# Analysis of Foods and Beverages:
# Modern Techniques

# FOOD SCIENCE AND TECHNOLOGY

## A SERIES OF MONOGRAPHS

### Series Editors

**Bernard S. Schweigert**
University of California, Davis

**John Hawthorn**
University of Strathclyde, Glasgow

### Advisory Board

**C. O. Chichester**
Nutrition Foundation, New York City

**Emil Mrak**
University of California, Davis

**J. H. B. Christian**
CSIRO, Australia

**Harry Nursten**
University of Reading, England

**Larry Merson**
University of California, Davis

**Louis B. Rockland**
Chapman College, Orange, California

**Kent Stewart**
USDA, Beltsville, Maryland

A complete list of the books in this series appears at the end of the volume.

# Analysis of Foods and Beverages: Modern Techniques

EDITED BY

## GEORGE CHARALAMBOUS
St. Louis, Missouri

1984

## ACADEMIC PRESS, INC.
(Harcourt Brace Jovanovich, Publishers)

Orlando   San Diego   San Francisco   New York   London
Toronto   Montreal   Sydney   Tokyo   São Paulo

ACADEMIC PRESS, INC.
Orlando, Florida 32887

*United Kingdom Edition published by*
ACADEMIC PRESS, INC. (LONDON) LTD.
24/28 Oval Road, London NW1 7DX

Library of Congress Cataloging in Publication Data
Main entry under title:

Analysis of foods and beverages.

(Food science and technology)
Includes index.
1. Food--Analysis. 2. Beverages--Analysis.
I. Charalambous, George, Date. II. Series.
TX541.A76 1984 664'.07 83-11783
ISBN 0−12−169160−8

PRINTED IN THE UNITED STATES OF AMERICA

84 85 86 87 9 8 7 6 5 4 3 2 1

# Contents

## 3  Quantitative Thin-Layer Chromatography of Foods and Beverages
*Dieter E. Jänchen*

## 4  Gas Chromatography
*Takayuki Shibamoto*

## 5  HPLC in the Analysis of Foods and Beverages (Including Ion Chromatography)
*Donald H. Robertson*

## 6  Mass Spectrometry
*Ian Horman*

## 7  NMR Spectroscopy
*Ian Horman*

# 8 Minicomputers and Robotics
### Gerald F. Russell

# 9 On-Line Methods
### Manfred Moll

# 10 Molecular Structure–Activity Analyses of Artificial Flavorings
### A. J. Hopfinger, R. H. Mazur, and H. Jabloner

# 11 Bioassays
### Chandralekha Duttagupta and Eli Seifter

# 12 X-Ray Analysis
### Gene S. Hall

## 13  Scanning Electron Microscopy in the Analysis of Food Microstructure: A Review
*Samuel H. Cohen*

## 14  Atomic Spectrometry for Inorganic Elements in Foods
*James M. Harnly and Wayne R. Wolf*

## 15  Quality Assurance for Atomic Spectrometry
*James M. Harnly and Wayne R. Wolf*

## 16  Near-Infrared Reflectance Analysis of Foodstuffs
*T. Hirschfeld and Edward W. Stark*

## 17  Applications of Fourier Transform Infrared Spectroscopy in the Field of Foods and Beverages
*R. A. Sanders*

## 18 Quantitative Quality Control of Foods and Beverages by Laser Light-Scattering Techniques
*Philip J. Wyatt*

## 19 Automated Multisample Analysis Using Solution Chemistry
*Gary R. Beecher and Kent K. Stewart*

# Contributors

Numbers in parentheses indicate the pages on which the authors' contributions begin.

**Gary R. Beecher** (625), Nutrient Composition Laboratory, Beltsville Human Nutrition Research Center, United States Department of Agriculture, Beltsville, Maryland 20705

**Samuel H. Cohen** (421), Science and Advanced Technology Laboratory, United States Army Natick Research and Development Laboratories, Natick, Massachusetts 01760

**Chandralekha Duttagupta** (353), Department of Gynecology/Obstetrics, Albert Einstein College of Medicine, Yeshiva University, Bronx, New York 10461

**Gene S. Hall** (397), Department of Chemistry, Rutgers, The State University of New Jersey, New Brunswick, New Jersey 08903

**James M. Harnly** (451, 483), Nutrient Composition Laboratory, Beltsville Human Nutrition Research Center, United States Department of Agriculture, Beltsville, Maryland 20705

**T. Hirschfeld** (505), Lawrence Livermore National Laboratory, Livermore, California 94550

**A. J. Hopfinger** (323), Searle Research and Development, Skokie, Illinois 60077

**Ian Horman** (141, 205), Nestlé Research Laboratories, CH-1814 La Tour de Peilz, Switzerland

**H. Jabloner** (323), Technical Center, Hercules Incorporated, Wilmington, Delaware 19899

**Dieter E. Jänchen** (69), CAMAG, CH-4132 Muttenz, Switzerland

**R. H. Mazur** (323), Searle Research and Development, Skokie, Illinois 60077

**Manfred Moll** (293), Centre de Recherches TEPRAL, F-54250 Champigneulles, France

**Howard R. Moskowitz** (13), Moskowitz/Jacobs, Inc., Scarsdale, New York 10583

**Donald H. Robertson** (117), Science and Advanced Technology Laboratory, United States Army Natick Research and Development Laboratories, Natick, Massachusetts 01760

**Gerald F. Russell** (265), Department of Food Science and Technology, University of California, Davis, California 95616

**R. A. Sanders** (553), Food Product Development Division, Winton Hill Technical Center, The Procter and Gamble Company, Cincinnati, Ohio 45224

**Eli Seifter** (353), Departments of Biochemistry and Surgery, Albert Einstein College of Medicine, Yeshiva University, Bronx, New York 10461

**Takayuki Shibamoto** (93), Department of Environmental Toxicology, University of California, Davis, California 95616

**Edward W. Stark** (505), Technicon Industrial Systems, Tarrytown, New York 10591

**Kent K. Stewart** (625), Department of Food Science, Virginia Polytechnic Institute and State University, Blacksburg, Virginia 24061

**Roy Teranishi** (1), Western Regional Research Center, United States Department of Agriculture, Berkeley, California 94710

**Wayne R. Wolf** (451, 483), Nutrient Composition Laboratory, Beltsville Human Nutrition Research Center, United States Department of Agriculture, Beltsville, Maryland 20705

**Philip J. Wyatt** (585), Wyatt Technology Company, Santa Barbara, California 93105

# Foreword

In the last decade there has been a rapid development in analytical techniques to detect and measure trace or micronutrients of foods. Microanalytical methods that were once seldom employed have now become routine. Detailed information on these techniques is widely available in the literature, but there is no single source of reference on their application to food systems. The publication of this book, which contains selected analytical methodologies with emphasis on those more recently developed, is a valuable contribution to food scientists, because it gives comprehensive presentation of the principles, techniques, and applications in food systems. The applications cover volatile and nonvolatile components, micronutrients, and other constituents of foods, including agricultural residues and contaminants. This book also complements traditional texts of food analysis, which are often based on methods that are cumbersome, methods whose sensitivity and accuracy have been surpassed by modern and improved instrumentation.

Subsections and a special chapter on sample preparation are appropriate in a food analysis book, because validity and reliability of results are greatly dependent on proper sample preparation for the type of analysis employed. The chapter on sensory analysis is presented with a more modern approach to sensory evaluation as opposed to existing differential, hedonic, and profile testing. An interesting counterpoint to sensory analysis is offered by the chapter on quantitative quality control by laser light scattering. The recently developed fields of scanning electron microscopy, X-ray microanalysis, differential laser light scattering, near-infrared reflectance and Fourier transformations, continuous-flow and flow-injection analysis, as well as mass spectrometry, nuclear magnetic resonance, and bioassay, have found wide applications in food analysis and are elaborated on in this volume. The introduction of molecular analysis of synthetic flavors and the automation of food analysis by use of computers, robotics, and other on-line methods are indeed timely considering that one may find rapid growth and development of these fields in the next decade. More traditional techniques, such as the various forms of chromatography (including thin layer, which is experiencing a rebirth), are also reviewed.

The analytical techniques provide a tool for producers and manufacturers to develop new and improved products while regulatory laboratories can update methods to monitor the safety and quality of our food supply. The recent consciousness of consumers and their demands for a "safe" food supply possessing high nutritional and aesthetic qualities provides an impetus for producers, manufacturers, and regulators to monitor the safety of our foods, that is, the absence or minimal presence of toxic substances as additives, contaminants, or indigenous components of foods. Such substances, along with other constituents of foods present in micro quantities, were not easily detectable 10 years ago but can now be measured with rapidity and accuracy.

For these reasons, this book will be valuable to students, teachers, researchers, and all food analysts, providing within two covers an up-to-date overview of these modern analytical techniques and helping them to keep abreast of a rapidly advancing technology, while at the same time pointing to future directions.

Marjorie B. Medina
Eastern Regional Research Center
United States Department of Agriculture
Philadelphia, Pennsylvania

# Preface

Because of the complexity of their chemical composition, foods and beverages confront the analyst with remarkably difficult problems—not least of which is the correlation of analytical results with the appearance and organoleptic factors that actually help market the product. On the other hand, nutrition, safety considerations, and shelf life are quite as important as palatability and appearance.

In all cases, the recent developments in the analysis of foods and beverages have resulted in the determination (detection and measurement) of ever-decreasing concentrations of their constituents with greater accuracy and, often, more simply.

Over the past few decades we have seen not only the evolution of existing techniques but also the development of new methodology: both events have led to greater accuracy and precision. Also, the initial need for highly skilled specialists is mitigated by simplification and automation, enabling many routine analysis techniques to be carried out by technicians—with interpretation of the results still often being the province of the specialist.

This book is the result of a need to present the very latest developments not only in the time-honored chromatographic techniques but also in several newer ones. Scanning electron microscopy, X-ray microanalysis, near-infrared reflectance and Fourier transformations, atomic spectrometry, bioassay, nuclear magnetic resonance, laser techniques, and automated multisample techniques using solution chemistry are all covered by experts in these fields. The introduction of molecular analysis of synthetic flavors is discussed, as are the use of computers and robotics and sample preparation techniques.

An attempt was made to reach a balance in presenting an explanation of principles, practical applications in the field of foods and beverages, and an informed forecast of new developments for the near future.

The editor wishes to thank all of the contributors to this treatise and in particular Dr. K. K. Stewart for many valuable discussions on scope and content, Marjorie B. Medina for contributing the foreword, and last but certainly not least the publishers for their unfailing advice and assistance.

# Abbreviations

| | |
|---|---|
| AAS | Atomic Absorption Spectroscopy |
| AES | Atomic Emission Spectroscopy |
| AFS | Atomic Fluorescence Spectroscopy |
| ATR | Attenuated Total Reflectance |
| CFA | Continuous-Flow Analysis |
| COSY | Correlated Spectroscopy |
| DAWN | Dual-Angle Weighted Nephelometry (™Wyatt Technology) |
| DCP | Direct-Current Plasma |
| DEPT | Distortionless Enhancement by Polarization Transfer |
| DLS | Differential Light Scattering |
| EDXRF | Electronic Detector X-Ray Fluorescence |
| ET | Extinction Time |
| FIA | Flow-Injection Analysis |
| FID | Flame Ionization Detector |
| FTIR | Fourier Transform Infrared Spectroscopy |
| GC | Gas Chromatography |
| HCL | Hollow Cathode Lamp |
| HOMO | Highest Occupied Molecular Orbital |
| HPLC | High-Performance Liquid Chromatography |
| HPTLC | High-Performance Thin-Layer Chromatography |
| ICP | Inductively Coupled Plasma |
| LC | Liquid Chromatography |
| LM | Light Microscope |
| LUMO | Lowest Unoccupied Molecular Orbital |
| MS | Mass Spectrometry |
| MTR | Multiple Total Reflectance |

| NIRA | Near-Infrared Reflectance Analysis |
| NMR | Nuclear Magnetic Resonance |
| OT | Onset Time |
| PAS | Photoacoustic Spectrometry |
| PHA | Pulse Height Analyzer |
| PIXE | Proton-Induced X-Ray Emission |
| QSAR | Quantitative Structure–Activity Relationships |
| SCOT | Support-Coated Open Tubular Columns |
| SDXRF | Scintillation Detector X-Ray Fluorescence |
| SEE | Standard Error of Estimate |
| SEM | Scanning Electron Microscopy |
| SEP | Standard Error of Prediction |
| SNR | Signal-to-Noise Ratio |
| SP | Sweetness Potency |
| SR | Sucrose Reference |
| TEM | Transmission Electron Microscope |
| TLC | Thin-Layer Chromatography |
| TRS | Time-Resolved Spectroscopy |
| WDXRF | Wavelength Detection X-Ray Fluorescence |
| XRF | X-Ray Fluorescence |

# 1

# Sample Preparation

ROY TERANISHI

*Western Regional Research Center*
*United States Department of Agriculture*
*Berkeley, California*

## I. INTRODUCTION

The main reasons for analyzing foods and beverages are to ascertain quality and to determine properties to which we respond favorably or unfavorably. A *property* can be defined as an effect a material has on another material or an effect a material has on one or more of the senses of an observer. The former type of property is used in isolating and purifying a material so that physicochemical effects can be measured by instruments. The latter type of property is determined by various biological methods, bioassays, and panel sensory analyses.

In analyses of macroscopic properties, if a sample is to represent a given food or beverage, it must be assumed that the properties of the sample actually do reflect those of the food or beverage being analyzed, instrumen-

ANALYSIS OF FOODS AND BEVERAGES
ISBN 0-12-169160-8

tally or biologically. With heterogeneous materials, it is not a simple accomplishment to obtain a statistically valid, representative sample of a food or beverage. Even homogeneous materials must be thoroughly mixed before taking an aliquot or samples may have differerent concentrations of constituents.

In handling samples for flavor qualities, the time factor must also be considered. For example, if one lingers over a mug of beer or a cup of coffee, the quality will change from the first sip to the last. Some compounds are very labile, due to volatility or decomposition or both. Concentrations of some compounds can change markedly within minutes and certainly within hours. Sampling methods, as well as analysis methods, should attempt to accommodate such situations.

Success in any analysis of molecular properties, instrumental or biological, depends on a sufficient supply of pure materials, or evaluations cannot be made with much certainty. The definition of purity, constancy of properties after various repeated purification methods, implies that enough material has been isolated for several purity tests, such as melting points, high-resolution mass spectrometry (MS), high-resolution nuclear magnetic resonance (NMR), etc. How often has an observer found that a single spot or a single peak did not represent a pure material but was composed of several compounds? Most experimental difficulties are related not to physicochemical methods, since proficiency in such methods has been highly developed, but usually to preparation of samples not worthy of presentation to highly trained and skilled scientists utilizing expensive, complicated, sophisticated equipment.

The usual goal in food and beverage analysis is to relate properties measured by instruments to biological properties. In order to do this, the sample requirements of such instruments must be met. Sample size and physicochemical conditions vary considerably according to the method being used. Usually some sample preparation is necessary to concentrate the important constituents or to eliminate interfering materials. It is obvious that the various steps or manipulations must be accomplished without loss of the properties of interest. Physicochemical properties, predominantly vapor pressure, solubility in various solvents, and adsorptivity, are utilized in the isolation and concentration of desired samples.

In the preparation of samples into usable mixtures, preferably into their molecular components, various separation methods are used. All separations involve redistribution of molecules between two or more phases. In crystallization, usually not used in odor chemistry but used in taste chemistry, the redistribution is between liquid and solid phases. In distillation, the redistribution is between vapor and liquid phases. In chromatography, the redistribution is between gas and liquid and/or solid or between liquid and liquid and/or solid. In the redistribution processes, different chemical and physical properties are used to effect separation and isolation of desired samples.

The major purpose of these processes is to isolate the materials in a pure form to establish their physical and sensory properties. In an ideal situation, there would be enough starting material for distillation into fractions of very small boiling point ranges, for liquid chromatography (LC) into different functional groups, and for preparative gas chromatography (GC) for purification and isolation of pure compounds. Purity of all fractions should be checked with high-resolution capillary GC. Since there is rarely sufficient material to be so thorough, methods must be carefully chosen.

To obtain materials contributing to the biological characteristics of interest is no simple or easy task. If the analyses involve the gross constituents, most of the problems relate to valid statistical sampling. If the analyses involve biologically important constituents, such as trace elements, vitamins, or flavor compounds, it is often necessary to concentrate the important constituents to within the detection limits of the instrumental or biological methods being used. The task, then, is to obtain from a food or beverage a concentrate that, when reconstituted to the concentration found in the original product, exhibits the biological activity of interest. If the biological activity is lost in the sample preparation step, there is no need to proceed with the analyses.

To accumulate certain components and to discard others, some procedure employing the different properties of the constituents, such as distillation, extraction, or adsorption, is used. Because of the major disruption caused in foods and beverages by sample preparation, selection of the proper procedure and control of the conditions are critically important. Mistakes made in sample preparation cannot be remedied or overcome no matter how elegant and sophisticated the analytical method used.

## II. DISTILLATION

Water is the major constituent in most foods and beverages (see Table 1). Ethyl alcohol is the next major constituent in some beverages. Thus, distillation is most often selected as the initial concentration step [see Sugisawa (1981) and Weurman (1969)].

All liquid substances have a vapor pressure that is constant at a given temperature. When the temperature is raised so that the vapor pressure of the substance equals that of the external pressure, the substance boils, and

TABLE 1   Food and Beverage Composition

| | |
|---|---|
| Water | to 95% |
| Carbohydrates | to 80% |
| Lipids | to 40% |
| Proteins | to 25% |
| Alcohol | to 20% |
| Minerals | to 5% |
| Vitamins | ca. $10^{-4}$% |
| Flavor compounds | to less than $10^{-10}$% |

this temperature is defined as the boiling point of the substance. Thus, to accurately define a boiling point, temperature and external pressure must be specified [see Weissberger (1951) for detailed discussion].

Distillation is divided into two categories: fractional and steam. In either case, two criteria must be satisfied if distillation is to be used. There must be sufficient starting material, and the desired constituents must be stable to the conditions of distillation.

## A. Fractional Distillation

Distillation is a powerful method for obtaining samples of very narrow boiling point ranges. Factors such as theoretical plates, throughput, column hold-up, ease of approaching equilibrium conditions, pressure drop across column, and boil-up rate must be considered for satisfactory operation [see Weissberger (1951)].

First of all, the type of column must be chosen. There are many types of columns, as described by Weissberger (1951), but only three will be discussed.

1. The packed column, usually with glass or metal helices, is usually used for purification of solvents. The size of the sample available usually limits the use of such columns in flavor chemistry to purification of solvents, because the column hold-up of packed columns is usually in the 10-g range.

2. Spinning band columns can be made with metal or Teflon bands. The metal spinning band column is good for preparing samples of relatively broad boiling point ranges. The Teflon spinning band column consists of a glass tube containing a solid Teflon rod around which is wound a helix of Teflon. The band is tightly fitted to the bore of the glass tube and is rapidly rotated. The helical pumping action of the band forces the refluxing liquid down with an intimate contact of the vapor going up the column. Because of this downward pumping action of the liquid, flooding of the column is easily avoided, and because of the open path between the solid rod and glass tube, hold-up and pressure drop are minimized. An 8-mm-bore, 91-cm-long

Teflon spinning band column has approximately 200 theoretical plates at atmospheric pressure with a hold-up of 0.4 ml and a throughput of up to 20 ml/hr.* Equilibrium can be established in about $\frac{1}{2}$ hr, a very desirable feature and one much appreciated by anyone who does a great deal of distillation.

3. In the concentric tube column the vapor going up is in close proximity to the liquid going down the very narrow annular space between the walls of the concentric tubes.† Because of its construction, this type of column is much more easily flooded than the other types mentioned and is therefore more difficult to bring to equilibrium conditions. The concentric tube column can be used more effectively at very low pressures and with smaller amounts of starting mixture than the spinning band column because of the latter's lower pressure drop and lower hold-up. The simpler construction with no moving parts makes the smaller concentric tube columns attractive for regular laboratory use.

All three types of fractional distillation columns should be available for sample preparations because each has its own advantage and usefulness.

## B. Steam Distillation

In the fractional distillation of a homogeneous mixture, consisting of substances soluble in each other, Raoult's law prevails. This is expressed by

$$P_A = P'_A N_A$$

where $P_A$ is the partial pressure of A from the solution, $P'_A$ is the vapor pressure of pure A, and $N_A$ is the mole fraction of A in the liquid. [See Weissberger (1951) for detailed discussion of fractional distillation.]

In the distillation of a heterogeneous mixture, consisting of substances not soluble in each other, if one of the substances is water, the term *steam distillation* is used. Water and volatile organic matter not soluble in water distill together when the total vapor pressure adds up to the external pressure, expressed by

$$P_{total} = P_{water} + P_{organic}$$

where the pressure terms are self-explanatory.

---

* See the products of Perkin-Elmer Spinning Band Systems (Perkin-Elmer and Company, Main Avenue, Norwalk, Connecticut 06856) and B/R Instrument Corporation (P.O. Box 7, Pasadena, Maryland 21122) as examples.

† See the products of Precision Distillation Apparatus Company (Woodlands Hills, California 91366) and Fisher Spaltrohr Columns (Rinco Instruments Company, Greenville, Illinois 62246) as examples.

Steam distillation is used to isolate volatile, water-insoluble organic materials from nonvolatile materials such as carbohydrates and proteins. This is a crude separation, because the vapor pressure of the mixture is entirely independent of concentration. This method cannot be used for fractionation of the constituents. In fact, water cannot be separated from the organics distilled by this method, and solvent extraction must be used for isolation of the organics. Isolation can sometimes be accomplished simply by separating the organic material floating to the top of the aqueous layer, as in harvesting essential oils such as mint oils, or by extracting the voluminous steam distillate with a very volatile solvent such as ethyl ether, or by using a combination steam distillation – continuous extraction apparatus.

The combination of steam distillation and continuous extraction has been known to organic chemists for many years [see Vogel (1956)]. This method became popular with flavor chemists after Likens and Nickerson (1964) used it for isolating beer volatiles. Flath and Forrey (1977) designed a head with a large condensing capacity (see Fig. 1) to take care of the large heat of vaporization of water. Schultz *et al.* (1977) evaluated the recovery of 11 compounds in model solutions and proposed the name of simultaneous distillation – extraction (SDE) for this system and method. The advantages of SDE are the following:

1. Desired volatile compounds can be concentrated, about 1 : 10,000, from a dilute mixture in a single operation within a short time (1 – 3 hr).

2. Small volumes of water and solvent are used to minimize artifact introduction from solvent and water.

3. Thermal degradation and artifact formation can be reduced because the SDE system can be used at diminished pressure and temperature.

A dry-ice cold-finger condenser is usually used at lower pressures and temperatures to prevent excessive losses of solvent. Severe solvent losses are encountered if the sample is being prepared from material that has a lot of surface area, such as leaves or other botanical sources. The dry-ice cold-finger condenser is necessary to condense solvent and to prevent desired volatiles from being carried out by entrainment with nitrogen and oxygen evolved from outgassing from the surface of the starting material. Figure 2 shows the side and top views of the SDE head; a cold-finger condenser equipped with a ball joint can be easily added to the SDE system without causing any stress from alignment problems.

Teranishi *et al.* (1977) designed a system for isolation of volatiles from fats and oils by utilizing a metal spinning band column and an extraction system. The oil mixture is pumped down the distillation column by the action of the spinning band while steam is distilled up. The distillate, composed of water and volatiles from the oil mixture, is extracted by a

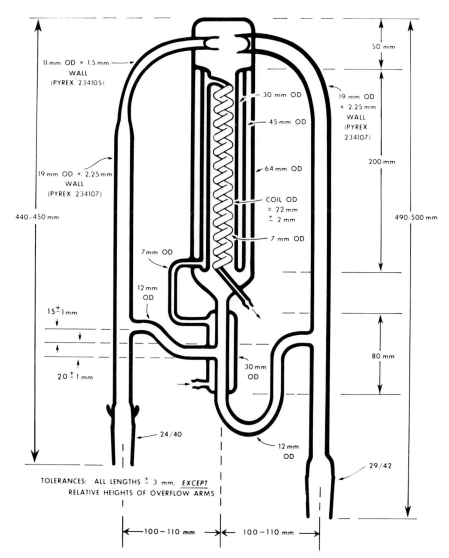

**Fig. 1.** Front view of simultaneous distillation–extraction (SDE) head used in concentration methods under vacuum and at atmospheric pressure. [Reprinted with permission from Flath and Forrey (1977). Copyright 1977 American Chemical Society.]

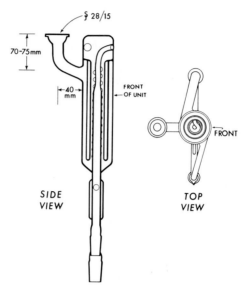

**Fig. 2.** Side and top views of simultaneous distillation–extraction (SDE) head. [Used with permission from T. H. Schultz, private communication (1983).]

solvent. Water and solvent are recycled continuously as in the standard SDE systems, while the oil mixture is pumped through the system and stripped of the volatiles. This system is similar to that described by Weurman (1969), but instead of a stationary glass spiral, the spinning band ensures a thin film of oil. The simultaneous extraction plus continuous pumping of the oil mixture permit the processing of a large amount of an oil mixture, thereby yielding sufficient sample size for subsequent characterization analyses.

## III. EXTRACTION

Extraction is the second most frequently used method for isolation and concentration of samples [see Weurman (1969) and Sugisawa (1981)]. Although extraction is a very powerful sample preparation method, other methods should be considered first for maximum effectiveness. Interference, such as formation of intractable emulsions with the presence of water-soluble proteins and/or carbohydrates, or extraction of compounds that mask properties of interest, may be encountered. Simultaneous extraction in combination with steam distillation has already been discussed, and this combination is an excellent example of how effective extraction is with a treatment before its use.

Many types of liquid–solid, liquid–liquid extractors are well known and commercially available. If extraction must be the first step and if emulsions may be formed, then the emulsion formation should be avoided by use of very slowly moving parts and very slow countercurrent movement of the liquids. In order to improve efficiency by using a very large surface area, Flath [see Stevens et al. (1969)] designed a system consisting of a series of concentric perforated stainless steel tubes (designed for maximum surface area) rotated very slowly as two immiscible liquids are forced to move countercurrently. With this type of extractor, even fruit juices containing particulates, sugars, and proteins were extracted without serious emulsion formations.

Extraction solvents should be chosen on the basis of selectivity for compounds of interest [see Weurman (1969) for a list of solvents]. Ethyl ether is used extensively, because its extraction efficiency is high for most organic compounds. Hydrocarbons and fluorocarbons have a low affinity for ethanol; therefore, these solvents are useful in preparing samples from alcoholic beverages.

Schultz et al. (1967) compared extraction efficiencies of various methods for isolating volatiles from apple essence. Charcoal adsorption was the most efficient in removing $C_1 - C_5$ alcohols, while isopentane and tetrafluoroethane did not extract the alcohols as expected but were best for extracting aldehydes and esters. Ethyl ether and liquid carbon dioxide were similar in nonselective extraction.

Supercritical gases, such as ammonia, carbon dioxide, nitrous oxide, ethylene, and some fluorocarbons, have been used to extract thermolabile natural products. A symposium was held on the topic of extraction with supercritical gases [see Anonymous (1978)]. The solubility behavior of various natural products was examined by Stahl et al. (1978). The most attractive of the gases listed is carbon dioxide.

Liquid carbon dioxide is nontoxic and has the advantage of having a low boiling point and thus producing little heat damage, but this advantage becomes a disadvantage in experimentation because of the need for high-pressure equipment. However, a laboratory system is now commercially available for preparing small samples of adsorbed materials for analytical purposes on porous polymers or activated charcoal.* Schultz and Randall (1970) have studied liquid carbon dioxide as a solvent, and Sugisawa (1981) has listed some miscible, partial, and insoluble compounds.

* See the product of J & W Scientific, Inc. ( 3871 Security Park Drive, Rancho Cordova, California 95670 ) as an example.

## IV. ADSORPTION

All materials have some affinity for binding on certain solid surfaces. Common adsorbents are alumina, silica gel, charcoal, molecular sieves, and porous polymers [see Dressler (1979) for available adsorbents]. The adsorptive capacity of a given material depends in part on the treatment or manufacturing conditions and on the composition of the adsorbent. Selective desorption can be controlled by the solvent used.

The phenomenon of adsorption can be exploited in separating mixtures into functional groups in conjunction with partition chromatography. Liquid chromatography is discussed by Robertson (Chapter 5, this volume).

Adsorption was used in about 8% of studies of food flavors in the period of 1900–1967 (Weurman, 1969) but increased to 14% in the period 1970–1977 (Sugisawa, 1981). This increased use of adsorbents is probably due to the manufacture of better adsorbents, improved sample manipulation techniques, and smaller sample-size requirements for analytical instruments such as nuclear magnetic resonance and mass spectrometers [see the excellent review by Dressler (1979)].

Desorption is usually accomplished by heat, especially for GC [see Shibamoto (Chapter 4, this volume)]. Desorption by use of solvents is becoming more popular because of ease of manipulation of samples [see Dressler (1979)]. Novel use of liquid carbon dioxide for a minimum of heat damage has been mentioned in Section III.

## V. MISCELLANEOUS SAMPLE PROCEDURES

After distillation, extraction, and adsorption, only a few miscellaneous sample preparation procedures remain, all of which are based on crystallization.

In freeze concentration, a selective partial removal of water can be obtained by crystallizing out the water in the form of ice [see Weurman (1969) and Sugisawa (1981)]. Some of the problems of this method include the partial loss of volatiles by evaporation during long periods of freeze concentration and losses due to occlusion in the ice crystals if the concentration is carried too far.

Zone melting is not used very much but is a procedure that should be kept in mind for special sample preparations. This procedure depends on the principle that a solvent freezing slowly tends to exclude impurities. If the entire mixture is frozen and a melted zone is repeatedly moved in one direction, desired materials can be concentrated and moved to one end of the frozen mixture.

Derivatization is sometimes used as a means of preparing samples for the easier analysis techniques, such as conversion of carbonyls to 2,4-dinitro-phenylhydrazones, conversion of acids to esters, conversion of alcohols to amines and to trimethylsilyl derivatives, etc.

## VI. SAMPLE MANIPULATIONS

Some sample preparations and analytical mixtures lend themselves to automation, for example, filtration, simple extractions, derivatizations, GC and LC, MS, ultraviolet (UV)–visible spectrometry, and infrared (IR) spectroscopy. Computer control and on-line methods are discussed by Russell (Chapter 8, this volume) and Moll (Chapter 9, this volume).

For batch methods, such as NMR, X-ray analysis, and scanning electron microscopy, individual samples must be isolated. Material in the gaseous state is excellent for on-line systems, but for batch sample manipulation the material is easier to manage in the liquid or solid state. Often solids are dissolved in suitable solvents for transfer to special spectrometer cells. For techniques in transferring small amounts of liquids in capillary tubes, see Teranishi (1981).

For isolation of materials for batch analysis, preparative gas chromatography is still very useful. The fundamental problem is to quantitatively condense purified materials from the gaseous state to manageable liquids. To prevent losses due to aerosol or fog formation, excessive supersaturation conditions must be avoided. That is, condensation at the dew point must be attained without excessive nucleation, which is the cause of aerosol or fog formation. If aerosols are formed, the percentage of material trapped is very small. The simplest, most effective, and most commonly used technique for condensation is to reach the dew point gradually with a thermal gradient [see Schlenk and Sand (1962)]. For small samples, the most common device is a glass melting point capillary or a small-diameter Teflon tube inserted into a heated gas chromatograph exit. Gradient cooling is established quickly, and condensation occurs with a minimum of fog formation. For a discussion of large-sample collectors, see Teranishi (1981).

## VII. SUMMARY

Because of the range and variety of samples encountered in food and beverage analysis and because of the diversity of analytical methods used, no general sample preparation procedure can be outlined. Therefore, the experimentalist must rely on fundamental principles for guidance as to what procedures must be used under what conditions. The importance of sample preparation cannot be overemphasized because most of the experimental difficulties encountered result from improperly prepared samples presented for highly developed and expensive instrumental or biological analyses.

Some innovations, such as combinations of steam distillation and simultaneous extraction, have been made recently. The greatest recent innovation in experimentation and data manipulation has been the application of computer methodology [see Russell (Chapter 8, this volume)]. Computer automation can be used in proven, routine sample preparations but usually is not feasible for research samples.

## REFERENCES

Anonymous (1978). *Angew. Chem. Int. Ed. Engl.* **17,** 701–754.
Dressler, M. (1979). *J. Chromatogr.* **165,** 167–206.
Flath, R. A., and Forrey, R. R. (1977). *J. Agric. Food Chem.* **25,** 103–109.
Likens, S. T., and Nickerson, G. B. (1964). *Proc. Am. Soc. Brew. Chem.* pp. 5–13.
Schlenk, H., and Sand, D. M. (1962). *Anal. Chem.* **34,** 1676.
Schultz, T. H., Flath, R. A., Black, D. R., Guadagni, D. G., Schultz, W. G., and Teranishi, R. (1967). *J. Food Sci.* **32,** 279–283.
Schultz, T. H., Flath, R. A., Mon, T. R., Eggling, S. B., and Teranishi, R. (1977). *J. Agric. Food Chem.* **25,** 446–449.
Schultz, W. G., and Randall, J. M. (1970). *Food Technol. (Chicago)* **24,** 94–98.
Stahl, E., Schilz, W., Schultz, E., and Willing, E. (1978). *Angew. Chem. Int. Ed. Engl.* **17,** 731–738.
Stevens, K. L., Flath, R. A., Lee, A., and Stern, D. J. (1969). *J. Agric. Food Chem.* **17,** 1102–1106.
Sugisawa, H. (1981). *In* "Flavor Research" (R. Teranishi, R. A. Flath, and H. Sugisawa, eds.), pp. 11–51. Dekker, New York.
Teranishi, R. (1981). *In* "Flavor Research" (R. Teranishi, R. A. Flath, and H. Sugisawa, eds.), pp. 53–82. Dekker, New York.
Teranishi, R., Murphy, E. L., and Mon, T. R. (1977). *J. Agric. Food Chem.* **25,** 464–466.
Vogel, A. I. (1956). "Practical Organic Chemistry." Wiley, New York.
Weissberger, A., ed. (1951). "Distillation, Technique of Organic Chemistry," Vol. IV. Wiley (Interscience), New York.
Weurman, C. (1969). *J. Agric. Food Chem.* **17,** 370–384.

# 2

# Sensory Analysis, Product Modeling, and Product Optimization

HOWARD R. MOSKOWITZ

*Moskowitz/Jacobs, Inc.*
*Scarsdale, New York*

ANALYSIS OF FOODS AND BEVERAGES
ISBN 0-12-169160-8

## I. INTRODUCTION TO SENSORY ANALYSIS

Sensory analysis plays a key role in the development of foods and beverages for human consumption. People eat what they like. Often consumers reject foods from which they could derive nutritional benefits because the food tastes bad or because these particular consumers have not had previous experience with the specific item.

Over the past 50 years we have seen the emergence of a science of sensory analysis. Looking back in the scientific literature of a half century ago, we find scattered references to sensory analysis of foods, with the information concentrated in technical journals devoted to physiology and psychology. The sensory analysis research of a half century ago consisted principally of descriptions of the taste and texture of foods, but little else. From time to time one might have read of approaches to measuring perceived intensity of a specific flavor characteristic or of attempts at correlating sensory perceptions with more objective physical measurements. By and large, however, little existed in the way of a useful body of knowledge in either the basic scientific realm or the applied domains (Amerine *et al.*, 1965).

World War II and its consequences brought food science and sensory analysis leaping forward into the twentieth century. Preservation of foods became paramount, and consideration of soldiers' food preferences emerged as a key factor in the development of nutritious and storage-stable foods (Guth and Mead, 1943). The United States Army Quartermaster Food and Container Institute pioneered techniques for measuring likes and dislikes. The army's efforts spawned a generation of researchers armed with scientific techniques and schooled in the pragmatics of real-world problems. Those researchers laid the foundations of present-day sensory analysis.

This review covers the state of sensory analysis of foods in the 1980s. It provides an overview of techniques to accomplish the following objectives:

1. Describe the perceptions that accompany foods and flavors.

2. Measure the perceived intensities of these private sensory experiences, in a reproducible and usable fashion.

3. Measure the degree of acceptance — likes and dislikes — for foods and flavors.

4. Integrate the foregoing information with formulation, processing, and financial information to optimize the food product for acceptance and to achieve other marketing and financial objectives.

## II. DESCRIBING THE SENSORY ATTRIBUTES OF FOODS

### A. Introduction

Descriptive analysis lies at the heart of all research in sensory analysis, because it provides the lexicon of terms appropriate for the sensory responses. The earliest work in sensory analysis consisted of series of terms for the senses of taste, smell (Harper *et al.*, 1968a), and touch (Boring, 1942).

A review of the scientific literature will quickly reveal that most investigators have had to develop their own sets of terms to describe sensory attributes associated with foods. Those researchers who have worked with color have had a relatively easy time, because they could use the XYZ or CIE color systems (LeGrand, 1957). However, no general qualitative or quantitative system existed (or exists) for the other sensory dimensions of appearance. In fact, researchers still do not have an adequate lexicon for surface appearance, and must rely upon the research of psychologists to help them define its attributes [e.g., Katz (1925)].

### B. Developing a Language for Taste, Odor, Texture, and Appearance

Faced with the lack of a general system for descriptive analysis, researchers have had to develop their own sets of terms. They have done so by a variety of procedures, as follows:

1. Introspective method. The introspective method derives from the research of structural psychologists such as E. B. Titchener (1916), who felt that introspection alone could isolate the building blocks or units of percep-

tion. With the introspective method the researcher (or the panelist) thinks about the basic units of perception and intuitively arrives at the set of attributes. The introspective method has merit because it allows a rich language to emerge. However, practiced by one researcher with only himself or a limited number of panelists as resources, the method inevitably generates a biased and limited set of terms. Many of the descriptor systems for odor have developed out of introspection by scientists over the past four centuries who relied upon this as their sole method.

2. Group method. The group, or discussion, method relies upon interaction between panelists to generate a list of terms for description. Individuals sit around a table and discuss the characteristics of the product. A moderator may lead the discussion but not force it into a specific sequence. The group dynamics enhance the flow of terms from the private thoughts of the panelist to the public forum of the group. The primary benefit of the procedure comes from its ability to elicit many terms from consumers. Its principal problem arises from the same source—the consumer panelists use many terms that actually have little or no meaning, except idiosyncratically.

3. Repertory grid method. The repertory grid method, developed by Kelly (1955), puts the panelist into a situation optimally designed to elicit new ideas and enhance creativity. The interviewer presents the panelist with pairs of stimuli. For each pair the interviewer elicits first those characteristics that make the members of the pair similar to each other and then those characteristics that differentiate the items in the pair from each other. The panelist has to provide terms for both. Next the interviewer removes the items, replacing one (or even both) with a new item or items. Again the panelist has to provide characteristics or attributes that illustrate similarity and difference. The panelist must continually change frames of reference, a task that shatters complacency in evaluation and enhances creativity.

## C. Descriptive Systems for Odor, Flavor, and Texture

An overview of food science will quickly reveal the existence of different procedures to describe the nuances of flavor and texture. Two schools of thought predominate in food research. The first includes those researchers who use a system to develop their own descriptors uniquely tailored to a specific need [e.g., the Flavor Profile as promoted by the Arthur D. Little Company (Caul, 1957; Cairncross and Sjostrom, 1950)]. The second includes those researchers who provide a list of prespecified terms that they feel are appropriate to flavor or texture [e.g., the Texture Profile as presented by Szczesniak and her colleagues (Szczesniak et al., 1963)]. This section provides the reader with historical background for both approaches, an assessment of the state of the art regarding those techniques in the early 1980s, and an analysis of the strengths and weaknesses of each method.

## 1. Procedures That Show Panelists How to Profile but Do Not Provide Terms

Procedures that show panelists how to profile but do not provide specific terms have historically grown out of the applied research laboratories, which are occupied with the evaluation of actual samples. Looking back at the history of food science we can see the emergence in 1948 of the Flavor Profile technique, first reported by researchers at the Arthur D. Little Company in Cambridge, Massachusetts (Cairncross and Sjostrom, 1950). These researchers undertook the task of developing procedures that would "capture" the elusive sensory nuances of aroma and flavor. Taste and flavor do not easily lend themselves to description. Attempts to capture these perceptions often generate descriptions that possess little generality.

The Arthur D. Little group recognized the elusiveness of these descriptors. They pioneered a system whereby panelists first discuss the quality notes of a product and through the discussion develop a descriptive language for the different characteristics. The complexity of foods and the large differences between various foods require that the panelists form different panels, depending upon the food being evaluated. A beer panel might develop a language quite different from the language developed by a spaghetti sauce panel.

Intuitively, we can liken the procedure to bootstrapping. Working as a team the group of panelists eventually develop a consensus set of terms to represent the qualities of the food. Over the series of meetings, the individuals on the panel have the opportunity to exchange viewpoints so that they all agree about the meaning of the terms. Along with the specific terms the panelists use a semiquantitative scale, as shown in Table 1. The panelists meet as a group, discuss the specific notes present in the product (which the same group has previously developed), and generate a profile of intensities. In all cases the group members work as a committee, with consensus voting.

**TABLE 1    The Intensity Scale for the Flavor Profile[a]**

| |
|---|
| 0 = not present |
| )( = threshold |
| 1 = weak |
| 2 = moderate |
| 3 = strong |

[a] Adapted from Caul (1957).

The outcome of the panel becomes a profile of descriptive terms specifying the degree or intensity of each term as applied to the food under consideration.

For over three decades researchers have used the Flavor Profile technique as a method to develop a language. Although not often subjected to scientific scrunity, the method has gained wide acceptance by applied researchers in the food industry and allied industries because it generates reproducible data. Most importantly, it quantifies the elusive senses of odor, taste, and texture.

Recently other quantitative methods have appeared that supplement the Flavor Profile procedure, augmenting it by statistical analyses. For instance, one glaring problem with the Flavor Profile concerns its relatively nonquantitative properties. With the advent of statistical analysis in sensory testing researchers eventually demanded a modification of the profiling procedure that would allow the sensory analyst to test differences between products on a statistical basis. One modification of the procedure, known as the quantitative descriptive analysis, or QDA (Stone *et al.*, 1974), uses a similar initial phase of language development. Afterward the panelists rate the product on the attributes, using a line scale. The researcher then converts the line-scale data to numbers and analyzes the numbers statistically. Both the Flavor Profile and the QDA method generate pictorial representations, but the QDA has the added benefit of backup statistical tests for substantiation.

## 2. Fixed Descriptor Systems

**a. Descriptive Systems for Odor.** Over the years researchers have developed a variety of systems to describe odors. Table 2 shows one system developed by Harper and his colleagues at Norwich, England, in the Food Research Institute. These researchers attempted to develop a systematized list of attributes that researchers around the world could use to describe the sensory nuances of odor and flavor perception. Their training in both experimental psychology and food science led them quickly to recognize that the list they developed could only partly capture the myriad quality notes that the researcher and consumer panelist might encounter in the evaluation of food. In a sense the list served as a vade mecum to guide the sensory panelist in his or her attempt to verbalize the notes of a food.

The Harper scale comprises 44 different attributes that the authors culled from many hundreds of different descriptors, found in a variety of lists through the ages (Harper *et al.*, 1968b). The panelist uses the list in a simple, straightforward manner. After smelling (or tasting) the stimulus, the panelist simply rates it on the degree to which it possesses the particular characteristic. To aid in distinguishing between levels, the panelist uses a category scale

**TABLE 2  Odor Quality Scale[a]**

| Q | | | | | | | S | | | | | | |
|---|---|---|---|---|---|---|---|---|---|---|---|---|---|
| Fragrant | 0 | 1 | 2 | 3 | 4 | 5[b] | Oily, fatty | 0 | 1 | 2 | 3 | 4 | 5 |
| Sweaty | 0 | 1 | 2 | 3 | 4 | 5 | Like mothballs | 0 | 1 | 2 | 3 | 4 | 5 |
| Almondlike | 0 | 1 | 2 | 3 | 4 | 5 | Like gasoline, solvent | 0 | 1 | 2 | 3 | 4 | 5 |
| Burnt, smoky | 0 | 1 | 2 | 3 | 4 | 5 | Cooked vegetables | 0 | 1 | 2 | 3 | 4 | 5 |
| Herbal, green, cut grass etc. | 0 | 1 | 2 | 3 | 4 | 5 | Sweet | 0 | 1 | 2 | 3 | 4 | 5 |
| Etherish, anesthetic | 0 | 1 | 2 | 3 | 4 | 5 | Fishy | 0 | 1 | 2 | 3 | 4 | 5 |
| Sour, acid, vinegar, etc. | 0 | 1 | 2 | 3 | 4 | 5 | Spicy | 0 | 1 | 2 | 3 | 4 | 5 |
| Like blood, raw meat | 0 | 1 | 2 | 3 | 4 | 5 | Paintlike | 0 | 1 | 2 | 3 | 4 | 5 |
| Dry, powdery | 0 | 1 | 2 | 3 | 4 | 5 | Rancid | 0 | 1 | 2 | 3 | 4 | 5 |
| Like ammonia | 0 | 1 | 2 | 3 | 4 | 5 | Minty, peppermint | 0 | 1 | 2 | 3 | 4 | 5 |
| Disinfectant, carbolic | 0 | 1 | 2 | 3 | 4 | 5 | Sulphidic | 0 | 1 | 2 | 3 | 4 | 5 |
| P | | | | | | | R | | | | | | |
| Aromatic | 0 | 1 | 2 | 3 | 4 | 5 | Fruit (citrus) | 0 | 1 | 2 | 3 | 4 | 5 |
| Meaty | 0 | 1 | 2 | 3 | 4 | 5 | Fruity (other) | 0 | 1 | 2 | 3 | 4 | 5 |
| Sickening | 0 | 1 | 2 | 3 | 4 | 5 | Putrid, foul, decayed | 0 | 1 | 2 | 3 | 4 | 5 |
| Musty, earthy, mouldy | 0 | 1 | 2 | 3 | 4 | 5 | Woody, resinous | 0 | 1 | 2 | 3 | 4 | 5 |
| Sharp, pungent, acid | 0 | 1 | 2 | 3 | 4 | 5 | Musklike | 0 | 1 | 2 | 3 | 4 | 5 |
| Camphorlike | 0 | 1 | 2 | 3 | 4 | 5 | Soapy | 0 | 1 | 2 | 3 | 4 | 5 |
| Light | 0 | 1 | 2 | 3 | 4 | 5 | Garlic, onion | 0 | 1 | 2 | 3 | 4 | 5 |
| Heavy | 0 | 1 | 2 | 3 | 4 | 5 | Animal | 0 | 1 | 2 | 3 | 4 | 5 |
| Cool, cooling | 0 | 1 | 2 | 3 | 4 | 5 | Vanillalike | 0 | 1 | 2 | 3 | 4 | 5 |
| Warm | 0 | 1 | 2 | 3 | 4 | 5 | Fecal (like manure) | 0 | 1 | 2 | 3 | 4 | 5 |
| Metallic | 0 | 1 | 2 | 3 | 4 | 5 | Floral | 0 | 1 | 2 | 3 | 4 | 5 |

[a] From Harper *et al.* (1968b).
[b] 0 = absent, 5 = extremely strong.

from 0 to 5, with 0 representing "none at all" and 5 representing "an extreme amount" (see Table 2).

Analytically, the researcher can secure a wide number of profiles from consumers or experts, each of whom has the opportunity to taste or smell a variety of different stimuli. The profiling scheme provides a plethora of data, in terms of the number of products one can test and the variety of different attributes that the panelist can use to register the quality notes.

Table 3 presents some published profiles of different pure chemicals

TABLE 3 Examples of Attribute Profiles for Different Odorants[a]

| | Expert | | | Inexpert | | |
|---|---|---|---|---|---|---|
| **Musk xylol (solid)** | | | | | | |
| | Fragrant | 2·6 | (13/15) | Light | 2·3 | (12/20) |
| | Sweet | 1·8 | (12/15) | Fragrant | 2·0 | (13/20) |
| | Floral | 1·3 | (12/15) | Sweet | 1·9 | (11/20) |
| | Musklike | 1·0 | (6/15) | Floral | 1·8 | (11/20) |
| | Dry, powdery | 0·7 | (7/15) | Dry, powdery | 1·2 | (8/20) |
| | | | | Aromatic | 1·0 | (7/20) |
| **Naphthalene (1% in DNP[b])** | | | | | | |
| | Like mothballs | 4·1 | (15/15) | Like mothballs | 4·3 | (20/20) |
| | Camphorlike | 1·7 | (9/15) | Camphorlike | 2·4 | (13/20) |
| | Cool, cooling | 0·7 | (8/15) | Cool, cooling | 1·6 | (11/20) |
| | Etherish, anesthetic | 0·6 | (8/15) | Heavy | 1·5 | (10/20) |
| | | 0·6 | (8/15) | Aromatic | 1·1 | (11/20) |
| | | | | Sicky | 0·8 | (9/20) |
| **Octanol (0.5% in DNP)** | | | | | | |
| | Soapy | 2·3 | (9/15) | Soapy | 2·4 | (12/20) |
| | | | | Light | 1·8 | (13/20) |
| | | | | Cool, cooling | 1·6 | (11/20) |
| | | | | Sweet | 1·4 | (10/20) |
| | | | | Oily, fatty | 1·3 | (9/20) |
| | | | | Sickly | 1·0 | (11/20) |
| | | | | Fragrant | 1·0 | (7/20) |
| **Phenyl acetic acid (1.5% in DNP)** | | | | | | |
| | Sickly | 1·3 | (9/15) | Sickly | 1·7 | (14/20) |
| | Sharp, pungent, acid | 1·1 | (10/15) | Sweaty | 1·7 | (13/20) |
| | Floral | 1·1 | (7/15) | Putrid, foul, decayed | 1·4 | (13/20) |

| | | | Sweaty | 0·9 | (9/15) | Heavy | 1·4 | (10/20) |
| | | | Sweet | 0·9 | (8/15) | Sharp, pungent, acid | 1·4 | (9/20) |
| | | | Fragrant | 0·9 | (7/15) | Warm | 1·3 | (10/20) |
| | | | Sour, acid, vinegar | 0·8 | (7/15) | Sour, acid, vinegar | 1·3 | (9/20) |
| | | | | | | Animal | 1·2 | (8/20) |
| | | | | | | Sweet | 1·1 | (9/20) |
| | | | | | | Rancid | 1·1 | (9/20) |
| | | | | | | Fecal (like dung) | 1·0 | (8/20) |
| **Phenylethanol (0.5% in DNP)** | | | | | | | | |
| | | | Floral | 2·7 | (11/15) | Floral | 3·1 | (17/20) |
| | | | Fragrant | 2·2 | (10/15) | Fragrant | 3·0 | (19/20) |
| | | | Sweet | 2·0 | (13/15) | Sweet | 2·4 | (17/20) |
| | | | | | | Light | 2·2 | (12/20) |
| | | | | | | Herbal, green, grass, etc. | 1·4 | (11/20) |
| | | | | | | Aromatic | 1·3 | (12/20) |
| | | | | | | Cool, cooling | 1·2 | (12/20) |
| **Nitrobenzene (10% in DNP)** | | | | | | | | |
| | | | Almondlike | 3·1 | (13/15) | Almondlike | 3·7 | (17/20) |
| | | | Aromatic | 1·2 | (9/15) | Sweet | 1·9 | (13/20) |
| | | | Petrol, chemical solvent | 0·9 | (8/15) | Heavy | 1·8 | (10/20) |
| | | | Vanillalike | 0·9 | (7/15) | Oily, fatty | 1·6 | (11/20) |
| | | | Etherish, anesthetic | 0·7 | (7/15) | Sickly | 1·5 | (14/20) |
| | | | | | | Fragrant | 1·5 | (12/20) |
| | | | | | | Aromatic | 1·1 | (8/20) |
| | | | | | | Like mothballs | 1·0 | (8/20) |

[a] From Harper et al. (1968b).
[b] DNP = dinonyl phthalate, a liquid solvent (presumed odorless).

reported by Harper and colleagues in the earliest published accounts of the scale. Note that they analyzed the data in several ways to learn about flavor perception. Specifically, they answered questions such as the following:

1. What do the profiles look like for the different chemicals?
2. Do experts differ in the profiles they generate, as opposed to consumers? (They do. Harper and colleagues reported that experts tended to use fewer terms than did consumers, suggesting that the experts had refined their abilities to rate attributes to the point where they jettisoned the less relevant notes and concentrated on the more important quality descriptors.)
3. Do profiles differ in the number of attributes used, independent of the levels assigned to the attributes? (They do, suggesting that some stimuli evoke complex perceptions whereas others evoke simpler, more unidimensional perceptions.)

We can contrast this empirical system of description with a list of different systems reported by John Amoore (1969). Amoore had an entirely different aim in mind when he developed his system for describing fragrance quality perceptions. According to Amoore, some of the terms that researchers used to describe perceptions might actually reflect "basic primaries." In contrast to their colleagues in visual perception, researchers in chemical perception had not developed a list of basic sensory primaries. Amoore focused his efforts toward finding these primaries, which he reduced to a limited number. Food scientists will find his list interesting primarily as a contrast to the more pragmatic technique of profiling to obtain a sensory signature of the product. (See Table 4 for Amoore's categories compared to others.)

Researchers have modified the different profiling systems to accommodate their specific needs. For instance, in the evaluation of fruit juices by aroma only ("nose") and by aroma plus taste ("mouth"), the researcher has to incorporate taste-relevant terms, as well as modify and fine tune the list of terms to slant more toward the estery, fruity notes of fruits and less toward the fatty and nitrogenous notes of meats and vegetables. We see an expansion of the Harper scale in this direction in Table 5, which presents profiles of fruit juices. Note the free use of terms from the Harper list along with relevant terms too specific for a general aroma list but exceedingly relevant for fruit smells.

**b. Descriptive Systems for Taste.** In contrast to odor we possess a much more limited set of terms to describe taste. Even though we think of the "taste of food" as its primary sensory component, in reality taste sensations limit themselves to those mediated by the tongue.

Scientifically we have only a few taste primaries: principally sweet, salty, sour, and bitter. We perceive taste by means of taste buds on the tongue. All other perceptions probably reflect odor or texture characteristics. Since, however, we experience most of the odor and texture by mouth, we often ascribe sensations properly belonging to odor and texture as related to taste.

**c. Descriptive Systems for Texture.** Parallel to research on descriptive systems for odor, investigators have tried to develop systems to describe the attributes of texture. To some extent researchers have had an easier time with texture, because the dimensions of texture appear simpler than those of odor. Whereas thousands of chemicals exist, each having its own unique nuances, we seem not to have that problem with texture. Introspectively we can narrow down the attributes or dimensions of texture to a limited number. Furthermore, odor attributes generally go hand in hand with specific stimuli embodying those attributes, whereas texture dimensions have an existence independent of the actual stimuli. We seem more easily able to extract the essence of a texture dimension and leave the physical stimulus that embodies it behind.

Historically, investigators in food science have used *texture* as a catchall word to describe the entire response to the mechanical properties of the food. Until the late 1950s the only primary research on texture, such as the introspective studies reported and collected by E. G. Boring and illustrated by Table 6, had emerged from psychology laboratories. The table indicates, as far as a table can, the analysis of thirteen cutaneous perceptions.

In the late 1950s Alina Szczesniak and her colleagues at the General Foods Corporation in Tarrytown, New York, began work to isolate the fundamental subjective dimensions of texture. Szczesniak attempted to develop a rational scheme to describe the characteristics of texture from a sensory point of view (Szczesniak et al., 1963) and where possible relate these to instrumental measurements of the forces generated by deforming the product under known test conditions.

After a number of years of careful work the investigations paid off. Table 7 shows the General Foods Texture Profile. The Texture Profile provides the researcher with the following:

1. A list of basic texture dimensions.
2. Gradations or levels of each dimension.
3. Specific reference stimuli that represent the levels of that texture dimension and that can act as exemplars.
4. Where possible, correlations between those perceived texture levels and instrumental measures.

The Texture Profile provides a means whereby researchers can unambiguously report the attributes of texture. The panelists first spend an initial training period learning the profile, and afterward they describe the texture characteristics in a reproducible and meaningful fashion.

Subsequent research on texture tended to use the Texture Profile as a fixed

**TABLE 4  Categories of Odor Perception[a]**

| | | | | Odor classes: general | | | | |
|---|---|---|---|---|---|---|---|---|
| Zwaardemaker (1895) 30 (sub)classes | Linnaeus (1756) 7 classes | Henning (1915) 6 classes | Crocker and Henderson (1927) 4 classes | Amoore (1952) 7 classes | Schutz (1964) 9 classes | Wright and Michels (1964): 8 classes | Harper et al. (1968): 44 classes | Miscellaneous additional |
| 1 Fruity | | | | | | Hexylacetate | Fruity | |
| 2 Waxy | | | | | | | Soapy | |
| 3 Ethereal | | | | Ethereal | Etherish | | Etherish-solvent | |
| 4 Camphor | | | | Camphor | | | Camphor; mothballs | |
| 5 Clove | Aromatic | | | | | | Aromatic | |
| 6 Cinnamon | | Spicy | | | | Spice | Spicy | |
| 7 Aniseed | | | | | | Benzothiazole | | |
| 8 Minty | | | | Minty | | | Minty | |
| 9 Thyme | | | | | | | | |
| 10 Rosy | | | | | | | | |
| 11 Citrous | | Fruity | | | | Citral | Citrous | |
| 12 Almond | | | | | Spicy | | Almond | |
| 13 Jasmine | | Flowery | | | | | Floral | |
| 14 Orangeblossom | Fragrant | | Fragrant | Floral | Fragrant | | Fragrant | |
| 15 Lily | | | | | | | | |
| 16 Violet | | | | | | | | |
| 17 Vanilla | | | | | Sweet | | Vanilla, sweet | |
| 18 Amber | | | | | | | Animal | |
| 19 Musky | Ambrosial | | | Musky | | | Musk | |

| No. | (Nonolfactory) | | | (Pungent) | (Trigeminal) | (Pungent and five others) |
|---|---|---|---|---|---|---|
| 20 | Leek | Alliaceous | | | | Garlic |
| 21 | Fishy | | | | | Ammonia; fishy |
| 22 | Bromine | | | | | |
| 23 | Burnt | Burnt | Burnt | Burnt | Burnt | Burnt |
| 24 | Phenolic | | | | | Carbolic |
| 25 | Caproic | Hircine | Caprylic | | | Sweaty |
| 26 | Cat urine | | | | | |
| 27 | Narcotic | Repulsive | | | | |
| 28 | Bedbug | | | | | |
| 29 | Carrion | Nauseous | | | | |
| 30 | Fecal | | | | | Sickly |
| 31 | | Resinous | | | Resinous | Fecal |
| 32 | | Foul | | Putrid | Unpleasant | Resinous; paint |
| 33 | | | Acid | | | Putrid; sulfurous |
| 34 | | | | | Sulfurous | Acid |
| 35 | | | | Oily | Oily | Oily |
| 36 | | | | Rancid | Rancid | Rancid |
| 37 | | | | Metallic | Metallic | Metallic |
| 38 | | | | | | Meaty |
| 39 | | | | | | Moldy |
| 40 | | | | | | Grassy |
| 41 | | | | | | Bloody |
| 42 | | | | | | Cooked vegetable |
| 43 | | | | | | Sandal |
| 44 | | | | | | Watery |
| | | | | | | Urinous |

Affective (Trigeminal)

ª From Amoore (1969).

**TABLE 5   Descriptor System for the Smell and Taste of Fruit Juices**[a]

| | Grape | | | | Apple | | | |
|---|---|---|---|---|---|---|---|---|
| Attribute | "Nose" n = 146[b] | | "Mouth" n = 142 | | "Nose" n = 153 | | "Mouth" n = 153 | |
| Total odor strength | 5.99 | (0.16)[c] | 5.40 | (0.17) | 5.53 | (0.14) | 4.75 | (0.14) |
| Applelike | 1.07 | (0.09) | 1.36 | (0.11) | 5.46 | (0.18) | 5.11 | (0.19) |
| Musty, moldy | 1.81 | (0.13) | 1.68 | (0.11) | 2.10 | (0.11) | 1.92 | (0.10) |
| Sweet | 5.13 | (0.13) | 4.83 | (0.13) | 4.31 | (0.13) | 4.02 | (0.13) |
| Spicy | 1.80 | (0.11) | 1.77 | (0.10) | 1.84 | (0.10) | 1.72 | (0.09) |
| Fermented, winelike | 2.68 | (0.14) | 2.51 | (0.15) | 1.99 | (0.12) | 2.04 | (0.12) |
| Blueberrylike | 1.67 | (0.18) | 1.73 | (0.19) | 0.13 | (0.03) | 0.23 | (0.05) |
| Estery (hard candy) | 2.37 | (0.16) | 2.54 | (0.17) | 2.11 | (0.13) | 2.08 | (0.13) |
| Aromatic | 4.23 | (0.17) | 4.07 | (0.17) | 3.81 | (0.15) | 3.67 | (0.14) |
| Sharp, pungent | 2.31 | (0.13) | 2.11 | (0.13) | 2.61 | (0.12) | 2.39 | (0.12) |
| Cranberrylike | 0.33 | (0.06) | 0.42 | (0.06) | 0.33 | (0.05) | 0.54 | (0.08) |
| Woody, sawdustlike | 0.81 | (0.08) | 0.75 | (0.07) | 1.15 | (0.10) | 1.09 | (0.11) |
| Floral | 3.21 | (0.16) | 3.20 | (0.15) | 3.18 | (0.14) | 3.05 | (0.13) |
| Grapelike | 6.33 | (0.15) | 5.89 | (0.15) | 0.50 | (0.06) | 0.63 | (0.07) |
| Resinous | 1.54 | (0.12) | 1.43 | (0.11) | 1.54 | (0.11) | 1.49 | (0.11) |
| Fruity, berrylike | 3.57 | (0.19) | 3.75 | (0.19) | 2.32 | (0.17) | 2.33 | (0.16) |
| Green, cut grass | 0.57 | (0.07) | 0.56 | (0.07) | 1.13 | (0.10) | 1.04 | (0.10) |
| Fragrant | 4.84 | (0.14) | 4.61 | (0.14) | 4.13 | (0.15) | 3.85 | (0.12) |
| Earthy | 0.82 | (0.09) | 0.87 | (0.09) | 1.19 | (0.12) | 1.33 | (0.13) |
| Vinegarlike | 1.50 | (0.12) | 1.49 | (0.13) | 2.16 | (0.11) | 2.26 | (0.13) |
| Etherish, anesthetic | 0.73 | (0.08) | 0.73 | (0.09) | 1.31 | (0.12) | 0.94 | (0.11) |
| Pleasantness in odor | 6.47 | (0.09) | 6.38 | (0.09) | 5.58 | (0.12) | 5.59 | (0.10) |

[a] From von Sydow et al. (1974).
[b] n = number of ratings from the panel.
[c] Numbers in parenthesis represent the standard error of the mean.

system, with few changes. From time to time the researcher might add a couple of terms to capture individual texture characteristics. By and large, however, the system remains as Szczesniak and her colleagues originally presented it. The stability (and rigidity) of the system stands in sharp contrast to the unceasing, continual modification of the flavor and odor descriptors. Perhaps with texture research there exist only a limited number of dimensions, whereas for odor we feel compelled to modify the set of descriptors to accommodate new chemicals or new odors that arise in the research. Texture and odor thus differ fundamentally in terms of their susceptibility to descriptive systems, with texture much more susceptible to description by a limited set of terms and odor (and flavor) less susceptible and more protean and changing.

**TABLE 6  Sensory Impressions Involved in Cutaneous Perception and Texture**[a]

| Perception | Author | Date | Pressure | Pain | Warmth | Cold | Movement | Other characteristics |
|---|---|---|---|---|---|---|---|---|
| 1. Hot (cf. 2) | Alrutz | 1897 | — | — | — | Cold | — | |
| 2. Cool (cf. 1) | Rubin | 1912 | — | — | — | Strong cold | — | |
| 3. Burning hot | Knight | 1922 | — | Pain | — | Weak cold? | — | |
| 4. Burning cold | Von Frey | 1904 | — | Pain | — | Cold | — | |
| 5. Wet (cf. 6) | Bentley | 1900 | Even | — | — | Strong cold | — | |
| 6. Oily (cf. 5) | Cobbey and Sullivan | 1922 | Weak | — | — | — | Movement? | |
| 7. Hard (cf. 8) | Sullivan | 1927 | Even | — | — | Cold | — | Good boundary |
| 8. Soft (cf. 7) | Sullivan | 1927 | Uneven | — | — | — | — | Poor boundary |
| 9. Smooth (cf. 10) | Meenes and Zigler | 1923 | Even field | — | — | — | Movement | |
| 10. Rough (cf. 9) | Meenes and Zigler | 1923 | Uneven field | — | — | — | Movement | |
| 11. Sticky (cf. 10) | Zigler | 1923 | Variable | — | — | — | Movement | |
| 12. Ticklish (cf. 11) | Murray | 1908 | Variable spreading | — | — | — | Movement (reflex) | Tendency to withdraw; unpleasant |
| 13. Clammy (cf. 8) | Zigler | 1923 | Even | — | — | Cold | Movement | Poor boundary; unpleasant imagery |

[a] Adapted from Boring (1942).

**TABLE 7   General Foods Texture Profile: Standardized Rating Scale with Generalized Standards**[a,b]

A. Mechanical Properties

| Hardness | Fracturability | Chewiness |
|---|---|---|
| Cream cheese | Corn muffin | Rye bread |
| Velveeta cheese | Egg Jumbos | Frankfurter |
| Frankfurters | Graham crackers | Large gumdrops |
| Cheddar cheese | Melba toast | Well-done round steak |
| Giant stuffed olives | Bordeaux cookies | Nut Chews |
| Cocktail peanuts | Ginger snaps | Tootsie Rolls |
| Shelled almonds | Treacle brittle | |
| Rock candy | | |

| Adhesiveness | Viscosity |
|---|---|
| Hydrogenated shortening | Water |
| Cheese Whiz | Light cream |
| Cream cheese | Heavy cream |
| Marshmallow topping | Evaporated milk |
| Peanut butter | Maple syrup |
| | Chocolate syrup |
| | Cool 'n Creamy pudding |
| | Condensed milk |

B. Geometrical Properties

| Related to particle size and shape | | Related to particle shape and orientation | |
|---|---|---|---|
| Property | Example | Property | Example |
| Powdery | Confectioners sugar | Flaky | Flaky pastry |
| Chalky | Tooth powder | Fibrous | Breast of chicken |
| Grainy | Cooked Cream of Wheat | Pulpy | Orange sections |
| Gritty | Pears | Cellular | Apples, cake |
| Coarse | Cooked oatmeal | Aerated | Whipped cream |
| Lumpy | Cottage cheese | Puffy | Puffed rice |
| Beady | Cooked tapioca | Crystalline | Granulated sugar |

[a] From Szczesniak et al. (1963).
[b] Intensity of parameters increases downward.

## III. MEASURING INTENSITY OF PERCEPTION

Intensity measurement of perception plays an important role in sensory evaluation, because it allows the researcher to quantify perceptions rather than merely classifying them.

### A. Thresholds versus Sensory Intensity

The history of measurement in perception begins with psychologists and physiologists who in the early part of the nineteenth century began to measure the lowest intensity of a stimulus that an individual could perceive. In those early days the researchers did not conceive of the human senses as anything more than mindless physiological mechanisms that, in the great empiricist tradition, would lend themselves to quantification. The human being could not act as a measuring system directly, or so at least these researchers thought. On the other hand, the researcher could measure the threshold or the lowest intensity that a person could perceive and compare threshold values across different chemicals or across individuals. The relation between perception (at the threshold level) and stimulus intensity (at the objectively measured level of concentration) provided an initial foray into sensory measurement.

The historian of science who delves into the study of perceived intensity measurement will find many articles written in the nineteenth and early twentieth centuries that deal with the values of threshold for simple chemicals [see Boring (1942)].

Thresholds provide a means of assessing the sensitivity of the perceptual system, because they indicate the physical stimulus level lying at the bottom of the sensory scale. Below a certain point (the detection threshold or the recognition threshold, respectively) the panelist cannot detect the stimulus or recognize the stimulus. Compilations of representative threshold values appear in Table 8.

Keep in mind that, although we speak of a threshold, in actuality the threshold represents the average of a range of levels of the stimulus. Across the population of panelists we will find some individuals who exhibit low thresholds for a stimulus (e.g., some individuals can recognize saccharin at very low concentrations). The same population distribution will cast up individuals who exhibit low sensitivity and who have a high threshold for the same stimulus.

Many methods exist to measure the threshold experimentally. Table 9 presents a brief compilation of these procedures. Bear in mind that measures of threshold simply provide that physical stimulus level lying at the bottom of the sensory scale.

**TABLE 8   Representative Threshold Values for Some Compounds and Chemicals**

A. Odor Thresholds for Some Chemicals in Vegetable Oil[a,b]

| Compound | Threshold |
| --- | --- |
| Hexanal | 120 |
| Heptanal | 250 |
| Nonanal | 1000 |
| Hex-2-enal | 850 |
| Hept-2-enal | 1500 |
| Oct-2-enal | 500 |
| Non-2-enal | 150 |
| Dec-2-enal | 2100 |
| Deca-2,4-dienal | 135 |
| Pent-1-3n-3-one | 5.5 |
| 2,5-Dimethylpyrazine | 2600 |
| 2-Ethyl-5-methylpyrazine | 320 |
| 2,5-Diethylpyrazine | 270 |
| 2-Ethyl-3,6-dimethylpyrazine | 24 |

B. Representative Absolute Thresholds for Taste[c,d]

| Chemical | Threshold |
| --- | --- |
| Sourness | |
|    Hydrochloric acid | 0.0009 |
|    Formic acid | 0.0018 |
| Saltiness | |
|    Sodium chloride | 0.03 |
| Sweetness | |
|    Sucrose | 0.01 |
|    Sodium saccharin | 0.000023 |
| Bitterness | |
|    Quinine sulfate | 0.000008 |
|    Caffeine | 0.0007 |

[a] From Buttery et al. (1973).
[b] Threshold values given in parts per trillion.
[c] From Pfaffmann (1959).
[d] Threshold values given in molarity.

**TABLE 9   Methods Used to Measure Thresholds**

| Method | Procedure | Question asked |
|---|---|---|
| 1. Single stimulus | Present stimuli at low levels to panelist. Ask panelist to report when he or she detects a stimulus. | When do you detect the stimulus? Do you detect and clearly recognize the quality? |
| 2. Paired comparison | Present two samples to panelist. One contains the stimulus. The other contains a "blank." | Which sample has the stimulus? |
| 3. Sorting (e.g., triangle method) | Present three samples to the panelist. One contains the stimulus. The others contain "blanks." | Which sample has the stimulus? |

All too often researchers conclude that stimuli that exhibit low threshold levels (i.e., those that panelists detect at low concentrations) must exhibit strong sensory intensities. In reality we cannot draw this conclusion. No evidence exists that links perceived sensory intensity to threshold level. Thus, a stimulus such as saccharin, which we can taste at concentrations hundreds of times lower than the threshold for sucrose, does not really taste many times as sweet. Rather the concentrations for matching sweetnesses lie in the order of several hundred to one.

## B. Using the Panelist to Measure Intensity by Category Scales

Suppose we allow the panelist to do more than simply report when he or she perceives the stimulus (as the experiment might require for threshold testing). Rather, we allow the panelist to rate perceived intensity using a fixed point scale (e.g., of 11 points, with 0 representing "none" and 10 representing "extreme"). In such a case we say that the panelist has used the category scaling method to rate perceived intensity.

The category method provides the panelist with a set of categories that represent graded degrees of perceived intensity. The psychological differences between category points (i.e., differences between adjacent category values, or the size of the categories) presumably equal the numerical difference. Thus, in theory, the difference between a stimulus rated 7 and a stimulus rated 6 is one category point. The perceived difference between these two stimuli should by the nature of the scale equal the perceived difference between a pair of stimuli rated 4 and 3 on the same scale.

## C. Using Category Data in Sensory Analysis

It should come as no surprise that food technologists and sensory analysts have used the category scaling procedure quite extensively in the evaluation of sensory attributes. As we noted previously, Szczesniak and her colleagues, who developed the Texture Profile, adopted the category scale as the measuring instrument.

Typical instructions for using the category scale follow:

> Please evaluate the following products on the attribute of flavor intensity. Use the scale provided. A 1 on the scale represents "no perceived flavor intensity." In contrast, a 9 on the scale (the highest point) represents "extreme amount of perceived flavor." Use the scale to show how strong the flavor seems to you, by choosing that particular category which best reflects your feeling about the strength of flavor.

Note that the panelist has to know the number of available categories and often may have to look at the extremes of the products to ensure that the category scale ratings lie within the range. It does little or no good for the panelist to use the top end of the scale for most of the products if during the evaluation the investigator presents a stimulus that seems far more intense than the stimuli already tested. Where on the scale should the panelist put the stimulus, since he or she has already used the top of the scale?

Some investigators in sensory analysis familiarize the panelist with the set of products ahead of time and thus ensure that the panelist does not exceed the upper or lower boundaries of the scale. This procedure biases the data because it implicitly (or even explicitly) tells the panelist to shrink or stretch the range of ratings in order to match the range of the stimuli.

Typical results from the category scale appear in two forms:

1. Researchers who use the scale to evaluate the perceived intensity of stimuli, undergoing known quantitative variation, generate dose–response or intensity curves, such as the curve shown in Fig. 1. The aim becomes the correlation or functional relation between the stimulus (e.g., processing variable, or formulation level) and panelist sensory response. Figure 2 shows the curve for the relation between the General Foods Texturometer reading and perceived texture intensity, using the Texture Profile scale (Jeon *et al.,* 1973).

2. Researchers often use the scale as a convenient measuring device to determine the change in a panelist's response as a function of treatment, product type, etc. Thus, the data in Table 10 represent the category rating scale on a variety of dimensions for several products that vary in many ways. The investigator using the category scale in this fashion uses it purely to determine differences between products and generally does not look at the quantitative relation between stimulus level and response level.

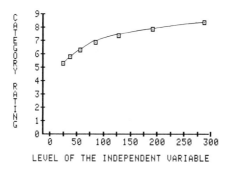

**Fig. 1.** Schematic category scale. The abscissa shows the level of physical intensity, whereas the ordinate shows the rating in terms of the fixed-point category scale.

## D. Types of Category Scales Currently Used

A glance at the food science literature will reveal a plethora of different scales for sensory analysis. Researchers develop their own category scales and vary them systematically depending upon their particular goals. Some of the variations include the following:

1. Scales that have no words ( just numbers) versus scales that have words alone versus scales that intermix both words and numbers. Sometimes investigators shy away from using numbers on the scale and present the panelist with words alone (see Table 11 for an example). The panelist chooses the right intensity rating from the words. The researcher codes the words numerically and analyzes the numerical ratings.

2. Scales that vary in the number of points. Investigators have not settled

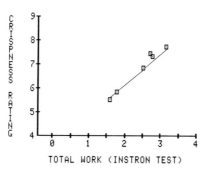

**Fig. 2.** Relation between the independent measure of texture (total work, measured by the Instron Universal Testing Machine) and the category rating of crispiness from the General Foods Texture Profile.

**TABLE 10**　Mean Values for Sensory Scores and Physical Measurements of Two Types of Beef Roasts Derived from Two Oven Temperatures[a,b]

| | Biceps femoris roasts | | | Semimembranosus roasts | | |
|---|---|---|---|---|---|---|
| | 135°C (n = 32) | 163°C (n = 32) | $s_{\bar{x}}$[c] | 135°C (n = 32) | 163°C (n = 32) | $s_{\bar{x}}$ |
| Sensory score[d] | | | | | | |
| Flavor intensity | 4.9 | 5.1** | 0.1 | 4.7 | 4.8 | 0.1 |
| Flavor acceptability | 5.4 | 5.4 | 0.1 | 5.2 | 5.2 | 0.1 |
| Juiciness | 5.0 | 5.1 | 0.1 | 4.3 | 4.7** | 0.1 |
| Tenderness | 5.1 | 5.1 | 0.1 | 4.6 | 4.7 | 0.1 |
| Chew count (number of chews) | 22.6 | 22.6 | 0.4 | 24.1 | 23.9 | 0.4 |
| Overall acceptability | 5.2 | 5.3 | 0.1 | 4.9 | 4.9 | 0.1 |
| Physical measurements | | | | | | |
| Shear force : Warner–Bratzler (kg) | 6.8 | 7.4 | 0.2 | 8.5 | 8.2 | 0.4 |
| Shear force : Allo–Kramer (kg) | 16.4 | 17.1 | 0.4 | 15.6 | 15.4 | 0.5 |
| Press juice (%) | 16.6 | 16.7 | 1.3 | 19.2 | 25.0** | 1.2 |
| Total cooking loss (%) | 25.2 | 24.8 | 0.7 | 27.1 | 26.2 | 0.4 |
| Drip loss (% of total loss) | 25.0 | 18.8** | 1.0 | 24.9 | 19.3*** | 0.9 |
| Cooking time (min/kg) | 107.1 | 82.5** | 2.1 | 125.2 | 88.2*** | 2.8 |
| Weight (kg) | 2.3 | 2.3 | 0.0 | 1.9 | 1.9 | 0.0 |

[a] From McCurdy et al. (1981).
[b] Pairs of means within a beef roast type followed by asterisk(s) are significantly different: ** indicates significance at the 1% level; *** indicates significance at the 0.5% level.
[c] Standard error of the mean.
[d] Based on an eight-point scale with 8 the highest score, except for chew count.

**TABLE 11**　Examples of Category Scales for "Intensity"

| | |
|---|---|
| 3 Strong | 4 Extremely strong |
| 2 Moderate | 3 Strong |
| 1 Weak | 2 Weak |
| 0 Not perceptible | 1 Extremely weak |

5 Extremely strong
4 Very strong
3 Moderate
2 Weak
1 Very weak
0 Not perceptible

on the best number of points to use in a category scale. We know that limiting the scale to 3 points reduces discrimination. We also know that the greater the number of points the higher the discrimination, but beyond a certain number of points we might just as well have an infinite scale. Some investigators prefer to use 5-point scales, others 9- or 10-point scales, and others 17-point or larger scales (Lundgren et al., 1978).

## E. Benefits and Drawbacks of Category Scales for Applied Research

Category scales have benefits and drawbacks. Among their benefits we find the following. They are easy to use, easy to analyze, and portable; they require little training; and they are easy to understand when cited in research reports and amenable to standardization. Among their drawbacks are the following: data are often biased because panelists shy away from using the entire scale; the scale does not always have equal-interval properties (i.e., the difference between adjacent scale points does not remain constant up and down the scale); researchers in different laboratories use different types of scales and do not agree on the number of or the verbal accompaniments for the category values; and the scale does not have ratio properties (so we cannot compute the ratios of scale values to conclude that one stimulus seems three times stronger than another).

## F. Ratio Scales of Perception in Sensory Analysis

According to Stevens (1946), researchers should attempt, whenever possible, to erect validated ratio scales of measurement rather than interval scales. The category scale typifies an interval scale, much like the Fahrenheit scale of temperature. In contrast stand ratio scales, such as the Kelvin scale of temperature, which can be used to establish valid relations between temperature levels and physical reactions.

For sensory analysis the analogy holds. Although category scales provide useful information, the researcher can nonetheless improve the quality of the results by developing ratio scales of perceived intensity, similar to the Kelvin scale of temperature. The ratio scale possesses a meaningful absolute zero (the interval scale does not) and the ratio scale allows the researcher to compute ratios of ratings (the interval scale does not).

In order to erect a valid ratio scale of perceptions the researcher has to allow the panelist to match numbers to perceived sensory intensity and not constrain the ratings within prespecified boundaries. (In contrast, the category scale prespecifies the boundary limits of the scale and forces the panelists to abide by these artificially constrained limits.)

The following instructions allow the panelist to assign ratio scale ratings to perceptions:

> Please evaluate the following products on the attribute of flavor intensity. Assign the first sample any convenient positive number that you wish. Small numbers on your scale will represent "low flavor intensity," and large numbers will reflect "strong flavor intensity." As you taste the remaining samples, match numbers to flavor intensity, so that the relative sizes of the ratings you assign best reflect the relative intensities of the flavors as you perceive them.
>
> You may go as high or as low as you wish in order to make your matches. Make sure, however, that the sizes of your ratings really reflect the perceived intensity of the flavor.

These instructions result from several decades of research experience obtained first by psychologists interested in perception and later by applied researchers. Note that the panelists can freely assign numbers to perceptions so that ratios of ratings represent ratios of perceptions.

By assigning ratio scale values to perceptions, the panelist can indicate to the researcher information such as whether one product tastes "twice as sweet" as another or whether one sample feels a third as tender as another. Magnitude estimation scaling, a variant of ratio scaling, has begun to find increased acceptance among researchers in sensory evaluation as a method for erecting the ratio scales (see the instructions just presented). Since each panelist can use his or her own scale, the researcher may wish to anchor the unit of the scale by presenting the panelist with a "standard" or reference product and requiring the panelist to assign this product a specific first rating (called the modulus or starting number). Then the panelist rates all of the remaining samples with reference to this standard, assigning numbers so that ratios of ratings reflect ratios of perceptions. During the course of the evaluation panelists soon become very comfortable rating the samples relative to the reference standard and eventually internalize the scale so that they need not return to the standard prior to rating each new sample.

Normalizing the scale ratings to ensure comparable numbers from panelist to panelist follows a simple procedure shown in Table 12. The panelist simply has to provide a rating that corresponds to the "top" of his or her scale.

## G. Relations between Stimulus Measures and Ratings Using Intensity Rating Scales

If the researcher has systematically varied the stimulus continuum (i.e., by changing the formulation level or by changing the processing time or

**TABLE 12  Example of Calibration of Magnitude Estimation Ratings**

| Before calibration | | After calibration[a] | |
| --- | --- | --- | --- |
| Panelist 1 | Panelist 2 | Panelist 1 | Panelist 2 |
| Top of scale = 50 | Top of scale = 200 | Top of scale = 100 | Top of scale = 100 |
| Product rating | Product rating | Product rating | Product rating |
| Greasiness = 20 | Greasiness = 60 | Greasiness = 40 | Greasiness = 30 |
| Aroma strength = 40 | Aroma strength = 160 | Aroma strength = 80 | Aroma strength = 80 |
| Purchase intent = 10 | Purchase intent = 100 | Purchase intent = 20 | Purchase intent = 50 |

[a] New magnitude estimate = 100 × (old magnitude estimate top of scale).

temperature or both), it becomes possible to relate the ratings assigned by panelists to the actual stimulus level, objectively measured.

Typical applications of this subjective–instrumental correlation include the following:

1. Relation between concentration in the product and perceived sensory intensity.

2. Relation between processing time and sensory intensity of specific characteristics.

3. Relation between objective physical measurements (e.g., Instron measures of fundamental mechanical properties) and sensory measures.

In all of these cases the researcher can relate the sensory ratings to the formulation or physical measures by means of statistical procedures, including correlation and regression.

### 1. Correlation

Correlation statistics show the researcher the degree of linear relation between two variables. Table 13 shows some typical correlation results between formulation or objective measurements and sensory responses. Keep in mind that the correlation coefficient does not allow the researcher to predict the likely sensory rating. Rather, it shows the degree of relatedness between two variables assuming that the underlying relation follows a straight line.

### 2. Regression

Regression or curve fitting (Draper and Smith, 1966) allows the researcher to develop a predictive and descriptive equation relating the rating assigned by the panelist to the physical level or processing variable. The typical equation follows a straight line of the form

$$\text{Rating} = A + B(\text{stimulus level})$$

For category ratings researchers have found that the underlying relation exhibits a substantial degree of curvature, so that the relation does not follow a simple straight line (although as we increase the stimulus level we will increase the rating of perceived intensity). Researchers have found that the rating follows a logarithmic curve, so that the relation becomes a well-behaved straight line once we plot the rating against the logarithm of stimulus level:

$$\text{Rating} = A + B[\log(\text{stimulus level})]$$

(We see this type of relation in the General Foods Texture Profile scale, in

**TABLE 13**   **Example of the Use of Correlation Coefficients to Relate Sensory Ratings of Tenderness and Marbling to Instrumental Measures**[a]

| Fiber parameter[d] | A maturity[b] | | | Combined maturity[c] | | |
|---|---|---|---|---|---|---|
| | Marbling score | Shear force | Tenderness rating | Marbling score | Shear force | Tenderness rating |
| Percent $\beta$R fibers | 0.24 | −0.07 | 0.15 | 0.33** | −0.22 | 0.23 |
| Percent $\alpha$R fibers | 0.30 | −0.36 | 0.32 | 0.31* | −0.22 | 0.23 |
| Percent $\alpha$W fibers | −0.46*[e] | 0.36 | −0.40* | −0.50** | 0.35** | −0.36** |
| Percent $\beta$R area | 0.26 | −0.11 | 0.17 | 0.32** | −0.22 | 0.24* |
| Percent $\alpha$R area | 0.45* | −0.46* | 0.44* | 0.40** | −0.25* | 0.24 |
| Percent $\alpha$W area | −0.45* | 0.36 | −0.38 | −0.48** | 0.31* | −0.32** |
| $\beta$R:$\alpha$W | 0.42* | −0.27 | 0.29 | 0.46** | −0.31* | 0.30* |
| $\beta$R:$\alpha$R | −0.06 | 0.22 | −0.14 | −0.04 | 0.03 | 0.00 |
| $\alpha$W:$\alpha$R | −0.44* | 0.43* | −0.42* | −0.43* | 0.32* | −0.32* |
| $\beta$R:($\alpha$R + $\alpha$W) | 0.30 | −0.14 | 0.19 | 0.32** | −0.24 | 0.25* |
| $\alpha$W:($\beta$R + $\alpha$R) | −0.37 | 0.30 | −0.34 | −0.43** | 0.29* | −0.31* |

[a] From Calkins et al. (1981).
[b] $n = 25$.
[c] $n = 65$.
[d] Percentage of a given kind of fiber is the percentage of the total fiber number attributed to that individual fiber type. Percent area is the percentage of the total area attributed to that individual fiber types. Fiber ratios are based on comparisons of the fiber areas.
[e] Asterisks mark significant scores: * indicates significance at the 5% level; ** indicates significance at the 1% level.

which ratings increase logarithmically with the readings from the General Foods Texturometer, the instrument used to measure texture properties.)

For magnitude estimates versus physical intensities the relation exhibits curvature. Figure 3 shows some of the relations. The type of curvature varies with the attribute that the panelist judges and with the physical stimulus level measured. In most cases we find a concave downward curvature. This indicates that increasing the stimulus level generates increased perceived intensity, but at a decreasing rate. This relation applied to perceived viscosity versus apparent physical viscosity (Moskowitz, 1972) and to perceived hardness versus the modulus of elasticity (Harper and Stevens, 1964).

In logarithmic coordinates the curved functions become straight lines. Those functions that exhibit the concave downward curvature generate straight lines whose slope lies between 0 and 1.0. For those functions in which the relation appears to accelerate (i.e., with perceived intensity growing more rapidly than we would expect on the basis of physical magnitude), the slope in log–log coordinates exceeds 1.0. An example is the

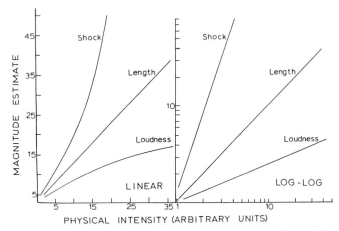

**Fig. 3.** Relation between physical intensities and magnitude estimates. The left panel shows the functions in linear coordinates. The right panel shows the functions in log–log coordinates, in which power functions become straight lines.

perceived grittiness of sandpaper, with the physical measure equaling the "grit" value (Cussler *et al.,* 1979). Other representative functions exhibiting this accelerating behavior include the perceived sweetness of sucrose (along with sweetness of other carbohydrate sweeteners) and the saltiness of sodium chloride.

As a unifying principle we write the relation between magnitude estimates and physical magnitudes in two forms. In the logarithmic form,

$$\text{Log magnitude estimate} = A + B(\text{log physical magnitude})$$

In the linear form,

$$\text{Magnitude estimate} = A(\text{physical magnitude})^B$$

Over the past 30 years researchers have used the magnitude estimation method to develop basic relations between physical magnitudes and perceived intensities. In many cases they have uncovered power functions that relate the two domains. Table 14 presents representative power function exponents for a wide variety of continua as presented by S. S. Stevens (1962), who discovered the basic power function relation, and by Moskowitz and Chandler (1977).

## H. An Overview of Intensity Measurements

This section has illustrated some of the different approaches to intensity measurements in sensory evaluation. The reader should keep in mind that

**TABLE 14  Representative Exponents of the Power Functions Relating Psychological Magnitude to Stimulus Magnitude**

A. General Array of Exponents[a]

| Continuum | Exponent | Stimulus condition |
|---|---|---|
| Loudness | 0.60 | Binaural |
| Loudness | 0.54 | Monaural |
| Brightness | 0.33 | 5° target — dark-adapted eye |
| Brightness | 0.50 | Point source — dark-adapted eye |
| Lightness | 1.20 | Reflectance of gray papers |
| Smell | 0.55 | Coffee odor |
| Smell | 0.60 | Heptane |
| Taste | 0.80 | Saccharin |
| Taste | 1.30 | Sucrose |
| Taste | 1.30 | Salt |
| Temperature | 1.00 | Cold — on arm |
| Temperature | 1.50 | Warmth — on arm |
| Vibration | 0.95 | 60 cps — on finger |
| Vibration | 0.60 | 250 cps — on finger |
| Duration | 1.10 | White-noise stimulus |
| Repetition rate | 1.00 | Light, sound, touch, and shocks |
| Finger span | 1.30 | Thickness of wood blocks |
| Pressure on palm | 1.10 | Static force on skin |
| Heaviness | 1.45 | Lifted weights |
| Force of handgrip | 1.70 | Precision hand dynamometer |
| Vocal effort | 1.10 | Sound pressure of vocalization |
| Electric shock | 3.50 | 60 cps — through fingers |
| Tactual roughness | 1.50 | Felt diameter of emery grits |
| Tactual hardness | 0.80 | Rubber squeezed between fingers |
| Visual velocity | 1.20 | Moving spot of light |
| Visual length | 1.00 | Projected line of light |
| Visual area | 0.70 | Projected square of light |

(*continued*)

**TABLE 14** *(Continued)*

B. Psychophysical Exponents for Real Foods[b]

| Food/attribute | Exponent |
|---|---|
| Appearance | |
| 1. Perceived size of grind in hamburger | 0.55 |
| Texture in mouth | |
| 1. Hardness of space cubes (vs. modulus of elasticity) | 0.41 |
| 2. Hardness of space cubes (vs. ultimate strength) | 0.61 |
| 3. Crunchiness of space cubes (vs. modulus of elasticity) | 0.55 |
| 4. Crunchiness of space cubes (vs. ultimate strength) | 0.72 |
| Aroma | |
| 1. Coffee odor | 0.55 |
| 2. Pork | 0.46 |
| 3. Cheddar cheese | 0.20 |
| 4. Grapefruit | 0.44 |
| 5. California orange | 0.52 |
| Flavor in mouth | |
| 1. Amount of mayonnaise | 1.2 |
| 2. Amount of cherry flavor | 1.2 |
| 3. Coffee flavor | 0.75 |
| Taste in mouth | |
| 1. Amount of sweetness in cherry beverage | 1.2 |

[a] Adapted from Stevens (1962).
[b] Adapted from Moskowitz and Chandler (1977).

proper intensity measurement of perception obeys the same rules whether one studies real foods or model systems (e.g., sugar and water). It should come as no surprise, therefore, that the greatest advances in sensory measurement have emerged from those laboratories that have taken sensory measurement down from the pedestal of the impossible and incomprehensible to the realm of the actionable, comprehensible, and meaningful.

## IV. MEASURING LIKES AND DISLIKES

### A. Introduction

The third basic area of sensory analysis concerns likes and dislikes. In addition to registering the sensory qualities and intensities of stimuli we make judgments about "good–bad" or "like–dislike." More than any other characteristic of a product, these hedonic characteristics determine our behavior when it comes to choosing foods the first time and repeating an initial choice upon continued exposure.

Sensory analysts, keenly aware of the importance of hedonics for the evaluation of food, have developed a variety of techniques to measure liking. They have developed different scales and a variety of measures.

## B. Attitudes versus Behavior

People say things, but when it comes to behavior they may or may not follow what they say. An individual who says he or she likes a food may not consume the food when given the opportunity. We have to distinguish between attitudes toward a food and behavior. Most of the research on food acceptance and sensory analysis has concerned attitudes: panelists rate a food for liking but do not necessarily have the opportunity to consume the food in the manner that they would if they actually purchased it.

## C. Measuring Likes by Classifying Responses

The earliest forms of hedonic measurement involved counts of the number of panelists who said that they liked or disliked particular foods (or stimuli) in laboratory test situations. From these counts researchers could determine whether panelists agreed with each other. The research from both the applied realm of food acceptance and the sensory scientific realm of odor and taste showed some fairly interesting results, which researchers obtained simply by polling the population.

1. For simple odors, no single aroma or fragrance received universal positive votes for acceptability. Rose fragrance received the greatest number of positive votes (Kenneth, 1924).

2. The votes agreed with each other more for the unpleasant than for the pleasant fragrances. Individuals agreed that the fragrance of carbon bisulfide smelled unpleasant, and they agreed more with each other for this (and similar rank odors) than for the pleasant ones. Apparently, the rules governing which odors smell "good" differ more from individual to individual than the rules governing which odors smell "bad."

3. For simple tastes, most individuals rated sweet as pleasant, salt as pleasant at low and middle levels but unpleasant at high levels, and sour and bitter as unpleasant at most concentrations (Engel, 1928).

4. For taste, the votes for pleasantness changed depending upon concentration.

5. For foods, no food received universally "pleasant" or "liked" ratings. Individuals differed. Unfortunately, we do not know whether individuals differ more on their hedonic classification of foods or on their hedonic classifications of simple tastes and aromas.

Some typical results from counts of votes of "like" versus votes of "dislike" appear in Table 15 for simple chemical stimuli that researchers would test in the laboratory and in Table 16 for actual foods.

Although in the formative years of sensory analysis researchers used headcounts or classification into likes and dislikes, they have recently either abandoned this simplistic method, or at least augmented it, for the following reasons:

1. Researchers now recognize that individuals can show varying levels or degrees of liking and do not simply like or dislike a food. By obtaining only an overall classification into one of two hedonic categories, the researcher loses quite a lot of information,

2. Simple incidence statistics (i.e., the number of panelists saying that they dislike a stimulus) permit weaker analysis than measurements of the degree of liking. With incidence statistics the researcher has to poll many individuals in order to obtain a stable estimate of the proportion who like or dislike a stimulus. With ratings of degree of liking the researcher needs to poll far fewer panelists, because even a single panelist can show "degree" of liking.

## D. Measures of Degree of Liking

Over the past 30 years researchers have developed a variety of scaling procedures to measure degree of liking. The scientific and pragmatic needs that impelled researchers to develop category and ratio scales of perceptions spilled over to the development of scales for degree of liking. Indeed, the history of sensory analysis will show more scales for measuring likes and dislikes than for measuring intensity of perception. In most applied research, overall "liking" represents the "bottom line" or key evaluative criterion against which the researcher judges all other variables.

Table 17 presents some category scales used by researchers in food science to measure the degree of liking of simple "model stimuli" (e.g., sucrose in water) as well as actual foods and even menus (Meiselman, 1978). Note that each researcher tends to select that set of attributes for liking with which he or she feels comfortable. Depending upon the researcher's particular needs the scale may have an odd number of categories or an even number, may have more liking categories or an equal number of categories for liking and

**TABLE 15**  Percentage Liking and Disliking the Taste of Simple Solutions of Chemicals[a,b]

| Level sucrose | | | Level NaCl | | | Tartaric acid | | | Quinine sulfate | | |
|---|---|---|---|---|---|---|---|---|---|---|---|
| (%) | Pleasant | Unpleasant | (%) | Pleasant | Unpleasant | (%) | Pleasant | Unpleasant | (%) | Pleasant | Unpleasant |
| 1 | 5 | 95 | 0.5 | 48 | 52 | 0.05 | 50 | 50 | 0.0001 | 0 | 0 |
| 3 | 76 | 24 | 1 | 81 | 19 | 0.11 | 94 | 94 | 0.0003 | 50 | 50 |
| 5 | 100 | 0 | 2 | 78 | 22 | 0.17 | 85 | 15 | 0.0005 | 52 | 48 |
| 7 | 100 | 0 | 3 | 62 | 38 | 0.22 | 77 | 23 | 0.0008 | 50 | 50 |
| 9 | 100 | 0 | 4 | 19 | 81 | 0.28 | 78 | 22 | 0.0011 | 52 | 48 |
| 11 | 100 | 0 | 5 | 9 | 91 | 0.34 | 70 | 70 | 0.0013 | 16 | 84 |
| 13 | 100 | 0 | 6 | 14 | 86 | 0.39 | 52 | 52 | 0.0015 | 2 | 98 |
| 15 | 96 | 4 | 8 | 4 | 96 | 0.45 | 55 | 45 | 0.002 | 0 | 100 |
| 20 | 96 | 4 | 10 | 0 | 100 | 0.56 | 33 | 67 | 0.0025 | 0 | 100 |
| 30 | 96 | 4 | | | | 0.67 | 25 | 75 | 0.003 | 0 | 100 |
| 40 | 95 | 4 | | | | 0.78 | 25 | 75 | 0.0035 | 0 | 100 |
| | | | | | | | | | 0.004 | 0 | 100 |

[a] Data from Engel (1928).
[b] Data reflect only those panelists who either liked or disliked the stimuli and not those panelists who voted "neutral."

**TABLE 16  Percentage Likes and Dislikes of Foods by College Students[a]**

| Items liked by 66% or more | | Items disliked by 33% or more | | Items unfamiliar to 5% or more | |
|---|---|---|---|---|---|
| Item | % "like" responses | Item | % "dislike" responses | Item | % "don't know" responses |
| French fried shrimp | 80.3 | Chicken giblets on rice pilaf | 50.5 | Southern ham shortcake | 14.6 |
| Chili and crackers | 78.2 | Yogurt fruit plate | 48.0 | Hunter's dinner | 13.8 |
| Grilled steak | 74.5 | Sole almondine | 44.7 | Veal cordon bleu | 13.3 |
| Roast turkey | 72.3 | Fried rabbit | 44.5 | Boston baked beans | 13.2 |
| | | Peach of a dairy salad plate | 42.7 | Cheese souffle | 12.8 |
| Bacon, lettuce, tomato sandwich | 71.2 | Brighten-a-blustery-day plate | 41.6 | Cheese balls on pineapple | 12.4 |
| Tacos | 70.4 | Boston baked beans | 39.7 | Snowdrift squares | 12.1 |
| Swiss steak with gravy | 69.8 | Corned beef hash | 38.1 | Cheese rarebit on toast | 11.9 |
| Roast ham | 68.6 | Salmon patties | 37.4 | Chicken crepes | 10.2 |
| Beef french dip | 68.5 | Omelet with mushroom sauce | 36.8 | Fried rabbit | 9.9 |
| Country fried chicken | 67.9 | Hunter's dinner | 35.4 | Beef biscuit roll | 9.4 |
| Grilled minute steak | 67.9 | Six-layer dinner | 35.1 | Yogurt fruit plate | 9.4 |
| Barbecued beef on bun | 67.8 | Salad greens, cottage cheese, fruit | 34.9 | Chicken antoine | 8.6 |

| | |
|---|---|
| Red snapper | 34.0 |
| Cold sliced meat loaf and turkey | 33.7 |
| Cheese souffle, with cheese sauce | 33.4 |
| Cheese rarebit on toast | 33.2 |
| Meat salad, cup of soup | 8.3 |
| Omelet | 8.1 |
| Sole almondine | 7.8 |
| Beef cutlet wrapped around dressing | 7.8 |
| Chicken giblets on rice pilaf | 7.5 |
| Dutch treat plate | 7.3 |
| Beef birds with gravy | 7.0 |
| Shrimp Louis salad bowl | 6.5 |
| Brighten-a-blustery-day plate | 6.4 |
| Trio luncheon plate | 6.2 |
| Chicken ala king | 6.2 |
| Salmon patties, cream sauce | 6.1 |
| Red snapper | 5.7 |
| Egg salad on lettuce | 5.7 |
| Hospitality plate | 5.7 |
| Pork ribs, sauerkraut | 5.6 |
| Plum delicious plate | 5.6 |
| Salad greens, cottage cheese, fruit | 5.4 |
| Peach of a dairy salad bowl | 5.4 |

[a] Adapted from Johnson and Vaden (1979).

**TABLE 17   Hedonic Scales of Food Preference**[a]

| Number of points | Scale categories |
|:---:|:---|
| 2 | Dislike, unfamiliar |
| 2 | Acceptable, dislike, not tried |
| 3 | Like a lot, like, dislike, do not know |
| 3 | Well liked, indifferent, disliked, seldom or never eaten |
| 5 | Like very much, like moderately, neutral, dislike moderately, dislike very much |
| 5 | Very good, good, moderate, tolerate, dislike, never tried |
| 5 | Very good, good, like moderately well, tolerate, dislike |
| 5 | Very good, good, moderate, dislike, tolerate |
| 9 | Like extremely, like very much, like moderately, like slightly, neither like nor dislike slightly, dislike moderately, dislike very much, dislike extremely |

[a] Summarized by Meiselman (1978).

disliking, may have words attached to all of the categories or only to some, and may have a neutral point to allow panelists the freedom to express "no hedonic response" or may not.

## E. Ratio Scales (Magnitude Estimation)

As we saw previously for the evaluation of perceived intensity, researchers have begun to use ratio scales to measure the degree of liking. Moskowitz and Sidel (1971) reported that magnitude estimation provides ratios of acceptance, allowing the researcher to conclude that a panelist liked one product several times more than another. Subsequent studies by McDaniel (1973) in her Ph.D. thesis at the University of Massachusetts showed that magnitude estimation scales for overall liking of foods generated more discrimination between beverages (whiskey sours) than did conventional category scales. McDaniel and Sawyer (1981) concluded that the magnitude estimation scale would thus provide a more sensitive measuring instrument for assessing likes and dislikes as compared to the category scale. Moskowitz (1982) also reported that, in the evaluation of foods systematically varied on different formulation characteristics, the magnitude estimation method generated greater sensitivity to product differences for overall liking than did the 9-point hedonic scale developed by Peryam and Pilgrim (1957). These findings suggest that for applied product testing researchers may not have used adequately sensitive scales to measure product differences and that the ratio scaling procedure might generate better data.

## F. Time-Preference Measures of Liking

We do not select foods in a vacuum. Much of food acceptance varies as a function of the time that has passed since the individual last ate the food. Looking at results from food acceptance surveys, we might think that consumers never accept such meats as liver. Yet observations of everyday behavior will quickly reveal that from time to time consumers would like to eat liver. In many panel tests consumers prefer steak to hamburger, but if we eat steak day in and day out, we tire of it, and eventually find it unacceptable (Moskowitz *et al.*, 1979a). These observations introduce the concept of time preference into sensory evaluation of acceptance.

Time preference refers to the observation that selection varies according to the intervening time since the panelist last ate the food as well as to the intrinsic acceptability of the food to be panelist.

As proposed by Joseph Balintfy and his colleagues at the University of Massachusetts, the concept of time preference allows the researcher to represent acceptance by these two independent factors. Then, it becomes possible to arrange foods in a selective menu so that the food appears sufficiently often that any more frequent or less frequent appearance would fail to maximize its overall acceptability over time (Balintfy *et al.*, 1974).

In terms of its implications for the sensory analysis of foods, we should note that most basic and applied research in sensory analysis takes place in the laboratory rather than in the home. Thus, many of the acceptance measures that researchers obtain for foods may not reflect true acceptance, because the panelist has not had a chance to consume the food in the natural environment. Rather, the panelist consumes only small portions of the food and tends thus to "uprate" the food in terms of liking. Testing the food at home and consuming it over several occasions will cause the acceptance level to change. This time-preference factor will come into play, and the data on acceptance will reflect both the intrinsic acceptability of the food and the degree to which panelists become bored with it.

## V. PUTTING IT ALL TOGETHER: USING SENSORY ANALYSIS FOR PRODUCT OPTIMIZATION

## A. Introduction

Over the past decade sensory analysis of foods has moved ever so gradually from a passive science that reports consumer reactions to food into a proactive tool by which product developers can determine the direction and magnitude of change to make in a product to increase its acceptance. The coupling of sensory analysis techniques (which provides a communication vehicle to consumers) with product modeling (which relates formulations to responses) naturally leads to product optimization and to the implementation of sensory analysis for actionable feedback to product developers (rather than just information reported back to them with little or no concrete guidance).

### B. The Approach of Optimization

Product optimization by sensory analysis strives to accomplish the following goals:

1. Develop an integrated data base of formulation and processing variables, panelist sensory ratings, and panelist acceptance measures.
2. Use that integrated data base to develop equations interrelating panelist ratings, formulations, and processing variables.
3. Use those equations to maximize acceptance ratings, subject to constraints (either cost, processing limitations, or sensory attributes constrained to lie within specified levels).

Over the past decade researchers have begun to use experimental designs along with optimization techniques as an efficient way to maximize consumer acceptance of ratings.

### C. Experimental Designs

Experimental design refers to arraying products so that they vary in a predesignated way on specific formulation variables. For instance, if the researcher wishes to investigate the effect of frying time on the perceived flavor intensity of potato chips, an experimental design might dictate the preparation of five different potato chips, representing systematically increasing frying times from a very short fry time to a very long fry time.

Experimental designs allow the researcher to investigate combinations of variables without having to prepare an undue number of different prototypes. The experimental design creates an efficient set of prototypes that can substitute for the many dozens or hundreds of samples that the researcher might make without this directed guidance. Table 18 presents some different experimental designs that researchers use in optimization studies.

### D. Sensory Evaluation Tests

The experimental design provides the efficient product set. The product evaluator then subjects these formulations (along, perhaps, with existing

**TABLE 18  Representative Experimental Designs: Three Formula Variables**[a]

A. Full Factorial (Two Levels)

|   | Variable 1 | Variable 2 | Variable 3 |
|---|---|---|---|
| 1 | H | H | H |
| 2 | H | H | L |
| 3 | H | L | H |
| 4 | H | L | L |
| 5 | L | H | H |
| 6 | L | H | L |
| 7 | L | L | H |
| 8 | L | L | L |

B. Central Composite Design

|   | Variable 1 | Variable 2 | Variable 3 |
|---|---|---|---|
| 1 | H | H | H |
| 2 | H | H | L |
| 3 | H | L | H |
| 4 | H | L | L |
| 5 | L | H | L |
| 7 | L | L | H |
| 8 | L | L | L |
| 9 | H | M | M |
| 10 | L | M | M |
| 11 | M | H | M |
| 12 | M | L | M |
| 13 | M | M | H |
| 14 | M | M | L |
| 15 | M | M | M |

C. Full Factorial (Three Levels)

|   | Variable 1 | Variable 2 | Variable 3 |
|---|---|---|---|
| 1 | H | H | H |
| 2 | H | H | M |
| 3 | H | H | L |
| 4 | H | M | H |
| 5 | H | M | M |
| 6 | H | M | L |
| 7 | H | L | H |
| 8 | H | L | M |
| 9 | H | L | L |
| 10 | M | H | H |
| 11 | M | H | M |
| 12 | M | H | L |
| 13 | M | M | H |
| 14 | M | M | M |

*(continued)*

**TABLE 18** (*Continued*)

C. Full Factorial (Three Levels)

|    | Variable 1 | Variable 2 | Variable 3 |
|----|-----------|-----------|-----------|
| 15 | M | M | L |
| 16 | M | L | H |
| 17 | M | L | M |
| 18 | M | L | L |
| 19 | L | H | H |
| 20 | L | H | M |
| 21 | L | H | L |
| 22 | L | M | H |
| 23 | L | M | M |
| 24 | L | M | L |
| 25 | L | L | H |
| 26 | L | L | M |
| 27 | L | L | L |

[a] H = high, M = medium, L = low.

competitors) to panel testing. The researcher might wish to use only consumers to evaluate the products or might wish to secure both expert panel and consumer panel data [see Moskowitz *et al.* (1979b)]. In either case the researcher treats these products as test formulations, evaluating them in the same way as one might evaluate any food product.

The output of the evaluation is a data matrix, similar to that shown in Table 19, which systematically varies three formula variables for pasta sauce. Note that the formulations involve wide ranges of three variables, so that it becomes possible to generate products that have quite different sensory characteristics and that vary (hopefully) in acceptability to the consumer.

## E. Developing the Product Model

By itself the matrix of data in Table 19 simply presents information that a sensory analyst or product developer could use to select the best product. However, because the product developer systematically varied the product formulations and processing conditions in known ways it becomes possible to tie the data together by means of equations that summarize the relations between variables in a quantitative form. The researcher can fit equations to the ratings and by so doing develop a "shorthand" summary of how formula variables relate to perceptions and acceptance as rated by consumers.

The equations that the researcher uses generally describe the data only.

TABLE 19  Data Base of Ratings for Pasta Sauce

| Product | Flavoring | Thickener | Tomato solids | Liking | Flavor intensity | Thickness | Spicy | Good by itself | All-purpose use |
|---|---|---|---|---|---|---|---|---|---|
| 1 | 10 | 16 | 1 | 57 | 55 | 76 | 47 | 53 | 30 |
| 2 | 10 | 16 | 4 | 27 | 34 | 79 | 33 | 46 | 45 |
| 3 | 10 | 1 | 1 | 16 | 54 | 15 | 37 | 28 | 21 |
| 4 | 10 | 1 | 4 | 1 | 34 | 15 | 49 | 31 | 31 |
| 5 | 2.5 | 16 | 1 | 51 | 38 | 69 | 43 | 49 | 41 |
| 6 | 2.5 | 16 | 4 | 44 | 44 | 84 | 52 | 67 | 57 |
| 7 | 2.5 | 1 | 1 | 8 | 25 | 23 | 37 | 28 | 33 |
| 8 | 2.5 | 1 | 4 | 17 | 32 | 16 | 34 | 28 | 32 |
| 9 | 10 | 8.5 | 2.5 | 23 | 37 | 19 | 40 | 36 | 40 |
| 10 | 2.5 | 8.5 | 2.5 | 22 | 26 | 16 | 25 | 26 | 26 |
| 11 | 6.25 | 1 | 2.5 | 49 | 57 | 78 | 53 | 66 | 57 |
| 12 | 6.25 | 16 | 2.5 | 14 | 39 | 11 | 63 | 38 | 37 |
| 13 | 6.25 | 8.5 | 1 | 21 | 38 | 23 | 35 | 30 | 26 |
| 14 | 6.25 | 8.5 | 4 | 33 | 51 | 21 | 41 | 45 | 53 |
| 15 | 6.25 | 8.5 | 2.5 | 29 | 45 | 33 | 48 | 43 | 52 |

They do not purport to model what physical processes really occur, but rather they summarize the data in an empirically based, mathematically tractable, and usable form. In effect the researcher simply searches for an equation that has two useful properties:

1. The equation describes the data (i.e., it "fits" the points reasonably well and accounts for a lot of the variation in the ratings).

2. The equation possesses a limited number of terms, sufficient to capture relations between formula variables and panelist ratings but not so many that in effect the equation really fits "random variability" or "noise." The researcher attempts to develop a reasonably parsimonious model that describes the ratings in an efficient fashion.

## F. Results: Modeling

Table 20 presents the parameters of the product model generated by the regression analysis. Along with the terms in the model we see measures of the "significance" of the entire model. To measure the significance of specific terms in the model we do a $T$ test of the coefficients and use only those coefficients that exhibit a significant $T$ value ($T > 1$ or $T < -1$). For the model as a whole we use two different statistics:

1. Multiple $R^2$. The square of multiple $R$ tells us the percentage of variability accounted for by the equation. Obviously, in developing a product model we would like an equation that explains as much of the variability as possible.

2. $F$ ratio. The $F$ ratio is a measure of the explained variability in the ratings to the unexplained variability, and it takes into account the number of predictors that we use. (As we use an increasing number of predictors or independent variables, we can account for more variability in the data, but the model becomes less parsimonious and "tight." The $F$ ratio will decrease even though the $R^2$ value increases.)

The reader should keep in mind that the equations involve linear, quadratic (second order, or squared), and interaction terms. Although we can easily visualize relations for two variables, once we have three independent variables and a dependent variable it becomes increasingly difficult to visualize the relations. The researcher must rely upon numerical methods to evaluate the equations.

## G. The Product Model

By developing equations for each attribute and then using the model for optimization the product developer can estimate the likely sensory and

**TABLE 20  Models Relating Formula Variations of Pasta Sauce to Attribute Rating**

| | Liking | Flavor intensity | Thickness | Spicy | Good by itself | All purpose |
|---|---|---|---|---|---|---|
| Constant | 12.68 | 12.84 | 11.43 | 3.28 | 18.46 | 15.12 |
| Flavoring | — | 11.23 | — | 13.76 | 9.12 | 9.39 |
| (Flavoring)$^2$ | −0.24 | −0.97 | — | −0.99 | −0.77 | −0.74 |
| Thickener | 3.99 | — | 2.22 | −5.30 | — | 1.55 |
| (Thickener)$^2$ | −0.15 | 0.08 | 0.15 | 0.37 | 0.13 | — |
| Tomato solids | — | −6.52 | — | 12.98 | −5.08 | — |
| (Tomato solids)$^2$ | −1.30 | — | 0.37 | −2.71 | — | −0.87 |
| Flavoring × thickener | — | −0.19 | — | −0.13 | −0.11 | −0.09 |
| Flavoring × solids | 1.03 | 1.29 | — | — | 0.51 | — |
| Thickener × solids | — | — | −0.31 | — | — | — |
| Multiple $R^2$ | 0.96 | 0.90 | 0.97 | 0.70 | 0.91 | 0.76 |
| F ratio | 36.37 | 11.95 | 10.99 | 2.32 | 14.22 | 5.89 |
| Degrees of freedom | 15,7 | 15,7 | 15,5 | 15,8 | 15,7 | 15,6 |

image–acceptance profile of any set of formulation variables. The researcher can do many things with the product model.

1. Optimize acceptance.
2. Estimate the likely sensory and image profile of the optimum.
3. Constrain the formulation variables to new limits.
4. Use cost or a sensory attribute as a constraint and find the best product
within formulation limits whose sensory profile or cost of goods also satisfies
preset constraints.
5. Explore the surface to get a feeling of which particular regions appear
most promising in terms of acceptance.

### H. Optimizing Acceptance

Optimization of a product can be considered in terms of the analogy of
"hill climbing." By envisioning the product model as a "hill" we see that to
optimize a product we must find the top of the hill. We cannot exceed the
boundaries in any dimension. Thus, the aim becomes to find the highest
point on the surface. Figure 4 shows a typical surface that one might find
with a well-defined optimum.

By the use of hill-climbing procedures it becomes possible to find the top
of the hill. Table 21 shows the results of steps or iterations. At each iteration
we ascend higher on the hill or acceptance surface. Early on in the hill
climbing we generate relatively large increments in acceptance for each

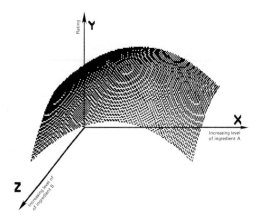

**Fig. 4.**    Schematic of a response surface showing how the variables interact to generate
ratings of acceptance. Optimization consists of going up to the top of the response surface.

**TABLE 21  Sequence of Iterations to Maximize/Optimize "Liking" of Pasta Sauce**

| | Formula variables | | | Ratings | | | | | | |
|---|---|---|---|---|---|---|---|---|---|---|
| Iteration | Tomato solids | Thick-ener | Flavoring level | "Liking" | Flavor | Thickness | Spicy | Good by itself | All purpose | Cost[a] of goods |
| 001 | 4.00 | 1.00 | 10.0 | 13 | 52 | 18 | 44 | 32 | 22 | 760 |
| 005 | 2.50 | 8.50 | 10.0 | 37 | 34 | 37 | 28 | 33 | 35 | 1330 |
| 010 | 4.00 | 16.00 | 2.50 | 50 | 35 | 71 | 45 | 50 | 41 | 1510 |
| 015 | 4.00 | 16.00 | 8.71 | 59 | 50 | 71.4 | 48 | 55 | 39 | 1882 |

[a] Cost of goods 20(tomato solids) + 80(thickener) + 60(flavoring). All concentrations are in relative units.

iteration or step. We go onward only if the next iteration generates an even more acceptable product. As one might expect, after several iterations we have gotten fairly close to the top of the hill. As a consequence, each iteration or next step generates only marginal improvements.

Table 21 shows the highest level achieved. The researcher can stop here and simply take the formulation variables as those levels that maximize acceptance. However, since in the previous step we developed a product model in which we related formulation variables to perceptions, it now becomes possible to estimate the likely attribute profile. This profile for the optimum also appears in Table 21.

## I. Changing Formulation Limits

From time to time product developers wish to modify the formulation limits of specific variables and evaluate the changes that might occur in the sensory characteristics as perceived by panelists. In the search for optimal products quite often the initial product development cycle puts no limitations on the levels of a formulation ingredient or the processing conditions. Nonetheless, later on, when it comes time to scale up the product manufacturing for commercial use, financial considerations dictate that the formula variables take on narrower ranges than those studied. The product-modeling technique allows the researcher to evaluate the change in sensory profile and in acceptance. Table 22 shows what occurs for our sauce product when we systematically reduce the range of formulation levels available for the product. Depending upon the specific product and the degree to which the optimum lies near an upper or a lower limit, the change in allowable formulation range will change acceptance and change the sensory profile.

## J. Constraining Cost and/or Sensory Attributes

Another use of the product model comes from developing the optimally acceptable product while at the same time ensuring that another perceptual attribute to an attribute such as cost of goods lies within predetermined limits. The aim of this approach consists of allowing the product formulation to vary within the original range tested. However, the optimization has two parallel goals: (1) to maximize acceptance (or some other criterion), and (2) at the same time to ensure that the product formulation generates a sensory profile whose values lie within specific predetermined guidelines.

We can see this approach for two types of constraints, as shown in Table 23. In the first set of constraints we systematically reduce the total cost of

TABLE 22  Examples of How Decreasing the Allowable Formulation
Range of Ingredients Decreases the Acceptance of the Product

|  | Wide range | Medium range | Narrow range |
|---|---|---|---|
| Formula variables |  |  |  |
| Tomato solids | 1–4 | 1.5–3.5 | 2–3 |
| Thickener | 1–16 | 4–12 | 4–8 |
| Flavoring level | 2.5–10 | 4–8 | 5–6 |
| Optimal formulas |  |  |  |
| Tomato solids | 4 | 3.5 | 3 |
| Thickener | 16 | 12 | 8 |
| Flavoring level | 8.71 | 7.44 | 6 |
| Ratings |  |  |  |
| Liking | 59 | 52 | 42 |
| Flavor | 50 | 48 | 45 |
| Thickness | 71 | 52 | 35 |
| Spicy | 48 | 42 | 40 |
| Good by itself | 55 | 49 | 42 |
| All purpose | 39 | 44 | 45 |
| Cost of goods | 1882 | 1476 | 1060 |

goods for the sauce. At each cost reduction phase we attempt to optimize
acceptance. The formulations can vary within the range. However, while
trying to maximize acceptance we cannot select those particular formulations whose costs exceed a specific dollar value. (We estimate cost of goods
by means of the equation relating cost of goods to formulation variables.) As
Table 23 shows, initially without any cost constraints we find a formulation
of intermediate thickener level to generate the maximal level of acceptance.
No matter how much we permit the cost of goods to increase, the maximally
acceptable product still has a finite, relatively moderate cost. Adding more
ingredients (such as thickener) will increase the total cost of goods but may
actually decrease acceptance. As we reduce cost of goods, however, we soon
begin to reduce the allowable formulation levels below their optimal point.
Eventually we reach formulations that exhibit reduced acceptance, as Table
23 shows.

Following the same logic, let us now develop a series of product formulations that exhibit prespecified sensory characteristics. The product model
provides a method for evaluating the likely sensory rating for each attribute
since the product model interrelates sensory attributes and formulation
levels. During the optimization process we try to maximize liking but at the
same time continue to check each possible formulation to ensure that its

**TABLE 23   Product Modeling with Two Types of Constraints**

|               | No cost constraint | Cost <1600 | Cost <1200 | Cost <800 | Low flavor (<40) | Low thickness (<40) |
|---------------|-------------------|------------|------------|-----------|------------------|---------------------|
| Tomato solids | 4.00              | 3.92       | 2.88       | 1.01      | 3.99             | 4.00                |
| Thickener     | 16.00             | 14.06      | 11.54      | 7.86      | 15.95            | 9.45                |
| Flavoring     | 8.71              | 6.58       | 3.65       | 2.51      | 3.21             | 10.00               |
| Cost          | 1882              | 1599       | 1200       | 800       | 1549             | 1435                |
| Liking        | 59                | 56         | 48         | 38        | 52               | 47                  |
| Flavor intensity | 50             | 51         | 38         | 33        | 40               | 43                  |
| Thickness     | 71                | 61         | 50         | 36        | 71               | 40                  |
| Moist spiciness | 48              | 47         | 38         | 21        | 49               | 21                  |

sensory profile lies within the range specified. Without any constraints on sensory profiles we will always find an optimal product. Placing a constraint on the sensory characteristics often decreases acceptance (just as cost reduction brought the product formulation below a certain level of acceptance). Finally we may place an infeasible constraint on the sensory characteristic by requiring an unachievably high or low level with the product variables and limits tested. In this case the technique of optimization will fail, because no product can yield a sensory profile satisfying the constraints set up ahead of time.

## K. Exploring the Response Surface

The final issue in optimization concerns the systematic exploration of the response surface to obtain an impression of which regions appear most promising in terms of acceptance or other relevant characteristics. Keep in mind that the product model, a series of equations, allows the researcher to evaluate the likely response profile of attributes for all regions lying within the limits of formulations tested. As a consequence the equations permit the researcher to search the space systematically, holding all but one ingredient or processing condition constant and then systematically varying that one ingredient. At each level in that systematic variation the researcher can estimate the likely sensory profile, by means of the equations relating formulation and processing variables to ratings.

To illustrate this procedure, consider the results shown in Table 24, where we have held all but one ingredient constant at the optimal level and then systematically varied that ingredient. We can obtain an idea of how changes in the ingredient manifest themselves in changes in responses. Such a

**TABLE 24  Exploring the Response Surface for Varying Tomato Solids**[a]

| Tomato solids | Liking | Flavor intensity | Thickness | Spicy | Cost of goods |
|---|---|---|---|---|---|
| 1.00 | 33 | 36 | 81 | 50 | 1822 |
| 1.37 | 38 | 38 | 79 | 53 | 1830 |
| 1.75 | 42 | 39 | 78 | 54 | 1838 |
| 2.13 | 46 | 41 | 76 | 55 | 1845 |
| 2.50 | 49 | 43 | 75 | 55 | 1853 |
| 2.88 | 52 | 45 | 74 | 55 | 1860 |
| 3.25 | 55 | 46 | 73 | 53 | 1868 |
| 3.63 | 57 | 48 | 72 | 51 | 1875 |
| 4.00 | 49 | 50 | 71 | 48 | 1882 |

[a] Tomato solids vary from 1.00 to 4.00; thickener = 16.00; flavoring = 8.71.

procedure would take much longer to accomplish if the researcher did not have a model that provided a generalized picture of relations between formulation levels and panelist responses.

## VI. MATCHING A PRODUCT TO A CONCEPT

Quite often applied sensory analysts try to match products to concepts. Product manufacturers develop product formulations that attempt to deliver what a concept or statement about the product promises. Traditional work in product–concept matching has required the panelist to rate the degree to which a product overdelivers, underdelivers, or delivers against a concept. The panelist reads the concept statement, which sets up expectations about the characteristics that the product will possess. Then, with the concept framework in mind, the panelist evaluates a product, rating it against the concept. On specific attributes the panelist has to state the degree of product–concept match.

Table 25 shows a typical ballot for this type of sensory study. Note that the results of this product–concept match provide the product developer with qualitative direction. The developer knows that the product fails to live up to expectations generated by the concept and has an idea of the dimensions or attributes in which the failure occurred, as well as some measure of the lack of fit. Nonetheless, the product developer cannot easily create a product formulation that fits the profile generated by the concept.

Recent work (Moskowitz *et al.,* 1977) deals with this question in the

**TABLE 25   Ballot for Concept–Product Test**

Please read the concept provided to you. It presents a new product for a low-calorie, dietetic margarine. Then taste the margarine. On each characteristic, check one of the three categories below, based upon how well the product lives up to the concept.

|  | The product | | |
|---|---|---|---|
|  | Delivers "too little" | Delivers "just right" | Delivers "too much" |
| Darkness | | | |
| Aroma strength | | | |
| Flavor strength | | | |
| Saltiness | | | |
| Margarine flavor | | | |
| Natural flavor | | | |
| Oiliness/texture | | | |
| Hardness | | | |
| Greasiness | | | |
| Mouth coating | | | |
| Aftertaste | | | |

framework of product optimization. These researchers developed a technique that accomplishes the following goals in product–concept matching:

1. To represent each product attribute by means of an equation. The equation relates each consumer-rated attribute to a linear or nonlinear (e.g., parabolic) combination of formula ingredients. Given the formula ingredients and their levels, the equation permits the researcher to estimate the likely sensory rating. A set of these equations makes up the product model, as we saw previously in Section V.

2. To provide a technique whereby the researcher can establish a "goal" or "target" profile of consumer-rated attributes. The profile may emerge from consumers themselves, who first evaluate existing products on their attributes, then read a concept, and finally profile the attributes of the concept just as if the concept represented another product. In other cases the profile may come from the sensory attributes of an existing product that the product developer wishes to reproduce with his or her own formulation variables.

3. To provide an algorithm or procedure by which one can evaluate different alternative formulation variables to determine which particular set of levels of the formulations generates the profile that comes "closest" (in a geometrical sense) to the desired goal profile.

Table 26 shows the goal profile and the formulation coming closest to that goal profile.


## VII. CONTRASTING AND RELATING EXPERT PANEL VERSUS CONSUMER PANEL DATA

In sensory analysis much controversy rages among practitioners regarding the appropriate type of panel to use in evaluation. Some practitioners aver that the product development team must use experts because experts know the characteristics of the product whereas consumers do not. Furthermore, experts by their very nature can better communicate nuances of the products that might totally escape the consumer. At the other end of the spectrum we see the researchers who work with consumers only, averring that consumers represent the target population. If a consumer does not perceive differences between products where an expert does, this simply means that the perceived differences do not amount to a relevant distinction. Despite what an expert may do, these researchers contend that the validity of evaluations lies primarily with the ultimate consuming population.

Recently researchers have begun to look at the data provided by experts and consumers as two different, parallel data sets for the same set of stimuli. Rather than choosing one type of data over another and ignoring either consumer or expert panel data, these researchers have begun to relate the two sets of data by equations and to predict the ratings from one group by the ratings for another.

Moskowitz *et al.* (1979b) published a key paper on the texture of bread, in which they developed equations relating formulation variables (e.g., sugar level, rye level) to the sensory attributes of bread (e.g., firmness, cohesiveness). In that study both expert panelists and consumer panelists participated. The results suggested that the two groups of panelists "looked at" the sensory attributes differently based upon the equation relating formulation variables to attribute ratings. Nonetheless, the data did provide a base with which to correlate the ratings of the two groups. Both groups may have assigned different ratings for the same bread, but when it came time to develop an optimally acceptable bread, both groups pointed to the same formulation. The regression analysis procedures used in that study showed that one could interrelate expert and consumer panel data. Similar types of interrelations emerged from a comprehensive study of texture by Cardello *et al.* (1982), who showed that the ratings of experts and consumers correlated with each other but that experts often tended to expand their scale of

**TABLE 26  Sequence of Steps for Developing a Product Having a Specific Profile**

Step 1: Develop a product model comprising equations which relate product attributes to formula variables. The equation can have linear, quadratic, and cross (interaction) terms

Step 2: Specify the goal profile (the desired attribute levels). These may represent levels of attributes for an existing product, which the product developer wants to copy with new formula ingredients

Step 3: Using the optimization approach, find that set of formula levels which generates a profile of attributes "closest" to the desired goal profile. Vary the formula levels in a systematic way, and compute the "distance" between the goal profile and the profile of the product corresponding to that formulation. Always choose the formula with the lowest or smallest distance

Example: Pasta sauce
Goal: A product with the following profile:

| | |
|---|---|
| Flavor intensity | 40 |
| Thickness | 50 |
| Spiciness | 40 |
| All-purpose use | 50 |

Ingredients varied: Flavoring level, thickener, tomato solids
Search sequence to find a product with a profile closest to the goal:

| | Formula variable | | | Expected attribute ratings | | | |
|---|---|---|---|---|---|---|---|
| Iteration | Flavoring level | Thickener | Tomato solids | Flavor intensity | Thickness | Spiciness | All-purpose use |
| 001 | 5.0 | 8.0 | 2.0 | 42.1 | 35.3 | 38.5 | 48.9 |
| 002 | 6.7 | 11.2 | 2.4 | 45.3 | 48.91 | 43.8 | 50.4 |
| 003 | 6.9 | 11.4 | 2.4 | 45.3 | 49.8 | 44.1 | 50.3 |
| 004 | 7.8 | 11.4 | 2.3 | 43.3 | 50.2 | 42.2 | 48.5 |

magnitude so that they would assign broader ranges of ratings for the same stimuli.

The area of interrelating consumer data, expert panel data, formulation data, and instrumental data remains largely unexplored in sensory analysis despite the intense investments by researchers in expert versus consumer panels. For the most part researchers do not know whether the differences that experts perceive translate to differences perceived by the consumer. Nor, in fact, in many cases, do researchers have any idea whether the products that experts direct the developer to generate will have any acceptance by the consumer until the sensory analyst or market researcher actually tests the prototype formulations.

## VIII. AN OVERVIEW OF SENSORY ANALYSIS

This chapter provides both a historical perspective and a description of the state of the art in sensory analysis. We can see a series of trends that have emerged from the history of sensory analysis and that have begun to propel the technology into the forefront of applied product development. These specific trends continue to appear:

1. Emphasis on better measurement of responses to stimuli.
2. Emphasis on a better language to describe perceptions and the development of that language by empirical means, using panelists.
3. Emphasis on measurement of consumer reactions, along with (or often in place of) expert panelists.
4. Cross-fertilization of sensory analysis by experimental psychology and statistics.
5. Search for models or equations relating ratings on attributes to either formulation variables or processing variables.
6. Use of those models to predict consumer ratings of products and/or to optimize consumer-rated acceptability.

The 1980s represent an exciting time for sensory analysis. Old procedures have begun to give way to new technology. The level of technical expertise among sensory analysts has increased and the actionability of their information for improved product development has increased in parallel fashion. With this continued growth we can expect to see more rapid development of consumer-acceptable products and a greater reliance by product developers on the data and guidance provided by sensory analysts. Certainly, the scientific literature reveals such development, with the applied sector following closely behind.

## REFERENCES

Amerine, M. A., Pangborn, R. M., and Roessler, E. B. (1965). "Principles of Sensory Evaluation of Food." Academic Press, New York.

Amoore, J. E. (1969). In "Olfaction and Taste III" (C. Pfaffmann, ed.), pp. 158–171. Rockefeller Univ. Press, New York.

Balintfy, J., Duffy, W. J., and Sinha, P. (1974). Oper. Res. 22, 771–727.

Boring, E. G. (1942). "Sensation and Perception in the History of Experimental Psychology." Appleton, New York.

Buttery, R. G., Guadagni, D. G., and Ling, L. C. (1973.) Agric. Food Chem. 21, 198–201.

Cairncross, S. E., and Sjostrom, L. B. (1950.) Food Technol. (Chicago) 4, 308–311.

Calkins, C. R., Dutson, T. R., Smith, G. C., Carpenter, L., and Davis, G. W. (1981). J. Food Sci. 46, 708–710, 715.

Cardello, A., V., Maller, O., Kapsalis, J. G., Segars, R. A., Sawyer, F. M., Murphy, C., and Moskowitz, H. R. (1982). J. Food Sci. 47, 1186–1197.

Caul, J. F. (1957). Adv. Food Res. 7, 1–40.

Cussler, E. L., Kokini, J. L., Weinheimer, R. L., and Moskowitz, H. R., (1979). Food Technol. (Chicago) October 1979, pp. 89–92

Draper, N., and Smith, H. (1966). "Applied Regression Analysis." Wiley, New York.

Engel, R. (1928). Arch. Gesamte Psychol. 64, 1–36.

Guth, C. E., and Mead, M. (1943). Report of the Committee on Food Habits. National Research Council,

Harper, R., and Stevens, S. S. (1964). Q. J. Exp. Psychol. 16, 204–215.

Harper, R., Bate-Smith, E. C., and Land, D. G. (1968a). "Odour Description and Odour Classification." Elsevier, New York.

Harper, R., Land, D. G., Bate-Smith, E. C., and Griffiths, N. M. (1968b). Br. J. Psychol. 59, 231–252.

Jeon, I. K., Breene, W., and Munson, S. T. (1973). J. Food Sci. 38, 334–337.

Johnson, K. E., and Vaden, A. G. (1979). J. Food Sci. 44, 457–262.

Katz, D. (1925). "Der Aufbau der Tastwelt," Z. Psychol. Monogr. Barth, Leipzig.

Kelly, A. T. (1955). "The Psychology of Personal Constructs." Norton, New York.

Kenneth, J. H. (1924). Perfum. Essent. Oil Rec. 15, 85–87.

LeGrand, J. Y. (1957). "Light, Color and Vision." Chapman and Hall, London.

Lundgren, B., Jonsson, B., Pangborn, R. M., Sontag, A. M., Barylko/Pikielna, N., Pietrzak, E., Dos Santos Garruit, R., Chabi Maroes, M. A., and Yoshida, M. (1978). Chem. Senses Flavour 3, 249–266.

McCurdy, S. M., Hard, M. M., and Martin, E. L. (1981). J. Food Sci. 46, 991–995, 998.

McDaniel, M. R. (1973). Unpublished Ph.D. dissertation, Univ. of Massachusetts, Food Science Dept., Amherst.

McDaniel, M. R., and Sawyer, F. M. (1981). J. Food Sci. 46, 182–185.

Meiselman, H. L. (1978). In "Encyclopedia of Food Science" (M. S. Peterson and A. H. Johnson, eds.), pp. 675–677. AVI, Westport, Connecticut.

Meiselman, H. L., Bose, H., and Nykvist, W. (1972). Percept. Psychophys. 12, 249–252.

Moskowitz, H. R. (1972). J. Texture Stud. 3, 89–100.

Moskowitz, H. R. (1982). ASTM Spec. Tech. Publ. STP 773, pp. 11–33.

Moskowitz, H. R., and Chandler, J. W. (1977). In "Sensory Properties of Foods" (G. G. Birch, J. G. Brennan, and K. J. Parker, eds.), pp. 189–209. Appl. Sci. Publ., Rippleside, United Kingdom.

Moskowitz, H. R., and Sidel, J. L. (1971). J. Food Sci. 36, 677–680.

Moskowitz, H. R., Stanley, D. W., and Chandler, J. W. (1977). *Food Prod. Dev.* **11**(2), 50–60.
Moskowitz, H., Fishken, D., and Ritacco, G. F. (1979a). *Food Technol.* (*Chicago*) August 1979, pp. 72, 74–75.
Moskowitz, H. R., Kapsalis, J. G., Cardello, A., Fishken, D., Maller, O, and Segars, R. A. (1979b). *Food Technol.* (*Chicago*) October 1979, pp. 84–88.
Peryam, D. R., and Pilgrim, F. J. (1957). *Food Technol.* (*Chicago*) **11**, 9–14.
Pfaffmann, C. M. (1959). *In* "Handbook of Physiology, Neurophysiology I" (J. Field, H.W., Magoun, and V. E. Hall, eds.), pp. 507–533. Am. Physiol. Soc., Washington, D.C.
Stevens, S. S. (1946). *Science* **103**, 667–670.
Stevens, S. S. (1962). Second Public Klopsteg Lecture, Northwestern Univ., Evanston, Illinois.
Stone, H., Sidel, J. L., Oliver, S., Woolsey, A., and Singleton, R. C. (1974). *Food Technol.* (*Chicago*) **28**, 24, 26, 28–29, 32, 34.
Szczesniak, A. S., Brandt, M. A., and Friedman, H. H. (1963). *J. Food Sci.* **28**, 397–403.
Titchener, E. B. (1916). "A Beginner's Psychology." Macmillan, New York.
Von Sydow, E., Moskowitz, H.R., Jacobs, H. L., and Meiselman, H. L. (1974). *Lebensm. Wiss. Technol.* **7**, 18–24.

3

# Quantitative Thin-Layer Chromatography of Foods and Beverages

DIETER E. JÄNCHEN

*CAMAG*
*Muttenz, Switzerland*

## I. INTRODUCTION

Since its introduction more than 25 years ago (Kirchner *et al.,* 1951; Stahl, 1962) thin-layer chromatography (TLC) has undergone not only

ANALYSIS OF FOODS AND BEVERAGES

different stages of methodological and instrumental development, as is usual, but also different degrees of recognition as an analytical method. Initially TLC had been considered a nondemanding technique appropriate for simple, qualitative bench tests. Around 1963 (Seiler *et al.,* 1963; Klaus, 1964) quantitative *in situ* evaluation methods (densitometry) were introduced and found appreciable interest, since in those days chromatographic analysis of nonvolatile substances was only possible by means of derivatization and gas chromatography or, without derivatization, by means of layer chromatography (i.e., TLC or paper chromatography).

When high-performance liquid chromatography (HPLC) became available and was rapidly accepted, quantitative TLC appeared to lose its justification. Only recently—after TLC had become an instrumental technique, which HPLC had always been—it regained recognition, and analysts became aware of the fact that both TLC and HPLC have their specific merits, often supplementing one another, and the decision which to prefer depends in the first place on the analytical application problem.

## A. Aspects for Choosing TLC

Thin-layer chromatography is characterized by a number of significant differences when compared with HPLC.

1. TLC is an open system. The three steps in the procedure—sample application, chromatogram development, and chromatogram evaluation—are carried out independently from one another. Many samples can be chromatographed side by side. Only those fractions that are of analytical interest need to be quantitatively measured. Consequently, TLC is particularly useful for handling a large number of samples (i.e., "mass production analysis"), and it has been shown (Jänchen, 1979) that the total analysis time per sample is in the order of magnitude of 2 min whereas labor time is about half a minute.

2. The TLC layer, that is, the stationary phase, is disposable, and this has a number of consequences: Lifetime considerations are immaterial. Therefore, sample preparation can be less critical, since irreversible contamination of the layer can be tolerated as long as artifacts with sample material of analytical interest can be precluded. For the same reason the selection of the developing solvent, the "mobile phase," can be made solely with respect to optimum chromatographic resolution. It should be noted that in HPLC one often resorts to reversed-phase systems for two reasons: (1) unacceptable lifetime of the stationary phase with a "straight system," and (2) compatibility of the RP mobile phase with the detection wavelength.

3. In TLC, detection is carried out in the absence of the mobile phase, and it can be repeated, with the same or with different parameters (wavelength as

well as scanning mode), for all or selected fractions, without repeating the chromatographic process. This makes TLC extremely selective and facilitates identification.

4. Of course, there are also considerations limiting the suitability of TLC. One is the limited separation distance. The number of theoretical plates in a TLC layer, preferably expressed by the separation number (Kaiser, 1977), limits the number of fractions in a sample to about 10 that can be quantitated and about 20 that, in total, can be tolerated. The migration speed of a developing solvent in a chromatographic layer, caused by capillary action, decreases with the square of the distance, while the speed of (nondirectional) diffusion of the fractions in the solvent remains constant. Because of this an optimum separation distance can be defined for a given layer chromatographic system that cannot be exceeded without loss of resolution [and unless external pressure is employed (Tyihak et al., 1979), a technique that has not become very common in TLC but that might extend its applicability to complex samples sometime in the near future].

## B. Review of Today's Instrumental TLC Techniques

### 1. Separation Media: Conventional or High-Performance TLC?

Including its modifications, such as RP derivatives, silica gel is the separation medium used on over 90% of all TLC applications. For several years the analyst has had a choice between conventional layers and the so-called high-performance (HPTLC) materials. Severe misunderstandings exist about the merits of the latter.

HPTLC silica gel has an average particle size of 7 $\mu$m compared to 12 $\mu$m in conventional layer material. HPTLC layers also have a more uniform particle size distribution. The immediate effect, apart from slightly improved optical properties, is the reduced optimum separation distance of about 50 mm compared with 100–120 mm of a conventional silica gel. Accordingly, chromatogram development takes place within the high-speed portion of the chromatographic flow behavior, so that development of an HPTLC plate is completed within 25% of the time required for a conventional plate. On the other hand, miniaturization of the plate dimensions requires that both the volume of the sample and its precise positioning be likewise adapted. Typical sample volumes in HPTLC are 100–200 nl, when applied spotwise and delivered at one time. The volume should not exceed 500 nl. More than with conventional layers, having the smallest possible area of the sample origin is essential for exploiting the better resolution of the layer. Automatic densitometric scanning of the chromatogram according to a preselected pattern requires that the precision of sample positioning be ±0.1 mm and that distortion during chromatography remain negligible.

This leads to the conclusion that true high-performance efficiency can only be achieved by combining high-resolution layers with appropriate techniques and instrumentation.

The gain in absolute sensitivity (detection limits) by HPTLC has been overestimated occasionally. Provided sufficient sample material is available, this gain is nearly compensated by the necessary reduction of the sample volume. But when sample application is combined with a sample enrichment step, for example, by employing a spray-on technique or by evaporating solvent during sample delivery at a controlled speed, then the lower detection limits in HPTLC offer direct advantages.

In any case, provided the number of samples to be analyzed is large enough to occupy at least the capacity of one HPTLC plate, economy is one reason for selecting HPTLC. Because many more samples can be chromatographed on a $10 \times 10$-cm HPTLC plate than on the conventional plate, which is four times larger, the HPTLC plate material costs less than the equivalent of conventional material. Also, the consumption of developing solvents is greatly reduced.

## 2. Sample Application

Two types of sample dosage devices are available for dispensing a certain volume of sample solution at one time and in the form of a spot. These are fixed-volume pipettes functioning by capillary action and syringe-type dosage devices.

Fixed-volume pipettes of a capacity smaller than 500 nl consist of a platinum–iridium tube of calibrated length, sealed into a glass holder. Glass capillary pipettes are available from 500 nl and up. They offer the advantage of being disposable and allow visual inspection of the content. It is advisable to lower all such pipettes, including those of 500 nl, onto the chromatographic layer by means of a precision device rather than manually. This technique avoids volume errors that may be caused by damaging the layer with such a narrow-bore tube, because the pipette is brought into contact with the layer compatible with the hardness of the layer surface. Also, precision positioning of the sample spot is a prerequisite for the automatic scanning of the chromatogram according to a preselected pattern. The Nanomat* spaces the samples 5.0 mm (or multiples thereof) apart and lowers the pipette onto the layer by controlled magnet action (see Fig. 1).

The Nano-Applicator is a micrometer-controlled syringe that serves for the application of variable sample volumes between 50 and 230 nl. The Micro-Applicator handles volumes exactly 10 times larger (i.e., $0.5-2.3 \mu l$)

---

* Products marked by an asterisk are available from CAMAG, CH-4132 Muttenz, Switzerland.

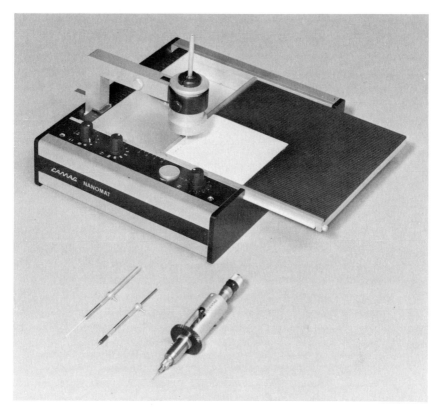

**Fig. 1.** The Nanomat with dosage devices: nanopipette, glass capillary in holder, and Nano-Applicator.

and is suitable for sample application onto conventional TLC layers. The syringe-type sample applicator is preferred when variable volumes are desirable or when the layer does not readily absorb liquid from a pipette, as may be the case with RP material. The Nano-Applicator and the Micro-Applicator can be positioned with the Nanomat too.

Most of the application examples of TLC analysis of foods and beverages described in Section II employ the spray-on technique for sample application with the Linomat* (Fig. 2). The spray-on technique, with nitrogen as a carrier gas, enables comparatively large volumes of sample solutions to be concentrated into narrow bands. The volume range of the Linomat is 1–99 $\mu$l. Sample application in the form of narrow bands ensures the best resolution that can be achieved with the respective chromatographic system and also reduces systematic errors occurring in quantitative evaluation by absorbance scanning.

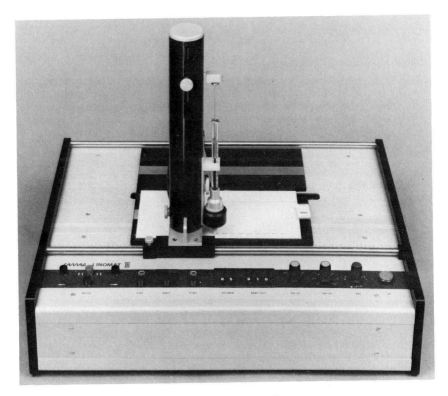

**Fig. 2.** The Linomat III.

The latest achievement for sample application is the Automatic TLC Sampler I* (Fig. 3). It has a volume range from 100 nl up to 20 $\mu$l and is freely programmable with respect to all parameters, including arrangement of samples and dispensing speed. Because of the latter it is also suited for the application of comparatively large sample volumes, which are concentrated into small spots. The main reason for employing the Automatic TLC Sampler, of course, is to save labor time — sample application takes up something like one-third of the total analysis time — and to make this step in the procedure independent of the skill and reliability of the laboratory personnel.

### 3. Chromatogram Development

The conventional way to develop a thin-layer chromatogram is to immerse the lower edge of the plate in the developing solvent contained in a suitable tank. Since the composition of the developing solvent can be significantly affected, either by preferential adsorption of a solvent compo-

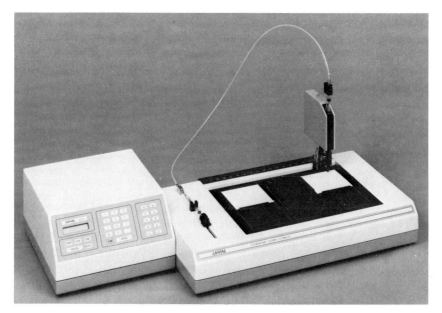

**Fig. 3.** The Automatic TLC Sampler I.

nent by the dry layer via the gas phase or by water introduced by the silica gel, it is recommended that fresh solvent be used for every plate. This is facilitated by using a Twin Trough Chamber,* which needs to be filled with only a little more solvent than is used up for chromatography (see Fig. 4a). Furthermore, the Twin Trough Chamber permits a certain degree of controlled pre-equilibration of the layer with solvent vapor prior to development, if that is desired (see Fig. 4b). Twin Trough Chambers are available in sizes for both conventional and HPTLC plates.

The HPTLC Linear Developing Chamber* permits the number of samples to be doubled. The chromatogram is developed from both sides after the samples have been applied parallel to both opposing edges. Development is performed in the sandwich configuration, that is, precluding vapor phase interaction during the actual run. If pre-equilibration of the layer is intended, this has to be done externally, prior to inserting the plate in the developing chamber. It should be noted that this type of developing chamber doubles the number of samples and thus more than doubles the ratio between unknowns and calibration standards, which makes this developing technique extremely attractive. On the other hand, not all multicomponent developing solvents appear to be suitable for the sandwich configuration. In all cases where, in the examples in Section II, a Twin Trough

a                                               b

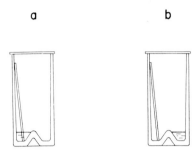

**Fig. 4.** A Twin Trough Chamber, schematic: (a) developing a chromatogram with a minimum of solvent; (b) conditioning the dry layer with solvent vapor.

Chamber was employed, development in the sandwich configuration of the HPTLC Linear Developing Chamber gave less favorable results.

Special developing techniques such as circular and anticircular chromatography, which offer advantages in particular cases, are not included in this survey, since these were not necessary to solve the separation problems of the analyses described in Section II.

### 4. Quantitative Chromatogram Evaluation

The recent methodology of quantitative chromatogram evaluation by *in situ* densitometry, including result computation via peak height or peak area has been described in sufficient detail (Schmutz, 1980; Ebel *et al.,* 1980a; Ebel and Geitz, 1981; Fenimore and Davis, 1981). Here only a few basic facts that are useful to remember will be presented. A light beam in the form of a slit selectable in length and width is moved over the sample zones to be evaluated. In order to avoid systematic errors, scanning should always be parallel to the direction of chromatography, not at right angles to it. The measured differences between the signal from the sample-free background and that from a sample zone can be correlated with the amount of the respective fraction of a calibration standard, usually chromatographed on the same plate. Instead of scanning with a slit, it is possible to move a light spot across the sample zones in a meander or zigzag pattern within a track. This offers certain advantages that, however, must be paid for by a drastically reduced signal-to-noise ratio, due to the lower light energy of the small spot. For *in situ* scanning the fractions to be measured must have inherent optical properties enabling them to be detected on the layer background. If this is not the case, they must be converted into suitable derivatives, preferably prior to chromatography. Generally speaking, scanning by fluorescence is preferable to scanning by absorbance, since sensitivity is usually 10–1000 times higher and the ratio between fluorescence signal and amount of substance is linear over a wide range. For scanning by absorbance a linear relationship can only be expected over a concentration range of $1:2$ to (at maximum) $1:5$.

The results of all of the following application examples (see Section II) were obtained with the CAMAG TLC/HPTLC Scanner* (Fig. 5), a slitscanner with a monochromator with selectable bandwidth and a wavelength range of 200 to 800 nm. It operates in the automatic scanning mode, that is, scanning the chromatogram according to a pattern preselected with respect to the length of the scan, the distance between tracks, and the number of tracks. This requires that the samples be positioned accordingly. In those cases where samples were applied in the form of narrow bands, the length of the scanning slit was chosen to be between one-half and two-thirds of the original band length. The only exception was the scanning of the aflatoxin chromatograms by fluorescence, in which case the slit length was set so as to overlap the fraction zones. This is recommended when fluorescing zones must be measured with a maximum of sensitivity and there is no background interference.

Registering the analog signal with a suitable recorder and measuring peak heights manually is the easiest form of quantitative readout. In routine

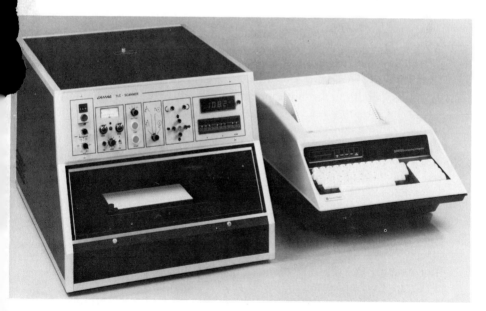

**Fig. 5.** The TLC/HPTLC Scanner with SP 4100 Spectra-Physics integrator.

analysis (i.e., when a large number of samples has to be handled), evaluation and presentation of results by means of an integrator and/or a computer is more convenient and therefore preferable. In most of the application examples the SP 4100 Spectra-Physics integrator was used for result computation.

## 5. Aspects of Automation

When a complex analytical procedure such as TLC is to be automated, care must be taken that none of the advantages based on the open system — the "off-line principle"—are sacrificed (Jänchen, 1982). Therefore it seems preferable to automate the three individual steps of sample application, chromatogram development, and chromatogram evaluation rather than the complete procedure. It should also be borne in mind what can be gained by automation: saving of labor and analysis time, precision, and reliability.

All of these can be expected from the automation of sample application, the step in the procedure that has hitherto required the largest part of tota' labor time and, accordingly, has been most prone to human errors. Automation of this step makes TLC more attractive for routine analysis.

Regarding chromatogram development, the simultaneous processing' many samples side by side can be accepted as an equivalent for automatior provided that the amount of manual handling is sufficiently small and th: the result does not depend on too high a level of skill. Gains in precisic (resulting in reproducible, nondistorted chromatograms suitable for au matic chromatogram evaluation) can be expected from employing develc ing devices that provide sufficient control of the gas phase prior to, duri and after chromatogram development. The large-volume tank, with a gen ous sump of developing solvent that is used over and again, is the le; favorable condition for development of reproducible chromatograms.

For the quantitative chromatogram evaluation the analyst today can already choose between three levels of automation (Schmutz, 1982a). The first two rely on the automatic scanning feature of the TLC/HPTLC Scanner, whereby result computation is achieved by a computing integrator such as the SP 4100. Results can be presented by the integrator in the form of a complete analysis report. However, this requires that the calibration standards be measured before the unknowns, which presupposes either that the standards are arranged suboptimally for chromatography or that the operator intervenes between measurement of the standards and the unknowns. The second level of automation includes a minicomputer for processing the raw data coming from the integrator. This increases flexibility with respect to arrangement of samples; use of random, complicated calibration functions; comparison of results with those from other chromatograms (stored on a flexible disk); etc. But this type of automatic chromatogram evaluation still requires that the scanning pattern correspond precisely with that of sample application and that no distortion have occurred during chromatogram development.

There is a third stage, namely, computer-controlled TLC evaluation (Ebel et al., 1980b), whereby a desk-top computer such as the HP 9826 controls all scanning movements of the TLC/HPTLC Scanner. The separation tracks on the chromatogram are automatically located and the densitometric measurement of each fraction is optimized. The computer, which has extended graphics capability, calculates results and presents them in the form of a complete analysis report. The computer-controlled system increases the reliability of TLC analysis.

## II. FOOD- AND BEVERAGE-RELATED APPLICATIONS OF INSTRUMENTAL TLC

The following examples are a selection of applications for which instrumental TLC appears attractive for various reasons. Most of the procedures employ the high-performance TLC variation.

### A. Aflatoxins B₁, B₂, G₁, G₂ in Peanuts, Almonds, Corn, and Similar Products

*Scope:* Sample preparation is started with 1 kg of material in order to obtain a reliable average. An aliquot of $\frac{1}{50}$ is processed to a final extract, of which $\frac{1}{200}$ to $\frac{1}{20}$ is chromatographed. Coextracted extraneous impurities are removed by prechromatography. Aflatoxins $B_1$, $B_2$, $G_1$, $G_2$ are separated and individually quantitated by *in situ* fluorescence scanning. Quantities as low as 25 pg per spot can be measured with good accuracy. This represents an aflatoxin concentration in the sample of 0.025–0.25 ppb, depending on the sample volume actually chromatographed. This procedure has been accepted as the official Swiss method (1983).

*Sample preparation:* Grind sample, making sure not to produce a pasty form or to leave coarse particles. Place 1000 g in a mixer and add 2500 ml methanol and 250 ml water; mix for 3 min, add 750 ml water, and continue mixing for another 3 min. Add some Celite, and filter through a plaited filter. Transfer a 70-ml aliquot to a separation funnel, and add 50 ml low-boiling petrol ether plus 55 ml water plus 5 g NaCl; shake for 2 min, and then discard the organic phase. Add 20 ml water to the aqueous phase and extract with 90 ml chloroform by shaking for 2 min. Dry the chloroform extract with $Na_2SO_4$ and evaporate the filtered extract in a rotating evaporator at 30°C to dryness. Take up residue with 1.0 ml chloroform; 50 $\mu$l of this extract represents 1 g of the original sample.

*Calibration standards:* Prepared from commercially available aflatoxin standards; stock solution containing 1 μg/ml each of $B_1$ and $G_1$ and 0.5 μg/ml each of $B_2$ and $G_2$ in benzene – acetonitrile 98 : 2. Refrigerated, this solution is stable for many months. From this stock solution the actual calibration standard is prepared by dilution with chloroform 1 : 50, so that 1 μl contains 20 pg each of $B_1$ and $G_1$ and 10 pg each of $B_2$ and $G_2$. It is stable (refrigerated) for several weeks.

*TLC layer:* HPTLC aluminium sheets, silica gel, MERCK 60 F 254, cut to size 10 × 10 cm.

*Sample application:* By spray-on technique with Linomat, band length 2 mm; distance between samples 8 mm, 40 mm(!) from edge of sheet (see Fig. 6); volume 5 – 50 μl of sample extract, 2 – 10 μl of calibration standard. Overspraying of a sample application of an extract that has been found to be aflatoxin-free with the calibration standard is recommended.

*Chromatography:* Special two-step developing technique in opposite directions. The first step removes lipoid impurities from the aflatoxins that remain at the origin. After the portion of the sheet on which the impurities are contained has been cut off, the sheet is turned 180° and the aflatoxins are chromatographed with chloroform – acetone 88 : 12. The prechromatography step is carried out with freshly dried ethyl ether into which the sheet is immersed 50 mm and developed to the upper edge. Caution: Moist ether could cause the aflatoxins to migrate into the area that is later cut off.

*Quantitative scanning:* By fluorescence; mercury lamp; excitation 366 nm; emission 400 nm; slit dimensions 0.3 × 4 mm; automatic scanning at 1 mm/sec; result computation, preferably via peak height, with suitable

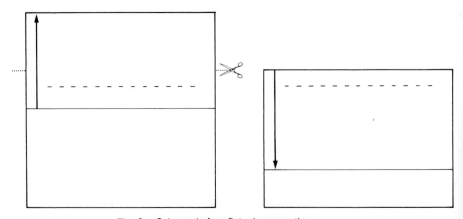

**Fig. 6.** Schematic for aflatoxin separation.

integrator or manually with analog recorder. A typical scanning curve is depicted in Fig. 7.

## Comments

1. The procedure includes an unusual clean-up step by prechromatography in the opposite direction.

2. The spray-on technique permits the application of comparatively large sample volumes chosen with respect to the aflatoxin content that is considered permissible in the product. If the analysis is aimed at the reliable detection of 0.5 ppb, it is sufficient to apply 5 $\mu$l of the extract. The critical quantity is then 100 pg, which is about four times the detection limit.

3. The linear correlation coefficients of peak heights versus aflatoxin quantities from 50 to 2500 pg ($B_2/G_2$) were found to be better than 0.995.

## B. Aflatoxin $M_1$ in Milk and Milk Products

A remarkable variation, suitable for milk aflatoxins in the ppt range, was described by Tripet *et al.*, (1981). The modified sample preparation includes precipitation of casein and lactose with acetone and elimination of other proteins and phospholipids with lead acetate. Comparatively large volumes of the final extract (i.e., 40–80 $\mu$l, depending on the legal limit of aflatoxin in the product) are sprayed onto TLC aluminum sheets, silica gel, MERCK 60. For extracts from milk, milk powder, butter, cream, and fresh cheese one-dimensional chromatography with chloroform–acetone–isopropanol 85 : 10 : 5 is sufficient. A "three-dimensional" chromatography procedure is employed for the determination of aflatoxins in extracts from complex material such as baby food and fermented cheese, as depicted in Fig. 8.

Unknown U and identification standard I are applied near the middle of the TLC sheet. Upon chromatography with ethyl ether–methanol–water 96 : 3 : 1 extraneous ingredients migrate to the upper area of the layer, which is then cut off. The TLC sheet is turned 180° and chromatographed in the opposite direction with chloroform–acetone–isopropanol 75 : 15 : 10. If a fluorescing zone corresponding with the identification standard is found in the track of the unknown, calibration standards of aflatoxin are applied as marked $C_1-C_3$ in the illustration. Chromatography in the third dimension, that is, at right angles, with ethyl acetate saturated with water separates the aflatoxin fractions. Quantitative scanning parameters are as described in Section II, A.

Table 1 compares the legal limits of aflatoxins in milk products with the detection limits of the one- or three-dimensional method, depending on the type of product.

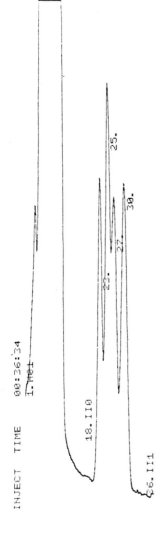

INJECT TIME   00:36:34

1.401

18.110

25.

23.

27.

30.

26.111

QUANT. HPTLC OF AFLATOXINS        00:36:34

FILE 1    METHOD 5.    RUN 9    INDEX 3    CALIB

| NAME | CONC | RT | PK HT BC | RF | RRT |
|------|------|-----|----------|-----|-----|
| G 2 | 500. | 23. | 194856 02 | 389.712 | 0.767 |
| G 1 | 1000. | 25. | 256232 02 | 256.232 | 0.833 |
| B 2 | 500. | 27. | 181624 02 | 363.248 | 0.9 |
| B 1 | 1000. | 30. | 191096 23 | 191.096 | 1. |
| TOTALS | 3000. | | 823808 | | |

**Fig. 7.** Analog curve (integrator print out) of aflatoxin $B_1$, $B_2$, $G_1$, $G_2$ chromatogram.

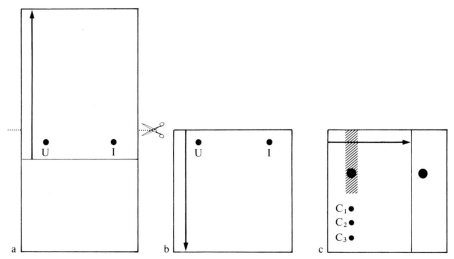

**Fig. 8.** Schematic for "three-dimensional" chromatography of aflatoxin $M_1$: (a) first dimension; (b) second dimension; (c) third dimension. U = unknown, I = identification standard, $C_1-C_3$ = calibration standards.

## C. Diethylstilbestrol in Veal and in the Urine of Fattening Calves

*Scope:* DES is extracted from meat, DES-glucuronides from urine; after derivatization with dansyl chloride the DES derivative is separated from complex by-products and quantitated by *in situ* fluorometry. The method (Wirz, 1982) allows safe detection of DES in the low ppb range.

*Sample preparation from meat:* Homogenize 1 g meat (veal) and 5 ml water; extract repeatedly with ethyl ether, combine extracts, and evaporate to dryness at 50°C under nitrogen; dissolve residue in 1 ml chloroform and extract three times with 1 ml 1 $N$ sodium hydroxide solution; adjust aqueous phase to pH 8, extract three times with 2 ml ethyl ether, and evaporate to dryness. The residue is ready for dansylation.

*Sample preparation from urine:* Adjust 5 ml filtered urine with 50%

**TABLE 1  Legal Limits of Aflatoxins in Milk Products** [a]

| Product | Legal limit of $M_1$ in Switzerland | FDA limit | Detection limit of the method |
|---|---|---|---|
| Cheese | 500 | 500 | 30 |
| Milk | 50 | 500 | 5 |
| Baby food (milk products) | 10 | 500 | 5 |

[a] In parts per trillion.

phosphoric acid to pH 1–2; extract DES-glucuronides in a separation funnel with 10 ml diethyl ether–ethyl acetate 1 : 1, and centrifuge if necessary; extract a second time with 5 ml of the same solvent; evaporate the combined extracts to dryness at 60°C under nitrogen. Dissolve residue in 1 ml 0.2 $M$ phosphate buffer of pH 7, wash two times with 2 ml ethyl ether, and discard the organic phases. Add 5 $\mu$l $\beta$-glucuronidase and incubate 1 hr at 37°C. Extract free DES three times with 2 ml ethyl ether, and evaporate combined ether phases to dryness. Dissolve residue in 1 ml chloroform, and proceed as described for the extract from meat.

*Dansylation:* Add 100 $\mu$l 5 $N$ NaOH to the residue, shake briefly, and boil for 5 min; when the mixture has cooled down to room temperature, add 100 $\mu$l of freshly prepared 0.125% solution of dansyl chloride in acetonitrile; shake for 15 sec, allow to stand for another 45 sec, then acidify with 600 $\mu$l 1 $N$ HCl, and boil for 10 min. Extract the cooled-down mixture two times with 2 ml ethyl ether; evaporate to dryness, and dissolve in 60 $\mu$l chloroform. This solution is ready for chromatography.

*TLC layer:* TLC precoated plates, silica gel, MERCK 60 F 254, 20 × 20 cm.

*Sample application:* Spray-on application by Linomat, 50 $\mu$l of solution; band length 10 mm; distance between bands 4 mm (12 samples per plate).

*Chromatography:* In Twin Trough Chamber with toluene–ethyl acetate 100 : 3 (without vapor phase saturation); separation distance 100 mm.

*Fluorescence scanning:* With TLC/HPTLC Scanner combined with SP 4100 integrator or with analog recorder; mercury lamp; excitation 366 nm; monochromator bandwidth 10 nm; emission > 400 nm; slit dimensions 0.6 × 10 mm; scanning speed 1 mm/sec. The quantity of DES in the sample is determined by comparison (via peak height) with calibration standards that have been added to the sample extract before dansylation (spiking method). Figure 9 shows typical scanning diagrams of samples from urine and from meat with and without DES spiking.

## Comment

The sample spray-on technique is employed for two reasons: a comparatively large sample volume has to be applied to achieve the desired very low detection limit; and band application enhances chromatographic resolution, which here is essential for the safe identification of DES in a urine extract.

### D. Quantitative Determination of Vitamin C in Fruit Juice

*Scope:* The procedure (Wirz, 1982) includes a two-step derivatization of vitamin C (dehydroascorbic acid hydrazone) and its chromatographic sepa-

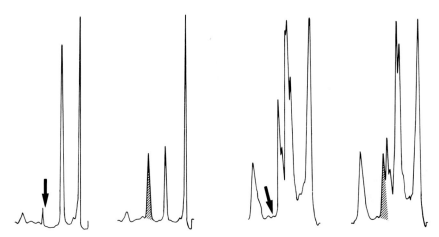

**Fig. 9.**   Analog curves of diethylstilbestrol chromatograms: left, meat extracts without and spiked with 25 ppb DES; right, urine extracts without and with 50 ppb DES.

ration from other fruit juice ingredients followed by *in situ* densitometry. The vitamin C derivative is identified by its absorption spectrum. Detection limit of the method is < 20 ng per sample zone.

*Sample preparation:* Pipette 2 ml fruit juice (or calibration standard solution containing 0.1 – 0.8 mg/ml L-ascorbic acid) into a 100-ml measuring flask. Add 25 ml 4% oxalic acid and 2 ml 0.5% aqueous solution of dichlorophenol – indophenol salt. Ascorbic acid oxidation is completed within 5 min. After that time add 10 mg thiourea to destroy the excess of oxidation reagent; fill up with water to 100 ml. Transfer a 20-ml aliquot into a 50-ml centrifuge vessel, and add 4 ml of a 2% solution of 2,4-dinitrophenylhydrazine in 75% sulfuric acid. Place the vessel in a 45 – 50°C water bath for 2 hr, and then let it cool down to room temperature. Extract the hydrazones three times with 6 ml ethyl acetate – acetic acid 98 : 2; centrifuging the vessel after vigorous shaking facilitates phase separation. Combine the extracts and fill up with same solvent to 20 ml. Aliquots of this orange-colored solution are chromatographed.

*TLC layer:* Precoated plates HPTLC, silica gel, MERCK 60 F 254, 20 × 10 cm.

*Sample application:* Spray-on application by Linomat, 5 μl of solution; band length 10 mm; distance between bands 5 mm. Twenty-two samples per plate are applied (11 on either side) in preparation for development in the HPTLC Linear Developing Chamber. Note: Since with the spray-on technique application of different sample volumes for calibration is permissible, it is sufficient to carry only one or two calibration solutions through the derivatization procedure.

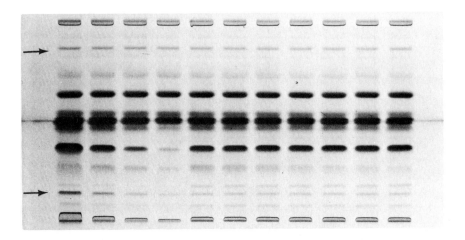

**Fig. 10.** Chromatogram of fruit juice extracts photographed under white transmitted light; ascorbic acid fractions marked with arrow; calibration standards with (from right) 160, 80, 40, and 20 ng per zone in upper right corner.

*Chromatography:* In the HPTLC Linear Developing Chamber, 20 × 10 cm (with the thinner, i.e., 2-mm, counter plate inserted) with chloroform–ethyl acetate 1 : 1; two runs with (careful) intermediate drying. The Rf of the ascorbic acid derivative is about 0.4 (see Fig. 10). The fraction of interest can be identified by its *in situ* absorption spectrum in comparison with that of the calibration–identification standard. It can be taken with the TLC/HPTLC Scanner, with tungsten lamp, monochromator bandwidth 10 mm, and wavelength changes from 400 to 570 nm at increments of 10 nm (see Fig. 11).

*Densitometry:* With TLC/HPTLC Scanner preferably in combination with integrator (e.g., SP 4100); scanning by absorbance either with mercury lamp at 254 nm or with tungsten lamp at 510 nm; monochromator bandwidth 10 nm in both cases; slit dimensions 0.3 × 6 mm; scanning speed 1 mm/sec.

## Comments

1. Again, the spray-on technique was chosen for the comparatively large sample volume, which, in turn, eliminates an additional preconcentration step. Another asset of band application is resolution, as can be seen from Fig. 10.

2. Employing the HPTLC Linear Developing Chamber doubles the number of samples that can be chromatographed on one plate, increasing

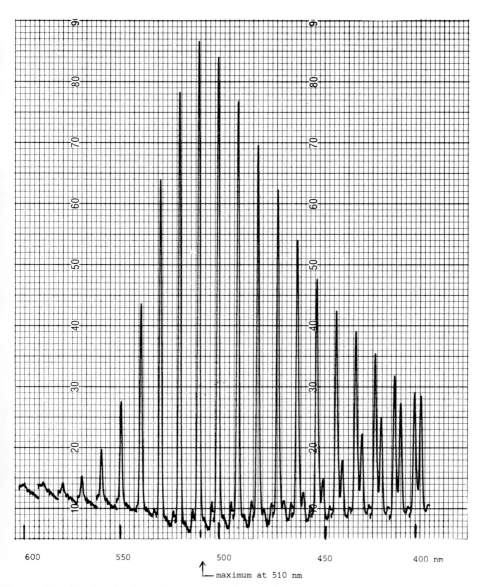

600             550             500             450             400 nm

maximum at 510 nm

**Fig. 11.** *In situ* absorption spectrum of L-ascorbic acid phenylhydrazone, taken with TLC/HPTLC Scanner.

the ratio of unknowns to calibration standards from 7 : 4 to 18 : 4, that is, by a factor of 2.5.

3. An overall reproducibility of 1.8% relative standard deviation was achieved by the method described.

## E. Caffeine in Foods and Beverages

*Scope:* Routine procedure (Wirz, 1982) for caffeine assay in roasted coffee, coffee extracts, caffeine-containing beverages, etc. Detection limit of the method is 10 ng per sample zone.

*Sample preparation from roasted coffee:* Shake 5 g ground coffee with 100 ml water for 20 min at 60°C; filter the suspension and dilute with methanol–water 9 : 1 to 1 liter. Instant coffee: Dissolve 1 g in 100 ml water at 60°C and dilute with methanol–water 9 : 1 to 1 liter. Cola drinks, etc.: Dilute 50 ml sample with methanol to 100 ml. Calibration standard: 20 mg caffeine per milliliter methanolic solution (= 20 ng/$\mu$l).

*TLC layer:* Precoated plates HPTLC, silica gel, MERCK 60 F 254, 20 × 10 cm.

*Sample application:* Spray-on application of extracts with Linomat, 5 $\mu$l of sample as 10-mm band; distance between samples 5 mm; 22 samples per plate (11 on either side) for development in HPTLC Linear Developing Chamber; different volumes of the calibration standard are sprayed on (e.g., 2, 4, 6, 8 $\mu$l).

*Chromatography:* In HPTLC Linear Developing Chamber, 20 × 10 cm, with chloroform-acetone 85 : 15; Rf of the caffeine fraction is about 0.25. The caffeine fraction can be identified by its *in situ* absorption spectrum (against that of the identification–calibration standard): TLC/HPTLC Scanner, with deuterium lamp, monochromator bandwidth 10 mm, wavelength changes from 200 to 300 nm in increments of 10 nm; the absorption maximum of caffeine is 270 nm.

*Densitometry:* Scanning by absorbance with mercury lamp; monochromator set to 275 nm with bandwidth 30 nm (in order to use the energy of the mercury lines at 265, 270, and 280 nm); slit dimensions 0.3 × 5 mm; scanning speed 1 mm/sec.

## Comments

1. With the method described an overall reproducibility of 1% relative standard deviation for pure standards and 2% for authentic samples was achieved.

2. The method can be simplified by preparing a concentration of the final extract five times higher than specified and applying only 0.5 $\mu$l spotwise.

Resolution is still sufficient; the loss of accuracy is about 30%, that is, 1.3% (2.5%) relative standard deviation.

## F. Determination of Sugars, Sugar Acids, and Polyols in Wine, Beverages, etc.

*Scope:* TLC separation followed by *in situ* derivatization and quantitative evaluation by fluorescence scanning; detection limit 20 ng per spot (for glucose and fructose). Method reported by Klaus (1979).

*Sample preparation:* Dilute the liquid sample (wine, beer, beverage) 1 : 50 with methanol–water 8 : 2. Prepare calibration standards of 0.05, 0.15, 0.25, and 0.5 mg/ml glucose and fructose in methanol–water 8 : 2.

*Sample application, chromatogram layer:* Precoated plates HPTLC, silica gel, MERCK 60, 20 × 10 cm. Two hundred nanoliters sample solution and calibration standards are applied with a fixed volume nanopipette or with a Nano-Applicator; positioning 5.0 mm apart is performed with a Nanomat.

*Chromatography:* In Twin Trough Chamber, 20 × 10 cm, with:

1. *n*-Butanol–isopropanol–0.5% boric acid–acetic acid 30 : 50 : 10 : 2, two runs 70 mm (note: this does not separate fructose from saccharose).

2. Ethyl acetate–water–acetic acid–propionic acid 50 : 10 : 5 : 0.5; solvent migration distance 70 mm (note: this does not separate glucose from fructose).

*Derivatization:* After careful drying in a stream of hot air the plate is immersed for 10 sec in a solution freshly prepared from 5 ml 20% lead(IV) acetate in acetic acid plus 5 ml 1% 2′, 7′-dichlorofluorescein in ethanol plus 200 ml toluene; the plate is dried carefully.

*Quantitative evaluation:* By fluorescence scanning with excitation at 313 nm (mercury lamp); monochromator bandwidth 10 nm; emission >400 nm; slit dimensions 0.3 × 4 mm; scanning speed 1 mm/sec; result computation with integrator SP 4100, quadratic regression function.

## G. Analysis of Malto-Oligosaccharides

*Scope:* Chromatographic separation and determination of malto-oligosaccharides for the purpose of characterizing starch hydrolysates used in food products. Method based on publications by Schweizer and Reimann (1982) and Würsch and Roulet (1982).

*Sample preparation:* Starch hydrolysates are dissolved in water so that the expected concentration of the sugars to be analyzed is between 0.05 and

0.5 $\mu$g/$\mu$l. If sugars of low molecular weight are expected, water – ethanol 1 : 1 is preferable.

*TLC layer:* Precoated plates HPTLC, silica gel, MERCK 60 F 254, 10 × 10 cm, prewashed with propanol – acetone – water 50 : 40 : 10 and dried at 120°C.

*Sample application:* With Linomat, 2 – 4 $\mu$l in form of bands 5 – 10 mm in length, 10 mm apart.

*Chromatography:* In Twin Trough Chamber, saturated with the solvent vapors; development with propanol – acetone – water 45 : 30 : 25; two runs followed by a third development with the same components but ratio 50 : 40 : 10; careful intermediate drying; final drying in vacuo prior to derivatization. *In situ* derivatization by dipping the plate rapidly into a solution of 4 ml aniline plus 4 g diphenylamine in 200 ml acetone plus 30 ml 85% phosphoric acid, leaving it immersed for about 10 sec and then heating it at 120°C for 15 min.

*Quantitative evaluation:* Scanning by absorbance at 546 nm with mercury lamp; monochromator bandwidth 10 nm; slit dimensions 0.3 × 5 mm or 0.3 × 2.5 mm, that is, half the band length of original sample application; scanning speed 0.5 mm/sec. Result computation with integrator via peak area, compared against calibration standards containing 1 $\mu$g maltose and maltotriose.

## Comments

The precision of the method was determined as 2.1% relative standard deviation for maltose and 0.6% for maltotriose. HPTLC is much faster than gel permeation chromatography, and compared with HPLC, it offers better detection (HPLC requires either refractometry or postcolumn derivatization).

## III. CONCLUSIONS

The previous examples show that instrumental thin-layer chromatography is suitable for a variety of food- and beverage-related analytical problems. Features that appear to be particularly attractive in this field of application are the following:

1. Comparatively uncomplicated sample preparation.

2. Positive identification of substances by cochromatography of authentic identification standards as well as by *in situ* derivatization (pre- and postchromatography), that is, "simultane" chromatography.

3. Application of comparatively large sample volumes by the spray-on technique, leading to a high sensitivity of the method.

## REFERENCES

Ebel, S., and Geitz, E. (1981). *Kontakte* 1/81, 44–48; 2/81, pp. 34–38.
Ebel, S., Geitz, E., and Klarner, D. (1980a). *Kontakte* 1/80, pp. 11–16; 2/80, pp. 12–16.
Ebel, S., Geitz, E., and Hocke, J. (1980b). *In* "Instrumental HPTLC" (W. Bertsch, S. Hara, R. E. Kaiser, and A. Zlatkis, eds.), pp. 55–80. Hüthig Verlag, Heidelberg, Basel, and New York.
Fenimore, D. C., and Davis, C. M. (1981). *Anal. Chem.* **53**, 252A–266A.
Jänchen, D. (1979). *Kontakte* 3/79, p. 9–13.
Jänchen, D. (1982). *In* "Instrumental HPTLC, Interlaken 1982" (R. E. Kaiser, ed.), pp. 220–232. Inst. for Chromatography, Bad Dürkheim, W. Germany.
Kaiser, R. E. (1977). *In* "High Performance Thin-Layer Chromatography" (A. Zlatkis and R. E. Kaiser, eds.), pp. 15–49. Elsevier, Amsterdam.
Kirchner, J. G. Miller J. M., and Keller G. J., (1951). *Anal. Chem.* **23**, 420.
Klaus, R. (1964). *J. Chromatogr.* **16**, 311–326.
Klaus, R. (1979). *Kontakte* 3/79, pp. 24–27.
Schmutz, H. R. (1980). *In* "Instrumental HPTLC" (W. Bertsch, S. Hara, R. E. Kaiser, and A. Zlatkis, eds.), pp. 315–344. Hüthig Verlag, Heidelberg-Basel-New York.
Schmutz, H. R. (1982a). *In* "Instrumental HPTLC, Interlaken 1982" (R. E. Kaiser, ed.), pp. 246–255. Inst. for Chromatography, Bad Dürkheim, W. Germany.
Schmutz, H. R. (1982b). *Labor-Praxis* **6**, 38–47. Preliminary note, based on private communication by W. Stutz, Kantonal Labor Liestal. Official Swiss method for the analysis of aflatoxins $B_1$, $B_2$, $G_1$, $G_2$, in peanuts, etc. "Schweizerisches Lebensmittelbuch," Vol. 2.
Schweizer, T., and Reimann, S. (1982). *Z. Lebensm. Unters. Forsch.* **174**, 23–28.
Seiler, N., Werner, G., and Wiechmann, M. (1963). *Naturwissenschaften* **50**, 643.
Stahl, E., ed. (1962). "Dünnschicht-Chromatographie." Springer-Verlag, Berlin and New York.
Tripet, F. Y., Riva, C., and Vogel, J. (1981). *Mitt. Geb. Lebensmittelunters. Hyg.* **72**, 367–379.
Tyihak, E., Mincsovics, E., and Kalasz, H. (1979). *J. Chromatogr.* **174**, 75.
Wirz, E. (1982). Kantonales Labor Solothurn. Private communication.
Würsch, P., and Roulet, Ph. (1982). *J. Chromatogr.* **224**, 177–182.

# 4

# Gas Chromatography

TAKAYUKI SHIBAMOTO

*Department of Environmental Toxicology*
*University of California*
*Davis, California*

## I. HISTORICAL ASPECTS

Tsvet first used column chromatography in 1906 to separate chlorophyll from a plant and tested more than 100 different adsorbents using his classic column chromatograph. He also interpreted his results correctly on the basis of chromatographic theory, which has been applied to all chromatographies since then. Various other chromatographic techniques followed the invention of column chromatography: Brown (1939) introduced paper chroma-

ANALYSIS OF FOODS AND BEVERAGES
Copyright © 1984 by Academic Press, Inc.

tography, which uses a filter paper and certain organic solvents, and Martin and Synge (1941) laid the theoretical foundations upon which liquid–liquid partition chromatography is based. The theory of partition chromatography advanced chromatographic technique considerably and led to the invention of gas–liquid chromatography (GLC).

Ten years after Martin and Synge predicted that partition chromatography would be applicable to gas–liquid systems, James and Martin (1952) invented the first instrument of GLC. They separated a series of fatty acids using an 11-ft stainless steel packed column. They also developed the technique for quantitative analysis of mixtures, another important application of gas chromatography (GC).

Once the gas–liquid system was devised, the new technique spread rapidly among analytical chemists. Pioneer works were accomplished by Dutch scientists who studied the separation of low-molecular-weight hydrocarbons, ethers, and alcohols on GLC during the early 1950s (Keulemans and Kwantes, 1955; Keulemans et al., 1955). Within a short time, several important studies set the seal on the technique (Cremer and Mueller, 1951; Zhkhoviskii et al., 1951; Janak, 1953; Ray, 1954; Paton et al., 1955).

At that time, however, the few instruments in use were located only in the central analytical research facilities of large corporations or universities. By the end of the 1950s, a commercial gas chromatograph had become available and was present in most laboratories. Today, the literature contains thousands of papers on GC. The analysis of natural plant components, air pollutants, pesticide residues, flavor and fragrance chemicals, and plasma cholesterols is now performed routinely. Its high-resolution ability allows tremendous numbers of chemicals to be isolated from a complex mixture.

## A. Instrumentation

The principal instrumentation has not changed since James and Martin devised the first gas chromatograph in 1952. It consists of an injector, a column, a detector, and a recorder. The injector vaporizes the sample to send it into a column, where the separation occurs. The only drawback of GC is that the sample must be vaporized under reasonably low temperature ($<450°C$). The vaporization occurs in a carrier gas stream so that a sample is vaporized at a temperature lower than its atmospheric boiling point. Therefore, GC is generally useful only for chemicals that have a boiling point below 600°C.

The active sites on the metal surfaces in the early injectors caused samples to alter at high temperatures. A glass inlet system is now used to prevent samples from contacting metal surfaces (Shibamoto, 1981). It is extremely important to prevent constituents from altering during flavor analysis, because the nature of flavor depends not only on the constituents of a sample but also on their relative quantities.

Recently, many investigators reported a method of on-column injection for capillary columns. The sample is introduced directly onto the column without prior vaporization, which has the following advantages (Knauss *et al.,* 1982):

1. Elimination of sample discrimination because of syringe effects.
2. Elimination of sample decomposition due to either thermal or catalytic (i.e., metal syringe needle) effects.
3. High analytical precision.
4. Excellent quantitation of individual solutes.

On the other hand, a couple of problems have been reported by those using on-column injection (Galli and Trestianu, 1981):

1. Incomplete and discriminative sample transfer from the syringe to the column owing to partial evaporation of the sample out of the syringe needle.
2. Back-ejection of part of the sample from the capillary column inlet, which causes sample loss, discrimination of sample components, and injector contamination.

Proske *et al.* (1982) studied routine pesticide analysis using an on-column injection system and concluded that on-column injection gave highly quantitative data of high accuracy (with relative standard deviations of $\leq 0.5\%$). Grob (1981) reported that using cold on-column injection and splitless sampling at low column temperature was advantageous with samples having a large quantity of solvent; on the other hand, the solvent spread out the less volatile sample components. On-column injection is a new field of GC study, and its applicability to routine analysis is still under investigation.

## B. Column Development

The most important part of a gas chromatograph is the column, since this is where separation occurs. Most pioneer work on food and beverage analysis was done using stainless steel packed columns.

In 1957, M. J. E. Golay devised an open tubular column, which increased column resolution significantly. Generally, the maximum resolution obtained by a packed column is 20,000 theoretical plates, but a capillary column can deliver more than 150,000 theoretical plates. Procedures for analyzing foods and beverages have evolved considerably since the invention of the capillary column.

The major problem posed by stainless steel columns is that the active metal surface promotes chemical changes in the sample components

(Hunter and Brogden, 1963; Wrolstad and Jennings, 1965). Glass tubing is now used to construct more inert columns. Although the superiority of glass capillary columns had been known since Desty *et al.* (1960) devised a glass capillary drawing machine, coating a stable liquid phase on the glass surface to obtain a reliable high-resolution column was a major difficulty. After intensive work on coating technique by many researchers, stable liquid phases on glass capillary columns became available during the late 1960s and early 1970s. In one of the earliest studies of GC analysis using a glass capillary column, Torline and Ballschmieter (1973) identified 38 compounds in the essential oil of freeze-dried guava puree using a glass capillary column (110 m × 0.3 mm inside diameter [i.d.]) coated with free fatty acid phase (FFAP).

The prominent position held by glass capillary columns did not last long: soon a flexible fused-silica capillary column appeared. In fused-silica columns, tailing is not observed, even with free fatty acids (Shibamoto, 1982). Glass columns have been replaced with fused-silica columns in many laboratories. In fact, many laboratories have switched from packed columns to fused-silica capillary columns without ever having used glass columns.

## C. Detectors for Food and Beverage Analysis

The details of detectors are available in many reference books (Purnell, 1962; Littlewood, 1962; Ettre, 1967; McNair and Bonelli, 1969). The most commonly used detector for food and beverage analysis is a flame ionization detector (FID). The FID operates on the principle that the electrical conductivity of a gas is directly proportional to the concentration of charged particles within the gas. The FID responds to most organic compounds and has an excellent stability. Recently, the thermionic detector, which is highly sensitive to nitrogen- and phosphorus-containing compounds (and which is also called the N/P detector), has received much attention for its use in the analysis of nitrogen-containing heterocyclic flavor chemicals such as pyrazines, thiazoles, and pyrroles.

Among several types of detectors, the thermal conductivity detector is unique in that it operates on the principle that a hot body will lose heat at a rate that depends on the composition of the surrounding gas. Its great advantage for flavor studies, which depend on GC and organoleptic analysis, is that the sample is not destroyed during analysis, so that one can sniff the individual unknowns at the exhaust outlet.

Because the human sense of olfaction is more sensitive than GC detectors in the analysis of certain chemicals, a more sensitive detector must be found for flavor research. Table 1 shows the sensitivity of the detector and the human nose to selected chemicals (Takagi, 1974).

TABLE 1 Comparison of the Odor Thresholds and GC Detector Sensitivities in the Study of Some Organic Compounds[a]

| Compound | Odor threshold (ppm) | GC sensitivity (ppm) |
|---|---|---|
| In aqueous solution | | |
| n-Propanol | 0.17 | 0.0025 |
| n-Butanol | 0.07 | 0.12 |
| n-Hexanol | 0.03 | 0.3 |
| Acetone | 500 | 0.03 |
| 2/Butanone | 50 | 0.017 |
| Dimethyl sulfide | 0.012 | 0.02 |
| Methyl mercaptan | 0.002 | 0.013 |
| In vapor phase | | |
| 2-Heptanone | $8.97 \times 10^{-4}$ | $6.5 \times 10^{-4}$ |

[a] From Takagi (1974).

## II. SAMPLE PREPARATION FOR GC ANALYSIS

### A. Headspace Sampling

Gas chromatography is an ideal technique for analyzing flavor volatiles of foods and beverages, because the basic principle of gas chromatography is the separation of mixtures in the vapor phase. When we enjoy flavors of foods or beverages, we are enjoying the vapor phase of those flavor compounds.

Direct analysis of vapor above the samples requires little sample preparation and provides less artifact formation. Most commonly, the sample is placed in a closed container, and the headspace from the closed container is injected directly into a gas chromatograph. Manning *et al.* (1976) investigated the role of hydrogen sulfide, methanethiol, and dimethyl sulfide in the headspace of cheese samples using a 30-ml bore hole.

One drawback to a headspace sample is that concentrations of objective compounds are very low. Sometimes the concentration is too low to be detected by the existing detectors. To compensate for this, many techniques for concentrating headspace samples have been developed. Schreyen *et al.* (1976) collected a condensed headspace sample using a container equipped with a cold finger above the collecting flask: the condensed headspace sample, containing mostly water, was subsequently extracted with an organic solvent. Trapping methods have been developed also.

The most widely used techniques for concentrating headspace samples involve entraining headspace vapor on adsorbents such as silica gel, active alumina, charcoal, or porous polymers. Tenax, a porous polymer based on *p*-2,6-diphenylene oxide, has been used successfully in many flavor studies. Buckholz *et al.* (1980) investigated a method for collecting, characterizing, and quantifying headspace volatiles from roasted peanuts using Tenax: compounds with low boiling points were trapped sufficiently with the nitrogen flow rate at 10 ml/min, and compounds with high boiling points were trapped more efficiently at a higher nitrogen flow rate of 60 ml/min. Doi *et al.* (1980) used Tenax to investigate changes in headspace volatile components of soybeans during roasting: more than 100 volatile chemicals, including alcohols, aldehydes, ketones, pyrazines, hydrocarbons, and a phenol, were trapped and recovered for gas chromatography. Porapack Q, a styrene polymer, offers advantages for trapping samples with high alcohol or water content. Because water and ethanol have considerably shorter retention times on Porapack Q at room temperature than the other organic compounds with low boiling points (Jennings, 1980), Porapack Q has been used to recover organic compounds from aqueous or alcoholic solutions (Jennings *et al.,* 1972). Figure 1 shows a gas chromatogram of the headspace sample obtained from Pepsi-Cola using Porapack Q.

Singleton and Pattee (1980) modified the injection port of the gas chromatograph in order to allow insertion of the polymer trap. In their method, the trapped headspace samples are introduced directly into the gas chromatograph, thereby eliminating the need for a microsyringe to transfer the

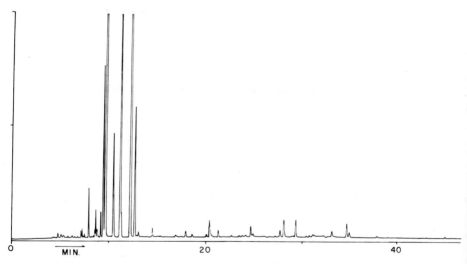

**Fig. 1.** Gas chromatogram of Pepsi-Cola headspace sample: 50-m × 0.23-mm-i.d. fused-silica capillary column coated with Carbowax 20M and programmed from 80 to 220°C at 2°C/min. (Courtesy of Dr. K. Yamaguchi, Ogawa and Company, Ltd., Tokyo, Japan.)

samples. More simplified headspace GC with glass capillaries has been developed and tested by Perkin-Elmer and Company (Kolb *et al.,* 1979). The actual application of this system for analyzing various samples such as water pollution, flavor, and volatiles in crude oil has been reported by the Perkin-Elmer research group (Kolb, 1980).

## B. Solvent Extraction

Solvent extraction may be the most commonly used method for preparing samples for GC. Samples for GC must possess a reasonably low boiling point. A "volatile compound" is sometimes interpreted to mean an organic-solvent-extractable compound. The most commonly used solvents are organic compounds that have a low boiling point (pentane, hexane, methanol, ethanol, dichloromethane, chloroform, ethyl acetate, ethyl ether, and acetone). If solubility is not a problem, pentane, hexane, and dichloromethane are generally preferred because of their purity and low boiling point. Although methanol and ethanol may be used for polar samples, the water content of methanol and ethanol is quite critical, because water destroys most liquid phases. Ethyl ether was the most commonly used solvent for GC samples for many years, but after the high-resolution capillary column was developed, ethyl ether was found to contain many impurities. Because high-purity ethyl ether has been too difficult to obtain, the use of ethyl ether as a solvent is diminishing. Dichloromethane has become more and more popular as a solvent for GC and may be the solvent most widely used today. One advantage to the use of chlorinated hydrocarbon for extractions is that the solvent locates in the lower layer so that the extract can be removed without removing the upper aqueous layer. The liquid–liquid continuous extractor provides more effective extractions than does the method of the conventional separatory funnel. Figure 2 shows a gas chromatogram of the dichloromethane extract obtained directly from Pepsi-Cola using a liquid–liquid continuous extractor.

## C. Steam Distillation

One annoying problem of solvent extraction is that extraction efficiency is reduced by the presence in the samples of high-molecular-weight materials such as proteins, carbohydrates, and fats. Steam distillation has been used to recover organic volatile components from foods and beverages containing those high-molecular-weight substances. Organic components are recovered subsequently from the distillates by either headspace or solvent extraction methods. Because heat treatment such as steam distillation alters the

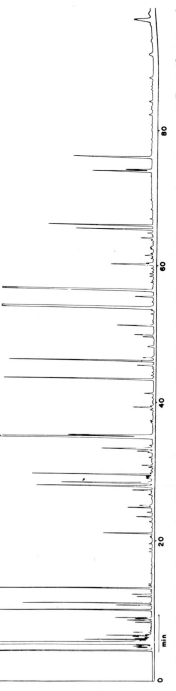

**Fig. 2.** Gas chromatogram of a dichloromethane extract obtained from Pepsi-Cola under the same circumstances as in Fig. 1. (Courtesy of Dr. K. Yamaguchi, Ogawa and Company, Ltd., Tokyo, Japan.)

constituents, it is recommended that distillation be performed under reduced pressure and at lower temperature. Figure 3 shows a gas chromatogram of volatiles recovered from whole milk by steam distillation.

A convenient simultaneous steam-distillation–solvent-extraction apparatus was invented by Likens and Nickerson (1966). Numerous food and beverage analyses have been conducted using this novel apparatus (Buttery et al., 1968; Cronin and Ward, 1971; Schultz et al., 1977; Yamaguchi and Shibamoto, 1979).

## D. Preparation of Derivatives

Although nonvolatile chemicals such as amino acids, sugars, and steroids cannot be analyzed satisfactorily using a gas chromatograph because of their high boiling point, several methods have been developed to analyze those nonvolatiles by GC. The most commonly used method is to prepare derivatives in order to reduce the polarity of samples; examples are methyl esters for fatty acids (Quin and Hobbs, 1958), N-trifluoroacetyl methyl esters for amino acids (Cruickshank and Sheehan, 1964), and trimethyl silyl esters for sugars (Wells et al., 1964).

## III. EXAMPLES OF APPLICATIONS

### A. Coffee Flavors

The most comprehensive analysis of coffee flavor was reported by Stoll et al. (1967), who found 202 constituents using the combined methods of preparative GC, mass spectrometry (MS), and infrared (IR) spectrometry. They also reported the mass spectra of 190 flavor compounds, which contributed significantly to the evolution of flavor analysis. Approximately 500 flavor chemicals have been identified in coffee to date (van Strater and de Vrijer, 1973; Vitzthum and Werkhoff, 1976). The major components are heterocyclic compounds such as pyrazines, thiazoles, pyrroles, and furans. Bondarovich et al. (1967) identified 22 pyrazine derivatives in coffee volatiles using 8-ft × 0.25-in.-i.d. 25% SE-30 on 60/100-mesh Chromosorb W and 13-ft × 0.25-in.-i.d. 20% Carbowax 20M on 60/80-mesh Diatoport S column. Additional information of the aroma constituents of roasted coffee, including GC retention data, was reported by the same research group several years later (Friedel et al., 1971).

Shibamoto et al. (1981a) used high-performance liquid chromatography

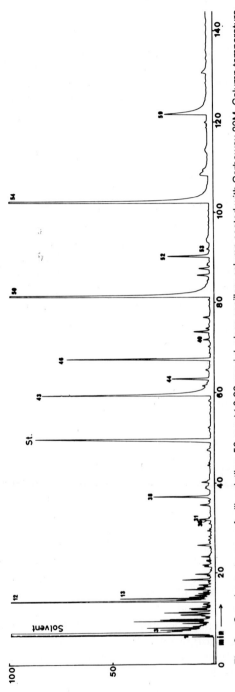

**Fig. 3.** Gas chromatogram of milk volatiles: 50-mm × 0.28-mm-i.d. glass capillary column coated with Carbowax 20M. Column temperature was programmed from 80 to 220°C at 2°C/min. (Courtesy of Dr. S. Mihara, Ogawa and Company, Ltd., Tokyo, Japan.)

(HPLC) to fractionate coffee volatiles for capillary GC. Five fractions from HPLC were analyzed using a 50-m × 0.28-mm-i.d. glass capillary column coated with Carbowax 20M. They identified 61 coffee volatiles. Figure 4 shows the gas chromatogram of coffee volatiles.

Flavor profiles of sulfur-containing compounds in coffee were investigated using a specific sulfur detector, the flame photometric detector (Nurok *et al.*, 1978). In further studies of the role of sulfur-containing compounds in coffee flavors, Tressel and Silwar (1981) identified 23 sulfur-containing compounds using a combination of preparative GC, adsorption chromatography, capillary gas chromatography, and capillary gas chromatography/ mass spectrometry (GC/MS). They found that mercaptan derivatives such as furfurylmercaptan and 5-methyl furfurylmercaptan increased in roasted coffee beans over storage time.

Although sulfur-containing compounds such as furfurylmercaptan play an important role in coffee flavor, stainless steel columns tend to trap the sulfur-container compounds. The early studies done on stainless steel columns did not provide sufficient information on the sulfur-containing compounds. One can predict that the development of inert fused-silica capillary columns will allow us to detect even more sulfur-containing constituents in coffee flavors.

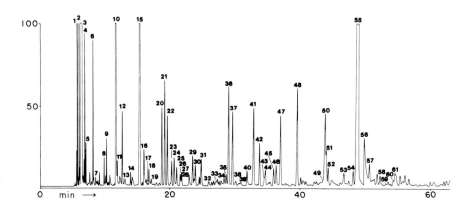

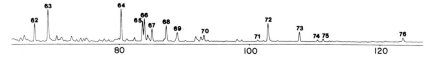

**Fig. 4.** Gas chromatogram of coffee flavors obtained under the same circumstances as in Fig. 3. (Courtesy of Dr. S. Mihara, Ogawa and Company Ltd., Tokyo, Japan.)

## B. Essential Oils

The more than 200 kinds of commercially available essential oils are widely used to create flavors for food and fragrances for cosmetics. Generally, essential oils, which consist mainly of volatile chemicals, give the characteristic odor to the plant in which they are found. These essential oils are obtained by extraction, expression, or distillation.

Essential oils had been analyzed before the gas chromatograph was invented. The acid values, ester values, alcohol, and aldehyde and ketone constants of nearly 200 essential oils had been reported before James and Martin succeeded in making a prototype gas chromatograph in 1952 (Guenther, 1952).

Since the development of GC, the number of essential oil constituents identified has increased significantly. Debrauwere and Verzele (1975) identified 51 compounds in the oxygenated fraction of pepper oil using glass capillary columns coated with SE-30 (70-m × 0.9-mm i.d.) and Carbowax 20M (75-m × 0.5-mm i.d.).

In laboratories, headspace sampling is used often to analyze the constituents of certain essential oils. Yabumoto et al. (1977) investigated the essential oil of cantaloupe using three 100-m × 0.25-mm-i.d. glass capillary columns coated with FFAP, Carbowax 20M, and SE-30. They proved that a glass capillary offers advantages for trace analysis of sulfur-containing compounds by identifying dimethyl disulfide in the oil, because sulfur-containing compounds are very susceptible to the metallic surfaces of stainless steel columns.

Gas chromatography/mass spectrometry is also a powerful technique for identifying constituents of essential oils. Since the early 1970s, numerous reports have revealed newly discovered essential oil constituents. The GC retention index coupled with the MS fragmentation pattern is recognized as the most conclusive method for positively identifying unknowns. The retention index and MS fragmentation patterns of unknowns are compared with those of authentic compounds. Determining retentions on two high-resolution capillary columns of different polarity offers high confidence of the identity of a compound. Yamaguchi and Shibamoto (1981) identified more than 60 constituents in green tea essence using this method. Figures 5 and 6 show gas chromatograms of green tea essence on two different polarity columns. Taskinen and Nykanen (1975) used two capillary columns of different polarity (25-m × 0.28-mm i.d., SP-1000 and OV-101) for analyzing angelica root oil and identified 70 volatile constituents.

Identification of sesquiterpenes has been one of the most difficult matters in essential oil analysis. A modern high-resolution capillary column can separate many sesquiterpenes that give similar or almost identical MS fragmentation patterns. Because of the difficulty of obtaining authentic

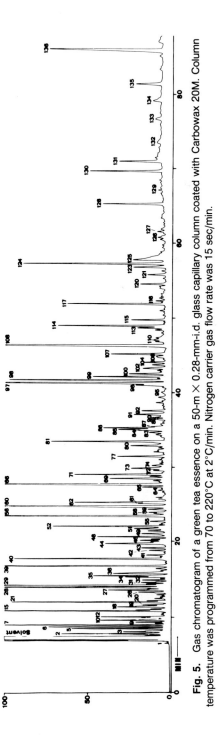

**Fig. 5.** Gas chromatogram of a green tea essence on a 50-m × 0.28-mm-i.d. glass capillary column coated with Carbowax 20M. Column temperature was programmed from 70 to 220°C at 2°C/min. Nitrogen carrier gas flow rate was 15 sec/min.

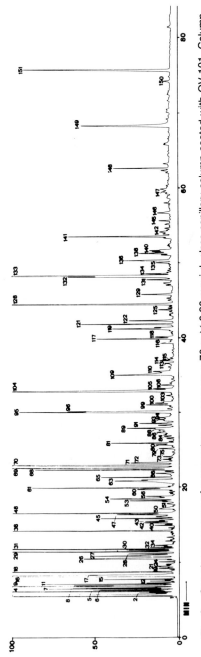

**Fig. 6.** Gas chromatogram of a green tea essence on a 70-m × 0.28-mm-i.d. glass capillary column coated with OV-101. Column temperature was programmed from 70 to 250°C at 2°C/min. Nitrogen carrier gas flow rate was 17 sec/min.

sesquiterpene compounds, positive identification of many sesquiterpenes in various essential oils is still pending. Shibamoto and Jennings (1977) observed the presence of more than 100 components in the oil of California juniper; 42 compounds were identified on the basis of their mass spectra and GC retention indices obtained from Carbowax 20M and SE-30 columns (100-m × 0.25-mm-i.d. glass capillary). However, 17 monoterpenes and 38 sesquiterpenes were left as only tentatively identified, because authentic compounds could not be obtained.

In spite of the lack of information on sesquiterpenes, GC is the most powerful or perhaps the only method that can be used for the modern quality control of essential oils, because both horticultural and crop-to-crop variations change the composition and relative proportions of individual constituents of an essential oil. The flavor and fragrance industries demand the consistency of each oil. Figure 7 shows a typical gas chromatogram of the peppermint oil used as the standard for chewing gum flavor.

## C. Fruit Flavors

Early research in fruit flavor was done using packed columns (Anderson and von Sydow, 1964; Heinz and Jennings, 1966). Some excellent studies were reported before the GC/MS technique was developed. Tang and Jennings (1967) analyzed apricot flavor using a preparative GC technique and the IR spectra. Initial separations were performed on a 13-ft × 0.25-in. stainless steel column packed with 10% Carbowax 20M on 40/60-mesh Gas Pak F. Fractions were trapped in thin-walled glass capillaries. Cochromatographing components were separated with dissimilar columns such as Apiezon L, and final collections were made from 500-ft × 0.03-in.-i.d. stainless steel capillary columns coated with SF-96-50 or Carbowax 20M. The isolated pure components were identified using the IR spectra. Fifteen constituents were isolated, including cis and trans isomers of an epoxyhydrolinalool; generally, cis and trans isomers are almost impossible to separate by conventional fractional distillation.

The GC/MS technique was first applied to identify volatile constituents of banana (Wick et al., 1969). Several columns were used to analyze banana flavor: 6-ft × 0.25-in.-i.d. 20% Carbowax 4000 on 60/80-mesh Chromosorb W for preparative separation of alcohol and neutral fractions; 10-ft × 0.125-in.-i.d. 10% DEGS on 35/80-mesh Chromosorb W for preparative separation of methyl ester fraction; and 500-ft × 0.02-in.-i.d. SF-96 open tubular column for the original oil. Ismail et al. (1981a,b) identified 73 volatile components in plum juice and 33 components in plum headspace using glass capillary in a gas chromatograph.

In the early 1970s, long, open tubular columns were used to increase

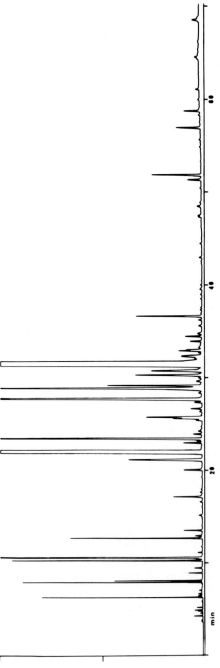

**Fig. 7.** A typical gas chromatogram of peppermint oil: 50-m × 0.25-mm-i.d. fused-silica capillary column coated with Carbowax 20M and programmed from 80 to 220°C at 2°C/min.

resolution. Flath and Forrey (1970) used 500-ft × 0.02-in.-i.d., 500-ft × 0.03-in.-i.d., and 1000-ft × 0.03-in.-i.d. columns, each coated with OV-101 to isolate pineapple flavors. Schreyen *et al.* (1976) used a 600-ft × 0.03-in.-i.d. column coated with OV-101 for leek flavor analysis. These long columns are almost extinct today, since satisfactory high resolution became obtainable using 30- to 50-m glass or fused-silica capillary columns. Shorter columns are advantageous because they reduce analysis time.

Recently, a microprocessor-controlled GC terminal became available to quantitate complex mixtures on the basis of normalization and internal standard methods (Wilson and Shaw, 1980). Microprocessor or computer calculation capabilities are a powerful quantitative analytical method. Wilson and Shaw (1980) determined certain chemical contributions to the overall flavor profile of grapefruit oil using a capillary column GC microprocessor. Figure 8 shows a typical gas chromatogram of peach volatiles.

## D. Meat Flavors

Gas chromatography has been used in the analysis of meat flavors since the early 1960s. The volatile constituents of cooked meat in the first report were methyl mercaptan, acetaldehyde, methyl sulfide, and acetone (Kramlich and Pearson, 1960). Hornstein *et al.* (1960) used gas chromatography to separate methyl esters of fatty acids in beef and pork fat. Wick *et al.* (1965) obtained nearly 30 peaks on the gas chromatogram of the irradiated beef flavor using a 2-m × 4-mm-i.d. stainless steel packed column (20% Carbowax 20M on 60/65-mesh Chromosorb P). The GC/MS technique became a major method for identifying volatile mixtures in the late 1960s. Nonaka *et al.* (1967) isolated 227 constituents from cooked chicken volatiles using a prototype GC/MS and identified 62 compounds. In the 1970s, numerous studies on cooked meat flavors were done using the GC/MS technique. Most of the comprehensive identification of meat flavors was completed by the late 1970s. Mussinan and Walradt (1974) identified 179 compounds in pressure-cooked pork liver using GC/MS; they used a 1000-ft × 0.03-in.-i.d. stainless steel column coated with either SF-96 or Carbowax 20M.

In the late 1970s, glass capillary columns began to be used. The meat volatiles produced in an L-rhamnose–hydrogen sulfide–ammonia browning model system were analyzed using a 100-m × 0.25-mm-i.d. glass capillary column coated with Carbowax 20M (Yamaguchi *et al.,* 1979). Figure 9 shows the gas chromatogram of the volatiles produced in the browning model system. Shibamoto *et al.* (1981b) identified 44 compounds in beef broth using a 50-m × 0.28-mm-i.d. glass capillary column coated with either Carbowax 20M (see Fig. 10) or OV-101. Recent studies of meat

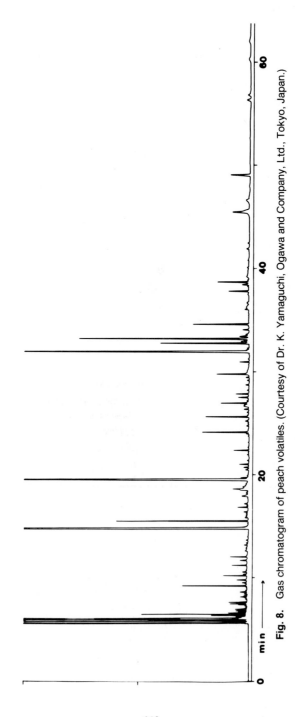

**Fig. 8.** Gas chromatogram of peach volatiles. (Courtesy of Dr. K. Yamaguchi, Ogawa and Company, Ltd., Tokyo, Japan.)

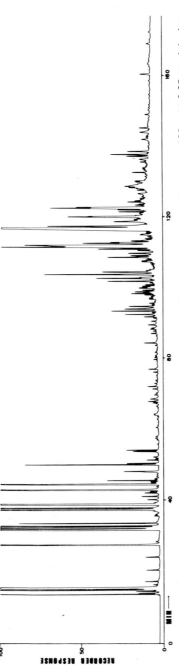

**Fig. 9.** Gas chromatogram of volatiles produced in an L-rhamnose–$H_2S$–$NH_3$ browning model system: 100-m × 0.25-mm-i.d. glass capillary column coated with Carbowax 20M and programmed from 70 to 170°C at 1°C/min.

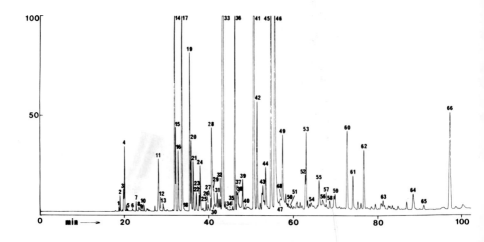

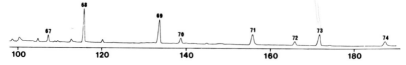

**Fig. 10.**   Gas chromatogram of volatiles in beef broth obtained from sukiyaki: 50-m ×
0.28-mm-i.d. glass capillary column coated with Carbowax 20M. Column temperature was
programmed from 70 to 220°C at 2°C/min.

flavors have focused on the isolation and identification of trace constituents
using GC techniques.

## IV. FUTURE VIEW OF GC

Gas chromatography has been used in almost every kind of food analysis.
A dynamic sampling method (a commercially available on-column injection system) is beginning to be used for food flavor analysis (Phillips, 1982).
The concentration in the trapping tube is still not 100% reproducible, and
the samples prepared by this purge-and-trap sampler are sometimes contaminated by unknown impurities, but once the drawbacks of this superior
method have been investigated and removed, conventional headspace sampling should become obsolete.

The column development seems to be an area of continued research. The
greatest drawback of GC is that it cannot be applied to nonvolatiles or to
high-molecular-weight compounds. The use of inert fused-silica capillary
columns and the on-column injection technique may solve part of this

deficiency. There is also interest in using GC in the direct analysis of amino acids and sugars in foods. The current goal is to delete the derivatization steps.

The most striking feature of GC is that the isolated compounds can be analyzed by a mass spectrometer. This method depends on the fast scanning ability of the mass spectrometer and its high sensitivity. Numerous compounds have been found in foods and beverages by means of this GC/MS technique. There are still some difficulties in identification with GC/MS alone. Confirmations with other instruments, such as ultraviolet (UV), IR, and nuclear magnetic resonance (NMR) are often required. Although GC and IR spectroscopy may solve this problem, there are still two technical difficulties in this method. One is the sensitivity limitations of IR detectors compared with GC detectors. Another is the possible time difference between the elution of a GC peak and the scan time required for IR. Fast Fourier transform IR spectroscopy may solve the latter problem (Rossiter, 1982). Increasing the sensitivity of IR spectroscopy may be most essential in devising a good GC/IR system. Application of a high-resolution capillary column requires high sensitivity, because the capacity of a capillary column is low compared with that of a low-resolution packed column. A high-resolution capillary GC/IR and a high-resolution capillary GC/NMR are expected to be available in the near future.

## REFERENCES

Anderson, J., and von Sydow, E. (1964). *Acta Chem. Scand.* **18**, 1105.
Bondarovich, H. A., Friedel, P., Krampl, V., Renner, J. A., Shephard, F. W., and Gianturco, M. A. (1967). *J. Agric. Food Chem.* **15**, 1093.
Brown, W. G. (1939). *Nature (London)* **143**, 377.
Buckholz, L. L. Jr., Withycombe, D. A., and Daun, H. (1980). *J. Agric. Food Chem.* **28**, 760.
Buttery, R. G., Seifert, R. M., Guadagni, D. G., Black, D. R., and Ling, L. C. (1968). *J. Agric. Food Chem.* **16**, 1009.
Cremer, E., and Mueller, Z. (1951). *Elektrochem.* **55**, 217.
Cronin, D. A., and Ward, M. K. (1971). *J. Sci. Food Agric.* **22**, 477.
Cruickshank, P. A., and Sheehan, J. C. (1964). *Anal. Chem.* **36**, 1191.
Debrauwere, J., and Verzele, M. (1975). *J. Sci. Food Agric.* **26**, 1887.
Desty, D. H., Haresnip, J. N., and Whyman, B. H. F. (1960). *Anal. Chem.* **32**, 302.
Doi, Y., Tsugita, T., Kurata, T., and Kato, H. (1980). *Agric. Biol. Chem.* **44**, 1043.
Ettre, L. S. (1967). *In* "The Practice of Gas Chromatography" (L. S. Ettre and A. Zlatkis, eds.), p. 402. Wiley (Interscience), New York.
Flath, R. A., and Forrey, R. R. (1970). *J. Agric. Food Chem.* **18**, 306.
Friedel, P., Krample, V., Radfor, T., Renner, J. A., Shephard, F. W., and Gianturco, M. A. (1971). *J. Agric. Food Chem.* **19**, 530.
Galli, M., and Trestianu, S. (1981). *J. Chromatogr.* **203**, 193.
Grob, K., Jr. (1981). *J. Chromatogr.* **213**, 3.

Guenther, E. (1952). "The Essential Oils," Vols. 1–6. Van Nostrand-Reinhold, Princeton, New Jersey.

Heinz, D. E., and Jennings, W. G. (1966). *Food Sci.* **31,** 69.

Hornstein, I., Crowe, P. F., and Sulzbacker, W. L. (1960). *J. Agric. Food Chem.* **8,** 65.

Hunter, G. L. K., and Brogden, W. B., Jr. (1963). *J. Org. Chem.* **28,** 1679.

Ismail, H. M. M., Williams, A. A., and Tucknott, O. G. (1981a). *J. Sci. Food Agric.* **32,** 498.

Ismail, H. M., Williams, A. A., and Tucknott, O. G. (1981b). *J. Sci. Food Agric.* **32,** 613.

James, A. T., and Martin, A. J. P. (1952). *Analyst (London)* **77,** 915.

Janak, J. (1953). *Chem. Listy* **47,** 464, 817, 1184.

Jennings, W. G. (1980). "Gas Chromatography with Glass Capillary Columns," 2nd ed. Academic Press, New York.

Jennings, W. G., Wohleb, R., and Lewis, M. J. (1972). *J. Food Sci.* **37,** 69.

Keulemans, A. I. M., and Kwantes, A. (1955). *Proc. 4th, World Pet. Congr.,* London, England.

Keulemans, A. I. M., Kwantes, K., and Zaal, P. (1955). *Anal. Chim. Acta* **13,** 357.

Knauss, K., Fullemann, J., and Turner, M. P. (1982). Hewlett-Packard Technical Pap. No. 94.

Kolb, B. (1980). "Applied Headspace Gas Chromatography." Heyden, London.

Kolb, B., Pospisil, P., Borath, T., and Auer, M. (1979). *J. High Res. Chromatogr. Chromatogr. Commun.* **2,** 283.

Kramlich, W. E., and Pearson, A. M. (1960). *Food Res.* **25,** 712.

Likens, S. T., and Nickerson, G. B. (1966). *J. Chromatogr.* **21,** 1.

Littlewood, A. B. (1962). "Gas Chromatography." Academic Press, New York.

McNair, H. M., and Bonelli, E. J. (1969). "Basic Gas Chromatography." Varian, Berkeley, California.

Manning, D. J., Chapman, H. R., and Hosking, Z. D. (1976). *J. Dairy Res.* **43,** 313.

Martin, A. J. P., and Synge, R. L. M. (1941). *Biochem. J.* **35,** 91.

Mussinan, C. J., and Walradt, J. P. (1974). *J. Agric. Food Chem.* **22,** 827.

Nonaka, M., Black, D. R., and Pippen, E. L. (1967). *J. Agric. Food Chem.* **15,** 713.

Nurok, D., Anderson, J. W., and Zlatkis, A. (1978). *Chromatographia* **11,** 188.

Patton, H. W., Lewis, J. S., and Kaye, W. I. (1955). *Anal. Chem.* **27,** 170.

Phillips, R. (1982). *Am. Lab. (Fairfield, Conn.)* **14,** 110.

Proske, M. G., Bender, M., Schomburg, G., and Humbinger, E. (1982). *J. Chromatogr.* **240,** 95.

Purnell, H. (1962). "Gas Chromatography." Wiley, New York.

Quin, L. D., and Hobbs, M. E. (1958). *Anal. Chem.* **30,** 1400.

Ray, N. H. (1954). *J. Appl. Chem.* **4,** 2182.

Rossiter, V. (1982). *Am. Lab. (Fairfield, Conn.)* **14,** 144.

Schreyen, L., Dirinck, P., van Wassenhore, R., and Schamp, N. (1976). *J. Agric. Food Chem.* **24,** 1147.

Schultz, T. H., Flath, R. A., Mon, T. R., Eggling, S. B., and Teranishi, R. (1977). *J. Agric. Food Chem.* **25,** 446.

Shibamoto, T. (1981). *In* "Applications of Glass Capillary Gas Chromatography" (W. G. Jennings, ed.), p. 469. Dekker, New York.

Shibamoto, T. (1982). *In* "Chemistry of Food and Beverages: Recent Developments" (G. Charalambous, ed.), pp. 73–99. Academic Press, New York.

Shibamoto, T., and Jennings, W. G. (1977). *Proc. 7th Int. Cong. Essent. Oils, 1977, Kyoto,* Abstr. No. K-2, p. 414.

Shibamoto, T., Harada, K., Mihara, S., Nishimura, O., Yamaguchi, K., Aitoku, A., and Fukada, T. (1981a). *In* "The Quality of Foods and Beverages: Chemistry and Technology" (G. Charalambous, ed.), p. 311. Academic Press, New York.

Shibamoto, T., Kamiya, Y., and Mihara, S. (1981b). *J. Agric. Food Chem.* **29,** 57.

Singleton, J. A., and Pattee, H. E. (1980). *JAOCS* December, p. 405.

Stoll, M., Winter, M., Gautschi, F., Flament, I., and Willhalm, B. (1967). *Helv. Chim. Acta* **50**, 628.

Takagi, S. (1974). *Kyukaku Hanashi, Iwanami Shoten, Tokyo*, p. 24.

Tang. C. S., and Jennings, W. J. (1967). *J. Agric. Food Chem.* **15**, 24.

Taskinen, J., and Nykanen, L. (1975). *Acta Chem. Scand.* **B29**, 757.

Torline, P., and Ballschmieter, H. M. B. (1973). *Lebensm. Wiss. Technol.* **6**, 32.

Tressel, R., and Silwar, R. (1981). *J. Agric. Food Chem.* **29**, 1078.

Tsvet, M. (1906). *Ber. Dsch. Bot. Ges.* **24**, 316, 384.

van Strater, S., and de Vrijer, F. (1973). "Lists of Volatile Compounds in Food Research," pp. 72.1–72.14. Naarden, Zeist, Netherlands.

Vitzthum, W. W., Sweeley, C. C., and Bentley, R. (1964). "Biomedical Applications of Gas Chromatography" (H. A. Szymanski, ed.), pp. 169–223. Plenum, New York.

Wick, E. L., Koshika, M., and Mizutani, J. (1965). *J. Food Sci.* **30**, 433.

Wick, E. L., Yamanishi, T., Kobayashi, A., Valenzuela, S., and Issenberg, P. (1969). *J. Agric. Food Chem.* **17**, 751.

Wilson, C. W., III, and Shaw, P. E. (1980). *J. Agric. Food Chem.* **28**, 919.

Wrolstad, R. E., and Jennings, W. G. (1965). *J. Chromatogr.* **18**, 318.

Yabumoto, K., Jennings, W. G., and Yamaguchi, M. (1977). *J. Food Sci.* **42**, 32.

Yamaguchi, K., and Shibamoto, T. (1979). *J. Agric. Food Chem.* **27**, 847.

Yamaguchi, K., and Shibamoto, T. (1981). *J. Agric. Food Chem.* **29**, 366.

Yamaguchi, K., Mihara, S., Aitoku, A., and Shibamoto, T. (1979). *In* "Liquid Chromatographic Analysis of Food and Beverages" (G. Charalambous, ed.), p. 303. Academic Press, New York.

Zhkhoviskii, A. A., Zolotareva, O. V., Sokolov, V. A., and Turkeltaub, N. M. (1951). *Dokl. Akad. Nauk. USSR* **77**, 435.

# 5

# HPLC in the Analysis of Foods and Beverages (Including Ion Chromatography)

DONALD H. ROBERTSON

*Science and Advanced Technology Laboratory*
*United States Army Natick Research and Development Laboratories*
*Natick, Massachusetts*

ANALYSIS OF FOODS AND BEVERAGES
Copyright © 1984 by Academic Press, Inc.
All rights of reproduction in any form reserved.
ISBN 0-12-169160-8

## I. INTRODUCTION

Especially in work with natural systems, such as those encountered in foodstuffs, a very heavy demand is placed upon an analytical technique because of the heterogeneity of the component materials. Chromatography in general and high-performance liquid chromatography (HPLC) in particular have responded well in providing the means whereby separations of diverse materials can be effected with a remarkable specificity. Extensive commercial development in HPLC has already led to a wide range of choices in basic instrumentation, column technology, and detection that can be selected by a knowledgeable researcher with great success to achieve optimum results.

What is referred to as HPLC is in fact a combination of several distinct phenomena operating either alone or in combination to effect a separation of components in a mixture. For pragmatic purposes, the concern is not so much what specific phenomena predominate but what is the extent of the separation that is achieved. For completeness, these phenomena are mentioned briefly.

1. Adsorption. The components of the sample adsorb and desorb on the stationary phase at different rates, thereby causing separation to occur.

2. Partition. The components spend varying amounts of time in the stationary phase and in the mobile or liquid phase, also leading in the ideal case to separation.

3. Size exclusion. The separation phase or solid phase (stationary phase) provides very well controlled pore sizes that filter sample components on the basis of their physical sizes or dimensions.

4. Ion exchange. The stationary phase has a charged surface, the charge being opposite to that of the sample components to be separated. Differential affinity of components for the charged surface as they are eluted from the column (by a buffer solution flowing through it) effects separation.

Many textbooks (Karger et al., 1973; Scott, 1976; Bristow, 1976; Simpson, 1976; Hamilton and Sewell, 1978; Knox et al., 1978; Englehardt, 1979; Snyder and Kirkland, 1979; Horvath, 1980; Provder, 1980; Kirkland, 1971; Yau et al., 1979) contain more detailed treatment of these phenomena. A combination of the phenomena mentioned above can lead to a rather complex interaction within an HPLC column.

In practice, HPLC commonly is further categorized into normal- and reverse-phase chromatography. The former, normal phase, is used when the stationary phase is highly polar and the mobile phase is nonpolar. Thus, the more polar the sample, the longer it is retained by the column. Reverse phase, as the name implies, is a nonpolar stationary phase with a polar mobile phase. A range of solvent polarities may be chosen from a group of pure single-component mobile phases, or in certain systems a mix of solvents (usually two) may be chosen that provides the requisite intermediary polarity. With this feature, a wide range of separations can be achieved. If the composition of the mobile phase remains constant throughout the run, it is called isocratic elution; if it is changed during the analysis, it is referred to as gradient elution.

## II. LIQUID CHROMATOGRAPHY SYSTEMS

Most manufacturers of present-day liquid chromatographic (LC) equipment supply a complete package that consists of all the components needed to conduct a successful analysis. A basic system will consist of the column in which the separations are effected and its contents, a means of getting a sample onto the column, a driving force to propel the mobile phase through the column with proper control and a means of detecting the sample as it elutes from the column. In principle, judicious selection of the proper column and conditions of operating it will bring about the desired separations and make the individual components amenable to detection. In practice, there are general guidelines and caveats that govern most modern HPLC.

## A. The Mobile Phase

Unlike the situation with gas chromatography (GC), the mobile phase in LC has an active part in the separation. There is the possible use of a single phase during the entire analysis, as indicated above, in which case the terminology used is isocratic elution, or the use of more than one mobile phase, which is termed gradient elution. Gradient elution has special application to natural product chromatography because the sample components often have broadly differing polarities; change of the mobile phase during the course of the run is helpful in eluting with good separation, in particular for the compounds that elute late in the chromatogram; with the sophisticated pumping systems that are available with modern chromatographic systems it is possible to program linear or other changes in mobile-phase composition that will optimally enhance the separation desired. After a moderate amount of experience, it is possible for a chromatographer to design the mobile phase composition profile to effect the desired separations. Of course, much benefit arises from repeated dealings with similar sample types, and establishing optimum conditions for separation thereby becomes easier. As LC has continued to grow, the demand for ultrapure solvents has been satisfied by numerous manufacturers. The characteristics of the solvent or mobile phase are of different importance depending upon the specific type of chromatography being performed (see Sections IV – VII). These characteristics include polarity, viscosity, ultraviolet (UV) adsorbance, refractive index, and ability to serve as a solvent for specific materials.

Several of the characteristics, particularly spectrophotometric behavior and refractive index, are especially important as regards their performance in the detector being used. Some of the most commonly used phases are hexane, chloroform, tetrahydrofuran, and methanol.

## B. The Stationary Phase

The essence of chromatographic separation is the interaction of components of a sample with the stationary phase. Expanding use of HPLC has led to custom manufacture of materials with characteristics that optimize their function in effecting a particular type of separation. In the case of modern HPLC the stationary phase is actually composed of two entities, one being the inert support and the other being the functional component, which when chemically bonded to the inert support provides a stable medium for active sites or functional groups that exhibit differential affinity for materials in a mixture scheduled for analysis. Much of the success of the HPLC technique has derived from clever methods for manufacturing columns with very specialized packings capable of performing optimized separation of a special problem mixture, for example, fatty acids, bile acids, or amino acids.

## C. Flow Control

An obvious requirement in an LC system is the means for moving the mobile phase through the closely packed column, thereby distributing the sample components and contributing to separation. These so-called pumping systems were initially a weak component of the LC instrument; modern-day systems, however, have been significantly improved and provide adequate precision in both isocratic and gradient elution modes. Particle size reduction for routine column packings (designed to effect better separation) dictate high inlet pressures (in the order of 1500 – 2000 psi). Viscosity of the mobile phase is another factor. It should be pointed out that the use of high pressures is a consequence of small particle size and viscosity rather than a direct choice by the operator. High pressure does not improve the chromatography; in point of fact it can cause problems in operation, such as leaks. Most commercial equipment operates on one of two principles: constant pressure or constant flow. A good rule of thumb is to operate at the lowest pressure that will effect the requisite flow for the analysis one wishes to perform. Another rule of thumb says to operate at the lowest effective pressure, since it is observed that efficiency increases as flow decreases, all other factors being equal. A low flow rate also has an economic advantage; optical purity of mobile phases is a requirement for detectors, and this purity comes dear. In addition, the problem of storage and use of gross amounts of toxic and/or flammable materials in the laboratory is not desirable.

## D. Sampling

The topic of sampling is addressed by Teranishi (Chapter 1, this volume), and therefore only a cursory treatment is accorded the topic here.

Sampling in a rigorous sense can be divided into two broad categories: the preparation of samples for use with an HPLC system and the actual mechanism of introducing an aliquot of that sample into the equipment.

It would appear initially that the use of a syringe and septum inlet, which serves so well in GC, would be the simplest choice. However, the operation of the system under high pressure on the inside of the septum leads to difficulty. Thus, most commercial systems use an elaboration of valves to compensate for the pressure problem. Sample-size requirements have been continually decreasing as small-bore columns, packed with microparticles, are improved. Microflow detectors make possible the detection of smaller and smaller amounts of material. Colin *et al.* (1979) have discussed the theory of injection. Simpson (1979) has reviewed the injection process. Certainly nanogram levels of detection can be achieved in the majority of applications. Since LC analysis conditions rarely exceed a temperature of 70°C, there is much less chance than in GC for sample deterioration due to temperature. Liquid samples can be introduced directly; samples that are solids at room temperature can be first dissolved in an appropriate solvent. Since the solvent is a small volume of material vis-à-vis the bulk of the mobile phase, it is not necessary that one use the same solvent as used for the mobile phase. As usual, the chemistry of the circumstances determines possible limitations upon selection of the solvent: reaction with the sample or the components of the column or incompatibility with the detector in use, for example, adsorbance at the wavelength used for detection, are considerations.

Generally speaking, the amount of sample preparation time is less for LC than for GC because it is easier to eliminate unwanted material. Natural product samples are notoriously "dirty," coming from a multicomponent biological system as they do, and the treatment accorded the sample prior to chromatography depends upon what one wishes to emphasize in the sample versus what may be considered waste or interfering material.

## E. Columns

The column is the backbone, as it were, of the HPLC system, and into it are packed the coated particles that provide the separation. What is placed in the column is really the subject for Sections III–VII. The column per se is usually stainless steel to minimize interaction with reactive components and should be of a grade especially manufactured for HPLC. It has been found important that the column be of uniform bore and that the inner walls be a smooth, polished surface.

A typical standard column length is less than 50 cm with inside diameter of 0.5–1 mm. More emphasis is currently being placed on short columns (5–10 cm) that are packed with very small particles (as small as 3 $\mu$m). This emphasis results in faster separations. Precolumns, guard columns, and the like can be used to clean up the sample material before it reaches the main column, thus extending column life. With short, microdiameter columns, numerous extracolumn effects manifest themselves. Microparticulate packings have been discussed by Rabel (1980). Several good reviews cover the general utilization of columns, among which are Reese and Scott (1980), Ohmacht and Halasz (1981), and Chen et al. (1981). As the capability of HPLC to effect good separations increases, there is a strong tendency to produce columns that are optimized for a specific separation, for example, protein or carbohydrate columns. Specialized items such as the radially compressed columns from Waters Associates* give superior results by reducing wall effects (Fallick and Rausch, 1979). Especially with natural product work, metal columns can be a distinct drawback, mostly because of reactivity, and so development of the technology to allow the use of glass columns under high-pressure conditions has been welcomed (Frei and Schulten, 1979). Some of the relative merits of conventional and microbore columns have been discussed by Henion (1980). General use of columns has been covered by Majors (1980).

Generally speaking, it is advised that one purchase commercially packed columns. If the HPLC experimental program in a given laboratory is very extensive in depth and duration, it may be worthwhile to learn how to do it oneself. However, special efforts are required if the venture is to be successful.

## F. Detectors

The search is under way for a truly all-purpose or universal detector for use with HPLC. The detector is the weak link in the chromatographic system. The classic controversy pertains here between sensitivity and selectivity. Quite naturally, the requirement for greater sensitivity has arisen from

* Milford, Massachusetts 01757.

the emphasis on short columns packed with microparticulate material. For these columns, flows are small, leading to improved chromatographic separation but also generating a requirement for microvolume detectors.

Currently, the most widely used detectors measure absorbance in the UV range (Fell, 1980; Smythe, 1981); second are the fluorescence detectors; these are followed by the refractive index and electrochemical detection, which are about equal in use. Two good reviews of detector use are Mc-Dowell *et al.* (1981) and McKinley *et al.* (1980).

Refractive index detection finds application in the food and beverage industry for determination of sugars in particular. Fluorescence peak height ratios have been used for determination of polynuclear hydrocarbons in food, water, and smoke (Crosby *et al.*, 1981). Considerable applications for fluorescence detection have been created by pre- or postcolumn derivatization of compounds that do not normally fluoresce into those that do. Application of this technique is found in the areas of aromatic hydrocarbons as well as a host of biologically important compounds, especially metabolites. Fluorometer determination is possible at the nanogram level (Zelt *et al.*, 1980); poor resolution may be compensated for by taking the second derivative of the elution curve. Picogram levels of detection have been reported for anthracycline drugs (Sepaniak and Yeung, 1980).

Electrochemical detection offers a remarkably wide range of possibilities in the choice of an electrode reaction, which is tantamount to detection of the eluates from the LC column. Some sophisticated systems have used two electrodes in series, the first to convert an eluate to a new species and the second to detect the newly created species. Electrochemical detection has been used in the range of 3–15 pg for phenols. It has also been used in the detection of biogenic amines (Weisshaar *et al.*, 1981).

The classic dropping mercury electrode is proving to be of value with low-conductivity electrolytes such as those sometimes used in reverse-phase techniques. Sensitivity is in the nanomole range, which (for cholic acid, for example) is better than UV absorbance detection at 210 nm but not as good as refractive index (RI) detection. A general-purpose radioactivity detector has been produced (Kessler, 1982).

Potentiometric determination has application to foodstuffs, as exemplified by the detection of amino acids with reverse-phase HPLC (Alexander *et al.*, 1981).

The combination of HPLC with mass spectrometry (MS) was often criticized in its early days because of the great expense involved in acquisition and maintenance of the MS as a detector for a much less expensive device. The same criticism was leveled against the now classic combination of GC with MS. In both cases there is an inherent incompatibility to be overcome in interfacing a device that depends upon pressure for its operation with a device that operates under high vacuum. The advantages of this unlikely combination rapidly became apparent, and soon the literature was filled with articles disclosing yet another wrinkle in the fabric of GC/MS. It is obvious that LC/MS is going the same route followed a decade and more ago by the pioneers of GC/MS. Applications of a rather extensive nature have been made to peptides, saccharides, fatty acids, vitamins, and antibiotics in the range of 1–10 ng. Subnanogram quantities have been detected by single-ion monitoring techniques (Blakley et al., 1980). Improvements will continue apace until LC/MS is as commonplace and utilitarian as GC/MS.

Arpino et al. (1979) have discussed the optimization of LC/MS. A good example of on-line LC/MS is in the qualitation and quantitation of natural capsaicinoids (Englehardt, 1979). McFadden (1980) has written on the subject of LC/MS. Melera (1980) has described split of the LC effluent prior to detection with a chemical ionization MS source. Schulten and Soldati (1981) have described determination of sugar sequences with LC in combination with field desorption – mass spectrometry. Scott (1979) has described the moving-wire method of introducing samples in a mass spectrometer from a liquid chromatograph that has been commercially produced.

## III. REVIEW OF THE SEPARATION PROCESS

Chromatography considered as a whole is a method whereby a mixture of compounds is separated into individual components. Based either directly or indirectly on elution time, one basic identification is accomplished. Final identification usually follows with an ancillary method such as infrared (IR), UV, or RI analysis. The principles on which chromatographic separations are based are common to all types of chromatography. Liquid chromatography in fact encompasses four different areas of study, all of which have application in the food and beverage industry. These areas, as mentioned in the introduction, are adsorption (liquid–solid), partition (liquid–liquid), exclusion (gel permeation and gel filtration), and ion exchange and ion pairing. But they share certain general features, particularly concerning separation, which is usually the most important aspect for any practitioner of the art of HPLC.

Much effort has been devoted to development of chromatographic theory, particularly in the area of GC. Based on theory, it is usually possible to construct a model that will serve well to approximate what is taking place in the column.

At the most basic level there is a stationary phase and a mobile phase involved in each chromatographic separation. As a sample moves through the column with the mobile phase, it is involved in a partition between the mobile phase and the stationary phase that is the basis for separation. The partitioning effect causes components of the sample to be selectively retarded; that is, the amount of the retardation will be different for each sample component. A standard means of measuring this retardation for each component is the retention volume $V_R$, which is as the name implies the volume of mobile phase that is needed to elute the component from the column under the particular conditions of analysis. The use of the term *retention volume* is somewhat antiquated for use with modern HPLC because of the accurate control of mobile-phase volume that is possible with modern instrumentation. Therefore, in current terminology retention volume has been replaced with retention time, or $t_R$. By the same token what was referred to initially as interstitial volume $V_c$ now is commonly called mobile-phase hold-up time, or $t_M$.

The equation

$$t'_R = t_R - t_M$$

gives us the adjusted retention time or the time that the solute molecule spends in the stationary phase. For a mixture of components the time spent in the mobile phase will be the same for each component. But each of them will spend a different amount of time in the stationary phase. The extent to which they are separated from each other is a function of the different amounts of time spent in the stationary phase.

A useful measure, the capacity ratio, arises from Eq. (1) and is designated by $k$:

$$k = t'_R / t_M = (t_R - t_M)/t_M$$

If we solve the equation for $t_R$, or retention time, we get:

$$t_R = t_M{}^{(1 + k)}$$

Generally speaking, the range of values for capacity is from 1 for the first peak eluted in a chromatogram to 14 or 15 for the last peak eluted. When capacity factors exceed about 15, the accompanying increase in analysis time and peak width leads to poor resolution.

Ideally, the shape of a chromatographic peak is a Gaussian, or normal distribution, curve. As one might expect with a Gaussian distribution, the shape of the curve can be characterized by the $\sigma$ or standard deviation of the curve.

Three different measurements of width can be used to characterize the peak: (1) $w_b$, peak width at baseline; (2) $w_h$, peak width at half height; and (3) $w_i$, peak width at the inflection points of the curve. These three measurements of width are shown to be interrelated by the following equations: (1) $w_b = 2w_i$ and (2) $w_b = 1.699 \, w_h$. For quantitative purposes, the total area under the curve is proportional to the amount of material being detected. Again, under ideal conditions, one desires that the peak width be as small as possible; that is, that the peak be sharp, which suggests that the conditions of operation of the column are close to ideal.

It has become standard to define the ratio of $t_R/\sigma$ as the efficiency of the column:

$$n = (t_R/\sigma)^2$$

In terms of retention time and peak width, the number $n$ of theoretical plates in the column can be expressed by:

$$n = 16(t_R/w_b)^2$$

or

$$n = 5.545(t_R/w_b)^2$$

From classical chromatographic theory we take the term height equivalent to theoretical plate (HETP), which relates the column length to its efficiency:

$$\text{HETP} = L/h$$

An efficient column will have a small number for its plate height and a high number for its plate number. For example, consider two columns, 1 and 2, that have plate numbers of 6588 and 7920, respectively. Assume that both columns are the same length, say 20 cm or 200 mm; then the plate heights will be:

$$h_1 = 200/6588 = 0.0304 \qquad h_2 = 200/7920 = 0.0253$$

Column number 2 will be the more efficient of the two columns.

Yet another value that is frequently employed to characterize a column is that of separation factor, denoted by $\alpha$:

$$\alpha = t_R'(2)/t_R'(1) = k_2/k_1$$

where $t_R' = t_R - t_M$. Let $t_M = 45$ seconds and let $t_R(1) = 105$ sec, $t_R(2) = 200$ sec, and $t_R(3) = 318$ sec. Then $t_R'(1) = 105 - 45 = 60$ sec, $t_R'(2) = 200 - 45 = 155$ sec, and $t_R'(3) = 318 - 45 = 273$ sec. If we calculate the ratios for the first and second peaks, we get:

$$\alpha_1 = k_2/k_1 = 155/60 = 2.5833$$

and for the second and third peaks we get:

$$\alpha_2 = k_3/k_2 = 273/60 = 4.55$$

The larger ratio indicates the pair of peaks that will be easiest to separate in a given column under a given set of operating conditions. This factor will remain constant for any instrument that may be used for analysis, assuming constant operating conditions.

And finally we come to deal with peak resolution, which shows the actual separation of peaks realized by a given analysis. Resolution is expressed by:

$$R = \frac{\Delta t}{\dfrac{w_b(1) + w_b(2)}{2}} = \frac{2\Delta t}{w_b(1) + w_b(2)}$$

where $t$ is the time difference of two maximum peak values and $w_b$ is the peak width at baseline. When one is dealing with two peaks that elute in close sequence, it can be assumed that the widths are the same, that is, that $w_b(1) = w_b(2) = \Delta t/R$. When this value is 1.5, there is baseline resolution; numbers greater than 1.5 indicate a flat baseline existing between peaks, and numbers less than 1.5 show less than complete separation of the two peaks.

## IV. REVERSED-PHASE CHROMATOGRAPHY

It is quite necessary that the selection of mobile phase be given sufficient consideration. In the initial work with most liquid chromatography, the normal or standard arrangement utilized a polar stationary phase that was operated with a nonpolar mobile phase. As the evolution of the technique of HPLC has proceeded, the use of normal-phase chromatography has gradually given way, until today some 75% of the chromatography performed is with reversed-phase configuration, that is, nonpolar stationary phase with polar mobile phase. A typical mobile phase is a combination of water and another polar solvent that allows one to provide a polarity ranging from that of water, which in this terminology is the weakest solvent, to the more polar phases such as methanol or acetonitrile. Assuming a nonpolar stationary phase in place on the solid support, a nonpolar solute will be quite soluble in it and less soluble in, say, a mobile phase consisting of water and methanol together.

## V. GRADIENT ELUTION

The nature of a sample from a foodstuff or another biological material is usually such that all components elute from an HPLC column operated isocratically or with a single solvent with less than ideal separation and resolution. A very common means of improving the separations on reversed-phase HPLC in particular is gradient elution. Initially, the use of fixed blends throughout the chromatographic run allowed one to optimize the components in the middle portion of the chromatogram, but early and late components were handled unsatisfactorily. Upon the development of more sophisticated pumping systems for modern HPLC it was possible to set gradient elution patterns that are under automatic machine control. Working out the precise gradient program required for a particular job can be very time-consuming. However, most commercial equipment can operate a number of standard programs that provide satisfactory results for a wide variety of sample types. Depending upon the nature of the sample to be separated, one can start with a mobile phase, which is, for instance, 100% solvent A, and gradually increase the admixture of solvent B until the mobile phase is 100% solvent B. For nonlinear requirements, the composition of the mixture may be held constant for a period of time, which will result in optimum separation of the components of the sample.

Virtually all of the work on gradient elution has been performed with binary mobile-phase mixtures. Recently there have been increasing reports of the use of ternary or three-component mobile-phase systems. The programming of three separate solvents becomes very complex but can be of considerable usefulness if the routines are provided by the manufacturer or if the construction of an elution profile can be achieved with the help of automatic control. Often, if the polarity range of a sample is great enough to justify the involvement with ternary programming, it may prove advantageous to consider an ion chromatography technique.

## VI. SIZE EXCLUSION

*Size exclusion* is the present-day term for what was at one time in the past referred to as gel permeation or gel filtration. Currently it finds application in the separation of polymeric material, either the synthetic industrial type or the biological type, such as peptides and proteins. The technique can also be used to separate small molecules from polymer molecules.

The relatively rapid advances in this field in recent years have been possible because of the manufacture of a wide range of packing materials for size-exclusion columns that can be used with a variety of solvents, aqueous and organic. The leading material in this classification has been divinylbenzene.

Functioning more or less like a sieve, the hold-up in the column has its greatest effect upon the smaller molecules; hence, the elution pattern is from large to small. With a foreknowledge of the size of the pore openings in the column packing, the retention time of a particular component can be related to its molecular weight; in fact, molecular weight is the outstanding information gleaned via size-exclusion chromatography. A column can be tested with a mixture of known molecular weights very much as one uses a mixture of components to test the performance of a GC column or an HPLC partition column. The molecular weight of an unknown can then, of course, be deduced from its time of elution as bracketed by times for two known molecular weights.

It is logical to assume that, since the separation effected in the size-exclusion column is only a physical phenomenon, based on size, the mobile phases used in this technique should be only a medium for transporting the sample through the column, having no interaction with the sample, the column, or the column packing.

## VII. ION CHROMATOGRAPHY

As the name implies, ion chromatography involves the use of ion exchangers as the stationary phase. The sample will be able to displace the charges on the stationary phase in proportion to the strength of the charge it possesses. These charges can be either positive or negative for a given column; that is, the sample ions can exchange with cations or anions. A common starting point for a stationary phase is silica gel (polymeric materials are also used) to which ionic groups are bonded by chemical action. Common cation exchangers are sulfuric acid and carboxylic acid groups; the most common anion exchanger is the quaternary ammonium group.

As the liquid mobile phase passes through the column, it controls the retention of an ionic solute by dint of ionic strength, pH, and the possible presence of an organic modifier. A happy medium must be struck, so that the ionic strength of the sample will be enough for it to react with the stationary phase and thus move from site to site until it has passed through the column. If the ionic strength is too weak, the ions will remain on the column indefinitely. As we saw in our earlier discussion of partition and adsorption chromatography, we can call upon gradient elution techniques to deal with samples that have a range of charge from weak to strong; in ion chromatography gradient elution, one begins with a mobile phase of low molality and progressively increases the molality to the extent necessary to elute components of higher ionic strength.

Some common types of separations that are handled in ion systems are nucleic acids, carboxylic acids, and organic amines that cover a range from weakly charged to strongly charged samples.

The control of pH is important because of its influence on weak exchangers, both cationic and anionic. The effect is somewhat diminished for strong exchangers or for samples that are highly ionized. A systematic approach to control of pH involves the adjustment of $pK_a$ values.

A rapidly developing approach to ion chromatography has been showing promise for some time now; this is referred to as ion-pair chromatography. Its use provides for analysis of certain categories of sample that have heretofore been difficult to analyze by any method. This includes materials that are so highly ionized that they do not lend themselves to analysis by standard adsorption and partition techniques and that are not soluble enough in water to be studied by the more well known ion-exchange techniques. In brief, ion pairing provides for the addition of a second ion that combines with the sample ion to produce a neutral pair that is then separated by conventional column chromatographic means, such as a reversed-phase column. One can also use the second ion to modify the stationary phase so that ion exchange takes place with the sample. The choice of ion obviously depends upon the sample to a certain extent, but in general terms the ion should possess either a high or a low $pK_a$ value in order to ensure that weakly ionizable materials are completely suppressed and that all strongly ionizable materials are completely ionized. A general survey of ion chromatography has been provided by Pohl and Johnson (1980).

## VIII. QUALITATIVE AND QUANTITATIVE DETERMINATIONS

One tacitly assumes that the ultimate questions in HPLC are what compounds are present in the sample under analysis and how much of each compound is present therein. Although the science of chromatography is in general a technique for effecting separations, it does not necessarily indicate what has been separated. It is further possible to determine very precisely how much of something you have yet not have a clue about its qualitative aspects. In order to get a closer fix on identifications, retention data are customarily employed. When a given column is operated under standard reproducible conditions, it is possible to obtain replicate retention times that provide the basis for an identification. Let us start with the basic relationship:

$$t'_R = t_R - t_M$$

For a typical analysis $t'_R$ represents the adjusted retention time, $t_R$ is the retention time from the point of injection to the maximum of the peak of interest, and $t_M$ is the retention time of a compound that is totally unretained by the column. Multiplication of each of the terms in the above equation by the flow rate $F$ results in a volume term, that is, $t'_R \times F = V'_R$, adjusted retention volume; $t_R \times F = V_R$, the retention volume; and $t_m \times F = V_m$, the void volume or interstitial volume of the column.

Probably the most common way of representing retention is via relative retention ($r$) or separation factor.

We can write:

$$r_{p,st} = \frac{t'_R(p)}{t'_R(st)} = \frac{V'_R(p)}{V'_R(st)} = \frac{k_p}{k_{st}}$$

where p represents the peak of an individual component and st indicates the standard reference compound, which will vary with the nature of the material expected to be separated on the column in question. The $k$ values refer to the capacity factor, which has been addressed in Section III. Using $r_p$ values obviously ties each component in the sample to a common reference point. Furthermore, it is advisable to run the "standard" as an internal standard; that is, the standard component is added to the sample and is thus chromatographed along with the components of the sample.

For the best overall results a standard should be chosen that falls midway in retention between the first and last components to elute from the column; although the terminology is not applied in practical usage, it is in fact a median relative retention. Although most HPLC work is done with a single column for which the operational parameters are carefully controlled, there is no absolute assurance that a retention value on this single column is unique. Therefore, rigorous identifications will be made with two columns of fairly wide polarity difference or activity. Even with two columns there is far from complete lack of ambiguity. At worst, however, one may show the lack of any particular component by comparing chromatograms of the sample mixture with and without internal standard X. If the mixture when chromatographed alone does not show a peak in a given section of the chromatogram but does present a peak in the admixture of "standard" and mixture, one can safely assume that there is no "standard" present in the sample. This may seem a rather negative approach to analysis, but in actual applications in food and beverage analysis establishing the absence of a suspected contamination or additive to a product could be the major problem to be solved. The food industry, concerned with government regulations, is very often faced with proving that a certain component has not found its way into the product. One additional general counsel would be to use the smallest sample sizes that will give good response with the chromatographic detector system being employed. Overload conditions can lead to false retention times. Sometimes, deliberate overload of one component is elected to allow the production of good elution profiles for components that are present in very small concentrations. The potential distortion of retention times should be kept in mind, however.

A now classic application to chromatographic retention was made by Kovats (1958) while he was working in the field of gas chromatography. It is called the Kovats retention index system, and it produces a linear relationship when the log of adjusted retention time is plotted against the carbon number of a homologous series of organic compounds. In this system members of a homologous series of hydrocarbons (specifically alkanes) have basic values equal to $100n$, where $n$ is the carbon number of the alkane. When a plot is made for these data, a straight line results; similar linear relationships obtain for other homologous series and can be applied to LC separation as well. Retention data for unknown members of a series can be obtained by interpolation or extrapolation, and conversely identification of a member of the series can be effected by reading its carbon number from a linear plot of log of retention time versus carbon number. By manipulation of the operating parameters of a column, most especially the composition and nature of the mobile phase, the relative positions of the various components of a sample can be shifted. The wider the range of polarity exhibited by these components, the greater their spread will be. Therefore, the observed spreads resulting from various specific changes in mobile phase or other operating parameters will suggest the nature of the components being separated.

Retention data for LC separations are published from time to time in books and technical journals, but there has been little progress toward providing a uniform basis for universal utilization. Although published information about a particular type of sample may be useful in dealing more directly with successful separation of its components, the average researcher still works out the details for his or her own equipment. If nothing else agrees between two chromatographic runs in two different laboratories, usually the order of elution (or retention) of components will remain the same.

Once a component has been separated by the LC column it must be identified; a major feature of LC is the ease with which the peak can be collected for subsequent qualitative analyses. Such analysis can be accomplished with one of the numerous types of detectors mentioned in Section II, F, or the actual "peak" can be collected for derivatization or another chemical reaction that will disclose the functionality and/or basic physicochemical characteristics of the compound in question.

Another advantage in sampling arises from a feature offered by numerous manufacturers called stopped-flow. This provision allows flow to be stopped with the sample in the cell. Detection can become more sophisticated, and more successful qualitative results beyond those indicated by retention times alone can be obtained.

For the more or less routine analysis of foods and related substances there are still numerous problems to be solved, problems of which there is a very active pursuit. Among these are detection of low concentrations of material, most particularly those that do not exhibit UV absorption characteristics. In addition there are the problems of small peak capacity, limitations on solute solubility, and/or limited adsorptive capacity of the stationary phase.

Quantitative analysis of HPLC runs is finding greatly increased application in most areas of HPLC use. Peak height could be expected to be a convenient basis for quantitation, but in practice it turns out to vary too much with operating conditions to be very useful. Therefore, virtually all accurate work is based on peak areas. Most commercial HPLC equipment of recent vintage comes supplied with electronic digital integration, which currently provides the epitome of performance in quantitation. It is with this method that the best precision is obtained.

Two other methods, the disk integrator and the cut-and-weight method of standard curve creation, yield rather acceptable precision. Modern HPLC workers probably would not have any involvement with methods offering poorer precision.

## IX. APPLICATION TO FOODS AND BEVERAGES

Since some 70% of the world of chemical compounds is solid at room temperature, there is obviously strong need for techniques to analyze solids. This is especially true of natural products and foodstuffs. Although attention has been directed for many years to studying the volatile components of foods with such powerful techniques as GC, there is ever-increasing interest in the investigation of the higher-molecular-weight materials that are solids at room temperature, such as carbohydrates, lipids, and peptides/proteins, the three basic categories of food. The significant advances that have been made in LC column technology and instrumentation make it possible to perform an increasing number of analyses at high speed and resolution. A general application for foodstuffs is provided by Lawrence (1981).

In the area of natural components of foods one is concerned primarily with the effects on flavor that result from the entire range of treatments that foodstuffs undergo, from growth of the source material through various processing and preservation steps to the packaging of a final consumable product. Since HPLC analyses can usually be carried out at ambient temperature, they have been in the last decade probably the most important analytical techniques for dealing with polar and thermally unstable compounds; molecular weights of up to about 2000 have been involved. We have already mentioned some of the progress in detection of components separated by HPLC, many of which are concerned ultimately with the fact that large numbers of flavor constituents do not contain a good chromophoric group, and are thus incapable of being analyzed with the most common UV adsoprtion detectors.

The complexity of foodstuffs has, in general, taxed the ability of HPLC to effect good separation in one continuous run. Among the drawbacks have been restricted peak capacity and a lack of good range in $k'$ values. The lack of capability in this respect should not really be considered a serious drawback, especially in light of a similar problem that arose in the separation of complex flavor mixtures by means of capillary GC, a technique that has been widely heralded as a panacea for a flavor chemist intent on doing difficult separation.

The applications of HPLC to natural components of foods are proliferating at a dramatic rate. This makes it very nearly impossible to list completely the worthwhile investigations that have been carried out. As far as this reviewer is concerned, the first line of information is a list of journals in one's specific discipline that have the greatest likelihood of reporting items of interest on a timely basis.

Additionally there is a journal published by Marcel Dekker Journals,* which, however, deals with general as well as specific applications of HPLC. This is the *Journal of Liquid Chromatography.*

On a continuing basis the annual reviews published by the ACS journal *Analytical Chemistry* offer the most exhaustive general coverage of HPLC in all its ramifications as well as specific coverage of applications in food. These reviews are two in number:

1. Applications reviews appear every other year in odd years; the next edition is due in April of 1985. One of the chapters in this publication is devoted to analytical chemistry of food. In the last edition, appearing in 1983, the following section titles were used: additives; adulteration, contamination, decomposition; carbohydrates; color; enzymes; fats, oils, and fatty acids; flavors and volatile compounds; identity; inorganic; moisture; organic acids; nitrogen; and vitamins. Some 1300 references were cited among these categories. Although the material reviewed covers all aspects of food analysis, the representation of HPLC is considerable; the non-HPLC developments that are reported will serve to provide a general background in analysis of foodstuffs and thereby to suggest areas where HPLC may find new application.

2. Fundamental reviews appear every other year in even years; the next edition is due in April of 1984. The chapters in this review address the fundamental disciplines of analytical chemistry, and one chapter is devoted to column LC. The compilers of this biannual review provide outstanding coverage of the field in breadth and depth. The subject matter includes gen-

---

* P. O. Box 11305, New York, New York 10249.

eral reviews, specialty reviews, and sections on columns, instrumentation, detectors, liquid–solid chromatography, liquid–liquid chromatography, normal- (bonded-) phase chromatography, reversed-phase chromatography, ion-exchange chromatography, sample preparation, pre- and postcolumn techniques, preparative liquid chromatography, and process liquid chromatography. Of course the coverage here includes more than food and beverages, but the material is organized in a manner that makes it easy for the user to extract what he or she needs from the total mass of information.

An outstanding example of the current trends in LC be found in the following listing of subjects of papers presented at the Eastern Analytical Symposium in November of 1982 in New York City:

Steric Exchange Chromatography (SEC) of Complex Polymers
Recent Developments in HP Aqueous and Nonaqueous SEC Packing
Optimization of the Gel Permeation Chromatographic Separation
Advances in Detectors for SEC
Porous Glassy Carbon: A New Column Packing Material for HPLC and GC
Controlled Surface Interaction Utilizing Mobile-Phase Additives
Fluorine-Containing Bonded Phases for HPLC
Separations Using Dynamic Surfaces
Reproducible Stationary Phases for LC
Automated Precolumn Derivatization for Analysis of Peptides and Amino Acids
Solid-Phase Peptide Synthesis Using LC Equipment
High-Performance Ion-Exchange Chromatography of Proteins
Large-Scale Purification of Proteins Produced by Recombinant DNA
Optimization of Mobile-Phase Components for LC Separations.
On-Line Multidimensional LC
HPLSC of Polynucleotides
Emerging Trends in the Design of Novel Stationary Phases for HPLC Separations of Nucleic
    Acid Constituents
HPLC Assays of Enzymes in the Purine Metabolic Pathway
Kinetic Analyses of Phosphoribosyl Transferase Activities Using HPLC
Nutritional Studies Involving Nucleic Acid Metabolism
Application of Gradient Elution to the Separation of Macromolecules in Reversed-Phase
    Chromatography
High-Performance Liquid-Exchange Chromatography Using Chemically Bonded Phases
Displacement Chromatography
Computer-Assisted HPLC
Use of Solvent Programming for Optimum Conditions in Isocratic Analysis.

## A. Natural Components of Foods

General papers continue to appear on carbohydrate determinations, employing both old and new techniques. HPLC techniques continue to be applied for determination of individual sugars (Li and Stewart, 1979). In fact the extensive use of HPLC for carbohydrate-related work makes selection of representative examples rather difficult. One useful treatment of individual carbohydrates in foods has been published by Hurst *et al.* (1979), another by Macrae and Dick (1981).

HPLC has been used to determine sugars in molasses (Damon and Pettitt, 1980) and in high-fructose syrups (van Olst and Joosten, 1979). It has been reported (DeVries *et al.*, 1979) that HPLC results on carbohydrates in foods are comparable with results from chemical determination. A classic determination is that of ethyl vanillin in chocolate (Hurst and Martin, 1982).

Fatty acid determinations are a major portion of the analyses of lipid materials being reported since their profiles are so useful in identifying specific natural sources. The fatty acids are usually derivatized prior to chromatography; methyl ester and phenacyl derivatives are among the more common forms. Triglyceride analysis constitutes the second major area of lipid determination. Particularly noteworthy are analyses on silica gel (Biacs *et al.*, 1979) and with argentation (Shipe *et al.*, 1980).

Charalambous (1979) has edited a two-volume set on LC analysis of food and beverages. Dong and DiCesare have focused on improved food analysis by means of LC (Dong and DiCesare, 1983). Amino acid and small-peptide determinations are perhaps best effected with ion-exchange column technology because of their charged character. As usual, a requirement for less rigorous cleanup of samples prior to chromatography makes use of a column technique advantageous.

## B. Food Additives

An additive can conceivably be a natural product that is added to a food product to improve its characteristics but that by virtue of the fact that it was not present initially in the natural product becomes classified as an additive. A useful review on determination of food additives is attributed to Conacher and Page (1979). Among the most common applications of HPLC are determination of antioxidants in fat, using both regular and reversed phase; a full line of food preservatives including benzoic and sorbic acids; and other additives such as saccharin and caffeine.

## C. Contamination in Food

Contamination is combined by Sloman *et al.* (1981) with adulteration to form a separate subject heading in their review of food analysis. The categories that have been investigated are legion. The most frequently cited group of compounds to be determined by HPLC is the mycotoxins, especially in peanuts and corn; both regular and reverse phase have been used in approximately equal amounts; determination level has been reduced to approximately 5 ng.

Veterinary drugs constitute another area of intensive interest in HPLC determinations in foods, specific examples being chloramphenicol in chicken, nitrofurazone in milk, and sulfamethazine in beef.

Size-exclusion chromatography has been used for antioxidants and DDT. Ion-pair HPLC has been used to determine ethylenediaminetetraacetic acid.

## D. Miscellaneous Determinations

As usual, miscellaneous becomes a catchall category. In surveying the literature, the most frequently reported category turns out to be color. This category takes the form of anthocyanins (Camire and Clydesdale, 1979; Pohl and Johnson, 1980), glucosides and diglucosides of anthocyanidins (Williams *et al.,* 1978), carotenoids, $\beta$-carotenes, and synthetic acid dyes. Food dyes (Merle *et al.,* 1978, 1979; Altzetmueller and Arzberger, 1979) have been analyzed with ion-pair HPLC. One report on fruit and vegetable extracts (Galensa and Herrmann, 1980) is of particular interest because it makes use of four different solvents in conjunction with a silica gel column. Actual components analyzed were flavonoids. Postcolumn determination by means of immobilized enzymes has been prominent in Japanese reports (Arisue *et al.,* 1980, 1981). The technique has also been applied to determination of the oxidation products of cholesterol (Oegren *et al.,* 1980). A review of postcolumn design selection has been offered by Frei and Schulten (1979). Large-scale LC has been described by DiCesare and Vandemark (1982).

Quantitation (Qureshi *et al.,* 1979) has been effective with pyrimidines, purines, nucleosides, nucleotides, polyphenols, and pyrazines.

Studebaker (1979) has reported LC of thiols, disulfides, and proteolytic enzymes. Food emulsifiers have been determined with semipreparative columns (Sudrand, 1981), and Verzele and Geeraert (1980) have described preparative applications.

HPLC of capsaicinoids from capsicum fruits is also an active area of application.

## X. SUMMARY AND CONCLUSIONS

In concluding it is certainly appropriate to take a look at the trends that are expected to be followed in HPLC during the next couple of years. A good general review of instrumentation has been given by Henry and Sivorinovsky (1981).

The experts are not predicting any substantial breakthroughs in the actual technology of the LC separation. Rather, there is expected to be a general refinement and stabilization of what has been reported during approximately the last few years. Electronics will show the greatest innovation as applied to HPLC, as indeed it can be expected to be a major part of what is new in the entire field of analytical instrumentation. Microprocessors will surely come into more common usage for the total control of an analysis. Also, HPLC equipment will manifest more self-diagnostic capability, leading to less down time, inasmuch as problems will be sensed before they become significant enough to cause disruption in the operation of the system. Computers have been manifesting self-diagnostics or remote diagnostics, for which the computer goes on line via a modem device to a remote computer that analyzes performance of the system and provides a set of instructions for the technician to follow when he or she makes the site visit to do repairs; in some cases, replacement of modules can be carried out by nonservice personnel. All this leads to lower maintenance costs for rather involved systems, and HPLC systems are expected to join this number. At the present time some 32% of laboratories in the United States are equipped with some form of LC, thus providing fertile field for instrument manufacturers (Mosbacher, 1982). With a microprocessor as part of the instrumentation, it can be expected that liquid chromatographs will "learn" to perform their own experiments, change operational parameters, and rerun with more extensive optimization of conditions until a preset level of agreement is reached.

The refinement and expansion of techniques for both pre- and postcolumn derivatization will aid in the development of HPLC in two ways: (1) the range of possible separations will increase substantially as formerly intractable mixtures become subject to more facile analysis by dint of their new forms, and (2) detection will be enhanced by conversion of nonchromophores, for example, to chromophores, which can more readily be detected with existing optical detectors. Several manufacturers are expected to follow the lead of du Pont, which has introduced a quarternary or four-solvent system. Automated solvent mixing via sophisticated pumping systems makes this concept feasible. What little is new in HPLC equipment will constitute refinement in and development of more capable detectors that in concert with the expanded derivatization techniques mentioned earlier will expand the horizons of HPLC for all applications, including those in natural products, foods, and beverages.

## REFERENCES

Alexander, P. W., Hadda, P. R., Low, G. K. C., and Maitra, C. J. (1981). *J. Chromatogr.* **69**, 9–39.
Altzetmueller, K., and Arzberger, E. (1979). *Z. Lebensm.-Unters-Forsch.* **169**, 335.

Arisue, K., Ogawa, Z., Koda, K., Hayashi, C., and Ishida, Y. (1980). *Rinsho Kagaku* **9**, 104.
Arisue, K., Maru, Y., Yoshida, T., Ogawa, Z., Kohda, K., Hayashi, C., and Ishida, Y. (1981). *Rinsho Byori* **29**, 459.
Arpino, P. J., Guichon, G., Krien, P., and Devant, G. (1979). *J. Chromatogr.* **185**, 529.
Biacs, P., Erdelyi, A., Kurucy-Lusztig, E., and Hollo, J. (1979). *Acta Aliment. Acad. Sci. Hung.* **7**, 181; *Anal. Abstr.* (1979), 2F61.
Blakley, C. R., Carmody, J. C., and Vestal, M. L. (1980). *Clin. Chem. (Winston-Salem, N.C.)* **26**, 1467.
Bristow, P. A. (1976). "LC in Practice." HETP Publ. Handforth, Wilmslow, Cheshire.
Camire, A. L., and Clydesdale, F. M. (1979). *J. Food Sci.* **44**, 926.
Charalambous, G. (1979). "Liquid Chromatography Analysis of Food and Beverage," Vols. 1 and 2. Academic Press, New York.
Chen, H.-W. Bao, M.-S., Fang, S.-H., and Lu, P.-C. (1981). *Chromatographia* **14**, 129–134.
Colin, H., Martin, M., and Guichon, G. (1979). *J. Chromatogr.* **185**, 79.
Conacher, H. B. S., and Page, B. D. (1979). *J. Chromatogr. Sci.* **17**, 188.
Crosby, N. T., Hunt, D. C., Philp, L. A., and Patel, I. (1981). *Analyst (London)* **106**, No. 1259, 135.
Damon, C. E., and Pettitt, B. C., Jr. (1980). *Chem.* **63**, 476.
DeVries, J. W., Heroff, J. C., and Egberg, D. C. (1979). *J. Assoc. Off. Anal. Chem.* **62**, 1292.
DiCesare, J. L., and Vandemark, F. L. (1982). *Ind. Res. Dev.* **24**, 138.
Dong, M. W., and DiCesare, J. L. (1983). *Food Technol. (Chicago)* **37**, 58.
Englehardt, H. (1979). "High Performance Liquid Chromatography." Springer-Verlag, Berlin and New York.
Fell, A. F. (1980). *Anal. Proc. (London)* **17**, 512.
Fallick, G. H., and Rausch, C. W. (1979). *Am. Lab. (Fairfield, Conn.)* **11**, 87, 89–90, 92, 94, 97.
Frei, R. W., and Schulten, A. H. M. T. (1979). *J. Chromatogr. Sci.* **17**, 152, 79A.
Galensa, R., and Herrmann, K. (1980). *J. Chromatogr.* **189**, 217.
Hamilton, R. J., and Sewell, P. A. (1978). "Introduction to High Performance Liquid Chromatography." Chapman and Hill, London.
Henion, J. D. (1980). *J. Chromatogr. Sci.* **18**, 101, 112.
Henry, R. A., and Sivorinovsky, G. (1981). *Liq. Chromatogr. Clin. Anal.* pp. 21.
Horvath, C. (1980). "High Performance Liquid Chromatography Advances and Perspectives," Vols. I and II. Academic Press, New York.
Hurst, W. J., and Martin, R. A., Jr. (1982). *Am. Lab. (Fairfield, Conn.)* **14**, 74.
Hurst, W. J., Martin, R. A., Jr., and Zoumas, B. L. (1979). *J. Food Sci.* **44**, 892.
Karger, B. L., Snyder, L. R., and Horvath, C. (1973). "An Introduction to Separation Science." Wiley (Interscience), New York.
Kessler, M. J. (1982). *Am. Lab. (Fairfield, Conn.)* **14**, 52.
Kirkland, J. J., ed. (1971). "Modern Practice of Liquid Chromatography." Wiley (Interscience), New York.
Knox, J. H., Done, J. N., Fell, A. T., Gilbert, M. T., Pryde, A., and Wall, R. A. (1978). "High Performance Liquid Chromatography." Edinburgh Univ. Press, Scotland.
Kovats, E. (1958). *Helv. Chim. Acta.* **41**, 1915.
Lawrence, J. F. (1981). "Organic Trace Analysis by Liquid Chromatography." Academic Press, New York.
Li, B. W., and Stewart, K. K. (1979). *NBS Spec. Publ. (U.S.)* No. 519, p. 271; *Chem. Abstr.* (1979), **91**, 73238X.
McDowell, L. M., Barber, W. E., and Carr, P. W. (1981). *Anal. Chem.,* **53**, 1373.
McFadden, W. H. (1980). *J. Chromatogr. Sci.* **18**, 97.
McKinley, W. A., Popovich, D. J., and Layne, T. (1980). *(Fairfield, Conn.)* **12**, 37–38, 40, 42, 45–47.

Macrae, R., and Dick, J. (1981). *J. Chromatogr.* **210**, 138.
Majors, R. E., ed. (1980). *J. Chromatogr. Sci.* **18**, 393–486, Part I; **18**, 487–582, Part II.
Melera, A. (1980). *Adv. Mass Spectrom.* **8**, 1597.
Merle, M. H., Puerta, A., and Puerta, M. (1978). *Ann. Falsif. Expert. Chim.* **71**, 263; *Chem. Abstr.* (1979), **90**, 37686N.
Mosbacher, C. J. (1982). *Ind. Res. Dev.* **24**, 177.
Oegren, L., Csiky, I., Risinger, L., Nilsson, L. G., and Johansson, G. (1980). *Anal. Chim. Acta* **117**, 71.
Ohmacht, R., and Halasz, I. (1981). *Chromatographia* **14**, 216–226.
Ohta, H., Akuta, S., Okamoto, T., and Osajima, Y. (1979). *Kyushu Diagaku Nogakubu Gakugei Zasshi* **33**, 101; *Anal, Abstr.* (1980), **39**, 1F25.
Pohl, C. A., and Johnson, E. L. (1980). *J. Chromatogr. Sci.* **18**, 442.
Provder, T. (1980). *ACS Symp. Ser.* No. 138.
Qureshi, A. A., Prentice, N., and Burger, W. C. (1979). *J. Chromatogr.* **170**, 343.
Rabel, F. M. (1980). *J. Chromatogr. Sci.* **18**, 394.
Reese, C. E., and Scott, R. P. W. (1980). *J. Chromatogr. Sci.* **18**, 479–486.
Schulten, H. R., and Soldati, F. J. (1981). *J. Chromatogr.* **212**, 37.
Scott, R. P. W. (1976). "Contemporary Liquid Chromatography." Wiley (Interscience), New York.
Scott, R. P. W. (1979). *NBS Spec. Publ. (U.S.)* No. 519, 637.
Sepaniak, M. J., and Yeung, E. S. (1980). *J. Chromatogr.* **190**, 377.
Shipe, W. F., Senyk, G. F., and Fountain, K. B. (1980). *J. Dairy Sci.* **63**, 193.
Simpson, C. F. (1976). "Practical High Performance Liquid Chromatography." Heyden, London.
Simpson, C. F. (1979). *Proc. Anal. Div. Chem. Soc.* **16**, 222.
Sloman, K. G., Foltz, A. K., and Yeransian, J. A. (1981). *Anal. Chem.* **53**, 242R–273R.
Smythe, L. E. (1981). *Chem. Aust.* **48**, 249.
Snyder, L. R., and Kirkland, J. J. (1979). "Introduction to Modern Liquid Chromatography." Wiley (Interscience), New York.
Studebaker, J. F. (1979). *J. Chromatogr.* **185**, 497.
Sudrand, G., Constand, J. M., Retho, C., Cande, M., Rosset, R., Hagemann, R., Gaudin, D., and Vivelizier, H. (1981). *J. Chromatogr.* **204**, 397.
van Olst, H., and Joosten, G. E. H. (1979). *J. Liq. Chromatogr.* **2**, 111.
Verzele, M., and Geeraert, E. (1980). *J. Chromatogr. Sci.,* **18**, 559.
Weisshaar, D. E., Tallman, D. E., and Anderson, J. L. (1981). *Anal. Chem.* **53**, 1809–1813.
Williams, M., Hrazdina, G., Wilkinson, M. M., Sweeny, J. G., and Iacobucci, G. A. (1978). *J. Chromatogr.* **155**, 389.
Yau, W. W., Kirkland, J. J., and Bly, D. D. (1979). "Modern Size-Exclusion Liquid Chromatography." Wiley, New York.
Zelt, D. T., Owen, J. A., and Marks, G. S. (1980). *J. Chromatogr.,* **189**, 209.

# 6

# Mass Spectrometry

IAN HORMAN

*Nestlé Research Laboratories*
*La Tour de Peilz, Switzerland*

ANALYSIS OF FOODS AND BEVERAGES
Copyright © 1984 by Academic Press, Inc.
All rights of reproduction in any form reserved.
ISBN 0-12-169160-8

## I. INTRODUCTION

When J. J. Thomson (1913) was studying the motion of charged particles propelled through magnetic and electric fields, he probably thought of food only as something to eat, if indeed he consciously thought of it at all. And yet, Thomson's work provided the basis for mass spectrometry (MS) as we know it today, a technique that has contributed more than any other to our knowledge on a molecular level of food composition. Although at first MS was firmly in the hands of physicists, its potential was rapidly appreciated by organic chemists and thence by food chemists, who have since used it to analyze everything from dried apricots to boiled fish.

In the early days, the concept of analyzing foods by MS, rather than being taken seriously, would more likely have conjured up humorous mental pictures: for example, of an ionized fruitcake negotiating the bends in the flight tube of the mass spectrometer and perhaps being subsequently identified in terms of its constituents — namely, the flour, sugar, eggs, dried fruit, and spices of the original kitchen recipe.

For better or for worse, these culinary elements of flour, sugar, etc., are still generally considered to be the fundamental components of foods, in much the same way as Empedocles of Agrigentum, in Sicily, over 2000 years ago considered all matter to consist of the elements earth, air, fire, and water. The food chemist today has no such illusions, knowing that each of these basic food materials is a complex mixture of macromolecules such as proteins, polysaccharides, polyphenols, and lipids, along with a host of smaller molecules, together responsible for its different physical, organoleptic, and physiological qualities. It is this myriad of individual chemical entities that we must analyze and characterize.

Further, we are aware that the relative amount of a particular constituent varies in different samples of a given food material. In the example of the dried apricot, the quantitative chemical composition will depend not only on the variety of the tree on which it was grown but also on the tree's geographical location. Trace impurities may be introduced from the irrigating water or as metabolites of agricultural chemicals. The drying process itself will certainly be accompanied by chemical changes, either thermally or fermentation induced, as will the subsequent storage period. Cooking then produces a multitude of new compounds through Maillard reactions, Strecker degradations, and other interactions, and microbial spoilage at any time during growth or storage may give problems with mycotoxins or other metabolites. A total analysis of a food product would therefore read like a biographical account, and the objective of this chapter is to trace the significant role played by MS in establishing such qualitative and quantitative profiles.

From the point of view of instrumental development, MS progressed rapidly throughout the postwar years up to the early 1970s. A period of about 5 or 6 years of apparent stagnation then followed, but the past 3 years have seen an almost unbelievable upsurge of new and exciting techniques, promoted partly by the widespread introduction of computerized data acquisition and processing and partly by novel instrumentation. The potential of these new techniques in food analysis is great, although for some of them no food-related examples are described to date in the literature. In these cases, original results on appropriate samples have been compiled with the cooperation of different laboratories; these results are presented here to demonstrate the operation of each technique and the nature of the information obtained. Where earlier reviews exist, these are cited. Some publications are treated in detail where it is felt that the work described represents a significant milestone in the applications of MS to food analysis.

## II. THE MASS SPECTROMETER AND THE TYPE OF STUDIES IT PERMITS

Consider first the basic nature of the studies we can perform using MS. Their nature can be generalized in a few words, namely, the identification and quantification of individual chemical compounds. The one notable exception to this among food applications, where a whole food item rather than any individual constituent is assessed in a single measurement, involves isotope MS and its use in detecting food adulteration and misrepresentation (see Section II, I). Identification and quantification can be subclassified as follows. Identification can consist of (a) original identification of a totally unknown compound, that is, one never previously identified from any source, or (b) recognition of a compound known to exist but never previously identified as a constituent of the mixture or the source being investigated. Quantification can consist of (a) simple confirmation of the presence or absence of a known specific compound or compounds, or (b) accurate quantitative estimation in a sample of a specific compound that may be present at any level between a few parts per billion or less (trace analysis) and 100% (sample purity analysis).

Foods, and in particular cooked or prepared foods, which often contain spices and condiments in addition to the basic food materials themselves, are undoubtedly the most complex mixtures we might ever wish to find. Individual constituents covering almost the entire range of physical and chemical characteristics are present. Modern mass spectrometers are capable of analyzing routinely the great majority of these compounds, possible exceptions at present being intact natural polymers such as proteins or polysaccharides. Even here, progress is under way, with radioactive californium sources permitting ionization of molecules with molecular weights of around 10,000 (MacFarlane, 1981). Applications of MS to foods today are therefore limited more by our capacity to use optimally the volume of data we can create than by instrumental limitations in running spectra.

A mass spectrometer consists of three basic elements: an ion source, a mass-to-charge ratio analyzer, and an ion collector. These elements were already embodied in the first instruments built by Aston and by Dempster over 60 years ago, and Beynon (1981, 1982) has given interesting accounts of these early days. Today, several alternative forms of ionization are in current use, and the same is true for mass analyzers and ion collectors. This is seen in Table 1, where the list is certainly not exhaustive. In addition, mass spectrometers can be used with several different modes of sample introduction, analyzers can be made to transmit positive or negative ions formed in the source or by decomposition in the analyzer itself, and outcoming data can be treated by computer methods in a variety of different ways. By combination of these various elements, several hundred configurations of instruments could be built. Evidently, not all of these would be practically viable, but many are.

The sensitivity and specificity of MS and associated techniques in mixture analysis, coupled with an increasing simplicity and routine reliability, have made it an instrument of choice for many food analysis laboratories. A single instrument cannot be made to embody all possible techniques, and, indeed, care must be taken to ensure that different techniques to be performed on the same instrument are compatible. As a general rule of thumb, we take as compatible those techniques that can be switched from one to another without breaking the vacuum; in this way, problems due to air leaks are minimized, and operation time is maximized.

## A. Historical Aspects

As early as 1951, Turk *et al.* hinted at the potential of MS for the study of food isolates when they ran the electron-impact (EI) spectrum of mixed volatile components from apples. More than 15 years then elapsed before MS began to be used routinely in food analyses. In the late 1960s, spectra of chromatographically isolated individual volatile constituents were obtained, and high-resolution MS yielded information on elemental compositions of these compounds and their fragments. From the early 1970s, gas chromatography/mass spectrometry (GC/MS) permitted the direct analysis of mixed volatiles without prior fractionation, and involatile constituents such as monosaccharides, amino acids, polyphenolics, and long-chain fatty acids could also be analyzed after transformation into volatile derivatives.

**TABLE 1  Alternative Techniques and Modes of Operation of Mass Spectrometry**

| Modes of ionization | Analyzers | Collectors |
|---|---|---|
| Electron impact (EI) | Quadrupole | Photoplate |
| Chemical ionization (CI) | Triple quadrupole | Faraday cup |
| Field ionization (FI) | Time of flight | Electron multiplier |
| Field desorption (FD) | Scanning magnet | Daly collector |
| Fast-atom bombardment (FAB) | Electrostatic analyzer | |
| Laser desorption | plus magnet | |
| Californium plasma desorption | Reverse geometry | |
| Desorption chemical ionization (DCI) | Fourier transform | |

| Sample introduction | Analysis techniques | Data system |
|---|---|---|
| Hot box | Collision-activated | Mass assignment |
| Probe | decomposition | Elemental compositions |
| GC/MS | Linked scan meta- | Mass chromatograms/fragmentograms |
| LC/MS | stable analysis | Spectrum substraction |
| | Single-ion monitoring | Sample identification using spectrum |
| | Multipeak monitoring | libraries |
| | Isotope ratio measure- | Mixture profiling |
| | ment | |

By the late 1970s, quantitative MS had become a widely used technique. To date, most qualitative food applications of MS have involved the identification of naturally occurring compounds, while quantitative analysis has concentrated more on food additives, contaminants, and toxins. Little, if any, use has been made of ionization methods other than EI nor of sample-introduction methods other than GC/MS or the direct-insertion probe. There are still many possibilities to explore.

The growing importance of MS in food analysis is to be seen in the increasing numbers of review articles published on the subject. Horman (1979, 1981) has described the general qualitative and quantitative problems confronting the food analyst and has covered the literature from 1976 to mid-1978. Questions of flavor have been treated by Kolor (1979, 1980). Self (1979, 1980) has considered quantitative aspects, while Binnemann and Jahr (1979) and Jahr and Binnemann (1979) have looked more specifically at applications to food products. Winkler and Schmidt (1980) and Krueger and Reesman (1982) have reviewed uses of carbon-isotope MS in food products. The latter article has appeared in a new journal, *Mass Spectrometry Reviews (MSR),* which puts emphasis on practical aspects of MS and as such is a welcome addition to the libraries of practicing spectroscopists. Many other articles already published or scheduled for publication in *MSR* are of direct or indirect interest to spectroscopists engaged in food chemistry or technology.

## B. The State of the Art

MS instrumentation in food laboratories lags some 5 to 10 years behind the development of new techniques. For example, chemical ionization (CI) was first used around 1970 and has been readily available commercially since 1973. Even so, it is only during the past 2 years that a significant number of food-related applications have been described. Thus, in this review, EI GC/MS instruments still provide the large majority of the results reported. There are many reasons for this, including the sharp rise of instrument costs in recent years (Berlowitz *et al.,* 1981) and the routine nature of many analyses, for which standard methods have been developed on small, low-cost, dedicated EI GC/MS systems. These instruments are easy to operate, are reliable, and have comprehensive data-processing facilities along with extensive libraries of EI reference spectra. The newer techniques, although potentially interesting, must develop a reputation for routine applicability before they will be widely used by foods analysts, who are generally more interested in solving their own particular problem in food science than in researching esoteric applications for novel techniques.

## C. Gas Chromatography/Mass Spectrometry: The Workhorse

In identifying ethanol and acetaldehyde from the EI spectrum of a mixture of volatiles emanated from apples, Turk *et al.* (1951) recognized that their main limitation was their inability to separate the mixed volatiles into individual components. By 1952, chromatography had won respectability as a science through the Nobel Prize-winning efforts of Martin and Synge. In 1964, Watson and Biemann described a device that coupled the outlet of a gas chromatographic (GC) column directly to the source of a mass spectrometer, separating the carrier gas from the sample by diffusion: GC/MS was born, and it has since grown to become the major form of MS to date, not only in food analysis but also more generally. From the early 1970s, when GC/MS systems became commercially available, almost all known food constituents could be readily analyzed, either directly or after suitable derivatization.

In the early days of GC/MS, there were two schools of thought. Mass spectrometrists looked upon the GC as an elegant sample-introduction device, whereas gas chromatographers saw MS as a sophisticated GC detector. It took some time to realize that the column, the GC/MS interface, and the source of the mass spectrometer form a single integrated unit that must be regarded as a whole. Gas chromatography/mass spectrometry has now established itself as a technique in its own right, as witnessed by the ever-increasing number of articles summarized in the publication "Gas Chromatography – Mass Spectrometry Abstracts" (Brooks, 1970 – 1983), now in its fourteenth year, where many articles on food analysis are to be found. The fundamental aspects of GC/MS are comprehensively described in books by Gudzinowicz *et al.* (1977).

McFadden (1979) has described the various GC/MS interfaces used. Today, the most suitable interfaces for use in food studies are probably the jet separator, which can work with high-flow-rate columns, and the open coupling interface, which takes 1 – 5 ml/min of column effluent directly into the mass spectrometer source. A major advantage of the open coupling interface is that the GC column can be operated with the outlet at atmospheric pressure, that is, under the same conditions as in GC alone. This avoids the problem of the loss of GC resolution due to vacuum effects on the end of the column, an important feature in the analysis of food isolates, which, as shown in Fig. 1, can contain several hundred compounds. With the ever-increasing use of capillary columns in food laboratories, it seems reasonable to predict that open split coupling will become the most widespread interface in use. Further information on this interface is to be found in the original articles by Henneberg *et al.* (1975, 1978) and more recently by Koller and Tressl (1980) and Kenyon and Goodley (1981).

COFFEE HEADSPACE

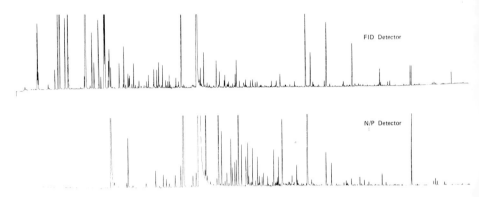

**Fig. 1.** Simultaneous flame ionization detector (FID) and nitrogen–phosphorus (N/P) detector chromatograms of a coffee headspace sample. (Courtesy of R. Liardon, Nestlé, La Tour de Peilz, Switzerland.)

Capillary columns were popularized almost overnight by the introduction of flexible silica tubing. Columns in this material are much more robust than their glass counterparts and contribute greatly to the continuing trend of eliminating metal surfaces on which decomposition is likely to occur in GC systems, interfaces, and transfer lines. Relative to the wide GC peaks obtained with conventional packed columns, those coming from capillary columns usually have peak widths of only a few seconds. This means that mass spectrometers must be capable of scanning rapidly, and whereas this is easy on quadrupole instruments, new magnet technology has been needed on magnetic sector instruments to get the same performance. Now, many instruments are capable of generating spectra at 0.1-sec/decade scan speeds, and it is possible in addition to change the mode of ionization from EI to CI in alternate scans. An example of this technique, applied to the analysis of the essential oil of geranium, is presented by Chapman and Wakefield (1980). This is an approach with many potential applications in food analysis, particularly where new compounds must be identified and where an EI spectrum alone is not enough. A homologous series of $\delta$-lactones, for example, gives basically the same EI mass spectrum, dominated by a large $m/z$ 99 peak. A mass chromatogram of the $m/z$ 99 ion can be used to identify individual members of this series from their GC retention indices (Cazenave *et al.*, 1974), but it would evidently be preferable to have confirmation of that identity. A CI spectrum can give this confirmation through the molecular weight information it contains, and it is of interest to have this CI spectrum in the same analysis as the EI spectrum to be sure that both spectra are run on the same peak.

Environmental Protection Agency (EPA) regulations now require the simultaneous monitoring of over 100 potentially noxious volatile contaminants in drinking water. For this type of repetitive analysis of specific target compounds, GC/MS lends itself readily to automation, as described by Smith (1981). Electron-impact spectra are compared with those of the target compounds in a library, and a quantitative report of the target compounds identified in a given sample can be automatically printed.

Precise mass measurement gives information on the elemental composition of ions, and an interesting assessment of results obtained with a high-resolution GC/high-resolution MS/real-time computer system for the analysis of multicomponent organic mixtures has been presented by Meili et al. (1979). However, it is no longer necessary to work at high resolution (i.e., 10,000 or higher) to get precise mass measurement. First, double-beam operation on magnetic sector instruments and, more recently, precise Hall probes that sense the magnetic field now permit mass measurement at resolutions of about 2500. Elemental composition information is of great value in identifying unknowns, and the interest for food analysts is that this information can be readily obtained in GC/MS operation.

Many examples of the use of GC/MS are described in Section III,A, which describes the analysis of volatiles, but it is worth noting specific examples of GC/MS of volatile derivatives of solids. A good example is the characterization of mixtures of dipeptides, such as might be obtained in protein hydrolysates. W. E. Seifert et al. (1978) analyzed over 120 dipeptides in the form of their $N,O$-perfluoropropionyl methyl esters, obtaining EI and CI spectra and GC retention indices. From the retention data, they were able to predict the retention indices for most of the 400 common dipeptides and to identify dipeptides in mixtures unambiguously.

The introduction of columns with chiral stationary phases (Frank et al., 1978) extends GC/MS to the analysis of optical isomers: this is of interest for amino acids, hydroxy acids, and sugars, etc., because D and L isomers can have different physiological activities in, for example, nutritional value or flavor. Liardon and Ledermann (1980) used such a column to separate derivatized D-amino acids formed by racemization of the naturally occurring L-amino acids during protein hydrolysis, assigning isomeric pairs by their EI mass spectra. Liardon et al. (1981) have since proposed a method for the simultaneous determination in a protein or peptide sample of the initial D-amino acid content and the amounts formed during hydrolysis. Kinetics of the racemization process for different aminoacids are also reported (Liardon and Jost, 1981).

There seems little doubt that GC/MS will remain the food analyst's workhorse for some time to come.

## D. Chemical Ionization

Chemical ionization is what is known as a soft ionization technique. To understand what is meant by this, and to appreciate its implications for food analysis, we must look at what happens to samples in the MS source. The problem may be stated simply: how do we create a population of ions in the gas phase that is as characteristic as possible of the sample we are analyzing? The basic possibilities are limited to two: (1) volatilize the sample and then ionize it, or (2) ionize the sample and then volatilize it. Traditional EI spectra use the former approach, but, unfortunately, many polar compounds decompose at the temperatures needed to volatilize them and many naturally volatile compounds fragment extensively at the high ionizing voltages needed, thus giving structurally uninformative spectra. Soft ionization techniques overcome these problems and extend MS to the study of larger and more polar molecules, thus opening up a whole new area in food application. Biological applications of CI and other soft ionization methods are described in a recent book edited by Morris (1981).

Like EI, CI uses the principle of volatilization followed by ionization. The sample is introduced into the source along with a reagent gas. The latter, in excess, is preferentially ionized and transfers its ionization to the sample in a low-energy process, usually by protonation to give positive ions 1 amu higher than the molecular ion, by electron transfer to give the negatively charged molecular ion, or by proton abstraction to give negative ions 1 amu below the molecular ion. Many substances, such as methane, isobutane, ammonia, water, methanol, and chlorine, alone or in mixtures, can be used as reagent gases, the choice depending on the nature of the sample to be analyzed. Generally, however, medium to high intensity peaks are obtained in the molecular ion region of the sample, giving information that is much more structurally representative than is the EI spectrum. This is illustrated in Fig. 2, which shows the EI and CI spectra of the S-methyl derivative of the ring-opened form of the vitamin thiamine.

Chemical ionization therefore lends itself to the study of structurally delicate or labile compounds. Thus, Hendriks and Bruins (1980) used OH⁻ as the reactant ion to characterize terpene alcohols and their esters in the essential oil of valerian. The same authors (1983) have since shown how mass chromatograms of specific ions generated from repetitive CI scans recorded during GC/MS analysis of essential oils can be used to identify esters in general. Chai and Harrison (1981) have used vinyl methyl ether as a reagent gas to locate double bonds in aliphatic chains, and Stan and

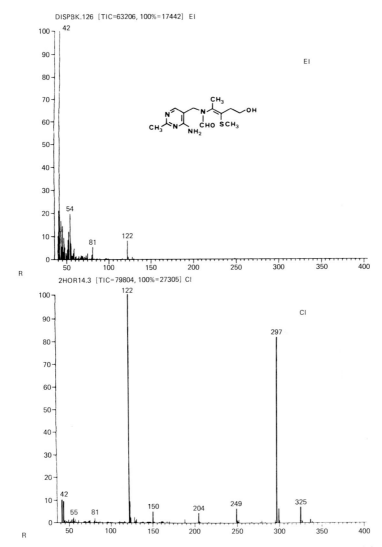

**Fig. 2.** The comparative EI and CI mass spectra of a stabilized ring-opened form of thiamine.

Scheutwinkel-Reich (1980) describe the use of methane or isobutane to detect hydroxy fatty acids in biological samples with positive- and negative-ion CI.

Food-related positive-ion CI studies include the determination of histamine in tuna fish as its trimethylsilyl (TMS) derivative, using methane as reagent gas (Henion *et al.*, 1981); the determination of the antibacterial drug sulfadimethoxin in the liver and kidneys of pigs and cattle (Garland *et al.,* 1980a); and the determination of carbofuran and its metabolites in vegetables (Robinson and Chapman, 1980).

Negative-ion CI, usually abbreviated to NCI, has been used by Hunt and Crow (1978) for trace analysis at levels of attomoles ($10^{-18}$) in selected ion-monitoring GC/MS methods. Dixon (1981) describes applications of NCI in environmental science, and Slayback and Nan (1979) report biomedical applications. Both areas have a broad overlap with food and food-related studies, so the general principles described are directly transferable. In identifying hazardous organic chemicals in fish, including chloro derivatives of hydrocarbons, alkylamines, styrene, norbornene, benzyl alcohol, and phenol, Kuehl *et al.* (1980) confirm that the advantages of NCI over EI are the former's increased sensitivity and selectivity for electronegative chemicals.

Desorption chemical ionization (DCI) couples the advantages of CI with a specific sample-introduction probe that permits the extremely rapid evaporation of a solid sample from a heated filament directly into the plasma of ionized reagent gas in the CI source. By this means, many labile materials can be evaporated as molecular species before decomposition occurs. Thus, the range of compounds that can be analyzed by MS is considerably extended. Wakefield *et al.* (1980) have reported the DCI analysis of underivatized natural products such as sugars, vitamins, glycosides, peptides, and antibiotics. In an interesting food application, Schulte *et al.* (1981) have shown how DCI can be used to determine the triglyceride patterns of fats. They applied this to the study of chocolate and used spectra such as those in Fig. 3 to detect the presence of a cocoa butter substitute (spectrum in Fig. 3c) in a chocolate sample (spectrum in Fig. 3b), which distinguishes itself from pure chocolate (spectrum in Fig. 3a) by the distribution of triglycerides at $m/z$ 950, $m/z$ 878, and $m/z$ 906. Summer and winter butter could be equally distinguished, as could beef tallow and lard.

### E. Field Desorption

Field desorption (FD) is another soft ionization technique, but one in which the sample first is ionized and then is desorbed into the gas phase. The principles of FD have been described in a book by Beckey (1977). Field desorption can be used to run spectra of highly polar and even ionic samples, and because it readily gives molecular ion information, it finds use in mixture analysis. Substances with molecular weights of 1500 amu or more are readily analyzed, and although the technique is less routine and less easy to use than EI or CI, it could make a useful contribution to the analysis of natural polymers.

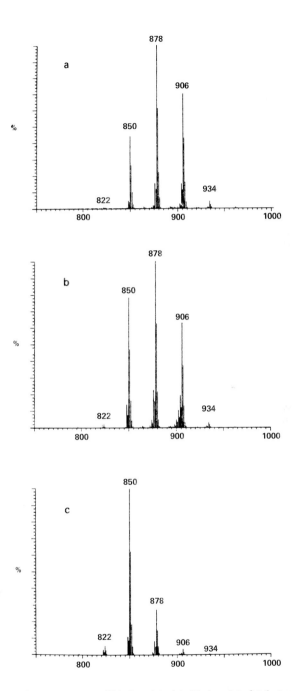

**Fig. 3.** The DCI mass spectra of (a) chocolate fat, (b) chocolate fat that contains a cocoa butter substitute, and (c) the cocoa butter substitute. Ammonia is used as the reagent gas. [Redrawn from Schulte *et al.* (1981).]

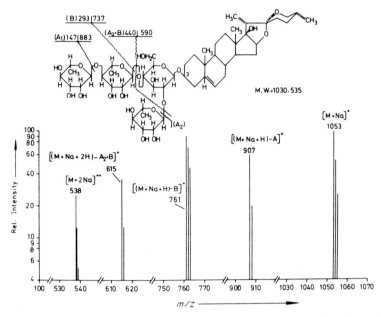

**Fig. 4.** The FD mass spectrum of pennogenin-3-O-α-L-rhamnopyrosyl-(1 → 4)-α-L-rham-nopyrosyl-(1 → 4)-[α-L-rhamnopyrosyl-(1 → 2)]-β-D-glucopyranoside. [Reprinted with permission from Schulten *et al.* (1978).]

   Schulten *et al.* (1978) have used FD to investigate glycosides bearing six or more monosaccharide units, getting information on molecular weight, and from fragment ions formed by the cleavage of successive sugar units also determining the monosaccharide sequences of the oligosaccharide units. Other food-related applications have involved vitamins (Schulten and Schiebel, 1978), mycotoxins in tomato extracts (Sphon *et al.*, 1977), and sugar sequences (Schulten *et al.*, 1982).

   A typical spectrum of an underivatized glycoside having four monosac-charide units is shown in Fig. 4. The base peak at *m/z* 1053 corresponds to an [M + Na]+ ion.

## F. Liquid Chromatography/Mass Spectrometry

   Gas chromatography/mass spectrometry catered to the analysis of com-plex mixtures of food volatiles and to some degree of lower-molecular-weight solids, but what about mixtures of involatile solids, which account for the bulk of food material? For reasons of incomplete reaction or perhaps undesirable side reactions, transformation into volatile derivatives followed by GC/MS was not always the best solution. Attention was turned to high-performance liquid chromatography (HPLC), and several groups set about developing combined HPLC/MS (usually further abbreviated to LC/MS) systems. Games (1981a,b) has recently published two reviews on general aspects of the subject. Controversy still rages as to whether LC/MS should be operated as an off-line technique with fraction collection of the HPLC column effluent and subsequent probe introduction into the mass spectrometer of individual fractions or, alternately, as an on-line technique with some kind of interface between the liquid chromatograph and the mass spectrometer. This is evidently a doubtfully realistic argument, in that the specific approach depends on the sample and on the type of ionization required. For example, EI and CI can be used in conjunction with on-line LC/MS systems, whereas FD cannot.

Controversy also exists as to the best type of interface to use in on-line systems. It was not without reason that Arpino characterized LC/MS as "the impossible marriage": the HPLC yields several milliliters per minute of liquid effluent, and the mass spectrometer operates under conditions of high vacuum. Thus, stringent requirements are imposed on the operation of the interface. Arpino (1983) has reviewed two different interfacing methods. The first is the transport interface, in which the sample is carried on a moving belt from the column outlet to the MS source, solvent being evaporated from the belt *en route;* the second is the direct liquid interface, in which a portion of the column effluent is nebulized directly into the MS source. The former has the advantage that it can be used with a wide variety of ionization methods, whereas the latter is essentially restricted to CI operation, the LC elution solvent becoming the CI reagent gas. On the other side of the coin, nebulization of the column effluent by the direct liquid interface leads to the formation in the MS source of a cloud of sample molecules dispersed in the vapor phase, and this is advantageous for compounds that may decompose during evaporation from a belt. Some manufacturers of MS equipment favor the one approach, others the other; as such, the controversy is perhaps of commercial rather than scientific origin. For food analysts, either approach significantly extends their analytical possibilities, although, ideally, it would be better to have an interchangeable probe system permitting both to be used.

No specific food applications of LC/MS can be cited, but the separation of a mixture of nucleosides will serve as an example. Thus, Fig. 5a shows the reconstructed total ion current trace obtained by ammonia CI during LC/MS of uridine, adenosine, and cytidine and reported by Games and Lewis (1980). The spectrum of uridine obtained is shown in Fig. 5b. This analysis was achieved with a moving-belt interface, and about 10 $\mu$g of sample was typically used. The authors reported a loss of molecular weight

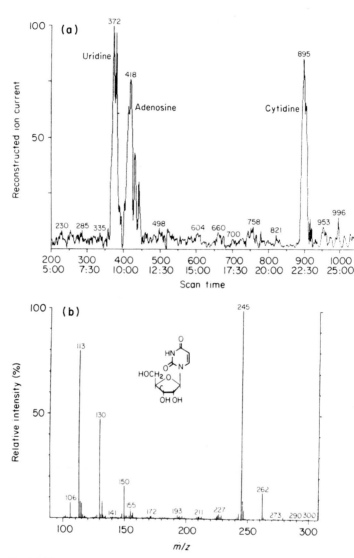

**Fig. 5.** (a) The total ion current trace (ammonia CI) taken during an on-line LC/MS separation of uridine, adenosine, and cytidine; (b) the mass spectrum of uridine obtained. [From D. E. Games and E. Lewis (1980), *Biomed. Mass Spectrom.* **7,** 433. Copyright 1980 by John Wiley and Sons. Reprinted by permission of John Wiley and Sons.]

information on very dilute solutions and suggested that this may impose a limit on sensitivity. The same article describes applications to glycosides, glucuronides, and sugars. Other food-related studies have involved monosaccharides (Arpino *et al.*, 1981), hydroxy acids (Voykanee *et al.*, 1982), flavonoid glycosides (Schuster, 1980), carbohydrates, esculin, and α-tocopherol (Kenyon *et al.*, 1981), peptides, nucleosides, and nucleotides using a new type of interface that combines the direct-introduction approach with features of the GC/MS jet separator (Blakley *et al.*, 1980), atrazine in soil (Evans and Williamson, 1981), organophosphorus pesticides (Parker *et al.*, 1982), and aromatic amines and chlorinated pesticides (Yoshida *et al.*, 1980). Further to this, Yorke and Crossley (undated) have described identification of geometrical isomers formed by derivatizing carbonyl compounds with 2,4-dinitrophenylhydrazine, and McFadden (1980) has summarized a wide variety of applications.

This impressive collection of examples suggest that LC/MS is now ripe as a technique, although we may still expect an improved facility of operation in the future. Once it is routinely applied in food analysis, it could have the same dramatic effect in extending our knowledge of the solid components of food as GC/MS has had for volatiles.

## G. Mass Spectrometry/Mass Spectrometry

By analogy with the terminology of GC/MS and LC/MS, mass spectrometry/mass spectrometry (MS/MS) is the in-series combination of two mass spectrometers. The analogy can be taken further in that, like the GC or the LC compound of the other combinations, the first mass spectrometer acts as a separator while the second plays the conventional role of mass analyzer. The main strength of MS/MS lies in its ability to analyze for specific target compounds, and in particular to do this directly and rapidly in complex biological mixtures without preseparation. The basic principles and uses of MS/MS have been reported by McLafferty (1982a), McLafferty and Lory (1981), Hunt *et al.* (1980), and Steiner *et al.* (1980), and a book on the subject has recently been edited by McLafferty (1982b).

Two basic types of food analysis can be envisaged: first, the rapid screening of a series of, say, intact fruits or vegetables for the presence of mycotoxins or pesticides; second, studies of the natural distribution of specific compounds throughout the matrix of a food product, such as amines in the rind and in the mass of ripening cheeses.

How MS/MS operates is best explained by the now classical example of determining the pesticide parathion directly on lettuce (Slaybach and Story, 1981). This first requires the introduction of a new concept, namely, the collision-activated decomposition (CAD) spectrum of a compound. The methane negative CI spectrum of parathion shows a medium-intensity molecular ion at $m/z$ 291. By means of tuning the analyzer of the first mass spectrometer, this ion is specifically selected and is passed through a collision chamber into the second MS analyzer. If the collision chamber is totally evacuated, the ion is transmitted intact to the collector of the second analyzer. If, however, an inert gas such as helium is leaked into the collision chamber, $m/z$ 291 ions collide with this gas and are broken down into smaller fragments characteristic of the structure of the ion. The resulting spectrum is called a CAD spectrum. The CAD spectrum of parathion is shown in Fig. 6a: in addition to the $m/z$ 291 peak, two fragment ions are seen at $m/z$ 169 and $m/z$ 154. Figure 6b shows the complete normal spectrum of the volatiles from a 10-mg sample of uncontaminated lettuce leaf introduced directly into the source of the mass spectrometer. A peak is seen at $m/z$ 291, but since the leaf is not contaminated, this cannot come from parathion. A second 10-mg sample of lettuce leaf to which 1 ng of parathion had been added gave a spectrum (not shown) almost identical to that in Fig. 6b and with an $m/z$ 291 peak insignificantly different in intensity from that in the uncontaminated sample. The CAD spectrum of this $m/z$ 291 peak selected from the veritable forest of other peaks is shown in Fig. 6c. The $m/z$ 291 peak is more intense than in Fig. 6a, but this is to be expected because of the background peak at the same mass coming from the lettuce leaf itself. The $m/z$ 169 and 154 peaks are, however, present and in approximately the correct ratio relative to Fig. 6a, thus confirming the presence of parathion and at the same time confirming the sensitivity that can be achieved by the method. To do the same analysis by conventional GC/MS would involve a multistep extraction and enrichment followed by a time-consuming chromatographic separation and spectral identification. Mass spectrometry/mass spectrometry is therefore much more rapid, and direct application on the matrix to be analyzed requires less sample.

A result similar to that described above can be had with a normal double-focusing magnetic sector mass spectrometer by running it in the $B/E$-linked scan mode. In a normal spectrum the electrostatic analyzer voltage $(E)$ is kept constant as the magnetic field $(B)$ is varied. In the linked scan mode, the ratio of $B$ to $E$ is kept constant as $B$ is scanned, and the resulting spectrum contains fragment peaks from a selected precursor ion. Warburton et al. (1981) have applied this principle to the study of egg yolk and lemon juice, showing that the $m/z$ 210 peak of lemon juice corresponds to citric acid and the $m/z$ 369 peak that is barely visible in the spectrum of egg yolk comes from cholesterol. The $B/E$-linked CAD spectrum of pure citric acid is shown in Fig. 7, along with the equivalent CAD spectrum from lemon juice.

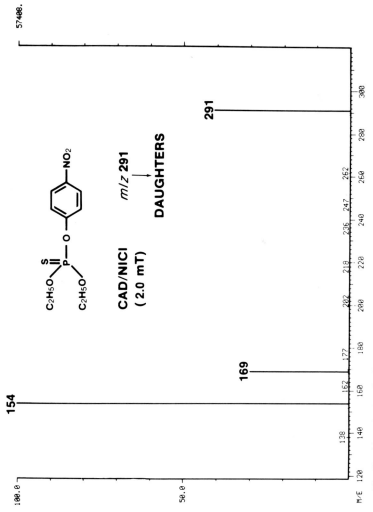

**(a)**

**Fig. 6.** Negative-ion CI (CH$_4$) mass spectra. (a) The collision-activated decomposition (CAD) spectrum from the $m/z$ 291 ion of pure parathion. (*Continued.*)

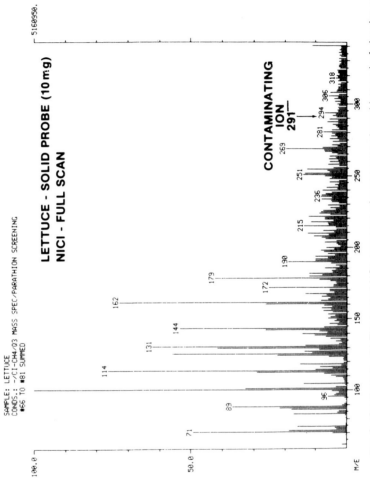

**Fig. 6.** (*Continued*) (b) The normal NCI spectrum of a noncontaminated lettuce leaf showing a large background at *m/z* 291.

160

(c)

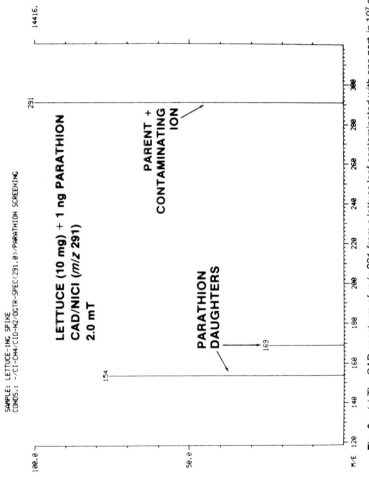

**Fig. 6.** (c) The CAD spectrum of $m/z$ 291 from a lettuce leaf contaminated with one part in $10^7$ of parathion. (Reprinted by permission of Finnigan-Mat, Palo Alto, California.)

161

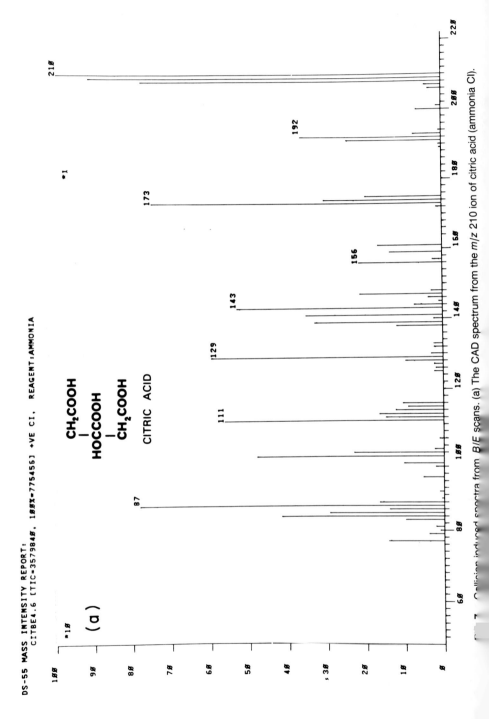

Collision induced spectra from B/E scans. (a) The CAD spectrum from the m/z 210 ion of citric acid (ammonia CI).

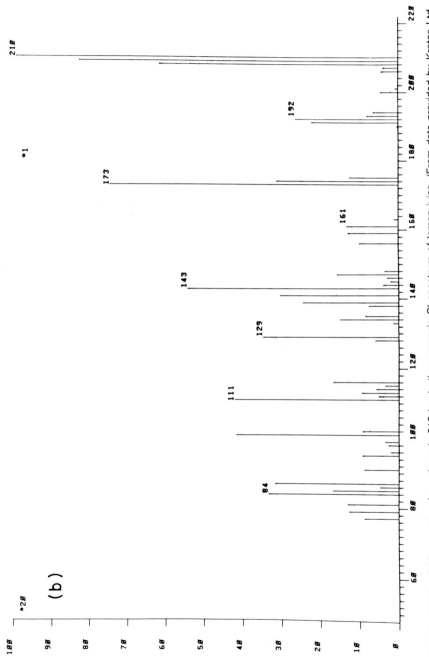

LEMBE4.33 [TIC=3482176, 100%=9394721] +VE CI, REAGENT:AMMONIA

**Fig. 7.** (b) The CAD spectrum from the $m/z$ 210 ion in the ammonia CI spectrum of lemon juice. (From data provided by Kratos Ltd., Manchester, England.)

The potential applications of MS/MS in food analysis must be taken seriously, particularly by those laboratories involved in quality control.

## H. Fast-Atom Bombardment

The group of compounds accessible to analysis by fast-atom bombardment (FAB) ionization is similar to that analyzed by field desorption, namely, involatile polar solids. First described by Barber *et al.* (1981), FAB is the most recent of the soft ionization methods. Ionization is accomplished by bombarding the sample with a beam of neutral atoms, typically argon at 2–8 eV, and classes of compounds that have been studied with success include oligosaccharides, oligopeptides, vitamins, glycolipids, porphorins such as chlorophyll, steroid conjugates such as cholesterol sulfate, and many others. Catalogs of typical spectra collected from several authors have been published by Kratos Ltd. (1981) and by VG Analytical Ltd. (1981).

A pertinent example is to be found in the FAB spectrum of the tomato glycoalkaloid $\alpha$-tomatine, shown in Fig. 8. This spectrum is typical in that it has an intense $(M + H)^+$ ion and also gives a considerable number of fragment ions that are useful for structure elucidation.

An interesting combination of FAB with $B/E$-linked scans has been provided by Taylor *et al.* (1982) in a study of the amino acid sequence in the oligopeptide Antiamoebin I. This compound contains 16 amino acids; the $(M + H)^+$ ion obtained by FAB ionization of the underivatized molecule gives a peak at $m/z$ 1670. A $B/E$ scan from this ion reveals the loss of the terminal proline·phenylaninol·H (Pro·PHOL·H) group to give a peak at $m/z$ 1422. A similar $B/E$ scan from the latter ion gives ions at $m/z$ 1337 and $m/z$ 1244 for the loss of aminoisobutyric acid (AIB) and hydroxyproline·aminoisobutyric acid (HYP·AIB), respectively. This process can be repeated throughout the entire sequence of 16 amino acids and is particularly useful in the terminal sequence, where fragment ions in the normal spectrum are present but are hidden by a high background. The entire amino acid sequence could be identified as

ACPHE·AIB·AIB·AIB·IVA·Gly·Leu·AIB·AIB·HYP·Gln·IVA·HYP·AIB·Pro·PHOL

where in addition to the abbreviations for amino acids already defined ACPHE = acetylphenylalanine, IVA = isovaline, Leu = leucine, Gly = glycine, and Gln = glutamine.

So far, compounds with molecular weights of up to about 4000 amu have been analyzed by FAB, and it may reasonably be expected that this is not the limit. Evidently, there is a degree of overlap between the fields of application of FAB and FD. If a classification can be made, FAB is probably more suited to the very polar solids and FD to compounds of lower polarity. For

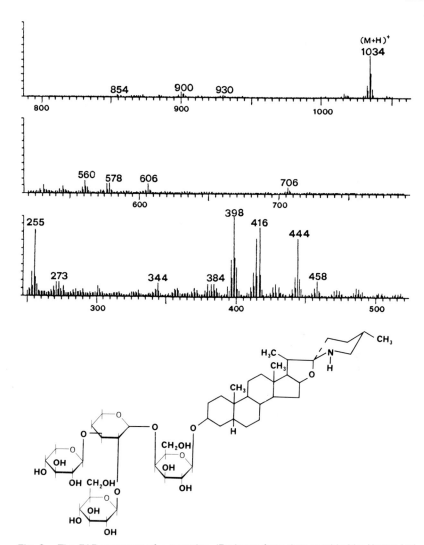

**Fig. 8.** The FAB spectrum of α-tomatine. (Redrawn from data provided by Kratos Ltd., Manchester, England.)

nonpolar solids, perhaps DCI provides a viable alternative, as in the triglyceride example of Fig. 3.

## I. Isotope Mass Spectrometry

For most MS applications involving isotopes, it is never necessary to consult anything more than a standard isotope table, which tells us that the relative abundance of $^{13}C$ relative to $^{12}C$ is about 1%, that $^{34}S$ is about 4% of $^{32}S$, and that any hydrogen isotope other than $^{1}H$ is insignificant. However, when studied systematically, it is found that accurately measured isotope abundances give information on the geographical and botanical origins of plant materials and the foods produced from them and also on whether specific compounds are of natural or synthetic origin.

Effectively, the relative amount of deuterium in rainwater is a function of local temperature and is higher at the equator than at the poles. Plants therefore grow with a D/H ratio in their tissues that is characteristic of the latitude at which they are produced. Preferential evaporation of $H_2O$ relative to HDO from the leaves of plants markedly increases the D/H ratio relative to the local rainwater. Measurements of D/H ratios have been applied to orange juices from crops grown in different Mediterranean regions. Juices from Morocco, Israel, Corsica, and Spain gave significantly different D/H ratios, and concentrated juices after redilution even in local tap water gave vastly different values (Bricout and Merlivat, 1971). This type of information can be used to control commercial purity claims.

A similar effect is noted for $^{13}C/^{12}C$ ratios. Some plants fix atmospheric $CO_2$ through a $C_3$ metabolite in what is known as the Calvin cycle, while others go through a $C_4$ metabolite in the Hatch–Slack cycle. The former class of plants is more deficient in $^{13}C$ than the latter, and this allows differentiation between the different botanical origins. For example, sugar from beets or grapes (Calvin plants) can be readily detected from that extracted from cane or corn (Hatch–Slack plants). Carbon-13 ratios thus can reveal the addition of inexpensive corn or cane sugar ($C_4$) to apple juice, maple syrup, or honey ($C_3$) (Doner et al., 1980). For further examples from food analysis and for more experimental details, the reviews by Krueger and Reesman (1982) and by Winkler and Schmidt (1980) should be consulted [see also Chapter 7, Section IV,E (this volume) for isotope determination by NMR].

### J. Fourier Transform Mass Spectrometry

Fourier transform mass spectrometry, FTMS as it is known, is based on a totally different principle from that of conventional mass spectrometers. The source, analyzer, and collector regions are all combined into a single region where ionization and mass analysis take place. This has the advantage that slits, focusing lenses, and flight tubes are eliminated and all ions formed are detected simultaneously. Instead of detecting the masses of ions by their trajectory in a magnetic or electric field, the mass analysis is performed by detecting the cyclotron frequencies corresponding to the different mass-to-charge ratios.

Analyses by FTMS can be run in EI and CI modes, but it is not out of the question that other modes of ionization could be used. In fact, there would be potential advantages in using FTMS in conjunction with any ionization technique in which ions are formed in short bursts, such as in the rapid evaporation of sample during FD or DCI operation. The reason for this is that the ions formed are kept in the observation chamber in the gas phase for as long as is desired and can thus be sampled with the FT pulse as often as is necessary to obtain a good spectrum. Moreover, the longer the observation time, the higher the resolution obtained, and resolutions up to 700,000 have been demonstrated (White *et al.,* 1980). Possibilities involving GC/MS and even MS/MS exist, as described by Wilkins and Gross (1981). For MS/MS measurements, for example, all ions except the ion in question can be removed from the measurement cell prior to collision with the background gas, thus eliminating all but the CAD ions.

Fourier transform mass spectrometry is relatively new in the commercial sector and is obviously still in a state of development. As such, the vast majority of examples to date are models to demonstrate the various likely areas of application. For the moment, concrete applications to foods are left to the imagination of potential users. It is of interest to imagine, for example, FAB ionization of a macromolecule in the mass range 5000 to 10,000, with MS/MS to detect the individual structural units through CAD fragmentation, and the possibility of doing this at high resolution and hence obtaining accurate mass measurements and elemental compositions on all fragments.

## III. APPLICATIONS

### A. Analysis of Volatiles

The flurry of activity in the identification and characterization of volatiles that resulted from the introduction of GC/MS shows no sign of abating. Developments in GC/MS instrumentation (see Section II,C) continue to provide new opportunities for discoveries, and it is likely that the most important of these are still waiting to be brought to light. However, if as Table 2 suggests the number of studies of volatiles and the number of different food materials investigated is increasing, this is certainly not true for the number of new compounds identified, and the time has come for us to question our strategy in approaching the problem.

Probably the major information to be obtained from the collective works on volatiles to date lies in the great differences between the qualitative and quantitative contributions of individual compounds to specific food aromas and flavors. For example, ethanol can be present in food volatiles in widely

**TABLE 2  Studies of Volatile Constituents of Foods and Beverages**

| Food or beverage | Study | Reference |
|---|---|---|
| Fruits and vegetables | | |
| Apple | Volatile flavor components of wood apple (*Feronia limonia*) and a processed product | MacLeod and Pieris (1981a) |
| Apricot | Characterization of additional volatile flavor components of apricot | Chairote *et al.* (1981) |
| Cranberry | The aroma of cranberries | Hirvi *et al.* (1981) |
| Litchi | Volatile constituents of litchi (*Litchi chinesis* Sonn.) | Johnston *et al.* (1980) |
| Orange, etc. | Aroma components in oranges and their changes during juice processing | Schreier *et al.* (1977) |
| | Aroma characteristics of citrus | Yamanishi *et al.* (1980b) |
| | Characterization of a new citrus component, *t,t,α*-farnesene | Moshonas and Shaw (1980) |
| | Isolation of α-farnesene isomers from dehydration of farnesol | Spencer *et al.* (1978) |
| Peach | Gas chromatographic and sensory analysis of volatiles from cling peaches | Flath and Takahashi (1978) |
| Prickly pear | Volatile constituents of prickly pear (*Opuntia ficus indica* Mill., de Castilla variety) | Ismail *et al.* (1980a) |
| Plum | The flavor components of plum: an investigation into the volatile components of canned plums | Ismail *et al.* (1981a) |
| | The flavor components of plums: an examination of the aroma components present in the headspace above four cultivars of intact plums | Ismail *et al.* (1981b) |
| | The flavor of plums (*Prunus domestica* L.): an examination of the aroma components of plum juice from the cultivar Victoria | Schreyen *et al.* (1979) |
| Quince | Flavor analysis of quince | Tsuneya *et al.* (1980) |
| | Isolation and identification of novel terpene lactones from quince fruit (*Cydonia oblonga* Mill., Marmelo) | |
| Raisins | 2-Hexyl-3-methylmaleic anhydride: an unusual volatile component in raisins and almond hulls | Buttery *et al.* (1980c) |
| Raspberries | The aroma of Finnish wild raspberries, *Rubus idaeus* L. | Honkanen *et al.* (1980) |
| Soursop | Volatile flavor components of soursop (*Annona muricata*) | MacLeod and Pieris (1981b) |
| Strawberry | Volatiles of wild strawberries, *Fragaris vesca* L., compared to those of cultivated berries, *Fragaria* × *ananassa* cv. Senga Senanga | Pyysalo *et al.* (1979) |
| | Subjective and objective evaluation of strawberry pomace essence. | Schen *et al.* (1980) |

| | | |
|---|---|---|
| Vanilla | Some benzyl ethers present in the extract of vanilla (*Vanilla planifolia*) | Galetto and Hoffman (1978) |
| Artichoke | Volatile aroma components of cooked artichoke | Buttery et al. (1978a) |
| Carrot | Identification of additional volatile constituents of carrot roots | Buttery et al. (1979) |
| | Characterization of some previously unidentified sesquiterpenes in carrot roots | Seifert and Buttery (1978) |
| Legumes | Volatile constituents of dried legumes | Lovegren et al. (1979) |
| Onion | Identification of a new volatile compound in onion and leek: 3,4-dimethyl-2,5-dioxo-2,5-dihydrothiophene | Albrand et al. (1980) |
| Potato | Chemistry of baked potato flavor: pyrazines and thiazoles identified in the volatile flavor of baked potatoes | Coleman and Ho (1980) |
| | Isolation and identification of volatile compounds from baked potatoes | Coleman et al. (1981) |
| | Analysis of the volatile constituents of baked "Jewel" sweet potatoes | Purcell et al. (1980) |
| | Odorous compounds from potato processing waste effluent irrigation fields: volatile acids | Buttery and Garibaldi (1980) |
| Tomato | Constituents of tomatoes: a review article | Hermann (1979) |
| Alcoholic beverages | | |
| Beer | Hop aroma in American beer | Peacock et al. (1980) |
| | Studies of the volatile composition of hops during storage | Tressl et al. (1978c) |
| | Determination of aroma substances in Spalter hops by GC/MS | Tressl and Friese (1978) |
| | Use of cation-exchange resin for the detection of alkylpyridines in beer | Peppard and Halsey (1980) |
| | Determination of thiolacetates and some other volatile sulfur compounds in alcoholic beverages | Leppanen et al. (1980) |
| | Identification and determination of S-methyl thioacetate in beer | Matsui et al. (1980) |
| | Isolation and identification of new staling-related compounds from beer | Williams and Wagner (1978) |
| Wine | Identification of volatile compounds of wine of *Vitis vinifera* cultivar Pinot Noir | Brander et al. (1980) |
| | Isolation and identification of volatiles from Catawba wine | Nelson et al. (1978) |
| | Wine aroma composition: red wine | Schreier (1980) |
| | Volatile phenol determination in wine | Etievant (1981) |
| | Differences in the formation of certain acetamides in grape and apple wines | Stackler and Ough (1979) |
| | 3,7-Dimethyl-octa-1,5-dien-3,7-diol — a new terpenoid component of grape and wine aroma | Rapp and Knipser (1979) |

(continued)

**TABLE 2** (*Continued*)

| Food or beverage | Study | Reference |
|---|---|---|
| Spirits | The flavor constituents of gin | Clutton and Evans (1978) |
| | Changes in the composition of neutral volatile components during the production of apple brandy | Schreier et al. (1978a) |
| | The influence of HTST-heating of the mash on the aroma composition during the production of apple brandy | Schreier et al. (1978b) |
| | Composition of neutral volatile constituents of grape brandies | Schreier et al. (1979) |
| | Aroma components in a distillate from fermented plum juice | Ismail et al. (1980b) |
| Cider | The neutral volatile components of cider apple juices | A. A. Williams et al. (1980) |
| | Volatile aroma components of fermented ciders: minor neutral components from the fermentation of sweet Coppin apple juice | Williams and Tucknott (1978) |
| Alcohol | Isovaleronitrile: a characteristic component of beet molasses alcohol | Strating and Westra (1982) |
| **Honey and floral notes** | | |
| Honey | Volatile constituents of some uniforal Australian honeys | Graddon et al. (1979) |
| Flowers | Studies on the constituents of flowers: on the components of the flower of *Ligustrum ovalifilium* Hassk | Kurihara and Kikuchi (1980) |
| | Volatile constituents of the Castanopsis flower | Yamaguchi and Shibamoto (1979) |
| | Qualitative and quantitative analysis of the essential oils of red and white clovers | Takayasu (1978) |
| | Volatile constituents of the chestnut flower | Yamaguchi and Shibamoto (1980) |
| **Cereals and grains** | | |
| Corn | *cis*-Hept-4-en-2-ol in corn volatiles: identification and synthesis | Buttery (1979) |
| Barley | Volatile flavor components of malt extract | Farley and Nursten (1980) |
| Oats | GC/MS investigations on the flavor chemistry of oat groats | Heydanek and McGorrin (1981a) |
| | GC/MS identification of volatiles from rancid oat groats | Heydanek and McGorrin (1981b) |
| Rice | Volatile flavor components of cooked rice | Yajima et al. (1978) |
| | Volatile flavor components of cooked Kaorimai (scented rice, *O. sativa japonica*) | Yajima et al. (1979) |
| | Volatile components in the steam distillate of rice bran: identification of neutral and basic compounds | Tsugita et al. (1978) |

| | Description | Reference |
|---|---|---|
| | Volatile components after cooking rice milled to different degrees | Tsugita et al. (1980) |
| | Isolation and identification of volatile components from wild rice grain (Zizania aquatica) | Withycombe et al. (1978) |
| Soya | Changes of headspace volatile components of soybeans during roasting | Doi et al. (1980) |
| **Eggs, meat, and fish** | | |
| Eggs | Thermally produced volatile basic components of egg white and ovalbumin | Kato et al. (1978) |
| Chicken | Formation of monocarbonyl compounds in chicken tissue | Moerck and Ball (1979) |
| Beef | Preliminary identification of volatile flavor compounds in the neutral fraction of roast beef | Min et al. (1979) |
| | Volatile flavor compounds from beef and beef constituents | Ching (1979) |
| | Identification of radiolytic compounds from beef | Vajdi et al. (1979) |
| Mutton | Nonacidic constituents of volatiles from cooked mutton | Nixon et al. (1979) |
| Meats, general | Isolation and identification of volatile compounds in cooked meats: sukiyaki | Shibamoto et al. (1981) |
| | Instrumental flavor analysis of meat products containing cottonseed protein | Fore et al. (1980) |
| Fish, general | GC/MS determination of volatile organic compounds in fish | Easley et al. (1981) |
| | Fatty acids of whitefish (Coregonus albula) flesh lipids | Kaitaranta and Linko (1979) |
| Krills | Cooked odor of Antarctic krills | Kubota et al. (1980) |
| Salmon | Volatile components of smoked salmon | Kasahara and Nishibori (1979) |
| Sardines | Undesirable odor of cooked sardine meat | Koizumi et al. (1979) |
| **Essential oils and spices** | | |
| Cardamon | GLC study of the essential oil of wild cardamom oil of Sri Lanka | Rajapakse-Arambewela and Wijese-kera (1979) |
| Cloves | Identification of some volatile components of ground cloves | Koller (1981) |
| Eucalyptus | Essential oil of Eucalyptus globulus in California | Nishimura and Calvin (1979) |
| Gossypium | Identification of the major monoterpenes in the leaf oil of Gossypium sturtianum var. nandewarense (Der.) Fryx | Kumamoto et al. (1979) |
| Juniper | The volatile terpenoids of Juniperus monticola f. monticola, f. compacta, and f. orizabnsis | Adams et al. (1980) |

(continued)

**TABLE 2** (Continued)

| Food or beverage | Study | Reference |
|---|---|---|
| Wild cumin | The volatile trepenoids of *Juniperus blancoi* and its affinities with other entire leaf margin junipers of North America | Adams *et al.* (1981) |
| | Investigation of the odoriferous principles of grains of *Nigella damascena* Lann | Paris *et al.* (1979) |
| Nutmeg | Analysis of nutmeg oil using chromatographic methods | Schenk and Lamparsky (1981) |
| Sage | Study of essential oil of *Salvia officinalis* growing in Egypt | Karawya and El-Hawary (1978) |
| Prepared foods, sauces, etc. | | |
| Peanut butter | Studies on roasted peanuts and peanut butter flavor | Lee (1980) |
| Mayonnaise | Analysis of neutral volatiles of mayonnaise by direct GC and MS | Fore *et al.* (1978) |
| Soy products | Shoyu (soy sauce) volatile flavor components: basic fraction | Nunomura *et al.* (1978) |
| | Analysis of flavor quality and residual solvent of soy protein products | Rayner *et al.* (1978) |
| Pastry | Changes of flavor compounds caused by addition of soy flour to short pastry | Heimann *et al.* (1980) |
| Sponge cake | Flavor components of sponge cake and production of 2,5-dimethyl-4-hydroxy-3(2H)-furanone in several model systems | Takei (1977) |

differing amounts up to several percentage points without markedly changing the organoleptic aroma quality, whereas methyl anthranilate confers richness to tea aroma at ppb levels in the cup (Cazenave and Horman, 1974). Almost all of the major volatiles of foods are like ethanol and along with most of the minor constituents contribute little more than the general background note of the aroma; typical general aroma descriptors would be "fruity," "fresh," "green," "fermented," and "roasted." In the average GC/MS run on a sample of food volatiles most of the spectral information generated evidently comes from this type of "ordinary" compound. One of the main difficulties consists of cutting through this forest of data to get at the few key compounds that characterize the source and that are known as flavor impact substances.

An excellent example of the identification of a powerful flavor impact compound has been reported by Delmole *et al.* (1982). Working with the total volatiles from grapefruit juice, they first separated the sample into five fractions by chromatography on silica gel. One of these fractions, representing little more than 1 ppm of the original juice, exhibited a sulfurous odor typical of grapefruit. This fraction was further analyzed by capillary column GC using three simultaneous detector systems: the first gave a regular GC trace showing the presence of over 30 components; the second was a "sniffing" port, allowing location of the compound of interest in the overall chromatogram by smell; the third was a specific sulfur detector. By sniffing of the column effluent, the compound of interest was detected in an essentially empty region of the regular chromatogram, but the sulfur detector gave a small peak. Further enrichment of the sample was necessary, and this was performed by preparative GC on a silicone column to give 17 fractions. One of these, amounting to about 1% of the previous fraction, was then separated into four fractions. The final sample thus analyzed by the GC/MS represented little more than 1 ppb of the original juice and consisted principally of 8-ethoxy-1-*p*-menthene, geranyl ethyl ether, two stereoisomers of vitispiranes, a bornyl acetate, and propyl octanoate. Along with these, there were trace amounts of two unknown sulfur compounds, one of which was the flavor impact compound that had been followed through the various stages of enrichment. After further elegant work in mass spectral interpretation and synthesis, this compound was identified as 1-*p*-menthene-8-thiol (**I**) and the second sulfur-containing compound as 2,8-epithio-*cis*-*p*-menthane (**II**).

I                                        II

One of the two enantiomers of (I) was found to have a taste detection threshold in water of $2 \times 10^{-5}$ ppb, making it the most powerful flavor impact compound ever identified. Compound (II) was about a million times less intense. This remarkable piece of work illustrates the successful combination of the human nose as detector along with modern instrumentation. Indeed, in this type of study, the human element cannot be underestimated. Had the authors not followed their noses, so to speak, they would never have looked specifically for a sulfur compound and the trace of (I) present may have been dismissed as an impurity.

In the drive to understand flavor chemistry, it would be instructive to have more studies of this nature and fewer studies that merely produce lists of the most predominant components in a particular aroma sample. Only such thorough methodology and careful sample preparation will ever allow us to identify from separations such as that shown in Fig. 1 powerful flavor components like (I) that are present in such low amounts that they are lost in the baseline noise. Trace analysis represents one of the major challenges in aroma chemistry today, and here the sensitivity of mass spectrometry, coupled with intelligent sample preparation, provides the only viable analytical approach.

Sensory evaluation of volatiles by sniffing can be carried out on an ordinary GC by blowing out the flame of the flame ionization detector (FID) just as a GC peak begins to emerge. This is clearly not as efficient as having a separate sniffing port, but using this method Godshall et al. (1980) noted the odors of molasses volatiles, remarking that none of the constituents they identified by GC/MS were likely to be responsible for the characteristic "green" or "grasslike" aroma of molasses.

Other reports include studies of the ingredients of smoke flavor preparations by Baltes and Soechtig (1979). Here, the authors looked specifically at the phenolic fraction, identifying some 70 compounds and, for the most characteristic of these, describing the aroma and flavor notes they impart. They also defined guaiacol and syringol and their 4-methyl and 4-ethyl derivatives as GC indicators to reveal admixture of smoke flavors in sausage meat. Phenols again were the subject of studies on coffee by Tressl et al. (1978a,b), who reported the effects of coffee variety and roasting conditions on the quantitative distribution of these compounds. Steam treatment of green coffee was shown by Rahn et al. (1979) to create 3,4-dihydroxystyrene as the principal phenolic compound, presumably through decomposition of chlorogenic acid. Partial removal of this styrene during roasting was reported to lower the contents of catechol, furfuryl alcohol, and 4-ethylcatechol in the final brew. Staying in the sphere of popular beverages, Yamanishi et al. (1980a) have identified the flavor constituents contributing to the floral aroma of pouchong tea, comparing its aroma pattern with that of jasmine tea. The latter is a low-grade pouchong tea scented by blending with jasmine flowers. Several of the key compounds introduced or reinforced by the addition of jasmine flowers are present naturally in much higher quantities in high-grade pouchong tea and seem to contribute to the superior floral aroma quality. Damascenone plays an important role in the flavor of alcoholic beverages, particularly in rum and malt whiskey. According to Masuda and Nishimura (1980), it appears to form during distillation, and they describe its occurrence in several alcoholic beverages. In an elegant investigation of nutmeg oil, Schenk and Lamparsky (1981) newly identified 19 compounds, including *(E)*-3-methyl-4-decen-1-ol and its acetate, described for the first time in nature. Two terpene lactones have been reported by Tsuneya *et al.* (1980) to be the main components responsible for the characteristic note of quince fruit. Other qualitative studies on volatiles are summarized in Table 2.

## 1. Multivariate Analysis

In addition to the identification of specific aroma compounds, we must consider what other information we can obtain from analysis of volatiles. By GC/MS, an enormous mass of data can be created from an aroma sample in a very short time. Computer-aided comparison of such data from different samples permits correlation between flavor profile and product characteristics. For example, we may use a taste panel to rank a series of coffee samples according to their acidity and then compare their GC/MS aroma profiles to identify the volatile constituents that vary proportionally with the organoleptically determined acidity levels. In this way, a set of compounds that can be used as acidity indicators will be defined. These compounds are not necessarily responsible of themselves for the acidic character, and whether they are identified or not is of little importance for this type of analysis. Two elements are, however, important; specifically, the method of collecting volatiles must give a totally representative sample, and the GC/MS analysis must be entirely reproducible with the best GC separation possible.

Using the above approach, Aishima *et al.* (1979) identified *trans*-2-hexen-1-ol, 4-ethylguaiacol, $\gamma$-butyrolactone, ethyl lactate, 2-ethyl-6-methylpyrazine, and 2,6-dimethoxyphenol as indicators for soy sauce quality, and more recently Aishima (1982) has used discrimination and cluster analysis to establish the relationships between quantitative characteristics (from capillary GC/MS) of different commercial brands of soy sauce and aroma quality as evaluated by a taste panel. A set of 19 sensorially significant compounds were determined in white wines by Noble *et al.* (1980), and these could be used for the varietal classification of different wine samples. Kwan and Kowalski (1980), also studying wines, showed how pattern recognition could be used to determine the geographical origins of different wines of the same variety. In the latter work, 1-hexanol and cyclohexane were reported to be the most important compounds for distinguishing French and American Pinot Noirs, and *p*-hydroxybenzaldehyde along with 2-phenylethanol further differentiated between California Pinot Noirs and those from Washington State. Cole and Phelps (1979) have described the use of canonical variate analysis in differentiating swede cultivars according to their hydrolysis products. These correlated with keeping qualities during storage and possibly with the susceptibilities of the different cultivars to attack by insect pests.

## 2. Insect Attractants

Insects are responsible for extensive crop damage every year. This has recently led to several studies with the aim of identifying volatile constituents of growing plants that may be insect attractants. Thus, Buttery *et al.* (1978b) analyzed corn husks and kernels to define compounds that might attract corn ear worm moths. By comparison with volatiles from tomato and cotton, plants that are also susceptible to the corn ear worm, they identified some 15 compounds (aldehydes, ketones, and terpenes) that were common to all three sources. These compounds appeared to show some insect-attracting capacity. Similar studies have been carried out on corn silk (Flath *et al.*, 1978), on alfalfa (Buttery and Kamm, 1980), on corn tassels (Buttery *et al.*, 1980a), and on almond hulls (Buttery *et al.*, 1980b). Generally speaking, the results to date are not conclusive, but insect control is an extremely important area on both health and economic grounds, and continued effort in this direction is justified.

## 3. Data Treatment and Library Systems

No account of the GC/MS investigation of volatiles would be complete without a mention of MS libraries and computer-assisted spectral search systems. Comprehensive spectral collections such as the "8 Peak Index" (Mass Spectral Data Centre, 1974) or "PMB/STIRS" (McLafferty *et al.*, 1980) exist, but many groups engaged in GC/MS of volatiles prefer to create their own more restricted libraries based, for example, on the excellent cumulative compilation of volatile compounds in foods edited by van Stratten *et al.* (1981) or on libraries provided by manufacturers of MS equipment, supplemented with their own spectra. Automatic identification proceeds by pairwise comparison of the spectrum of the unknown with each of the spectra in the library. For each pair, the probability that the two spectra are identical is expressed in terms of a matching factor, and it is usual to report the 10 "best hits," that is, the 10 library spectra giving the highest matching factors with the unknown. Many food volatiles belong to homologous series and give similar spectra. Here, problems can be minimized by basing identifications on GC retention data as well as on mass spectra.

Restricted libraries have their disadvantages in that they limit our possibilities to compounds that have been found in foods or that we think it might be reasonable to find in foods. This can be overcome to some degree by introducing data from model systems such as are described in the next section.

Computer-assisted spectral recognition by library search can indeed be very useful in studies of volatiles, where massive amounts of data must be assessed in a short time. But there is one major danger, namely that of believing too implicitly the results given in the computer output. No library system to date has ever provided an output indicating that a certain spectrum was not in the library, that a certain spectrum was a mixture of two, three, or four particular components, or that a certain spectrum was a random spectrum including several noise peaks, etc. In fact, any spectrum is always matched against the library contents, and the best hits are always output. Particularly for spectra obtained by subtracting background spectra from those for low intensity peaks, where the data are often almost random, this can lead to some interesting cases of almost perfect—but totally unrealistic—"identifications."

## B. Genesis of Aroma and Flavor Components

### 1. Biogenesis

Having considered food volatiles in the previous section, it is now of interest to discuss their origins. These are to be found in the biochemically, chemically, or thermally induced transformations of flavor precursors during ripening, storage, or cooking. Typical flavor precursors are sugars, amino acids, polyphenols, and fats, which form the basis of plant and animal tissues. Few studies of biosynthesis of volatiles have been published, and to find good examples of what might be done, we must go back some 6 years. Yamashita et al. (1976, 1977) reported the formation of aldehydes, alcohols, and esters in ripening strawberries and studied the interconversions between them by incubating whole strawberry fruits with mixtures of aldehydes and acids. Another investigation of ester formation in whole fruits, this time at natural levels, was described by Yabumoto et al. (1977). These authors measured the quantitative changes of 16 volatile constituents found in the cavity of the cantaloupe melon during postharvest storage. Headspace samples were taken at 24-hr intervals through a septum-covered glass tube inserted through the melon flesh and into the fruit cavity. In parallel with the findings on strawberries, quantities of some compounds (ethyl esters, acetaldehyde, and ethanol) continually increased, while others (such as acetate esters) remained relatively constant.

## 2. Thermally Induced Aroma Genesis

The thermal or oxidative degradation of precursors has received much more attention than biogenesis. An idea of the complicated mixtures that can be obtained during the cooking or roasting of a food can be given by the example of trigonelline (**III**). This compound is found in green coffee at a

III

level of about 1% and falls to about 0.1% after roasting. When it is heated alone at 180°C in a sealed tube for 15 min, over 60 volatile compounds are produced (Viani and Horman, 1974). Some of these are shown in Fig. 9, where those also described in coffee aroma itself are marked with a star. The presence of some and the absence of others in coffee is rationalized in terms of the mode of formation. Generally speaking, most of the starred compounds can be formed by monomolecular decomposition or rearrangement, whereas the others would require a bimolecular or higher-order reaction. In coffee, where trigonelline is dispersed throughout the bean, the probability of finding two trigonelline molecules close enough together to react is low, and there may be other more reactive species around. Even so, if we looked sufficiently carefully in coffee, we would expect to find minute quantities of all of the compounds of Fig. 9. Studies of model systems can therefore provide reference mass spectra of components that are likely to be present in trace amounts in real food samples. These spectra can be used to extend spectral library systems, as discussed in Section III,A.

In other studies of model systems, Sakaguchi and Shibamoto (1978) have identified 39 compounds produced during the reaction of hydrogen sulfide with glucose in aqueous solution, most of which had already been identified in roasted peanuts, coffee, beef, pork liver, chicken, filberts, popcorn, or bread. Nishimura *et al.* (1980) have investigated the products formed in model food browning systems involving cyclotene, ammonia, and hydrogen sulfide and have reported several compounds with aroma notes typical of roasted foods. Shibamoto and Bernhard (1978) and Shibamoto *et al.* (1979) have described the creation of pyrazines, pyrroles, and imidazoles in the reaction of rhamnose and glucose, respectively, with ammonia, and Spingarn and Garvie (1979) have reported on the mutagenic activity of the products of these and other sugar–ammonia systems. The mutagenicity of products from a model system containing cysteamine and glucose was also the concern of a study by Mihara and Shibamoto (1980). Several sugar–amino acid systems have been studied, including glucose, fructose, and

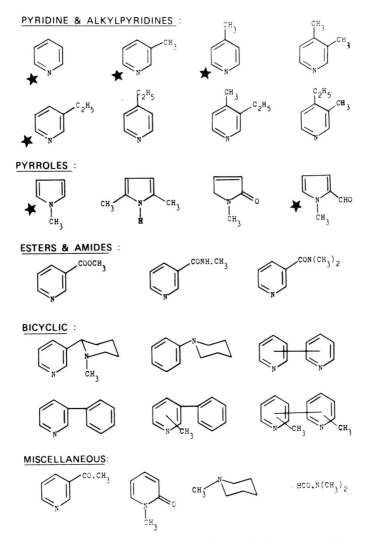

PYRIDINE & ALKYLPYRIDINES :

PYRROLES :

ESTERS & AMIDES :

BICYCLIC :

MISCELLANEOUS:

**Fig. 9.** Compounds formed during the roasting of trigonelline (III), a green coffee constituent. Compounds marked with a star have been described in the volatile constituents of coffee.

ascorbic acid with phenylalanine, where furfural and aromatic hydrocarbons were the major products (Seck and Crouzet, 1981). The thermal degradation products of Amadori compounds formed from sugars and amino acids as Maillard reaction intermediates have been reported by Shigematsu *et al.* (1977). In further investigations of the Maillard reaction, Otto and Baltes (1980) identified volatile products from the reaction of glucose with 4-chloroaniline, showing the formation of *N*-4′-chlorophenyl-pyrrole aldehyde and derivatives of quinoline and dihydroquinoline as well as the usual furan derivatives. In a similar study, Piloty and Baltes (1979) have described compounds formed by the reaction of amino acids with diacetyl; these included novel pyridines. The origin of alkylpyridines from the reaction of aldehydes with amino acids was reported by Suyama and Adachi (1980), while Henderson and Nawar (1981) went one stage further back in the sequence of precursors, showing that alkylpyridines are also formed when the doubly unsaturated linoleic acid is heated with the amino acid valine in the presence of oxygen. A total of 27 compounds, mainly heterocycles, are reported by Breitbart and Nawar (1979) as the thermal decomposition products of lysine. Szafranek and Wisniewski (1978) have studied compounds formed by the acid catalysis of mannitol, and Donnelly *et al.* (1980) have worked with pyrolysis products of oligosaccharides. The characteristic odor compound formed from thiamine is now thought to be a bicyclic disulfide (**IV**) (R. M. Seifert *et al.*, 1978).

IV

Several authors have reported the results of heating fatty acids and other lipids. *n*-Alkylbenzenes and ω-phenylcarboxylic acids form by cyclization during thermal treatment of fatty acids with kraft lignin (Traitler *et al.*, 1980), and according to Henderson *et al.* (1980), linoleate esters give larger amounts of alkyl ethers and cyclic hydrocarbons at 250°C than at 180°C, where hexanal and esters of keto acids were found in larger amounts. Alcohols, aldehydes, and esters are described as the major decomposition products of methyl oleate hydroperoxides and of trilinolein heated in air (Selke *et al.*, 1978, 1980).

Model studies are also carried out using food systems or food isolates. Thus, Ramey and Ough (1980) investigated the effects of temperature, ethanol concentration, and pH on the rate of formation or hydrolysis of volatile esters in wine: ethyl esters of lower-molecular-weight fatty acids tend to increase in quantity at the expense of other esters that may have been formed biosynthetically in the maturing or fermenting grape. P. J. Williams *et al.* (1980) have identified grape polyols (hydroxylated linalools) as precursors that generate volatile monoterpenes nonenzymatically in muscat grapes. Alkylthiomethylhydantoin-*S*-oxides were demonstrated by Tahara and Mizutani (1979) to act as nonenzymatic precursors of fresh flavors in garlic and onion plants. Swoboda and Peers (1978) cite *trans*-4,5-epoxyhept-*trans*-2-enal as the major volatile compound formed when butterfat is oxidized in the presence of copper and α-tocopherol.

## C. Water Analysis

Water has no inherent food value. As such, the problem of water purity seems at first sight to belong more in the hands of the environmental scientist or the ecologist rather than the food analyst, until we come to realize that water is the universal vehicle in the growth, cooking, consumption, and digestion of the food we eat.

In many countries, water of good quality is becoming scarce, and it is found, for example, that fish and birds living in or by the Rhine River in Holland are so badly polluted with chlorinated pesticides that their flesh can no longer be eaten (van Esch, 1978). In 1971, the EPA, already concerned with this problem, selected a computerized GC/MS system as its principal tool for the analysis of drinking water and of industrial and municipal effluents that are returned to the public water supply. Since then, 129 compounds termed "priority pollutants" have been selected as target compounds for routine analysis in water samples. These are aliphatic and aromatic hydrocarbons, chlorinated hydrocarbons, acrolein, acrylonitrile, nitrosamines, phenols, chloro- and nitrophenols, pesticides, and metals [see, for example, Keith and Telliard (1979) and Finnigan *et al.* (1979)]. Of these, 114 compounds are amenable to analysis by GC/MS with a high degree of automation in sample handling and with direct computer output of amounts of pollutants present. Techniques of sample preparation and details of instrumentation are published by several authors, who have emphasized the superior reliability, speed, and cost effectiveness of GC/MS as compared with pure GC for routine target compound analysis (Thomas *et al.,* 1980; Pereira and Hughes, 1980; Trussel *et al.,* 1980; Schnute, 1980; Beggs, 1978).

By mid-1984, all public water in the United States must be controlled using these or equivalent methods. The problem is no less acute in other countries, which may soon be expected to follow suit with equivalent legislation. A book on the mass spectrometry of priority pollutants has recently been published by Middleditch *et al.* (1981).

## D. Lipids

One of the main problems confronting the lipid analyst involves the location of sites of chain branching and unsaturation in the long-chain fatty acids and alcohols that are the principal components of food fats and oils. Effectively, after ionization and before fragmentation, the long hydrocarbon chains of these compounds undergo extensive molecular rearrangement such that the final mass spectra have lost most of the information characteristic of the original structure. Many suggestions to overcome this have been put forward, particularly for the study of double-bond positions. These have involved performing chemical reactions such as deuteration or oxidation to "fix" the double bond and to introduce diagnostic peaks in the mass spectra. In spite of this, spectra frequently were still difficult to interpret. More recently, Chai and Harrison (1981) have located double bonds by CI mass spectra using vinyl methyl ether as a reagent gas. In a new EI approach, Vetter and Meister (1981) have analyzed long-chain alcohols in the form of their nicotinates, and Harvey (1982) has analyzed long-chain acids as their picolinyl esters. In both cases, a series of medium-intensity ions is formed presumably by random hydrogen abstraction along the chain followed by $\beta$-cleavage to the newly formed radical site. These ions are diagnostic both for double-bond positions and for sites of chain branching. Both sets of derivatives can be handled by GC/MS, although they tend to tail. Perhaps derivatives having better GC properties but making use of the same principle in MS could be developed.

In the general analysis of lipids, Slover et al. (1980) have surveyed the lipid composition of fast foods as part of a larger study on the kinds and amounts of dietary lipids consumed by humans. Alexandridis and Lopez (1979) have reported that unsaturated lipids decrease significantly in sweet potato flakes during processing and storage whereas saturated lipids increase only slightly. The decrease in unsaturates suggested a link with the development during storage of off-flavors and off-odors. Bitter-tasting monoglycerides were identified in stored oat flour (Biermann and Grosch, 1979), and unusual minor carboxylic acids have been identified in Finnish tall oil (Hase et al., 1980). The composition of subcutaneous fat in lambs was studied as a function of diet (Miller et al., 1980). Corn-fed lambs had lower quantities of saturated fatty acids and softer fat than animals fed on corn silage. Lipids from fish have been investigated by Miyagawa et al. (1979), who observed a high percentage (26%) of odd-carbon-numbered fatty acids in snow crab, and by Kaitaranta (1980), who remarked on the exceptionally high amounts of polar lipids in whitefish roe. Bastic et al. (1978) has shown that the hydrocarbon fraction of several oils is composed of n-paraffins, isoprenoidal poly-olefins, squalene, diterpenes, and triterpenes. Some oxygen-containing compounds are also present. They suggest that the hydrocarbon fraction of oils may be used for characterization. Roderbourg and Kuzdzal-Savoie (1979) have shown how technological treatment can alter the hydrocarbon composition of anhydrous butterfat. Krulick and Osman (1979) have reported the metabolic production of the fatty alcohols, hexacosanol, and octacosanol in bruised potato tissue.

Sterols in sunflower and poppy-seed oils were investigated by Johansson (1979): sunflower oil was characterized by Δ7-sterols with almost 50% of the total sterols present as esters, whereas poppy-seed oil contained campesterol, stigmasterol, sitosterol, and Δ5-avenasterol with little more than 10% esterified. A detailed study of the minor sterols of the oyster is reported by Teshima *et al.* (1980). The presence of 31 desmethylsterols and at least eight 4-monomethylsterols was confirmed; most of these were identified. The origins of sterols in marine invertebrates being little known, it was hoped that these findings might provide insight into the biosynthesis and modification of exogenous sterols in the oyster and other bivalves.

## E. Food Solids

Before MS became widely used in organic analysis, the solid components of food had received much more attention than the volatiles. Today, the reverse is true, partly because of the instrumental limitations in separation and MS sample-handling techniques described earlier and partly because mass spectra of even relatively small solid molecules are often difficult to interpret. Schulten, speaking of his FD spectra of glycosides, such as that of Fig. 4, states that is easy to obtain information on molecular weights, isotopic distribution, and the nature of the individual sugar units present but that to say exactly how the sugars are linked together is impossible from a single spectrum.

### 1. Carbohydrates

A good example of the work required to identify simple structures is the identification of two galactopinitols, unusual disaccharides found in soya (Schweizer and Horman, 1981). These are formed by linking together galactose (**V**) and pinitol (**VI**), an *O*-methyl inositol. It was easy to demon-

| V | V I |

strate the presence of the two individual sugar units and to detect that the galactose was linked from the 1-position. The question was to determine the position of linking in the pinitol ring for the two isomers. The problem was solved by pertrideuteromethylation of the galactopinitols, followed by hydrolysis and acetylation. The spectrum of the pinitol moiety was readily obtained and corresponded to a compound in which free hydroxyl groups in the original compound had been replaced by —$OCD_3$ groups and that had

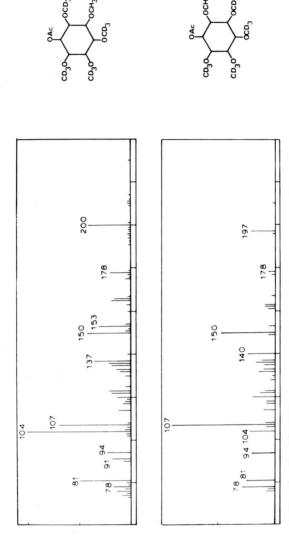

**Fig. 10.** The 70-eV EI mass spectra of the cyclitol acetates prepared from two galactopinitols [compounds with a galactosyl residue in position a or b of **(VI)**] by pertrideuteromethylation, hydrolysis, and acetylation.

an acetoxy group in the position initially occupied by the galactosyl residue. Two such compounds were prepared from the two galactopinitols, and their spectra, shown in Fig. 10, were compared with those of two model inositol derivatives also bearing a simple acetoxy group along with either five —OCH$_3$ groups or five —OCD$_3$ groups. The spectra of this limited series of compounds, limited because similar compounds of known structure were totally unavailable, could be rationalized in terms of three fragmentation mechanisms, shown in Schemes 1–3. This showed that in one of the galactopinitols the galactosyl and methoxy groups are on adjacent carbons and in the other they are separated by a CHOH group. This finding was confirmed by spectra of a second set of derivatives in which the acetoxy group in the previous derivatives was replaced by methoxy, and also by $^{13}$C nuclear magnetic resonance of the parent compound. This kind of multi-derivatization approach is frequently the only solution to problems in the structural elucidation of solids, particularly when samples are too small to obtain complementary data by other spectroscopic techniques.

Also in the field of carbohydrates, Aman (1979) has reported the distribution of carbohydrates in raw and germinated mung beans and chick-peas, noting a decrease in raffinose oligosaccharides during germination, and

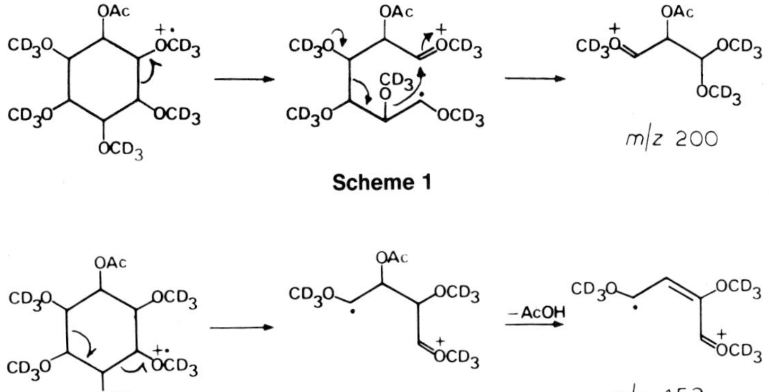

**Scheme 1**

**Scheme 2**

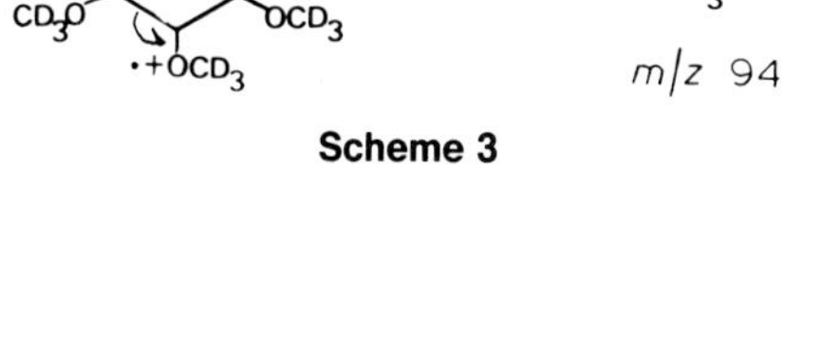

**Scheme 3**

Krusius and Finne (1981) have looked at branching patterns in glycopeptides and glycoproteins. It is worth noting here the work of Bjorndal *et al.* (1967), which exemplifies a classical MS method for obtaining information on the individual sugars constituting a polysaccharide and their mode of linkage in the macromolecular structure.

## 2. Polyphenols

Flavonoids in wines and their changes on aging have been the subject of a study by Wulf and Nagel (1980). The flavonoids were separated by HPLC and analyzed by MS in the form of their TMS derivatives. Also using TMS derivatives, Horvat and Senter (1980) have described a method for the GC/MS analysis of phenolic acids in plant materials, and Senter *et al.* (1980) have applied the method in establishing a correlation between phenolic acid content and stability of pecans on storage. Horman and Viani (1971) have published EI spectra of several phenolic acids from plants, showing the likely fragmentation pathways for their TMS derivatives. Hydroxybenzoic and hydroxycinnamic acids are described in spices (Schulz and Herrmann, 1980), and methyl and ethyl esters of phenolic acids are reported in vegetables (Sontag *et al.,* 1980). Hedin *et al.* (1979) have isolated juglone (5-hydroxy-1,4-naphthoquinone) from pecan and have shown it to vary seasonally. They report that this compound inhibits pecan scab disease.

## 3. Gibberellins

Kuroguchi *et al.* (1979) have used GC/MS to identify gibberellins in the rice plant and to follow changes in one of these, gibberellin $A_{19}$, throughout the life cycle of the plant. The same group has identified a new gibberellin (gibberellin $A_{57}$) and in this and the previous publication report detailed mass spectra (Murofushi *et al.,* 1980). Sponsel *et al.* (1979) report the identification of six gibberellins in immature seeds of broad bean, and Metzger and Zeevaart (1980) used ion-monitoring methods to follow changes in the levels of different gibberellins in spinach as a function of day length.

## 4. Miscellaneous

Among other food solids that have recently received attention are glucosinolates, which give isothiocyanates, thiocyanates, and cyanides in food plants (Olsen and Sorensen, 1980), cytokinins from cabbage hearts (Hashizume *et al.,* 1979), cardanols from cashew nuts (Strocchi and Lercker, 1979), and ginger constituents (Harvey, 1981). The last study gives a comprehensive account of the pungent principles of ginger, including newly identified gingerdiones and hexahydrocurcumin analogs. Mass spectral data and proposed biosynthetic pathways for the formation of the newly identified compounds are presented.

## F. Quantitative Determination of Natural Products

The principles of quantitative MS have been presented by Markey (1981) and by Millard (1978), and Hertz *et al.* (1978) have given excellent general accounts of the problems involved and measures needed to assure good quantitative trace organic analysis. In its simplest terms, determinations are made by comparing the height of a selected characteristic sample peak with a peak from a standard substance present in known quantity. The peaks chosen may be either molecular ions or fragment ions, and one, two, or more peaks may be used for the same analysis. Also, MS peak surfaces or the envelopes given by selected peaks in mass chromatograms may be used instead of peak heights to determine relative proportions. However, all methods described are refinements of the simple procedure.

The extreme natural complexity of food materials means that, when we carry out quantitative analyses, we must frequently find and estimate the amount of compounds present in minute quantities among hundreds of other compounds. This is a severe test for any instrumental technique, and the performance of MS must be judged against this background.

Most quantitative measurements of compounds in foods have been concerned with agricultural, environmental, or technological residues present as contaminants or impurities, and these are described in the following sections. A few quantitative measurements have been carried out on natural food constituents, and some of the recent applications are presented below.

Erucic acid (docosenoic acid; $22:1$ $n$-9) is a constituent of rapeseed oil, which is reported to cause fatty deposits around the heart if consumed in large amounts. Strains of rapeseed low in erucic acid have been developed, but these need to be monitored. Thus, Blomstrand *et al.* (1978) have described a method whereby rapeseed oil is first subjected to a standard transesterification to transform the triglycerides into methyl esters; then a known amount of $1$-$^{14}$C erucic acid methyl ester is added, and the mixture is and analyzed by GC/MS. The comparative heights of the $m/z$ 320 and $m/z$ 322 peaks, corresponding to the $(M - MeOH)^{+\cdot}$ ions from the normal and labeled erucic acids, give a direct measure of the amount of erucic acid in the sample with a relative standard deviation of less than 2%. This example uses a reference compound that cochromatograms with the compound to be analyzed, but this is not a necessity, as seen in the method of analyzing sorbic acid described by Stafford and Black (1978). Here, decanoic acid was added to a homogenate of dried prunes and was coextracted along with other acids. These were converted into *n*-butyl esters and analyzed by GC/MS, comparing surfaces of peaks in reconstructed chromatograms. The method works down to levels of 4 ppm sorbic acid. Using butyl behenate as an internal standard, Hasegawa and Oakamoto (1980) have described a method for the estimation of lysinoalanine in alkali-treated food proteins. The lysinoalanine is transformed into its *N*-trifluoracetyl-*O*-butyl ester before analysis. Yamamoto *et al.* (1980) have described how tyramine, in the form of its *N*-*O*-bis(ethyloxycarbonyl) derivative, can be estimated relative to 3,4-dimethoxyphenethylamine as an internal standard. Tohma *et al.* (1979) have determined cholestanols and their derivatives by GC mass fragmentography using deuterated cholestanols as standards. Iwai *et al.* (1979) have separated total capsaicinoids from red peppers by HPLC and have then analyzed the mixture after trimethylsilylation for the five individual capsaicin analogs by mass chromatography of characteristic ions at $m/z$ 365, 377, 379, 391, and 393, respectively, using 5-$\alpha$-cholestane as an internal standard. Routine analysis capabilities from microgram to nanogram levels of individual components are claimed. Two analyses involving quantitative estimation by HPLC with confirmation of the compound analyzed by MS are reported, one being the determination of 25-hydroxycholecalciferol in chicken egg yolks (Koshy and van der Slik, 1979) and the other being the determination of vitamin $D_3$ and 7-dehydrocholesterol in cows' milk (Adachi and Kobayashi, 1979).

## G. Agricultural Residues

Pesticides, insecticides, herbicides, and fungicides must be looked upon as necessary agricultural aids if we wish to combat our natural competitors for a common food supply. Equally, the use of drugs in livestock helps to keep our meat disease and infection free. But what is the fate of these products, and to what degree do they accumulate in plant and animal tissues?

Ideally, at the time of eating, the residual levels of these compounds present in the tissues of plants and animals would be zero. However, it is inevitable that small amounts are found, and for most of these compounds legal upper limits have been established. A large body of effort has been addressed to the question of measuring residual levels in routine food control, and the elements of regulatory MS have been presented by Brumley and Sphon (1981). The sensitivity and specificity of MS have played an all-important role in these studies, which have focused on several points: establishment of appropriate MS conditions for obtaining optimum spectra, estimation of the various compounds in plant and animal tissues, metabolism in plants and animals, impurities in commercial products, breakdown under conditions of oxidation and solar irradiation, residual levels in soil, etc. Specific examples of these studies are presented below. For further examples, a book by Safe and Hutzinger (1973) treats the mass spectrometry of pesticides and pollutants, including examples using EI, CI, NCI, and FD ionization.

Soft ionization techniques such as CI and NCI (Section II,D) have proved invaluable in the analysis of pesticides and other agricultural chemicals, many of which are thermally sensitive and decompose to give unrecognizable or uncharacteristic spectra in EI. Stan (1977) used CI GC/MS to identify and estimate 23 of the most frequently used organophosphorus pesticides in Germany, showing their unequivocal detection at the 40 ppb level, some 20 times lower than could be obtained for the same compounds by EI spectra. Garland *et al.* (1980b) demonstrated the power of NCI GC/MS in detecting the antihistomoniasis drug ipronidazole and a major hydroxylated metabolite at the regulatory levels of 2 ppb in turkey muscle. Deuterated analogs of the two compounds were used as standards. Sonobe *et al.* (1980) illustrated one of the problems associated with pure GC in pointing out the potential interference from long-chain fatty acids in the analysis of rice bran for pesticides using different detector systems. The acids have similar retention times on the GC column proposed in the official method, but these difficulties are overcome when MS is used as the detector. The purity of some agriculture chemicals can be cause for concern. Lamoureux and Feil (1980) have provided MS evidence for the presence of more than 50 impurities in technical methoxychlor, 25 of which they could identify. The presence of such impurities, albeit in trace amounts, cannot but complicate studies of metabolic transformations. Rusness and Lamoureux (1980) treated peanut plants with [14]C-labeled pentachloronitrobenzene, a fungicide, using the radioactive label to follow the evolution of chloroform-soluble metabolites. After 4 days, compounds in which the nitro group had been replaced by $-SCH_3$, $-NH_2$, $-SH$, $-SO \cdot CH_3$, $-SCH_2 \cdot COOH$, $-SCH_2 \cdot CH(OH)$, COOH, $-SCH_2 \cdot CH_2 \cdot COOH$, and $-SCH_2 \cdot CH(NH_2) \cdot COOH$ were identified along with the parent fungicide in the 59.2% of the original material that was recovered. Belasco and Harvey (1980) have investigated the rumen metabolism of the insecticide oxamyl to add metabolic data from animals used as food sources to earlier results obtained on crop plants, rats, and soil. The major metabolites produced by the cow and the rat were identical. Clark *et al.* (1978) have shown that trimefedon, active against powdery mildew fungus, is metabolized by the marrow to give two diastereoisomeric secondary alcohols. These compounds are also extremely active as fungicides. Alternatively, photolysis gives three compounds, none of which shows a fungitoxic activity. Argauer and Feldmesser (1978) have reported on the uptake of the nematocide ethoprop by vegetables grown in soil that had been treated with different amounts of this compound 1 week before planting. Onion, carrot, radish, and eggplant contained amounts of the order of 0.1 – 1 ppm, the exact figure increasing as the quantity of ethoprop used to treat the soil increased. Beet, cabbage, cantaloupe, pea, and tomato were apparently less susceptible, showing no uptake exceeding the detection limit of 100 ppb. Perhaps not surprisingly, skins and roots showed higher concentrations than did the whole vegetables.

Further studies have involved the development of methods for the determination of organochloro and organophosphorus pesticides, pyrethrins, and pyrethrinoids in cereal products (Mestres *et al.,* 1979); for the detection of methyl-4-chloroindolyl-3-acetate at levels of 0.2 ppm is canned and frozen peas (Heikes, 1980); for the preparation of samples from lipids contaminated with organochlorine residues (Veierov and Aharonson, 1980); for the characterization of an environmental alteration product of malathion in stored rice (Hansen *et al.,* 1981); for the identification and estimation of residues of the carbamate insecticide crotenon along with its sulfoxide and sulfone metabolites in citrus trees following soil treatment (Aharonson *et al.,* 1979); for the confirmation of endosulfan and its sulfate from CI data at levels down to 0.1 ppm in apples and carrots (Wilkes, 1981); for the analysis of the herbicide diquat in potato tubers at levels of 0.01 ppm after reduction with sodium borohydride to give a volatile derivative (King, 1978); and for the uptake of the herbicide prometryn from treated soil into oat plants (Khan, 1980).

Other metabolic or radiation-induced degradation studies have been the identification of products formed by the action of sunlight on pyrethrins and related compounds in kerosene solution (simulated commercial formulations) stored in clear glass bottles (Kawano *et al.,* 1980); the isolation and identification of two polar metabolites of methidathion occurring in tomato plants (Simoneaux *et al.,* 1980); the elucidation of metabolic products of the insecticide tetrachlorvinphos in goose and turkey liver, where it was shown that the initial demethylated metabolite is further transformed to give other metabolites (Akhtar and Foster, 1980); the mass spectral analysis of *s*-triazine metabolites (Lusby and Kearney, 1978); and the tentative structural identification of metabolites of the herbicide fluchloralin in soybean roots (Marquis *et al.,* 1979).

Drug studies that can be cited are the determination of arprinocid in chicken tissues analyzed at levels of 0.01 ppm using GC/MS and CI (Tway *et al.,* 1979) and the estimation of sulfamethazine in swine tissue by GC/MS and EI spectra in quantities lower than the permitted regulatory levels of 0.1 ppm (Suhre *et al.,* 1981). The control of progesterones used in livestock to synchronize estrus is also reported using selected ion-monitoring GC/MS techniques (Winkler *et al.,* 1977), and a further method for the determination of anabolic drugs in meat at levels of 1 – 5 ppb has been described (Stan and Abraham, 1980).

## H. Technological Residues

Nitrites occur naturally in foods and have also been added to foods for many generations, particularly to meat products via curing salts that prevent microbial spoilage and associated food poisoning. Along with food amines, they react to give trace amounts of nitrosamines, which have been reported to be carcinogens. Mass spectrometry has been used in nitrosamine investigations for over 10 years (e.g., Fazio *et al.,* 1971). An excellent recent text on volatile nitrosamines edited by Egan (1978) presents critical reviews of applications of low- and high-resolution GC/MS techniques to nitrosamine detection and estimation and concludes that GC combined with high-resolution MS provides the most reliable method. Here, mass fragmentography or chromatography of specific ions, usually the molecular ion or the $NO^+$ ion at $m/z$ 29.9980 common to all nitrosamines, has permitted estimation down to 1-ng levels (Gough, 1978). Other authors (Havery *et al.,* 1978; Sen *et al.,* 1979) have preferred to carry out quantitative measurements using a non-MS method based on the thermal energy analyzer (TEA), a detector designed for the specific analysis of nitrosamines (Fine *et al.,* 1975). However, the TEA can give false positive indications, particularly at the lower levels, and GC/MS is used as a backup technique to confirm identifications. Also, the relative merits of GC/MS methods have been compared with methods based on HPLC (Wolfram *et al.,* 1977) and thin-layer chromatography (TLC) (Cross *et al.,* 1978) along with fluorescence, or on chemiluminescence (Gough *et al.,* 1977). Condensed mass spectra are reported for 146 nitrosamines (Rainey *et al.,* 1978).

Among the results obtained, Sen *et al.* (1978) have used GC with high-resolution MS to determine *N*-nitrosodimethylamine (NDMA), *N*-nitrosodiethylamine (NDEA), and *N*-nitrosopiperidine (NPIP) in 27 commercial cheeses at levels of up to 20 ppb. The average amount of NDMA per sample was less than 3 ppb, with NDEA and NPIP present on average in much lower quantities. The same authors describe methods for MS analysis of nonvolatile nitrosamines and report traces of *N*-nitrosohydroxypyrrolidine (NHPYR) in six samples of cooked bacon: they found no significant levels of *N*-nitrosoproline (NHPRO) in the same samples. Using a TEA with GC and low-resolution MS as confirmation, Libbey *et al.* (1980) report levels of about 2 ppb NDMA in nonfat dried milk powders, and Havery *et al.* (1981) have surveyed the same compound in malt whiskeys and beers, finding generally less than 10 ppb in the former and less than 1 ppb in the latter. Using the TEA with GC and high-resolution MS, Janzowski *et al.* (1978) have found 1–7 ppb of NHPYR in cured meat products.

Factors controlling the formation of nitrosamines have also been studied: Spinelli-Gugger *et al.* (1980) have investigated the quantities of NDMA and NDEA formed as a function of the generation of free amino acids during the curing and frying of bacon and ham; Pensabene *et al.* (1980) have found that amounts of *N*-nitrosopyrrolidine (NPYR) formed on frying fresh bacon are less than those formed from bacon that has been refrigerated and thawed; and Spinelli-Gugger *et al.* (1981) have tied the formation of NPYR to reactions between the nitrite ion and the water-soluble fraction of pork belly adipose tissue. *N*-Nitrosamides and their fatty acid ester and free amine precursors have been studied by Kakuda and Gray (1980a,b), and a method for the GC/MS determination of nitrite in foods after reaction with 1,2-dia-minobenzene has been described (Tanaka *et al.,* 1980).

In studies directed toward the prevention of nitrosamine formation, pork bellies treated with $\alpha$-tocopherol alone or in admixture with sodium ascorbate were found to generate much lower amounts of NPYR than untreated samples (Fiddler *et al.,* 1978), and Sen *et al.* (1979), analyzing for NDMA, NDEA, NPYR, and NPIP in several cured meats, reported a continuously lowering trend in the total levels of volatile nitrosamines in these products.

The use of food additives in general is strictly controlled, and most of the problems associated with the incorporation of these compounds into foods have revolved around nitrates and the nitrosamines they create. However, other additives have been the subject of investigations of purity or of their interactions with other food constituents. Two such examples where MS has been employed concern the estimation of free $\alpha$- and $\beta$-naphthylamines at levels of a few parts per billion in the red food dye amaranth (Stavric *et al.,* 1979) and the identification of 4-methyl-1,3-dioxolanes in commercial beef flavors, probably formed by the reaction of carbonyl aroma constituents with propane-1,2-diol, a solvent commonly used for flavor preparations (MacLeod *et al.,* 1980).

## I. Mycotoxins

Mycotoxins are secondary metabolites that form in vegetable material as a result of fungal spoilage. By definition, any fungal metabolite that is known to be toxic to humans or animals is a mycotoxin, and this covers a wide range of chemical classes. The structures of several mycotoxins are to be found in an article by Scott (1978) on the origins of these compounds in food materials. They generally give well-defined, characteristic mass spectra that can be used for their identification (Kingston, 1976) or for their estimation through ion-monitoring methods (Salhab *et al.,* 1976), and a collection of about 100 such spectra have been published by Dussold *et al.* (1978) under the title "Mycotoxins Mass Spectral Data Bank."

For the estimation of patulin in apple juice, Price (1979) has critically compared the use of two methods based on GC/MS, namely GC combined with single ion monitoring at high-resolution (HRSIM) or with multiple-ion monitoring at low resolution (LRMPM). Effectively, he found that HRSIM of the molecular ion at $m/z$ 226.066 reliably measured 0.2 ppb patulin in apple juice whereas LRMPM of $M^+$ at $m/z$ 226 and fragment ions at $m/z$ 211 and 183 gave a detection limit of 5 ppb and gave reliable measurements only at 10 ppb or more. The outcome of this study is of interest not only for mycotoxins but also in determinations of nitrosamines, pesticides, or other contaminants where HRSIM or LRMPM methods may be employed. Chaytor and Saxby (1981), using GC with LRMPM, found detection limits of 20 ppb and 10 ppb, respectively, for patulin and penicillin in roasted cocoa beans. Collins and Rosen (1979, 1981) report methods for the estimation of T-2 toxin in milk and in wet-milled corn products, and Scott *et al.* (1978), Bennett and Shotwell (1979), and Marasas *et al.* (1979) report measurements of zearalenone in grains, corn, and corn products. The last study also includes the mycotoxin deoxynivalenol.

## J. Contaminants

In contrast to agricultural chemicals and food additives that are purposefully used on or in our foods, certain other compounds are introduced inadvertently by absorption from the environment. We live in contaminated surroundings, and it is in these same surroundings that our food is produced; hence the incorporation of environmental contaminants into the food chain. The processes by which this occurs and its impact on mammals has been outlined by Bowes (1981). He summarizes the magnitude of the problem by stating that contaminants are probably complex mixtures of chemicals at trace levels and that, apart from a few compounds, we have no idea what we are looking for. The question of contaminants has recently been highlighted by the Seveso affair in Italy, in which a large region was severely contaminated with 2,3,7,8-tetrachlorodibenzo-*p*-dioxin (TCDD), and the Love Canal situation, in which chemicals from an old dump site have infiltrated homes (Elder *et al.,* 1981).

The compound TCDD is highly toxic, and it has been necessary to develop analyses that can handle levels of parts per trillion ($10^{-12}$). This is several orders of magnitude below the normal limits required for pesticide analysis. Mass spectrometry provides the only solution to this problem, and a method has been described by Harless *et al.* (1980) with sub-ppt detection limits. In it, the sample is prepared using a rigorous cleanup procedure and is then separated on a capillary GC column that feeds directly into a high-resolution mass spectrometer (in the EI mode). Quantitation is effected by monitoring the molecular ion cluster of TCDD using $^{37}$Cl-TCDD as a standard. The precise methodology used is important because polychlorinated biphenyls (PCBs), which may also be present, give peaks in the same mass region. A similar method has been described by Cavallaro et al. (1980), who tested TCDD recoveries at the 0.5 and 1-ppt levels from spiked vegetables. Zabik and Zabik (1980) have measured TCDD levels in steaks and liver from cattle to which technical-grade pentachlorophenol (which contains traces of TCDD) had been administered; they reported that the liver showed the highest concentrations. Firestone et al. (1979), in a similar experiment, have reported on milk and blood levels of hexa-, hepta-, and octachlorodibenzo-p-dioxins, pointing out that levels drop to basal values within a few days of stopping the feeding of pentachlorophenol.

In other investigations, Kuehl et al. (1980) found chlorinated alkanes, alkenes, alkylamines, styrenes, norbornenes, pentachlorobenzyl alcohol, and pentachlorophenol in river fish; Wise et al. (1980) and Chesler et al. (1978) have reported trace-level aromatic hydrocarbons in mussels and in marine biota, respectively; Diachenko (1979) has described the presence in fish of industrial aromatic amines; and Ogata and Miyake (1980) identified organic sulfur compounds in fish and shellfish. Many other examples are presented in a review by Self (1979).

## IV. CONCLUSIONS

The scope of mass spectrometry in the analysis of foods and beverages is already vast but is still continuing to grow with the introduction of new instrumentation. The major areas of interest for the future must include the investigation of food solids, automated trace and ultratrace analysis for quality control, and the correlation of food composition with physiological responses. In these and other studies, MS must continue to prove its worth in improving our understanding of the food we eat.

## REFERENCES

Adachi, A., and Kobayashi, T. (1979). J. Nutr. Sci. Vitaminol. 25, 67.
Adams, R. P., von Rudloff, E., and Hogge, L. (1980). J. Nat. Prod. 43, 417.
Adams, R. P., von Rudloff, E., Hogge, L., and Zanoni, T. A. (1981). J. Nat. Prod. 44, 21.
Aharonson, N., Neubauer, I., Ishaaya, I., and Raccah, B. (1979). J. Agric. Food Chem. 27, 265.
Aishima, T. (1982). J. Food Sci. 47, 1562.

Aishima, T., Nagasawa, M., and Fukushima, D. (1979). *J. Food Sci.* **44**, 1723.
Akhtar, M. H., and Foster, T. S. (1980). *J. Agric. Food Chem.* **28**, 693.
Albrand, M., Dubois, P., Etievant, P., Gelin, R., and Tokarska, B. (1980). *J. Agric. Food Chem.* **28**, 1037.
Alexandridis, N., and Lopez, A. (1979). *J. Food Sci.* **44**, 1186.
Aman, P. (1979). *J. Sci. Food Agric.* **30**, 869.
Argauer, R. J., and Feldmesser, J. (1978). *J. Agric. Food Chem.* **26**, 42.
Arpino, P. J. (1983). *In* "Detectors in Liquid Chromatography" (T. M. Vickray, ed.), Dekker, New York (in press).
Arpino, P. J., Krien, P., Vajta, S., and Devant, G. (1981). *J. Chromatogr.* **203**, 117.
Baltes, W., and Soechtig, I. (1979). *Z. Lebensm-Unters.-Forsch.* **169**, 9, 17.
Barber, M., Bordoli, R. S., and Sedgwick, R. D. (1981). *In* "Soft Ionization Biological Mass Spectrometry" (H. R. Morris, ed.), pp. 137–152. Heyden, London.
Bastic, M., Bastic, L., Jovanovic, J. A., and Spiteller, G. (1978). *J. Am. Oil Chem. Soc.* **55**, 886.
Beckey, H. D. (1977). "Principles of Field Ionization and Field Desorption Mass Spectrometry." Pergamon, Oxford.
Beggs, D. P. (1978). Hewlett Packard Application Note, No GC-MS AN176-24.
Belasco, I. J., and Harvey, J., Jr. (1980). *J. Agric. Food Chem.* **28**, 689.
Bennett, G. A., and Shotwell, O. L. (1979). *J. Am. Oil Chem. Soc.* **56**, 812.
Berlowitz, L., Zdanis, R. A., Crowley, J. C., and Vaughn, J. C. (1981). *Science* **211**, 1013.
Beynon, J. H. (1981). *Biomed. Mass Spectrom.* **8**, 380.
Beynon, J. H. (1982). *Trends Anal. Chem.* **1**, 292.
Biermann, U., and Grosch, W. (1979). *Z. Lebensm.-Unters.-Forsch.* **169**, 22.
Binnemann, P., and Jahr, D. (1979). *Fresenius' Z. Anal. Chem.* **297**, 341.
Bjorndal, H., Lindberg, B., and Svensson, S. (1967). *Carbohydr. Res.* **5**, 433; *Acta Chem. Scand.* **21**, 1801.
Blakley, C. R., Carmody, J. J., and Vestal, M. L. (1980). *Anal. Chem.* **52**, 1636.
Blomstrand, R., Svensson, L., and Herslof, B. (1978). *Lipids* **13**, 283.
Bowes, G. W. (1981). *Biomed. Mass Spectrom.* **8**, 419.
Brander, C. F., Kepner, R. E., and Webb, A. D. (1980). *Am. J. Enol. Vitic.* **31**, 69.
Breitbart, D. J., and Nawar, W. W. (1979). *J. Agric. Food Chem.* **27**, 511.
Bricout, J., and Merlivat, L. (1971). *C. R. Acad. Sci. Paris* t273, 1021.
Brooks, C. J. W., ed. (1970–1983). "Gas Chromatography–Mass Spectrometry Abstracts." PRM Science & Technology Agency Ltd., London.
Brumley, W. C., and Sphon, J. A. (1981). *Biomed. Mass Spectrom.* **8**, 390.
Buttery, R. G. (1979). *J. Agric. Food Chem.* **27**, 208.
Buttery, R. G., and Garibaldi, J. A. (1980). *J. Agric. Food Chem.* **28**, 159.
Buttery, R. G., and Kamm, J. A. (1980). *J. Agric. Food Chem.* **28**, 978.
Buttery, R. G., Guadagni, D. G., and Ling, L. C. (1978a). *J. Agric. Food Chem.* **26**, 791.
Buttery, R. G., Ling, L. C., and Chan, B. G. (1978b). *J. Agric. Food Chem.* **26**, 866.
Buttery, R. G., Black, D. R., Haddon, W. F., Ling, L. C., and Teranishi, R. (1979). *J. Agric. Food Chem.* **27**, 1.
Buttery, R. G., Ling, L. C., and Teranishi, R. (1980a). *J. Agric. Food Chem.* **28**, 771.
Buttery, R. M., Soderstrom, E. L., Seifert, R. M., Ling, L. C., and Haddon, W. F. (1980b). *J. Agric. Food Chem.* **28**, 353.
Buttery, R. G., Siefert, R. M., Haddon, W. F., and Lundin, R. E. (1980c). *J. Agric. Food Chem.* **28**, 1338.
Cavallaro, A., Bartolozzi, G., Carreri, D., Bandi, G., Luciani, L., Villa, G., Gorni, A., and Invernizzi, G. (1980). *Chemosphere* **9**, 623.
Cazenave, P., and Horman, I. (1974). *Helv. Chim. Acta* **57**, 209.

Cazenave, P., Horman, I., Mueggler-Chavan, F., and Viani, R. (1974). *Helv. Chim. Acta* **57,** 206.

Chai, R., and Harrison, A. G. (1981). *Anal. Chem.* **53,** 34.

Chairote, G., Rodriguez, F., and Crouzet, J. (1981). *J. Food Sci.* **46,** 1898.

Chapman, J. R., and Wakefield, C. J. (1980). Kratos Data Sheet No. 117.

Chaytor, J. P., and Saxby, M. J. (1981). *J. Chromatogr.* **241,** 135.

Chesler, S. N., Gump, B. H., Hertz, H. S., May, W. E., and Wise, S. A. (1978). *Anal. Chem.* **50,** 805.

Ching, J. C. (1979). *Diss. Abstr., Int.* Order No. 8002343.

Clark, T., Clifford, D. R., Deas, A. H. B., Gendle, P., and Watkins, D. A. M. (1978). *Pestic. Sci.* **9,** 497.

Clutton, D. W., and Evans, M. B. (1978). *J. Chromatogr.* **167,** 409.

Cole, R. A., and Phelps, K. (1979). *J. Sci. Food Agric.* **30,** 669.

Coleman, E. C., and Ho. C.-T. (1980). *J. Agric. Food Chem.* **28,** 66.

Coleman, E. C., Ho C.-T., and Chang, S. S. (1981). *J. Agric. Food Chem.* **29,** 42.

Collins, G. J., and Rosen, J. D. (1979). *J. Assoc. Off. Anal. Chem.* **62,** 1274.

Collins, G. J., and Rosen, J. D. (1981). *J. Food Sci.* **46,** 887.

Cross, C. K., Bharucha, K. R., and Telling, G. M. (1978). *J. Agric. Food Chem.* **26,** 657.

Delmole, E., Enggist, P., and Ohloff, G. (1982). *Helv. Chim. Acta* **65,** 1785.

Diachenko, G. W. (1979). *Environ. Sci. Technol.* **13,** 329.

Dixon, D. J. (1981). Hewlett Packard Technical Paper, MS-12.

Doi, Y., Tsugita, T., Kurata, T., and Kato, H. (1980). *Agric. Biol. Chem.* **44,** 1043.

Doner, L. W., Krueger, H. W., and Reesman, R. H. (1980). *J. Agric. Food Chem.* **28,** 362.

Donnelly, B. J., Voigt, J. E., and Scallet, B. L. (1980). *Cereal Chem.* **57,** 388.

Dussold, L. R., Dreifuss, P. A., Pohland, A. F., and Sphon, J. A. (1978). "Mycotoxins Mass Spectral Data Bank." Assoc. Off. Anal. Chem., Washington, D.C.

Easley, D. M., Kleopfer, R. D., and Carasea, A. M. (1981). *J. Assoc. Off. Anal. Chem.* **64,** 653.

Egan, H., ed. (1978). *IARC Sci. Publ.* No. 18.

Elder, V. A., Proctor, B. L., and Hites, R. A. (1981). *Biomed. Mass Spectrom.* **8,** 409.

Empedocles of Agrigentum (460 B.C.). "The Four Elements and Four Qualities of Matter."

Etievant, P. X. (1981). *J. Agric. Food Chem.* **29,** 65.

Evans, N., and Williamson, J. E. (1981). *Biomed. Mass Spectrom.* **8,** 316.

Farley, D. R., and Nursten, H. E. (1980). *J. Sci. Food Agric.* **31,** 386.

Fazio, T., Howard, J. W., and White, R. H. (1971). *In* "Proceedings of the Heidelberg Meeting on Nitrosamines," pp. 16–24. IARC, Lyon.

Fiddler, W., Pensabene, J. W., Piotrowski, E. G., Phillips, J. G., Keating, J., Mergens, W. J., and Newmark, H. L. (1978). *J. Agric. Food Chem.* **26,** 653.

Fine, D. H., Rounbehler, D. P., and Oettinger, P. E. (1975). *Anal. Chim. Acta* **78,** 383.

Finnigan, R. E., Hoyt, D. W., and Smith, D. E. (1979). *Environ. Sci. Technol.* **13,** 534.

Firestone, D., Clower, M., Jr., Borsetti, A. P., Teske, R. H., and Long, P. E. (1979). *J. Agric. Food Chem.* **27,** 1171.

Flath, R. A., and Takahashi, J. M. (1978). *J. Agric. Food Chem.* **26,** 835.

Flath, R. A., Forrey, R. R., John, J. O., and Chan, B. G. (1978). *J. Agric. Food Chem.* **26,** 1290.

Fore, S. P., Legendre, M. G., and Fisher, G. S. (1978). *J. Am. Oil Chem. Soc.* **55,** 482.

Fore, S. P., Legendre, M. G., Cherry, J. P., Berardi, L. C., and Vinnett, C. H. (1980). *J. Food Sci.* **45,** 912.

Frank, H., Nicholson, G. J., and Bayer, E. (1978). *J. Chromatogr.* **146,** 197.

Galetto, W. G., and Hoffman, P. G. (1978). *J. Agric. Food Chem.* **26,** 195.

Games, D. E. (1981a). *Biomed. Mass Spectrom.* **8,** 454.

Games, D. E. (1981b). *In* "Soft Ionization Biological Mass Spectrometry" (H. R. Morris, ed.), pp. 54–68. Heyden, London.

Games, D. E. and Lewis, E. R. (1980). *Biomed. Mass Spectrom.* **7**, 433.

Garland, W. A., Miwa, B., Weiss, G., Chen, G., Saperstein, R., and MacDonald, A. (1980a). *Anal. Chem.* **52**, 842.

Garland, W. A., Hodshon, B. J., Chen, G., Weiss, G., Felicito, N. R., and MacDonald, A. (1980b). *J. Agric. Food Chem.* **28**, 273.

Godshall, M. A., Roberts, E. J., and Legendre, M. G. (1980). *J. Agric. Food Chem.* **28**, 856.

Gough, T. A. (1978). *IACR Sci Publ.* No. 18, pp. 25–42.

Gough, T. A., Webb, K. S., Pringner, M. A., and Wood, B. J. (1977). *J. Agric. Food Chem.* **25**, 633.

Graddon, A. D., Morrison, J. D., and Smith, J. F. (1979). *J. Agric. Food Chem.* **27**, 832.

Gudzinowicz, B. J., Gudzinowicz, M. J., and Martin, H. F. (1977). *Chromatogr. Sci.* **7**, Part 3.

Hansen, L. B., Castillo, G. D., and Biehl, E. R. (1981). *J. Assoc. Off. Anal. Chem.* **64**, 1232.

Harless, R. L., Oswald, E. O., Wilkinson, M. K., Dupuy, A. E., Jr., McDaniel, D. D., and Tai, H. (1980). *Anal. Chem.* **52**, 1239.

Harvey, D. J. (1981). *J. Chromatogr.* **212**, 75.

Harvey, D. J. (1982). *Biomed. Mass Spectrom.* **9**, 33.

Hase, A., Hase, T., and Holmbom, B. (1980). *J. Am. Oil Chem. Soc.* **57**, 115.

Hasegawa, K., and Okamoto, N. (1980). *Agric. Biol. Chem.* **44**, 649.

Hashizume, T., Sugiyama, T., Imura, M., Cory, H. T., Scott, M. F., and McCloskey, J. A. (1979). *Anal. Biochem.* **92**, 111.

Havery, D. C., Fazio, T., and Howard, J. W. (1978). *J. Assoc. Off. Anal. Chem.* **61**, 1374.

Havery, D. C., Hotchkiss, J. H., and Fazio, T. (1981). *J. Food Sci.* **46**, 501.

Hedin, P. A., Langhams, V. E., and Graves, C. H., Jr. (1979). *J. Agric. Food Chem.* **27**, 92.

Heikes, D. (1980). *J. Assoc. Off. Anal Chem.* **63**, 1224.

Heimann, W., Timm, U., Rapp, A., and Knipser, W. (1980). *Z. Lebensm.-Unters.-Forsch.* **171**, 35.

Henderson, S. K., and Nawar, W. W. (1981). *J. Am. Oil Chem. Soc.* **58**, 632.

Henderson, S. K., Witchwoot, A., and Nawar, W. W. (1980). *J. Am. Oil Chem. Soc.* **57**, 409.

Hendriks, H., and Bruins, A. P. (1980). *J. Chromatogr.* **190**, 321.

Hendriks, H., and Bruins, A. P. (1983). *Biomed. Mass Spectrom.* (in press).

Henion, J., Nosanchuk, J. S., and Bilder, B. M. (1981). *J. Chromatogr.* **213**, 475.

Henneberg, D., Henrichs, U., and Schomburg, G. (1975). *Chromatographia* **8**, 449.

Henneberg, D., Henrichs, U., Hussmann, H., and Schomburg, G. (1978). *J. Chromatogr.* **167**, 139.

Hermann, K. (1979). *Z. Lebensm.-Unters.-Forsch.* **169**, 179.

Hertz, H. S., May, W. E., Wise, S. A., and Chesler, S. N. (1978). *Anal. Chem.* **50**, 428A.

Heydanek, M. G., and McGorrin, R. J. (1981a). *J. Agric. Food Chem.* **29**, 950.

Heydanek, M. G., and McGorrin, R. J. (1981b). *J. Agric. Food Chem.* **29**, 950.

Hirvi, T., Honkanen, E., and Pyysalo, T. (1981). *Z. Lebensm.-Unters.-Forsch.* **172**, 365.

Honkanen, E., Pyysalo, T., and Hirvi, T. (1980). *Z. Lebensm.-Unters.-Forsch.* **171**, 180.

Horman, I. (1979). *In* "Specialist Periodical Report: Mass Spectrometry" (R. A. W. Johnstone, ed.), Vol. 5, pp. 211–233. Chem. Soc., London.

Horman, I. (1981). *Biomed. Mass Spectrom.* **8**, 384.

Horman, I., and Viani, R. (1971). *Org. Mass Spectrom.* **5**, 203.

Horvat, R. J., and Senter, S. D. (1980). *J. Agric. Food Chem.* **28**, 1292.

Hunt, D. F., and Crow, F. W. (1978). *Anal. Chem.* **50**, 1781.

Hunt, D. F., Shabanowitz, J., and Giordani, A. B. (1980). *Anal. Chem.* **52**, 386.

Ismail, H. M. M., Williams, A. A., and Tucknott, O. G. (1980a). *Z. Lebensm.-Unters.-Forsch.* **171**, 265.

Ismail, H. M. M., Williams, A. A., and Tucknott, O. G. (1980b). *Z. Lebensm.-Unters.-Forsch.* **171**, 24.

Ismail, H. M. M., Williams, A. A., and Tucknott, O. G. (1981a). *J. Sci. Food Agric.* **32**, 498.

Ismail, H. M. M., Williams, A. A., and Tucknott, O. G. (1981b). *J. Sci. Food Agric.* **32**, 613.

Iwai, K., Suzuki, T., Fujiwake, H., and Oka, S. (1979). *J. Chromatogr.* **172**, 303.

Jahr, D., and Binnemann, P. (1979). *Fresenius' Z. Anal. Chem.* **298**, 337.

Janzowski, C., Eisenbrand, G., and Preussmann, R. (1978). *J. Chromatogr.* **150**, 216.

Johansson, A. (1979). *Lipids* **14**, 285.

Johnston, J. C., Welch, R. C., and Hunter, G. L. K. (1980). *J. Agric. Food Chem.* **28**, 859.

Kaitaranta, J. K. (1980). *J. Sci. Food Agric.* **31**, 1303.

Kaitaranta, J. K., and Linko, R. R. (1979). *J. Sci. Food Agric.* **30**, 921.

Kakuda, Y., and Gray, J. I. (1980a). *J. Agric. Food Chem.* **28**, 580.

Kakuda, Y., and Gray, J. I. (1980b). *J. Agric. Food Chem.* **28**, 584.

Karawya, M. S., and El-Hawary, S. S. (1978). *Egypt. J. Pharm. Sci.* **19**, 301.

Kasahara, K., and Nishibori, K. (1979). *Bull. Jpn. Soc. Sci. Fish.* **45**, 1543.

Kato, Y., Watanabe, K., and Sato, Y. (1978). *Lebensm.-Wiss. Technol.* **11**, 128.

Kawano, Y., Yanagihara, K., Miyamoto, P., and Yamamoto, I. (1980). *J. Chromatogr.* **198**, 317.

Keith, L. H., and Telliard, W. A. (1979). *Environ. Sci. Technol.* **13**, 416.

Kenyon, C. N., and Goodley, P. C. (1981). Hewlett Packard Technical Paper, MS-14.

Kenyon, C. N., Melera, A., and Erni, F. (1981). *J. Anal. Toxicol.* **5**, 216.

Khan, S. U. (1980). *J. Agric. Food Chem.* **28**, 1096.

King, R. R. (1978). *J. Agric. Food Chem.* **26**, 1460.

Kingston, D. G. I. (1976). *J. Assoc. Off. Anal. Chem.* **59**, 1016.

Koizumi, C., Cao, tK.-T., and Nonaka, J. (1979). *Bull. Jpn. Soc. Sci. Fish.* **45**, 1307.

Koller, W. D. (1981). *Z. Lebensm.-Unters.-Forsch.* **173**, 99.

Koller, W. D., and Tressl, G. (1980). *J. High Res. Chromatogr. Chromatogr. Commun.* **3**, 359.

Kolor, M. G. (1979). *In* "Mass Spectrometry" (C. Merritt and C. N. McEwen, eds.), Part A, pp. 67–117. Dekker, New York.

Kolor, M. G. (1980). *In* "Biochemical Applications of Mass Spectrometry, First Supplementary Volume" (G. R. Waller and O. C. Dermer, eds.), pp. 821–854. Wiley, New York.

Koshy, K. T., and van der Slik, A. (1979). *J. Agric. Food Chem.* **27**, 180.

Kratos Ltd. (1981). "Fast Atom Bombardment Spectra," Publ. No. A300–0981. Manchester, England.

Krueger, H. W., and Reesman, R. H. (1982). *Mass Spectrom. Rev.* **1**, 205.

Krulick, S., and Osman, S. F. (1979). *J. Agric. Food Chem.* **27**, 212.

Krusius, T., and Finne, J. (1981). *Carbohydr. Res.* **90**, 203.

Kubota, K., Matsufusa, K., and Yamanishi, T. (1980). *Nippon Nogeikagaku Kaishi* **54**, 1.

Kuehl, D. W., Leonard, E. N., Welch, K. J., and Vieth, G. D. (1980). *J. Assoc. Off. Anal. Chem.* **63**, 1238.

Kumamoto, J., Waines, J. G., Hollenberg, J. L., and Scora, R. W. (1979). *J. Agric. Food Chem.* **27**, 203.

Kurihara, T., and Kicuchi, M. (1980). *Yakugaku Zasshi* **100**, 1161.

Kuroguchi, S., Murofushi, N., Ota, Y., and Takahashi, N. (1979). *Planta* **146**, 185.

Kwan, W. O., and Kowalski, B. R. (1980). *J. Agric. Food Chem.* **28**, 356.

Lamoureux, C. J., and Feil, V. J. (1980). *J. Assoc. Off. Anal. Chem.* **63**, 1007.

Lee, M.-H. (1980). *Diss. Abstr. Int.* Order No. 8013170.

Leppanen, O. A., Denslow, J., and Ronkainen, P. P. (1980). *J. Agric. Food Chem.* **28**, 359.

Liardon, R., and Jost, R. (1981). *Int. J. Pept. Protein Res.* **18**, 500.
Liardon, R., and Ledermann, S. (1980). *J. High Res. Chromatogr. Chromatogr. Commun.* **3**, 475.
Liardon, R., Ledermann, S., and Ott, U. (1981). *J. Chromatogr.* **203**, 385.
Libbey, L. M., Scanlan, R. A., and Barbour, J. F. (1980). *Fd. Cosmet. Toxicol.* **18**, 459.
Lovegren, N. V., Fisher, G. S., Legendre, M. G., and Schuller, W. H. (1979). *J. Agric. Food Chem.* **27**, 851.
Lusby, W., and Kearney, P. C. (1978). *J. Agric. Food Chem.* **26**, 635.
McFadden, W. H. (1979). *J. Chromatogr. Sci.* **17**, 2.
McFadden, W. H. (1980). *J. Chromatogr. Sci.* **18**, 97.
MacFarlane, R. D. (1981). *Biomed. Mass Spectrom.* **8**, 449.
McLafferty, F. W. (1982a). *Trends Anal. Chem.* **1**, 298.
McLafferty, F. W., ed. (1982b). "Tandem Mass Spectrometry." Wiley, New York.
McLafferty, F. W., and Lory, E. R. (1981). *J. Chromatogr.* **203**, 109.
McLafferty, F. W., Atwater, B. L., Hakari, K. S., Hosokawa, K., Mun, I. K., and Venkataraghavan, R. (1980). *Adv. Mass Spectrom.* **8B**, 1564.
MacLeod, A. J., and Pieris, N. M. (1981a). *J. Agric. Food Chem.* **29**, 49.
MacLeod, A. J., and Pieris, N. M. (1981b). *J. Agric. Food Chem.* **29**, 488.
MacLeod, G., Seyyedain-Ardebili, M., and MacLeod, A. J. (1980). *J. Agric. Food Chem.* **28**, 441.
Marasas, W. F. O., van Rensburg, S. J., and Mirocha, C. J. (1979). *J. Agric. Food Chem.* **27**, 1108.
Markey, S. P. (1981). *Biomed. Mass Spectrom.* **8**, 426.
Marquis, L. Y., Shimabukuro, R. H., Stolzenberg, G. E., Feil, V. J., and Zaylskie, R. G. (1979). *J. Agric. Food Chem.* **27**, 1148.
Mass Spectrometry Data Centre (1974). "Eight Peak Index of Mass Spectra." UKCIS, Nottingham.
Masuda, M., and Nishimura, K. (1980). *J. Food Sci.* **45**, 396.
Matsui, S., Yabuuchi, S., and Amaha, M. (1980). *Nippon Nogeikagaku Kaishi* **54**, 747.
Meili, J., Walls, F. C., McPherron, R., and Burlingame, A. L. (1979). *J. Chromatogr. Sci.* **17**, 29.
Mestres, R., Atmawijaya, S., and Chevallier, C. (1979). *Ann. Falsif. Expert. Chim.* **72**, 577.
Metzger, J. D., and Zeevaart, J. A. D. (1980). *Plant Physiol.* **66**, 844.
Middleditch, B. S., Missler, S. R., and Hines, H. B. (1981). "Mass Spectrometry of Priority Polluants." Plenum, New York.
Mihara, S., and Shibamoto, T. (1980). *J. Agric. Food Chem.* **28**, 62.
Millard, B. J. (1978). "Quantitative Mass Spectrometry." Heyden, London.
Miller, G. J., Kunsman, J. E., and Field, R. A. (1980). *J. Food Sci.* **45**, 279.
Min, D. B. S., Ina, K., Peterson, R. J., and Chang, S. S. (1979). *J. Food Sci.* **44**, 639.
Miyagawa, M., Miwa, T. K., and Spencer, G. F. (1979). *J. Am. Oil Chem. Soc.* **56**, 834.
Moerck, K. E., and Ball, H. R., Jr. (1979). *J. Agric. Food Chem.* **27**, 514.
Morris, H. R., ed. (1981). "Soft Ionization Biological Mass Spectrometry." Heyden, London.
Moshonas, M. G., and Shaw, P. E. (1980). *J. Agric. Food Chem.* **28**, 680.
Murofushi, N., Sugimoto, M., Itoh, K., and Takahashi, N. (1980). *Agric. Biol. Chem.* **44**, 1583.
Nelson, R. R., Acree, T. E., and Butts, R. M. (1978). *J. Agric. Food Chem.* **26**, 1188.
Nishimura, H., and Calvin, M. (1979). *J. Agric. Food Chem.* **27**, 432.
Nishimura, O., Milhara, S., and Shibamoto, T. (1980). *J. Agric. Food Chem.* **28**, 39.
Nixon, L. N., Wong, E., Johnson, C. B., and Birch, E. J. (1979). *J. Agric. Food Chem.* **27**, 355.
Noble, A. C., Flath, R. A., and Forrey, R. R. (1980). *J. Agric. Food Chem.* **28**, 346.
Nunomura, N., Saskai, M., Asao, Y., and Yokotsuka, T. (1978). *Agric. Biol. Chem.* **42**, 2123.

Ogata, M., and Miyake, Y. (1980). *J. Chromatogr. Sci.* **18**, 594.

Olsen, O., and Sorensen, H. (1980). *J. Agric. Food Chem.* **28**, 43.

Otto, R., and Baltes, W. (1980). *Z. Lebensm.-Unters.-Forsch.* **171**, 286.

Paris, M., Clair, G., and Unger, J. (1979). *Riv. Ital. Essenze Profumi.* **5**, 225.

Parker, C. E., Honey, C. A., and Hass, J. R. (1982). *J. Chromatogr.* **237**, 233.

Peacock, V. E., Deinzer, M. L., McGill, L. A., and Wrolstad, R. E. (1980). *J. Agric. Food Chem.* **28**, 774.

Pensabene, J. W., Fiddler, W., Miller, A. J., and Phillips, J. G. (1980). *J. Agric. Food Chem.* **28**, 966.

Peppard, T. L., and Halsey, S. A. (1980). *J. Chromatogr.* **202**, 271.

Pereira, W. E., and Hughes, B. A. (1980). *J. Am. Water Works Assoc.* **72**, 220.

Piloty, M., and Baltes, W. (1979). *Z. Lebensm.-Unters.-Forsch.* **168**, 374.

Price, K. R. (1979). *Biomed. Mass Spectrom.* **6**, 573.

Purcell, A. E., Later, D. W., and Lee, M. L. (1980). *J. Agric. Food Chem.* **28**, 939.

Pyysalo, T., Honkanen, E., and Hirvi, T. (1979). *J. Agric. Food Chem.* **27**, 19.

Rahn, W., Meyer, H. W., and Koenig, W. A. (1979). *Z. Lebensm.-Unters.-Forsch.* **169**, 346.

Rainey, W. T., Christie, W. H., and Lijinsky, W. (1978). *Biomed. Mass Spectrom.* **5**, 395.

Rajapaske-Arambewela, L. S., and Wijesekera, R. O. B. (1979). *J. Sci. Food Agric.* **30**, 521.

Ramey, D. D., and Ough, C. S. (1980). *J. Agric. Food Chem.* **28**, 928.

Rapp, A., and Knipser, W. (1979). *Vitis* **18**, 229.

Rayner, E. T., Wadsworth, J. I., Legendre, M. G., and Dupuy, H. P. (1978). *J. Am. Oil Chem. Soc.* **55**, 454.

Robinson, J. R., and Chapman, R. A. (1980). *J. Chromatogr.* **193**, 213.

Roderbourg, H., and Kuzdzal-Savoie, S. (1979). *J. Am. Oil Chem. Soc.* **56**, 485.

Rusness, D. G., and Lamoureux, G. L. (1980). *J. Agric. Food Chem.* **28**, 1070.

Safe, S., and Hutzinger, O. (1973). "Mass Spectrometry of Pesticides and Pollutants." CRC Press, Cleveland, Ohio.

Sakaguchi, M., and Shibamoto, T. (1978). *J. Agric. Food Chem.* **26**, 1260.

Salhab, A. S., Russell, G. F., Coughlin, J. R., and Hseih, D. P. H. (1976). *J. Assoc. Off. Anal. Chem.* **59**, 1037.

Schen, J. A., Montgomery, M. W., and Libbey, L. M. (1980). *J. Food Sci.* **45**, 41.

Schenk, H. P., and Lamparsky, D. (1981). *J. Chromatogr.* **204**, 391.

Schnute, W. C. (1980). Finnigan Corp. Application Report, No. AR8018, OWA.

Schreier, P. (1980). *J. Agric. Food Chem.* **28**, 926.

Schreier, P., Drawert, F., Junker, A., and Mick, W. (1977). *Z. Lebensm.-Unters.-Forsch.* **164**, 168.

Schreier, P., Drawert, F., and Schmid, M. (1978a). *J. Sci. Food Agric.* **29**, 728.

Schreier, P., Drawert, F., and Steiger, G. (1978b). *Z. Lebensm.-Unters.-Forsch.* **167**, 16.

Schreier, P., Drawert, F., and Winkler, F. (1979). *J. Agric. Food Chem.* **27**, 365.

Schreyen, L., Dirinck, P., Sandra, P., and Schamp, N. (1979). *J. Agric. Food Chem.* **27**, 872.

Schulte, E., Hoehn, M., and Rapp, U. (1981). *Fresenius' Z. Anal. Chem.* **307**, 115.

Schulten, H. R., Bahr, U., and Goertz, W. (1982). *J. Anal. Appl. Pyrolysis* **3**, 229.

Schulten, H. R., and Schiebel, H. M. (1978). *Naturwissenschaften* **65**, 223.

Schulten, H. R., Komori, T., Nohara, T., Higuchi, R., and Kawasaki, T. (1978). *Tetrahedron* **34**, 1003.

Schulz, J. M., and Herrmann, K. (1980). *Z. Lebensm.-Unters.-Forsch.* **171**, 193.

Schuster, R. (1980). *Chromatographia* **13**, 379.

Schweizer, T. F., and Horman, I. (1981). *Carbohydr. Res.* **95**, 61.

Scott, P. M. (1978). *J. Food Prot.* **41**, 385.

Scott, P. M., Panalaks, T., Kanhere, S., and Miles, F. (1978). *J. Assoc. Off. Anal. Chem.* **61**, 593.

Seck, S., and Crouzet, J. (1981). *J. Food Sci.* **46**, 790.

Seifert, R. M., and Buttery, R. G. (1978). *J. Agric. Food Chem.* **26**, 181.

Seifert, R. M., Buttery, R. G., Lundin, R. W., Haddon, W. F., and Benson, M. (1978). *J. Agric. Food Chem.* **26**, 1173.

Seifert, W. E., Jr., McKee, R. E., Beckner, C. F., and Caprioli, R. M. (1978). *Anal. Biochem.* **88**, 149.

Self, R. (1979). *Biomed. Mass Spectrom* **6**, 361.

Self, R. (1980). *Ann. Chim. (Rome)* **70**, 15.

Selke, E., Frankel, E. N., and Neff, W. E. (1978). *Lipids,* **13**, 511.

Selke, E., Rohwedder, W. K., and Dutton, H. J. (1980). *J. Am. Oil Chem. Soc.* **57**, 25.

Sen, N. P., Donaldson, B. A., Seaman, S., and Iyengar, J. R. (1978). *IARC Sci. Publ.* No. 19, pp. 373–393.

Sen, N. P., Seaman, S., and Miles, W. F. (1979). *J. Agric. Food Chem.* **27**, 1354.

Senter, S. D., Horvat, R. J., and Forbus, W. R., Jr. (1980). *J. Food Sci.* **45**, 1380.

Shibamoto, T., and Bernhard, R. A. (1978). *J. Agric. Food Chem.* **26**, 183.

Shibamoto, T., Akiyama, T., Sakaguchi, M., Enomoto, Y., and Masuda, H. (1979). *J. Agric. Food Chem.* **27**, 1027.

Shibamoto, T., Kamiya, Y., and Mihara, S. (1981). *J. Agric. Food Chem.* **29**, 57.

Shigematsu, H., Shibata, S., Kurata, T., Kato, H., and Fujimaki, M. (1977). *Agric. Biol. Chem.* **41**, 2377.

Simoneaux, B. J., Martin, G., Cassidy, J. E., and Ryskiewich, D. P. (1980). *J. Agric. Food Chem.* **28**, 1221.

Slayback, J. R. B., and Nan, M. N. (1979). Finnigan Application Report, No. AR8020, PPINICI.

Slayback, J. R. B., and Story, M. S. (1981). *Ind. Res. Dev.* February, p. 129.

Slover, H. T., Lanza, E., and Thompson, R. H., Jr. (1980). *J. Food Sci.* **45**, 1583.

Smith, D. E. (1981). Finnigan Corporation Technical Report, No. TR8019.

Sonobe, H., Carver, R. A., and Kamps, L. R. (1980). *J. Agric. Food Chem.* **28**, 265.

Sontag, G., Schaefers, F. I., and Herrmann, K. (1980). *Z. Lebensm.-Unters.-Forsch.* **170**, 417.

Spencer, M. D., Pangborn, R. M., and Jennings, W. G. (1978). *J. Agric. Food Chem.* **26**, 725.

Sphon, J. A., Dreifuss, P. A., and Schulten, H. R. (1977). *J. Am. Oil Chem. Soc.* **60**, 73.

Spinelli-Gugger, A. M., Lakritz, L., and Wassermann, A. E. (1980). *J. Agric. Food Chem.* **28**, 424.

Spinelli-Gugger, A. M., Lakritz, L., Wassermann, A. E., and Gates, R. A. (1981). *J. Food Sci.* **46**, 1136.

Spingarn, N. E., and Garvie, C. T. (1979). *J. Agric. Food Chem.* **27**, 1319.

Sponsel, V. M., Gaskin, P., and MacMillan, J. (1979). *Planta* **146**, 101.

Stackler, B., and Ough, C. S. (1979). *Am. J. Enol. Vitic.* **30**, 117.

Stafford, A. E., and Black, D. R. (1978). *J. Agric. Food Chem.* **26**, 1442.

Stan, H. J. (1977). *Z. Lebensm.-Unters.-Forsch.* **164**, 153.

Stan, H. J., and Abraham, B. (1980). *J. Chromatogr.* **195**, 231.

Stan, H. J., and Scheutwinkel-Reich, M. (1980). *Lipids* **15**, 1044.

Stavric, B., Klassen, R., and Miles, W. (1979). *J. Assoc. Off. Anal. Chem.* **62**, 1020.

Steiner, U., Solokow, S., and Slayback, J. R. B. (1980). Finnigan Topic, No. FT8004, TSQ.

Strating, J., and Westra, W. M. (1982). *J. Chromatogr.* **244**, 159.

Strocchi, A., and Lercker, G.(1979). *J. Am. Oil Chem. Soc.* **56**, 616.

Suhre, F. B., Simpson, R. M., and Shafer, J. W. (1981). *J. Agric. Food Chem.* **29**, 727.

Suyama, K., and Adachi, S. (1980). *J. Agric. Food Chem.* **28**, 546.

Swoboda, P. A. T., and Peers, K. E. (1978). *J. Sci. Food Agric.* **29**, 803.

Szafranek, J., and Wisniewski, A. (1978). *J. Chromatogr.* **161**, 213.

Tahara, S., and Mizutani, J. (1979). *Agric. Biol. Chem.* **43**, 2021.

Takayasu, K. (1978). *J. Agric. Food Chem.* **26**, 1194.

Takei, Y. (1977). *Agric. Biol. Chem.* **41**, 2361.

Tanaka, A., Nose, N., and Watanabe, A. (1980). *J. Chromatogr.* **194**, 21.

Taylor, K. T., Hazelby, D., and Wakefield, C. J. (1982). *Proc. Int. Symp. Mass Spectrom. 9th, 1982* (in press).

Teshima, S. I., Patterson, G. W. and Dutky, S. R. (1980). *Lipids* **15**, 1004.

Thomas, Q. V., Stork, J. R., and Lammert, S. L. (1980). *J. Chromatogr. Sci.* **18**, 583.

Thomson, J. J. (1913). "Rays of Positive Electricity and Their Application to Chemical Analysis." Longmans, London.

Tohma, M., Nakata, Y., and Kurosawa, T. (1979). *J. Chromatogr.* **171**, 469.

Traitler, H., Lorbeer, E., and Kratzl, K. (1980). *J. Am. Oil Chem. Soc.* **57**, 335.

Tressl, R., and Friese, L. (1978). *Z. Lebensm.-Unters.-Forsch.* **166**, 350.

Tressl, R., Grunewald, K. G., Koeppler, H., and Silwar, R. (1978a) *Z. Lebensm.-Unters.-Forsch.* **167**, 108.

Tressl, R., Daoud, B., Koeppler, H., and Jensen, A. (1978b). *Z. Lebensm.-Unters.-Forsch.* **167**, 111.

Tressl, R., Lothar, F., Fendesack, F., and Koeppler, H. (1978c). *J. Agric. Food Chem.* **26**, 1426.

Trussel, A. R., Lien, F. Y., and Moncur, J. G. (1980). *Identif. Anal. Org. Pollut. Water [Chem. Congr. North Am. Cont.] 1980.*

Tsugita, T., Kurata, T., and Fujimaka, M. (1978). *Agric. Biol. Chem.* **42**, 643.

Tsugita, T., Kurata, T., and Hiromichi, K. (1980). *Agric. Biol. Chem.* **44**, 835.

Tsuneya, T., Ishihara, M., Shiota, H., and Shiga, M. (1980). *Agric. Biol. Chem.* **44**, 957.

Turk, A., Smock, R. M., and Taylor, T. I. (1951). *Food Technol. (Chicago)* **5**, 58.

Tway, P. C., Wood, J. S., Jr. and Downing, G. V. (1979). *J. Agric. Food Chem.* **27**, 753.

Vajdi, M., Nawar, W. W., and Merritt, C., Jr. (1979). *J. Am. Oil Chem. Soc.* **56**, 611.

Van Esch, G. J. (1978). *In* "Aquatic Pollutants: Transformation and Biological Effects" (O. Hutzinger, I. H. van Lelyveld, and B. C. J. Zoeteman, eds.), Pergamon, Oxford.

Van Stratten, S., de Vrijer, F., and de Beauveser, J. C., eds. (1981). "Volatile Compounds in Foods: Cumulative Mass Spectral Collection." Division for Nutrition and Food Research, TNO, Zeist, The Netherlands.

Veierov, D., and Aharonson, N. (1980). *J. Assoc. Off. Anal. Chem.* **63**, 202.

Vetter, W., and Meister, W. (1981). *Org. Mass Spectrom.* **16**, 118.

VG Analytical Ltd. (1981). "Fast Atom Bombardment," Application Note No. 6.

Viani, R., and Horman, I. (1974). *J. Food Sci.* **39**, 1216.

Voykanee, R. D., Hass, J. R., and Bursey, M. (1982). *Anal. Lett.* **15**, 1.

Wakefield, C. J., Chapman, J. R., Evans, S., Bradford, J. H., and Done, D. (1980). *Br. Mass Spectrom. Meet. 1980.*

Warburton, G. A., Chapman, J. R., and Taylor, K. T. (1981). Kratos Data Sheet, No. 133.

Watson, J. T., and Biemann, K. (1964). *Anal. Chem.* **36**, 1135.

White, R. L., Ledford, E. B., Ghaderi, S., Wilkins, C. L., and Gross, M. L. (1980). *Anal. Chem.* **52**, 1525.

Wilkes, P. S. (1981). *J. Assoc. Off. Anal. Chem.* **64**, 1208.

Wilkins, C. L., and Gross, M. L. (1981). *Anal. Chem.* **53**, 1661A.

Williams, A. A., and Tucknott, O. G. (1978). *J. Sci. Food Agric.* **29**, 381.

Williams, A. A., Lewis, M. J., and Tucknott, O. G. (1980). *Food Chem.* **6**, 139.

Williams, P. J., Strauss, C. R., and Wilson, B. (1980). *J. Agric. Food Chem.* **28**, 766.

Williams, R. S., and Wagner, H. (1978). *J. Am. Soc. Brew. Chem.* **36**, 27.

Winkler, F. J., and Schmidt, H. L. (1980). *Z. Lebensm.-Unters.-Forsch.* **171**, 85.

Winkler, V. W., Strong, J. M., and Finley, R. A. (1977). *Steroids* **29**, 739.

Wise, S. A., Chesler, S. N., Guenther, F. R., Hertz, H. S., Hilpert, L. R., May, W. E., and Parris, R. M. (1980). *Anal. Chem.* **52**, 1828.

Withycombe, D. A., Lindsay, R. C., and Stuiber, D. A. (1978). *J. Agric. Food Chem.* **26**, 816.

Wolfram, J. H., Feinberg, J. I., Doerr, R. C., and Fiddler, W. (1977). *J. Chromatogr.* **132**, 37.

Wulf, L. W., and Nagel, C. W. (1980). *J. Food Sci.* **45**, 479.

Yabumoto, K., Jennings, W. G., and Yamaguchi, M. (1977). *J. Food Sci.* **42**, 32.

Yajima, I., Yanai, T., Nakamura, M., Sakakibara, H., and Habu, T. (1978). *Agric. Biol. Chem.* **42**, 1229.

Yajima, I., Yanai, T., Nakamura, M., Sakakibara, H., and Habu, T. (1979). *Agric. Biol. Chem.* **43**, 2425.

Yamaguchi, K., and Shibamoto, T. (1979). *J. Agric. Food Chem.* **27**, 847.

Yamaguchi, K., and Shibamoto, T. (1980). *J. Agric. Food Chem.* **28**, 82.

Yamamoto, S., Wakabayashi, S., and Makita, M. (1980). *J. Agric. Food Chem.* **28**, 790.

Yamanishi, T., Kosuge, M., Tokitomo, Y., and Maeda, R. (1980a). *Agric. Biol. Chem.* **44**, 2139.

Yamanishi, T., Fukawa, S., and Takai, Y. (1980b). *Nippon Nogeikagaka Kaishi* **54**, 21.

Yamashita, I., Nemoto, Y., and Yoshikawa, S. (1976). *Phytochemistry* **15**, 1633.

Yamashita, I., Iino, K., Nemoto, Y., and Yoshikawa, S. (1977). *J. Agric. Food Chem.* **25**, 1165.

Yorke, D., and Crossley, N. (undated). VG Organic Application Note No. 3.

Yoshida, Y., Yoshida, H., Tsuge, S., Takeuchi, T., and Mochizuki, K. (1980). *J. High Res. Chromatogr. Chromatogr. Commun.* **3**, 16.

Zabik, M. E., and Zabik, M. J. (1980). *Bull. Environ. Contam. Toxicol.* **24**, 344.

# 7

# NMR Spectroscopy

IAN HORMAN

*Nestlé Research Laboratories*
*La Tour de Peilz, Switzerland*

ANALYSIS OF FOODS AND BEVERAGES
Copyright © 1984 by Academic Press, Inc.
All rights of reproduction in any form reserved.
ISBN 0-12-169160-8

## I. INTRODUCTION

The 1952 Nobel Prize for physics was awarded to Felix Bloch of Harvard (1946; Bloch *et al.*, 1946) and to Edward Purcell of Stanford (1946) for their discovery of nuclear magnetic resonance (NMR) in solids. Today, some 30 years later, the gurus of pure NMR are still physicists who plumb the mysteries of angular momentum, vector moments, precession frequencies, and rotating frames to develop new instruments and new ways of using them. Fortunately for us lesser mortals who can measure in millieinsteins our ability to understand these intricacies, modern commercial spectrometers are relatively easy to use. Thus, with some general knowledge of the basic principles of NMR, chemists, biochemists, and biologists can now perform and interpret the most complex experiments with relative ease. The high sensitivity offered by present-day instruments makes NMR more and more attractive even in studies of multicomponent systems such as food or food extracts. The continuing increase in the number of food-related applications reported every year testifies to this. The present chapter briefly presents aspects of low-resolution NMR and treats more comprehensively high-resolution applications, for which no review exists in the literature to date.

### Theoretical Considerations

Nuclear magnetic resonance spectroscopy involves measuring the absorption of radio-frequency radiation by nuclei in a sample placed in a magnetic field. At a fixed field, different nuclei absorb different frequencies of radiation depending on their atomic characteristics and their position relative to other nuclei in the molecular structure or sample matrix. For groups of nuclei of the same kind, protons for example, the minor differences in frequencies of resonance observed reflect differences in the local charge density at the various atomic sites within the molecule. Because charge density is distributed in three-dimensional space, the information obtained relates to molecular shape, solvation, hydrogen bonding, and orientation relative to other molecules in solution. Alternatively, the frequency of the radiation to be absorbed, instead of the magnetic field, may be fixed: under these conditions, the various nuclei will resonate at different magnet field strengths. A spectrum may therefore be recorded at fixed field by scanning the frequency range, or at a fixed frequency by systematically varying the magnet field strength. Both principles have been applied in commercial spectrometers.

A nucleus, once excited by absorption of energy, must get rid of this excess energy and return to the ground state before the absorption process can be repeated. This dissipation of energy is known as relaxation. If it is prevented, the nuclei become saturated, no further energy is absorbed, and the corresponding signals will not be observed. Relaxation occurs by transferring energy either through space to other nuclei in what are known as spin–spin interactions or into the matrix of the sample, where it is converted into other forms such as kinetic energy or heat in so-called spin–lattice interactions. The rate of relaxation is not the same for all nuclei; it depends largely on the motional freedom of the individual atoms. This in turn depends on sample viscosity, concentration, molecular size, nature of solvent, and other factors. Relaxation rates are quantified in terms of first-order time constants $T_1$ and $T_2$ corresponding to spin–lattice and spin–spin interactions, respectively. Values of $T_1$ for different nuclei in the same molecule can vary from microseconds to minutes and can be exploited in studies of intramolecular mobility [e.g., Wehrli and Wirthlin (1976)].

Quantitative measurements use the fact that an NMR signal area is proportional to the number of nuclei giving rise to that signal: saturation must be minimized to avoid differential signal suppression. However, in qualitative studies involving sample identification, the feature of specific signal suppression can be used to advantage to simplify spectra. Examples of this are found in conventional proton-decoupled $^{13}C$ spectra and in DEPT spectra where $CH_3$, $CH_2$, and CH carbon signals may be selectively displayed and unambiguously recognized even in the most complicated $^{13}C$ spectrum.

Unlike the situation in mass spectrometry (MS), only two basic types of NMR instruments exist, namely low and high resolution. These are not, as might be inferred from their names, inferior and superior versions of the same but two distinct techniques. Low-resolution NMR is used to investigate the bulk properties of samples that are mixtures of liquid- and solid-phase material: effectively, the protons present in moisture and in molten fat have a higher degree of mobility than those in the solid phase, a difference that allows measurement of the proportions of liquid and solid present. High-resolution NMR permits more specific structural, kinetic, or equilibrium studies of individual compounds in the pure state or in mixtures. High- and low-resolution instruments can be made to work either at fixed frequency (i.e., continuous wave, or CW) conditions or at fixed field (i.e., pulsed Fourier transform, or PFT — or, more simply, FT) conditions. To avoid confusion, the two techniques are described separately in the succeeding pages.

More detailed information on the fundamental aspects and underlying theory of NMR is available in books, starting with the now classical opus of Pople *et al.* (1959). Several texts exist presenting the basic data needed to develop a working knowledge of spectral interpretation while minimizing the theoretical aspects, such as that of Williams and Fleming (1966). More recent texts on practical NMR include those of Wehrli and Wirthlin (1976) and Martin *et al.* (1980).

## II. LOW-RESOLUTION OR WIDE-LINE NMR

Bloch *et al.* (1946) and Purcell *et al.* (1946) furnished the first reports of bulk NMR of protons in wax and in water using instruments that were forerunners of present-day low-resolution spectrometers. Routinely usable CW spectrometers were developed by Richards and Smith (1951) and by Pratt and Richards (1953), and it was on an instrument such as these that the first food-related results were recorded by Chapman *et al.* (1959). These authors showed that the ability of NMR to distinguish between protons in liquid (and therefore mobile) fat and in solid (and therefore immobile) fat according to the line width of their respective signals permitted the determination of liquid–solid fat ratios in lipid mixtures. Effectively, the signal from solid fat protons was so broad that it was indistinguishable from the baseline, and only the sharper signals from protons in the mobile liquid phase were observed. The peak-to-peak height of the recorded derivative curve, compared to that for a whole liquid sample, gave a direct measure of the amount of liquid present with an apparent 2% error, much improved relative to the then existing dilatation methods. The principal advantages of this new method were threefold: samples could be examined in the state in which they were received, no prior knowledge of the triglycerides present was needed, and results were independent of the polymorphic form of the solids present. Shortly afterward, Chapman *et al.* (1960) reported NMR studies relating margarine consistency to liquid–solid fat contents. Wide-line NMR also found uses in the observation of the water signal to measure the rate of solution of dried milk powders (Samuelsson and Hueg, 1973) and, rather less conventionally, in the detection of red flour beetles and rice weevils in infested wheat (Anonymous, 1971). A comparison of direct and indirect wide-line methods for measuring solid fat content has recently been reported by Mills and Van de Voort (1981).

The progress achieved with CW wide-line instruments was considerable, but some limitations were apparent, and new instrumentation was explored. Thus, Van Putte and Van den Enden (1973) developed methods based on pulsed NMR. Their spectrometers operated at higher frequencies, typically 20 MHz instead of the 2–3 MHz of the earlier wide-line versions, and gave better definition. This allowed direct observation of protons in the solid phase as well as in liquids. Disadvantages of CW operations, such as long analysis times and influence of sample viscosity on signal intensity, that were prone to introduce inaccuracies were largely overcome. Further, the method did not require total melting of the sample and could be automated. As with the earlier wide-line instruments, liquid–solid fat ratios (solid fat indices) could be determined either indirectly by comparing the signal from the liquid only with that of a 100% liquid sample (Van Putte and Van den Enden, 1974) or directly by processing together the signals from both liquid and solid fat (Van Putte, 1975). These signals are clearly distinguished, because protons from solid fat have $T_2$ values around $10\,\mu sec$ and those from liquid fat around 100 msec. Desarzens et al. (1978) have described the use of this method to determine the temperature dependence of solid fat indices (melting curves), showing that NMR gives more complete information than dilatometry or calorimetry, through its capacity to detect liquid fat at low levels in the temperature range of $0-20°$.

Pulsed NMR opened the possibilities for many other experiments. Free liquid water in food materials could be distinguished from water that was bound, for example as water of crystallization, and the two types of water could be independently quantified. A comprehensive account of water determination by NMR methods and comparison with results from sorption isotherms and calorimetry has been presented by Bouldoires (1975). More recently, Weisser (1979) has reviewed NMR techniques for studying bound water in foods, and a comparative discussion of the possibilities and limitations of wide-line and pulsed NMR for studying water in starches and vegetables has been presented by Lechert et al. (1980). Lillford et al. (1980) have described the use of NMR techniques in relating the quality of codfish tissue to the state of water present, observing that structural reorganization occurring in the tissue during frozen storage persists into the cooked state.

An important advantage of pulsed NMR is its ability to observe spin–lattice relaxation, $T_1$, as well as spin–spin relaxation, $T_2$, observed in CW wide-line NMR. This permits selective determination with minimum interference of different mobile components, such as water and fat present in the same product (i.e., in products such as milk, meat, cheese, and mayonnaise). Currie et al. (1981) have described the use of $T_1$ measurements to investigate changes of muscle water in beef as a function of postslaughter time, Brosio has determined oil and water in olive husks (1981) and in emulsions (1982), and Trumbetas et al. (1978) have studied emulsion stability.

Total fats present in a food product can be readily analyzed at a temperature of $40°C$, which is sufficient to melt the fats but leaves the product's other major constituents, namely proteins, carbohydrates, and polyphenols, in their solid form. Even water, at the level of about 4% typically found in dehydrated products, is largely bound and does not interfere. An almost perfect correlation ($r^2 = 0.997$) is reported for fat contents in the range 1–30% between values derived from NMR and from standard solvent extraction methods (Lambelet, 1980). Desarzens and Michel (1982) have determined total fats in cocoa products using these methods, emphasizing the point that values obtained by pulsed NMR, unlike those from CW methods, are insensitive to metallic contamination. Oil contents of whole oilseeds are also readily estimated, as described by Daun and Biely (1980), who carried out a survey of the variation over a 1-year period in the oil content of shipments of rapeseed exported from Canada. A further extension of the possibilities of pulsed NMR involves the estimation of protein. Coles (1980) has presented a method based on changes in the relaxation rates of copper-based relaxation reagents, which are found to vary as a function of protein concentration. Effectively, the protein binds the copper, thus reducing its capacity to promote relaxation. The method was tested on animal and vegetable proteins, and interfering effects of carbohydrates were assessed.

The brevity of this section on low-resolution NMR compared to the following sections describing high-resolution applications should not be taken as indicative of the relative importance of the two techniques. In fact, the literature of the past decade shows that more than half of the NMR applications in food analysis are low resolution. However, most of these studies are described in detail in a recent review article by Coveney (1980) and in a literature survey by Weisser (1978). Also, Brennan (1982) has described how NMR can be used in food laboratories.

## III. HIGH-RESOLUTION NMR

### A. Historical Aspects

In 1951, Arnold *et al.* showed that the single peak observed with a wide-line instrument for the protons of ethanol could be separated by using a magnet with a more stable field into three distinct spectral lines for the methyl, methylene, and hydroxyl protons. With this discovery, high-resolution NMR was born. Organic chemists were quick to appreciate applications to their problems, and books by Jackman and by Roberts, both in 1959, testified that high-resolution NMR in organic chemistry was here to stay.

Almost exclusively, the early work involved proton NMR for two principal reasons. First, protons were the most abundant nuclei present in organic compounds, and second, as seen in Table 1, the signal intensity observed per unit quantity of nuclei was much higher for protons than for any other of the common nuclei. This was particularly the case when natural abundances were taken into account. Spectra were frequently run on pure samples, as

**TABLE 1    Nuclei of Interest in Food Analysis**

|  |  | Relative sensitivity[a,b] | |
| --- | --- | --- | --- |
| Isotope | Natural abundance (%) | Equal numbers of nuclei | At natural abundance levels |
| $^1H$ | 99.9844 | 1 | 1 |
| $^2H$ | 0.0156 | 104 | $6.66 \times 10^5$ |
| $^{13}C$ | 1.108 | 63 | $5.69 \times 10^3$ |
| $^{14}N$ | 99.635 | 990 | $9.93 \times 10^2$ |
| $^{15}N$ | 0.365 | 960 | $2.63 \times 10^5$ |
| $^{17}O$ | 0.037 | 34 | $9.19 \times 10^4$ |
| $^{19}F$ | 100.0 | 1.2 | 1.2 |
| $^{31}P$ | 100.0 | 15 | 15.0 |

[a] Calculated from data recorded by Varian Associates, Palo Alto, California.
[b] Number of times less sensitive than $^1H$ nucleus. Measured at constant field.

exemplified in Fig. 1a, the 40-MHz spectrum of eugenol, which is a major volatile constituent of clove oil. This approach worked for liquids, but solids had to be dissolved. Problems frequently arose because protons in the solvent molecules also absorbed and, in all but the most favorable cases, masked part of the spectrum. When solubility permitted, such difficulties could be overcome by using aprotic solvents such as carbon tetrachloride. Later, many perdeuterated solvents became readily available, and such problems were largely eliminated. Studies relating spectral fine structure to molecular shape were followed by observations of solvent dependency of peak positions, concentration effects, proton exchange, hydrogen bonding, tautomerism, and equilibria.

By the mid-1960s, in spite of these impressive advances in the under-

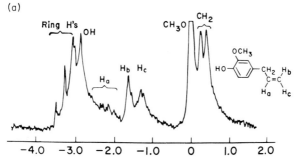

**Fig. 1.**    Four generations of NMR: proton spectra of eugenol at (a) 40 MHz, (b) 60 MHz, (c) 80 MHz, and (d) 400 MHz. (See pp. 212–214.)

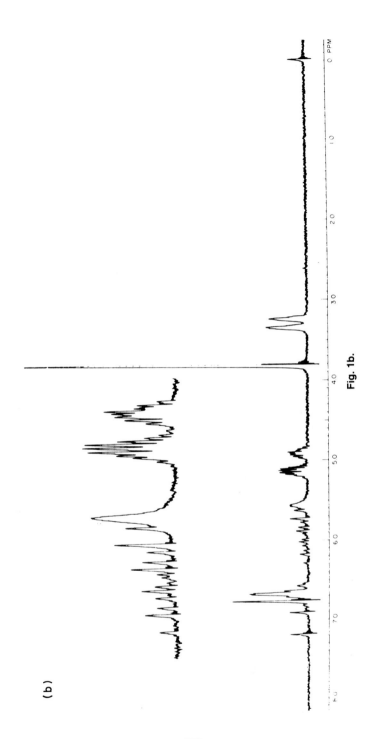

(b)

Fig. 1b.

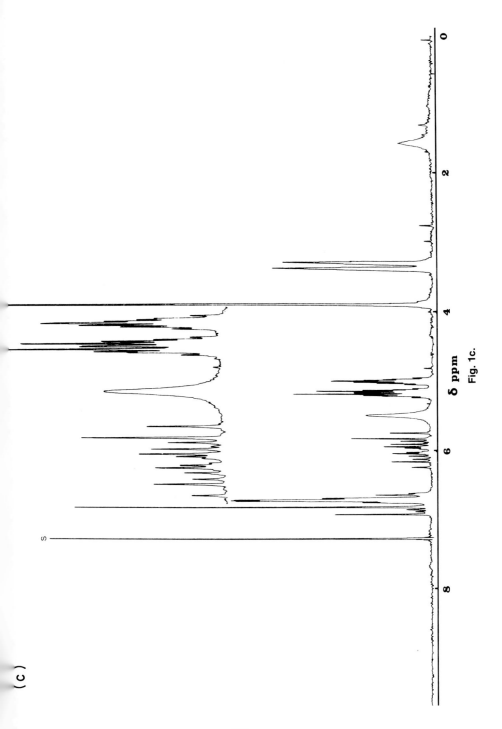

δ ppm

Fig. 1c.

(c)

(d)

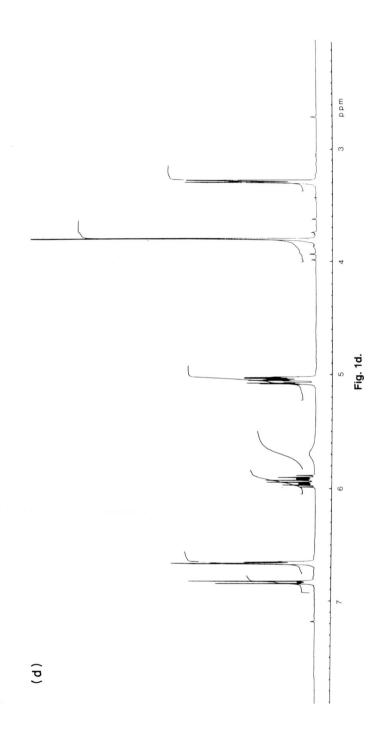

**Fig. 1d.**

standing of NMR spectra and their origins, the technique was still of limited use to food analysts. The instruments in widespread use at this time operated at 50–60 MHz instead of the 20–40 MHz of the previous decade and gave spectra such as that of Fig. 1b. However, resolution was still hardly adequate to permit direct analysis of complex mixtures, except when characteristic peaks were well separated from the bulk of the spectrum. A favorable example is that of Fig. 6, showing that trigonelline levels could be determined in coffee infusions as a function of degree of roasting (Horman and Viani, 1975). In addition to resolution, sensitivity was also a problem. As a general rule 10–50 mg of sample was needed to obtain good single-scan spectra like that of Fig. 1b with all signals clearly distinguishable above baseline noise. If such quantities presented few problems for organic chemists engaged in synthetic or mechanistic studies, this was certainly not the case for food analysts. For these, the isolation of such amounts of all but the most abundant food constituents represented — and still represents — a formidable task. Microcell methods offered some improvement in sensitivity, as did spectrum accumulation time-averaging devices, which gave results such as those in Fig. 2. However, it was not until the development of pulsed FT high-resolution instruments that spectacular advances in sensitivity were achieved. Samples could then be run at much lower concentrations, which minimized line broadening from solute–solute interactions such as dimerization and thus automatically enhanced resolution as well. This is clearly demonstrated by the comparison between the 60-MHz CW spectrum of eugenol in Fig. 1b and the 80-MHz FT spectrum in Fig. 1c, run at concentrations of about 7% and about 0.3%, respectively.

An approximate idea of NMR sensitivity routinely available over the past 25 years can be obtained from Fig. 3. The amount of sample needed to obtain an "acceptable" proton spectrum is also indicated. Progress is quantized, and therefore the joining of successive points in Fig. 3 to give a smooth curve is justified only in terms of artistic license. However, these curves highlight the effects of two highly significant developments in NMR technology that together open inestimable possibilities of application in food analysis. These are, first, the Fourier transform operations made possible by the integration into spectrometers of mini- and microcomputers and, second, the widespread introduction of superconducting high-field magnets.

Prior to pulsed FT instruments, the only common nucleus in addition to protons that could be readily observed was fluorine, of limited direct interest in foods. The other nuclei in Table 1 could be observed in isotopically enriched compounds, but only with great difficulty at natural levels. Computers and FT operation made possible observations on all of these nuclei in normal unenriched samples, stimulating interest in a field that up to the mid-1970s had been essentially inaccessible. The reason for this spectacular

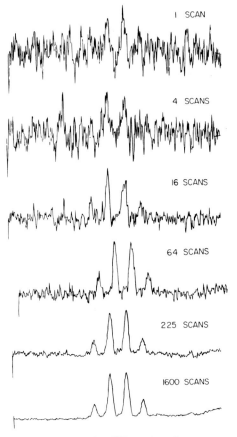

**Fig. 2.** Signal-to-noise enhancement in a $^{13}C$ spectrum by scan accumulation. [Reprinted with permission from Bovey (1969).]

breakthrough, which also reduced sample sizes required for proton spectra from milligram amounts to tens of micrograms, lies in the fundamental operating differences between CW and pulsed NMR spectroscopy. On CW instruments, spectra such as that of Fig. 1b were recorded typically in 5 to 10 min during a single sweep of the magnetic field. With pulsed NMR, typical single-scan spectra are obtained in 0.5 to 5 sec, and several such single scans are added to give the final spectrum. Figure 1c is the result of accumulating 150 scans of 2 sec each and was thus also recorded in 5 min. Relative to the limited success of scan accumulation on CW instruments, in which deterioration of resolution and peak shape was a problem, accumulation on pulsed FT instruments could be carried out over several hours or even days

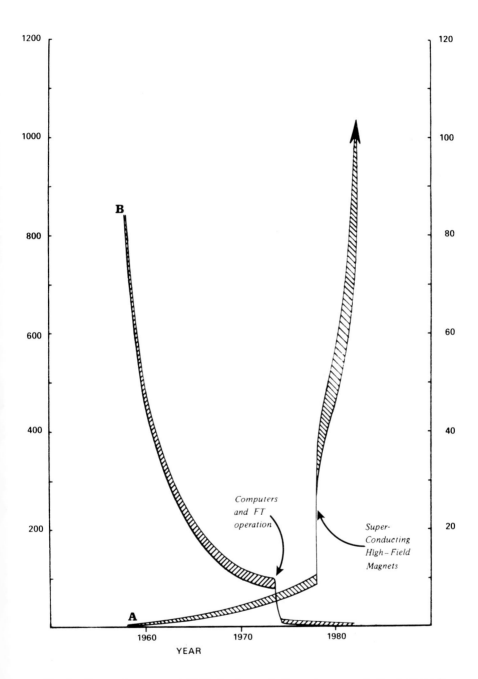

**Fig. 3.** Twenty-five years of NMR development. Curve A represents the increase in sensitivity for proton measurement expressed as signal-to-noise ratio (left vertical axis) for a single scan; curve B represents the decrease in the amount of sample (right vertical axis, in milligrams) needed to record a proton spectrum.

to give spectra of good quality: this meant that $^{13}C$ spectra could then be obtained on as little as 5 mg of cholesterol by accumulating about 5000 scans in 1 hr. A more spectacular example was the spectrum recorded from 500 µg of brucine by accumulating some 300,000 scans over a weekend, as shown in Fig. 4.

The ultimate stage in instrumental development has been the introduction of superconducting high-field magnets. The advantage of high fields is to be seen in Fig. 1d, which shows the 400-MHz proton spectrum of eugenol. Each proton now gives a distinctly separate group of signals, the shape of which is determined by coupling interactions with other protons in the molecule, and the whole spectrum is easier to interpret. The advantages of higher-field magnets were realized almost from the early days when the first commercial 30-MHz instruments were upgraded to 40 MHz in 1955. Johnson (1971) published several spectra at different fields up to 220 MHz, emphasizing the inherent gain in sensitivity and spectral simplification achieved. However, it was not until the late 1970s that high-field instrumentation became readily available. The impact of these super high-resolution spectrometers in food analysis can only be assessed in 10 years' time, but it is already clear from the preceding discussion that NMR now has enough resolution and sensitivity to be capable of analyzing complex mixtures having constituents at levels typical of those isolated from food materials.

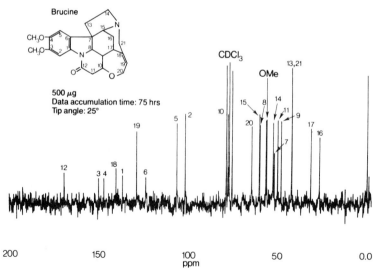

**Fig. 4.** A $^{13}C$ spectrum run on only 500 µg of brucine by data accumulation over a weekend. (Reprinted by permission of Varian Associates, Zug, Switzerland.)

## B. The Situation Today

A survey of the literature available to date on high-resolution NMR applications in foods and food-related areas reveals that in 42% of cases measurements have been made at 60 MHz or less, with 51% at 80–100 MHz and 7% at 200 MHz or higher. About 80% of examples have involved proton NMR, 18% $^{13}$C-NMR, and the remaining 2% other nuclei such as phosphorus. About 50% of applications are qualitative, dealing with identification of food constituents after isolation or with general observations of food systems. Of the 50% that are quantitative applications, some 10% are concerned with kinetics, equilibria, or other mechanistic studies. If a single subject has received more attention than any other, it must be edible oils and fats, accounting alone for most of the $^{13}$C applications.

Few, if any, of these applications use or demonstrate the full capacities of the modern ultrahigh-resolution NMR spectrometer in food analysis. The reason for this appears simple. Many innovations have been introduced within the past 2 years, some within only the past few months. Such a case is the MLEV-16 low-energy proton-decoupling method (Levitt *et al.*, 1982), which, as shown by Fig. 13, gives improved noise levels and peak shapes for $^{13}$C spectra. However, there is a more fundamental problem to be considered, namely conservatism, both in the approach to food analysis and in the subjects considered "suitable" as fields of application of high-powered instrumental techniques. This is partly a problem of philosophy or awareness, but too frequently in food analysis laboratories, where these techniques would be highly beneficial, they are suspected and rejected as unnecessarily sophisticated and costly research tools, while in most "pure science" laboratories where this instrumentation exists very few workers would ever consider using it to analyze foods. The result of this conflict of dogmas, where the universal loser is the discipline of food science, is that most applications of NMR to foods lag 5 to 10 years behind current instrumental possibilities. Section V therefore describes new applications of high-resolution NMR and hints at possible fields of utility in food analysis, this to accelerate where possible the passage of these techniques into the food analyst's armory.

The advances in today's NMR spectrometer are not only to be measured in terms of improved resolution and sensitivity, as depicted in Fig. 1d, but also in new methods and techniques. A wide range of nuclei can now be readily observed, as can the spectra of solids through the so-called magic angle operation. New pulse sequences permit simplification of the most complex spectra, and two-dimensional NMR (2D-NMR) not only facilitates interpretation of spectral fine structure but can now through carbon connectivity experiments directly depict the carbon skeleton of a compound using less sample than the amount required 30 years ago to run a simple proton spectrum such as that in Fig. 1a.

Before passing on to practical applications, a word on instrument sensitivity is in order. In principle, the intensity of a signal for a given nucleus is proportional to the square of the field or of the frequency at which it is observed. Thus, moving from the earliest commercialized instruments at 30 MHz to a present-day instrument of 300 MHz, we should theoretically observe a hundredfold increase in sensitivity for a single scan. Advances in electronics have increased this figure by a factor of 5 or more. Computerized data acquisition and processing further increase sensitivity limits almost to our own limits of patience and time in continuing to accumulate scans.

However, there is also a danger of which we must be aware: sensitivity increases only as the square root of the time spent in acquiring data. It is worth reflecting that, assuming the same analysis were possible at any two different frequencies (and hence two different magnet strengths), an analysis requiring a day at 60 MHz would take an hour or less at 300 MHz, and an analysis that would now take a day at 500 MHz would have taken almost a year at 30 MHz. Realistic data acquisition times are at most about 14 hr, corresponding to an overnight run, or about 60 hr, corresponding to a weekend run. For small amounts of sample, or for detailed spectra of complex mixtures, high-field operation is therefore more economic and efficient than lower fields. For routine work with larger amounts of sample, perhaps a lower-field instrument would do the job just as well as the higher-field instrument, and it would be more economical because of lower investment costs in instrumentation. The message here is that the optimum instrument is a function of the type of sample to be analyzed and the type of analysis required.

## IV. APPLICATIONS OF HIGH-RESOLUTION NMR

Applications can be broadly classified as qualitative or quantitative. For the sake of convenience these classes are here divided into subgroups treating general qualitative and quantitative aspects, cameo profiles of NMR applications in analyzing oils and fats, deuterium NMR studies, kinetics and equilibria, and finally new applications, including 2D-NMR.

### A. Qualitative Applications

#### 1. Volatiles

Deyama and Horiguchi (1971) have used NMR along with other instrumental techniques to identify volatile aroma constituents of cloves. The major constituent, accounting for 80% of the volatiles extracted, is eugenol.

This is evident from the comparison of the NMR spectrum of whole clove oil shown in Fig. 5 with the spectra of Fig. 1. Other compounds identified included $\beta$-caryophyllene, acetyl eugenol, methyl salicylate, $\alpha$- and $\beta$-humulene, $\alpha$-ylangene, and chavicol. Similarly, Maarse and van Os (1973) report the identification of 49 components in the volatile essential oil of oregano; Kim et al. (1973) have described smoked food flavors, reporting 98 smoky aroma constituents including alcohols, ketones, acids, furans, lactones, phenols, phenol ethers, and pyrocatechols; and Viani et al. (1969) have identified 46 components from tomato aroma.

Many other examples of the use of NMR in conjunction with mass and infrared spectroscopy to identify volatile substances are to be found in the literature around 1970. At that time, NMR could be used only for the major constituents, such as the case of eugenol in clove volatiles, and was of little practical value for the hundreds of minor volatiles that could be isolated only in submicrogram quantities. This type of work became the realm of GC–MS, as described in Chapter 6, Section II,C. However, NMR is much more structure specific than MS, and Flament et al. (1977) used this property to identify seven alkyl pyrrolo [1,2-$\alpha$] pyrazines (I) as flavor constituents of roast beef when MS alone had proved insufficient to distinguish these compounds from benzimidazoles, indazoles, cyclopenta[b]pyrazines, and dihydroquinoxalines.

I

Possibly, with the increased sensitivity of modern instruments, this kind of study may once more be undertaken by NMR for any compounds that can be isolated pure in quantities of a few micrograms. Such studies would undoubtedly be of value for clarifying doubtful or tentative identifications in the literature to date based on MS alone. More recently in this field, Formacek and Kubeczka (1982a) have published a study of the $^{13}$C spectra of a collection of essential oils. An example of the $^{13}$C spectrum of fennel oil is shown in Fig. 11.

Positive identification of volatiles frequently depends on the synthesis of reference compounds. Proton NMR spectra are reported for a series of alkylpyrazines (Kitamura and Shibamoto, 1981) and for methoxymethylpyrazines (Nakel and Haynes, 1972). Carbon-13 spectra were used to characterize and identify $\gamma$- and $\delta$-lactones (Pyysalo et al., 1975; Pyysalo and Enqvist, 1975).

80 MHz $^1$H
WHOLE CLOVE OIL

δ ppm

An mixture of the essential oil from cloves.

## 2. Polyphenols

The polyphenols of tea consist of a complex mixture of theaflavins and thearubigins. Here NMR has been used to advantage in identifying new compounds such as isotheaflavine (Ollis *et al.,* 1970) and three other flavanotropolones related to theaflavine (Takino *et al.,* 1967). Structures of bisflavonols proposed earlier were confirmed (Ferreti *et al.,* 1968), and theogallin was identified as the depside 3-galloylquinic acid (Stagg and Swaine, 1971). Other depsides are also described, for example from coffee, where the proton spectra of chlorogenic acids (caffeoylquinic acids) have been published by Corse *et al.* (1966), and from lettuce, where dicaffeoyl tartaric acid was identified as the principal polyphenol (Feucht *et al.,* 1971).

Ferulates are reported in rice-bran oil as their esters with campestrol, $\beta$-sitosterol, cycloartenol, and 24-methylenecycloartenol (Endo *et al.,* 1969), and the structures and formation of dimeric procyanidins from different fruits have been presented (Weinges, 1971), as have the structures of polyphenolic and terpenoid constituents of grapes (Piretti, 1975).

A study on thiamine stability in aqueous solution using proton NMR has settled a long-standing controversy in the literature on vitamins by showing that the thiamine structure remains unaltered in the presence of *ortho*-diphenols (Horman *et al.,* 1981). The bridge methylene group of thiamine [position d in structure (**VII**)] is adjacent to a positively charged nitrogen atom, and its position in the proton spectrum at $\delta$5.40 (see Fig. 12) is extremely sensitive to structural modifications. The presence of caffeic acid, which according to earlier literature is a potent antithiamine agent, leaves the $\delta$5.40 peak unchanged. It was thus concluded that the alleged strong antithiamine effect of *ortho*-diphenols, a widespread class of food polyphenols found in fruit, vegetables, coffee, etc., was unproved. Horman and Brambilla (1982) later showed that the thiochrome method used to establish the original evidence condemning *ortho*-diphenols was susceptible to dissolved oxygen, and they once more used NMR data to suggest that these earlier literature reports were erroneously based on an apparent disappearance of thiamine when oxygen was not excluded.

## 3. Sugars, Polysaccharides, Glycosides

The structural elucidation of three $\alpha$-D-galactopyranosylcyclitols isolated from soya and reported by Schweizer and Horman (1981) is described in Chapter 6, Section III,E. In the same publication, Schweizer and Horman assigned the $^{13}$C spectra of the three compounds along with those of galactose, galactinol, and related inositols and *O*-methyl inositols. Proton spectra have also found use in discriminating cell wall mannans from yeasts grown on different substrates. Spectral differences were related to different distribution patterns of individual mannose monosaccharide units in the polysaccharide structure (Hirata and Ishitani, 1978; Ishitani *et al.,* 1978). Mannans and galactomannans from yeasts isolated from bumblebee honey were identified with proton magnetic resonance, and the spectra were then used as an aid for characterizing the different yeast species (Spencer *et al.,* 1970). The $^{13}$C spectra of $(1 \rightarrow 6)$-$\alpha$-D-galactosyl-$(1 \rightarrow 4)$-$\alpha$-D-mannans from locust bean, guar, and fenugreek gums were recorded by Bociek *et al.* (1981), and the various peaks were assigned. Mannose–galactose ratios in the hydrolyzed gums as calculated directly from NMR data are comparable with gas chromatographic (GC) determinations on the same samples. In an extensive review article, Coxon (1980) has presented a discussion of $^{13}$C spectra of food-related di- and trisaccharides, including factors affecting peak positions and coupling constants; suggestions for the quantitative analysis of food oligosaccharides are put forward.

Two new sugar derivatives from green coffee beans have been isolated and identified. The first is a glucopyranosyl ester of kauran-18-oic acid, which was given the name cofaryloside (Richter *et al.,* 1977), and the second was a furokaurane glycoside named mascaroside (Richter and Spiteller, 1979). Glycyrrhetinic acid is the aglycone of glycyrrhizic acid, the sweet principle of licorice root and a disaccharidic glycoside. The $\alpha$- and $\beta$-isomers of glycyrrhetinic acid were separated by high-performance liquid chromatography (HPLC) in sufficient quantities for analysis by $^{13}$C-NMR (Tisse *et al.,* 1979).

## 4. Mycotoxins

The structures of mycotoxins, naturally occurring toxic compounds of fungal origin, are many and varied (Scott, 1978), and although most identifications have been effected by MS, a few assigned NMR spectra are reported. Proton spectra were used to confirm the structure of parasiticol (**II**), a metabolite structurally related to the aflatoxin family (Stubblefield *et al.,* 1970). Proton and $^{13}$C-NMR were used to identify the structure of satratoxin H as a macrocyclic dilactone of 12,13-epoxytrichothec-9-ene, and complete assigned proton and $^{13}$C spectra for this and related trichothecene mycotoxins of basic structure (**III**) are presented (Eppeley *et al.,* 1977).

II                                        III

Ellison and Kotsonis (1976) have reported the assigned $^{13}$C spectra of 8 further trichothecene mycotoxins, and Cox and Cole (1977) have shown the

$^{13}$C spectra of 12 other mycotoxins containing the fused bisdihydrofuran ring system characteristic of aflatoxins and sterigmatocystins and shown (for parasiticol) in structure (**II**). Carbon-13 spectra were also used to analyze tutin-related compounds in toxic honey; spectral assignments were effected for these compounds and also for a series of compounds related to hyenanchin and picrotin (Blunt *et al.*, 1979).

## 5. Fermented Beverages

Several groups of workers have used NMR to study beer constituents coming from hops and imparting bitter flavor. For example, Shannon *et al.* (1969) have studied the rates of biogenesis of alpha and beta acids in the growing hop cone. Results indicated an initial predominant formation of beta acids and 4-desoxyhumulones, with a subsequent increase in alpha acids as the cones came to maturity. The variation of other constituents such as humulone and lupulone and their adisomers was also described. Similarly, Kowaka *et al.* (1970) has identified and determined individual analogs of alpha and beta acids. Dadic (1977) also describes the use of NMR to characterize the three major alpha acids, to elucidate the structures of isohumulones, and to identify polyphenols. Belleau and Dadic (1977) have shown that low-field signals from aromatic protons in ethyl acetate extracts of beer, which differ according to beer type and age, can be used for fingerprinting and quality monitoring.

## 6. Miscellaneous Applications

Signals for hydroxyl groups in vinegars are seen to displace to lower field and to broaden with aging of the product, whereas signals from nonhydroxyl protons remain relatively unchanged. This is interpreted in terms of increased H-bonding and formation of molecular clusters, with a resulting reduction in proton mobility and exchange (Horike *et al.*, 1982). A meat soup stock was fractionated and individual fractions analyzed by taste and by $^{13}$C-NMR (Murai *et al.*, 1979). In this manner, a correlation could be established between variation of flavor profile and variation of individual substances. The fraction scoring highest in tasting was found to contain lactic acid, carnosine, anserine, and other compounds. Clearly, for this type of study, the specificity of $^{13}$C-NMR allows a more detailed analysis of the variation of individual components than does the more conventional ultraviolet (UV)–visible spectroscopy. Marmion (1974) has published the proton spectra of 27 certifiable food colors that are either pure compounds or secondary mixtures of pure colorants. The spectra permit a semiquantitative mixture analysis. Similarly, Turczam *et al.* (1977) report the proton spectra of 17 compounds listed in the Food Chemicals Codex that do not have analytically usable UV spectra. These are mainly amino acids, fatty acids, and hydroxy acids.

Proton, $^{13}C$- and $^{31}P$-NMR have been used for the study of plant and animal proteins. From proton spectra, Glickson *et al.* (1971) have studied oxidized and reduced forms of alfalfa and soya ferredoxins. Baianu (1981) has presented partially assigned $^{13}C$ spectra of wheat proteins, pointing out the advantages of NMR as a nondestructive method for protein character-ization relative to the more destructive current techniques. The chemical nature of phosphorous atoms in bovine $\alpha_s$-casein B and in egg yolk phosvitin has been investigated by Ho *et al.* (1969). Line multiplicities and variation of chemical shifts with pH suggested that most of the phosphate groups in both proteins were present as monoesters attached to seryl residues.

Biosynthetic pathways for the formation of individual compounds by plants, yeasts, or isolated enzymes can be studied by feeding with specifically $^{13}C$-labeled substrates, for example acetate or pyruvate, and then running $^{13}C$ spectra of the biosynthesized compounds after isolation. Signals corre-sponding to sites in the molecule where $^{13}C$ has been incorporated are more intense than in unlabeled material. In this manner, the biosynthesis of sesquiterpenes formed as stress metabolites in potatoes was investigated (Stoessl *et al.,* 1976, 1978), as were eugenin and 6-methoxymellein from carrot roots (Stoessl and Stothers, 1978).

## B. Quantitative Applications

Under appropriate recording conditions, the area under a given NMR signal is proportional within 1 or 2% to the number of nuclei giving rise to that signal. Continuous-wave instruments operated in their normal mode give good results for protons. More care must be taken with pulsed instru-ments, where signal irregularities can arise due to saturation, nuclear Over-hauser effects, and too much or too little power in the exciting pulses. With correct pulse intensities and durations, these problems can be largely over-come, even for nuclei other than protons. Quantitative information can be obtained from carbon spectra, for example, by recording data on samples to which a paramagnetic species has been added to accelerate relaxation. For further discussion, the excellent account of quantitative aspects of $^{13}C$-NMR by Shoolery and Jankowski (1973) should be consulted. Alternatively, this information is to be found in the book on practical aspects of NMR spectroscopy by Martin *et al.* (1980).

### 1. Proton Spectra

It is frequently the case that the monitoring of changes in the constituents of a food system as a function of technological parameters can be performed with only approximate quantitative measurements. Proton NMR is perfectly suited to this kind of task, as exemplified in its application to the degradation of trigonelline (**IV**) in coffee beans as a function of the degree of roasting (Viani and Horman, 1975).

I V

Trigonelline, present in green coffee at a level of about 1%, was shown to decrease with roasting in an S-shaped curve with a maximum rate of degradation around 14–15% weight loss (percentage of weight loss is a standard manner of indicating degree of roast). Proton spectra of coffee infusions at a concentration of 15 wt/vol % in $D_2O$ are shown in Fig. 6 for green coffee and two different levels of roasting. An approximate idea of the amount of trigonelline present can be got by measuring the height (or the area) of the characteristic signal at about $\delta 9.1$ relative to that of a simultaneously measured standard, in this case the trimethylsilane (TMS) used to position the spectrum and contained in a capillary tube that can be physically transferred from sample to sample. Other constituents, such as the chlorogenic acids, are also seen to decrease with roasting, but some, such as caffeine, apparently remain relatively constant. The advantage of NMR is that several constituents can be monitored simultaneously and that whole samples can be roughly analyzed without tedious separations.

Another application to the analysis of intact food products is the determination of turmeric, used as a yellow colorant in mustard preparations at levels of about 0.6%, by measuring the concentrations of curcumin and related compounds (Unterhalt, 1980). Further work in the area of spices and condiments involves the determination of relative amounts of capsaicins and dihydrocapsaicins in extracts of the pungent principle of red peppers by integration of their respective methyl signals (Müller-Stock *et al.,* 1973). Other individual plant constituents conveniently quantified are the alpha and beta acids of hops, which are readily isolated in hexane extracts. After evaporation and redissolution in deuterochloroform, these compounds give a group of peaks in the region of $\delta 19.0$ coming from protons in hydrogen-bonded hydroxyl groups. The intensities of six well-separated signals have been used to determine the relative amounts of the six different hop acids humulone, cohumulone and adhumulone (alpha acids), and lupulone, colupulone, and adlupulone (beta acids) in a range of hop varieties (Molyneux and Wong, 1975).

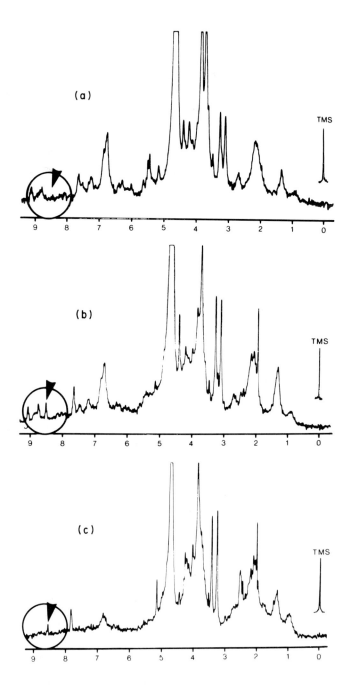

**Fig. 6.** The 60-MHz proton spectra of 15% coffee infusions in $D_2O$: (a) green coffee, (b) medium-roast coffee, and (c) high-roast coffee. The circled peaks come from trigonelline (**IV**). Peaks in the region $\delta 6.0 - 7.5$ are from chlorogenate, and the doublet at $\sim \delta 3.2$ from caffeine.

A rapid method for the determination of sucrose in sugar beet juices, based on preferential paramagnetic relaxation of water protons and subsequent data treatment to remove the water signal, has been described by Lowman and Maciel (1979). It works directly on solutions in ordinary water (as opposed to $D_2O$) over a sucrose concentration range of up to 26% wt/wt and gives a linear comparison with GC results. Stahl and McNaught (1970) have used proton NMR to determine the hydroxypropyl content of modified starches by first hydrolyzing the starch and then integrating the doublet for the terminal methyl of the hydroxypropyl group at about $\delta1.5$ relative to an acetic acid standard. A lower limit of 0.5% hydroxypropyl content could be detected, but conceivably with modern instrumentation much lower limits would be attainable.

A comparison of HPLC and NMR for the determination of phosphatidylcholine in soy lecithin is described by Press et al. (1981). The values from NMR are obtained by integration of the choline trimethylammonium signal at $\delta3.3$ and are systematically about 0.9 times the corresponding HPLC determinations. Similarly, the area of the trimethylammonium signal has been used to determine betaine in molasses and in sugar beet extracts (Merbach et al., 1975). The utility of the $\delta3.3$ signal for characterizing choline derivatives had already been established in one of the earliest applications of NMR to food analysis by Chapman and Morrison (1966), who published the spectra of phospholipids.

Further to the above, the determination of ethanol in liquors, wines, and other beverages has been described by Anders et al. (1976), and Sterk (1969) has determined small amounts of methanol in ethanol and small amounts of ethanol in water.

## 2. Other Nuclei

For nuclei other than protons, few quantitative studies are reported. There is, however, one major field of exception to this rule, namely that of using $^{13}C$ spectra to analyze edible oils and fats, and two other areas where promising results have been obtained: the $^{13}C$ analysis of essential oils, and the $^2H$ spectra of natural products. These applications are presented in more detail in Sections IV,C, IV,D, and IV,E, respectively.

In other applications, the specificity and sensitivity of $^{13}C$-NMR have been exploited to determine 4-hydroxy-L-proline in meat protein (Josefowicz et al., 1977; O'Neill et al., 1979). Hydroxyproline is found principally in the connective protein collagen, which imparts strength to tendon, skin, and muscle but which also increases the toughness of the meat. By comparing the signal intensity of the C-4 carbon of the hydroxyproline cycle with that of an internal standard, levels of about 3% could be measured with a standard deviation of ±0.16%. The total analysis time was about 30 min, and the method was at least as reliable as alternative procedures based on colorimetry or on amino acid analyzers.

Phytic acid is the hexaphosphate of myoinositol. It is an important plant constituent, particularly in grains and hence in cereal foods. A strong complexing agent for dietary metals, it can induce imbalances of calcium, zinc, iron, and magnesium in the body if consumed in large amounts. O'Neill *et al.* (1980) have described how $^{31}$P-NMR can be used to determine phytate in foods by comparing the area of the C-2 phosphate peak relative to an internal reference compound, cyclotetrametaphosphate. Sucrose was added to the samples to increase sample viscosity; this reduced the $T_1$ values for the different groups by a factor of about 10, speeding up the analysis and minimizing problems due to sample saturation. Some 21 cereal food samples were analyzed and found to have phytate contents of between 0.1 and − 3.5% with a standard deviation of 0.02%. A complete analysis takes about 95 min, of which only 25 min is operator time.

Finally, $^{19}$F-NMR has found use in the estimation of free $\epsilon$-amino groups of lysine in dairy products (Holsinger and Posati, 1974) and in other proteins (Ramirez *et al.,* 1975). The method is based on the conversion of free amino groups to their trifluoroacetates by reaction with *S*-ethyl trifluorothioacetate and is reported to be rapid, clean, and reasonably accurate and to offer a vast improvement in ease of operation relative to other methods. For five proteins, lysine contents determined by this method give a correlation coefficient ($r^2$) of 0.98 with values coming from an amino acid analyzer.

## C. Edible Oils and Fats

The main advantages of working with oils and fats are that they are normally plentiful, they are easy to extract as a discrete group of compounds, and they are miscible in all proportions with low-polarity solvents such as chloroform. This last characteristic has made them the favorite candidates to date for $^{13}$C-NMR spectroscopy.

### 1. Proton Studies

Johnson and Shoolery (1962) first used proton NMR for the accurate determination of unsaturation in edible fats and oils. Nielsen (1976) showed a correlation coefficient of 0.97 for iodine-value determinations of unsaturation as measured by NMR and by the classical Wijs's method. Pawlowski *et al.* (1972) used the cyclopropene methylene resonance at $\sim \delta 0.75$ to assay cyclopropenoid fatty acids in lipid mixtures, an analysis that was difficult to perform by the more conventional GC methods because of instability of cyclopropenoids on the column.

Heating oils and fats increases their chemical reactivity and can provoke the creation of new compounds. Here, Ferren and Seery (1973) used NMR along with gel permeation chromatography to determine the dimer content of corn oil by differences in the respective spectra of dimer and monomer forms at $\delta 5.5$, Sohde et al. (1974) determined hydroperoxides in methyl linoleate and in safflower oil, and Mel'nikov (1982) described a study of spectral changes occurring on oxidizing vegetable oils. Richter and Lakshminarayana (1979) have investigated the isomerization and rates of saturation of double bonds in cyclopentenyl fatty acids: they have shown an accumulation of trans double bonds, presumably a result of their faster formation and slower hydrogenation than cis double bonds. In a different type of study, Noda and Takahashi (1979) employed chemical shift reagents to resolve spectra of fatty acid methyl esters, remarking that praseodymium complexes gave better signal separation than europium complexes.

## 2. Carbon-13 Spectra

Carbon-13 spectra offer several advantages relative to protons in oil and fat analysis, largely because chemical shifts are some 20 times greater, resulting in higher resolution and better signal definition. This is seen in Fig. 7, which compares the 80-MHz proton spectrum of rapeseed oil with the corresponding 20-MHz $^{13}$C spectrum. Applications fall conveniently into four different categories, which are described separately below.

## 3. Oil and Fat Profiling

In the same way that a simple silhouette allows instant recognition of many everyday objects, oils and fats can be recognized from profiles in their $^{13}$C spectra. The distribution of unsaturated acids in the different edible oils and fats is characteristic of the specific animal or vegetable source. This is reflected in the distribution of peaks in the $\delta_C 130$ region of the spectrum coming from the different double-bond carbon atoms. In Fig. 7 this region is shown expanded in an insert for rapeseed oil, and in Fig. 8 for eight other oils and fats. The visual differences in these profiles, both qualitative and quantitative, are self-evident and are sufficient for the rapid screening of samples to control, for example, product labeling and integrity. These profiles were obtained at 20 MHz, and evidently at higher frequencies more detail is seen. To illustrate this, Fig. 9 shows once more the olefinic carbon region of rapeseed oil, this time at 100 MHz. At this higher resolution, peaks from individual carbons in oleic, linoleic, and linolenic acids are distinctly defined.

## 4. Composition of Oils and Fats

Following the procedure described by Shoolery and Smithson (1969), the $^{13}$C spectra of five oil samples were quantitatively analyzed. The percentages

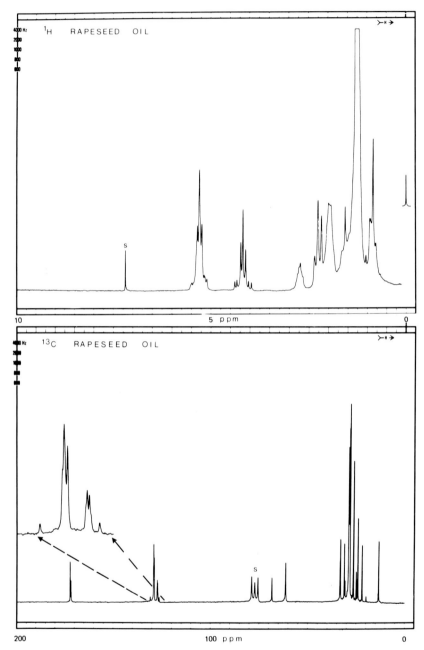

**Fig. 7.** The 80-MHz proton (upper) and 20-MHz $^{13}C$ (lower) spectra of rapeseed oil in CDCl$_3$, with the olefinic region of the $^{13}C$ spectrum expanded in the insert. Peaks labeled s are from the solvent.

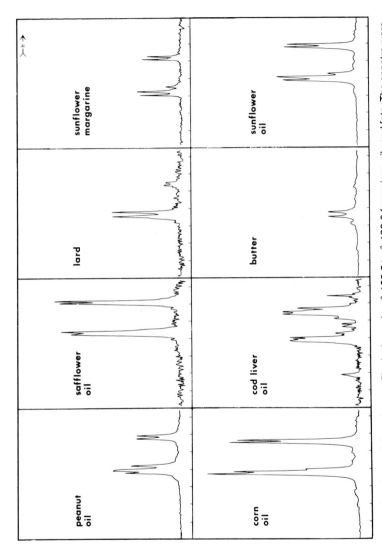

**Fig. 8.** The olefinic carbon profiles in the region $\delta_C 125.5$ to $\delta_C 133.0$ for various oils and fats. The spectra were run in CDCl$_3$ solution at 20 MHz.

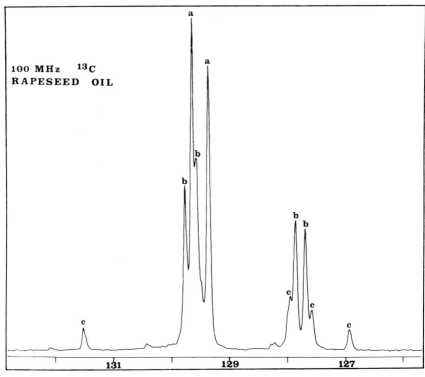

**Fig. 9.** The 100-MHz olefinic carbon profile of rapeseed oil showing good resolution on peaks arising from (a) oleic, (b) linoleic, and (c) linolenic acids, which in Fig. 7 at 20 MHz overlap extensively. (Courtesy of G. Wider, Bruker-Spectrospin, Zurich, Switzerland.)

of the different fatty acids present are given in Table 2. Here, 18:0, 18:1, 18:2, and 18:3 express acids with 18 carbon atoms in the fatty acid chain and with 0, 1, 2, or 3 double bonds respectively, namely stearic, oleic, linoleic, and linolenic acids. As can be seen, values compare well with those obtained by GC methods, with the advantage for the NMR determination being the fact that it is carried out on the intact oil sample whereas the classical GC determination involves a chemical transformation of the triglycerides to methyl esters. Modern GC methods with on-column injection (Traitler and Prévot, 1981) are beginning to make possible the direct analysis of triglycerides, but even here NMR would probably still have the edge in rapidity and in softness of analytical conditions for thermally sensitive samples. A complete description of the details of the method has been published by Shoolery (1975), and the appropriate recording conditions that minimize the differential effects of relaxation times on peak

**TABLE 2  Comparison of the Distribution of C$_{18}$ Fatty Acids in Different Edible Oils Determined by $^{13}$C-NMR and by Gas Chromatography[a]**

| | Fatty acid composition (%) | | | | | | | | |
|---|---|---|---|---|---|---|---|---|---|
| | NMR | | | | GC | | | | |
| Oil | 18:0 | 18:1 | 18:2 | 18:3 | 18:0 | 18:1 | 18:2 | 18:3 |
| Safflower | 10.7 | 14.1 | 74.2 | 0.0 | 10.4 | 13.0 | 76.3 | 0.3 |
| Soybean | 14.0 | 46.5 | 34.8 | 4.5 | 13.8 | 44.8 | 37.8 | 3.6 |
| Peanut | 18.4 | 49.5 | 32.0 | 0.0 | 18.6 | 48.7 | 32.6 | 0.0 |
| Cottonseed | 24.7 | 20.4 | 54.9 | 0.0 | 23.5 | 19.0 | 57.4 | 0.0 |
| Rapeseed | 6.3 | 76.0 | 11.0 | 6.7 | 6.8 | 76.9 | 10.0 | 6.4 |

[a] From data provided by J. N. Shoolery, Varian Associates, Palo Alto, California.

intensities and thus permit this kind of analysis are described by Shoolery and Jankowski (1973).

An excellent example of the use of the NMR method is described by Tulloch (1982), who has quantified 13 individual acids in 7 different seed oils. The acids included 6 isomeric conjugated octadecatrienoic acids, namely the cis,trans,cis- and trans,trans,cis-8,10,12-isomers, and the cis,trans,cis-, cis,trans,trans-, trans,trans,cis-, and trans,trans,trans-9,11,13-isomers, all of which give different characteristic peaks for double-bond carbons and for carbons near the double-bond systems. Although cis–trans isomerization during extraction could have been responsible for the forma-tion of the trans,trans,trans-isomer in the oil from *Centranthus ruber,* this isomer is thought as a result of this study to be naturally occurring. A further example involves a study by Bradbury and Collins (1982) of the lipids of the major histological components of rice. Specific compounds have also been determined, such as butyric acid and its isomers found at the 10.3 mol % level in butter oil (Pfeffer *et al.,* 1977) and cyclopropenoid acids in cotton-seed oil (Shoolery, 1982). The latter study suggests that the cyclopropene content of cottonseed oil can be determined by $^{13}$C-NMR at levels less than 0.1 mol %, that is, better than a tenfold reduction in the lower limit of measurability relative to earlier proton determinations (Pawlowski *et al.,* 1972). Further work by Shoolery (1982) reveals that almost all of the butyric acid in butter is present in the 1' or 3' position of the triglycerides concerned.

## 5. cis–trans Isomerisms

Catalytic hydrogenation of oils to give margarines can be accompanied by cis–trans isomerism and/or migration of double bonds, which occurs to a greater or lesser extent depending on the catalyst used. Bus and Frost (1974) reported that allylic carbon atoms such as $C_8$ and $C_{11}$ in oleic acid have chemical shifts differing by over 5 ppm for the cis and trans isomers. This was exploited by Pfeffer *et al.* (1977) to develop a method for determining cis–trans ratios in catalytically treated lipid mixtures. In butter, $11.2 \pm 0.5\%$ trans acids were detected relative to 12% by GC.

## 6. Degree of Unsaturation

As with proton spectra, carbon spectra can also be used to determine the degree of unsaturation of oils and fats. This is traditionally expressed in terms of an iodine value, namely, the number of centigrams of iodine that will react with 1 g of lipidic material. Shoolery and Jankowski (1973) doped corn oil with chromium triacetoacetate [Cr(acac)$_3$] and determined by integration an iodine number of 129, compared with 128.8 determined by titration. Correction factors were used to annul Nuclear Overhauser (NOE) intensity differences between olefinic carbon signals and those from carbonyl, glyceride, and remaining carbons. In our own experience, iodine numbers from NMR are generally within 5% of those determined by titration, but because titration is an easy operation it would make sense to use the nondestructive NMR technique only if limited sample was available.

## 7. Oil Composition in Whole Seeds

If for most quantitative measurements GC gives results comparable to those obtained by NMR, there is one area where NMR has no rival. This involves the detailed analysis of the oil content of intact oilseeds. Wide-line proton NMR has now been used for almost 20 years for the analysis of seeds and since the early 1970s has been significant in commercial corn-breeding programs. However, it can only give information on total oil content and as such is no use in breeding studies designed to alter the composition of individual fatty acids in the triglycerides. Carbon-13 NMR is not as susceptible as proton NMR to reduced sample mobility and field inhomogeneity induced by the presence of heterogenous solids. Being nondestructive, it can therefore be used to determine oil compositions of whole seeds, which can subsequently be planted and cultivated as normal. Seeds with desirable fatty acid distributions may thus be selected prior to planting, rather than after sowing, growing, harvesting, and subsequent detailed analysis of the extracted oil. Depending on the precision required, it can take 10–90 min analysis time per seed, which may be a limiting factor in the general application of carbon NMR in large-scale breeding programs. The spectrum of a sunflower seed is shown in Fig. 10, and it is seen to compare well in quality with the spectrum of rapeseed oil (Fig. 7). Results of analyses for five sunflower seeds are shown in Table 3, where significant differences in the relative quantities of oleic and linoleic acids are apparent.

Since the first reports of this approach by Shoolery (1973), it has been applied to selecting soybeans producing an oil depleted in the less desirable linolenic acid (Schaefer and Stejskal, 1974) and by the same authors (1975) to other seeds. The latter study also discusses the potential of NMR to determine protein and carbohydrate in the same seeds. Chen *et al.* (1979) have determined the fatty acid composition in three types of seed, including wild oats, presumably in preparation for sowing. Albornoz and Leon (1980) observed considerable differences in the characteristics of oils coming from 64 varieties of peanut, and Leal *et al.* (1981) report the analysis of 12 fatty acids in different seeds including soybean, castor bean, and tonka bean. Practical details for this and the other types of oil analysis here described are to be found in a recent publication by Shoolery (1982).

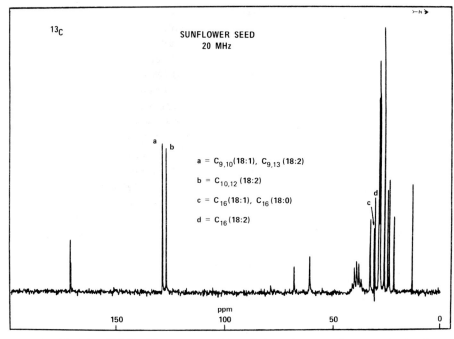

**Fig. 10.** The 20-MHz $^{13}$C spectrum of the oils in an intact sunflower seed, showing characteristic peaks for specific acids.

## D. Essential Oils

Essential oils extracted from herbs and spices are complex mixtures consisting principally of oxygenated and nonoxygenated mono- and sesquiterpenes along with a few other characteristic compounds such as phenylpropenes. Generally, a very limited number of compounds, perhaps up to five, account for as much as 98% of the total oil. The remaining 2% or less

**TABLE 3** Variation of Oil Composition in Intact Sunflower Seeds Determined by $^{13}$C-NMR[a]

| Seed | Stearic A | Oleic A | Linoleic A |
|------|-----------|---------|------------|
| 1 | 13.6 ± 1.3 | 29.0 ± 1.5 | 57.4 ± 0.4 |
| 2 | 11.3 ± 2.0 | 31.2 ± 2.0 | 58.1 ± 0.9 |
| 3 | 10.2 ± 1.8 | 33.1 ± 1.0 | 56.7 ± 2.1 |
| 4 | 12.0 ± 1.5 | 26.7 ± 2.5 | 62.4 ± 2.0 |
| 5 | 12.9 ± 2.4 | 37.3 ± 0.4 | 49.8 ± 2.0 |

[a] From data provided by J. N. Shoolery, Varian Associates, Palo Alto, California.

can, however, contain more than 100 compounds. Any analysis method for such materials must be capable of operating over a very wide dynamic range, and because we can expect to get a signal for almost every carbon in every compound present, the idea of attempting identifications and quantitative determinations in such a forest of peaks is almost frightening. In 1979, Formacek produced a thesis on the possibilities of applying $^{13}$C-NMR to the direct analysis of essential oils. He classifies oils according to the principal types of compound present and gives a spectral data bank of some 70 common constituents. Then, using fennel oil and patchouli oil as examples, he shows how individual compounds can not only be recognized but also be estimated in quantities down to less than 0.5% giving values in good agreement to those found by GC. To illustrate this, partial spectra of fennel oil in which the peaks for 7 compounds are assigned are shown in Fig. 11. This figure also presents the comparative compositional analysis values obtained from NMR and from GC. The qualitative and quantitative aspects of this work have been published by Formacek and Kubeczka (1979, 1982b).

It is also evident that $^{13}$C-NMR could be used for the rapid screening of essential oils and other aroma mixtures, comparing spectra of samples by simple pattern recognition and degree of fit with those in a library. This type of operation, common in MS, is little used in NMR.

## E. Deuterium NMR

In the preceding sections, several areas are described in which proton or $^{13}$C spectra provide elegant solutions to problems in food analysis. Examples are given in which the higher-resolution instruments and more powerful computer systems now available facilitate, or even make possible, analyses that were difficult, time-consuming, or even impossible with earlier generations of NMR equipment. There are, however, certain problems that cannot be tackled by observing protons or $^{13}$C nuclei, irrespective of the degree of instrumental sophistication. For these cases, other nuclei must be explored. Two totally different problems solved by deuterium NMR are therefore presented as examples, one involving isotope-enriched materials and the other measuring deuterium at natural abundance levels.

### 1. Oil–Water Emulsions

Oil and water are immiscible. On shaking together, droplets of water are dispersed in the lipid phase. Each droplet is surrounded by a lipidic monolayer, in which the triglycerides are oriented with their more polar triglyceryl ester moieties toward the bulk of the water and with the fatty acid chains radiating outward from the droplet surface. Thus, when two droplets come together, they are separated by their attached triglyceride layers in what are

Fennel Oil :
aliphatic C

| Compound | NMR % | GC % |
|----------|-------|------|
| tA | 25.2 | 21.4 |
| F | 16.7 | 14.6 |
| L | 25.0 | 26.8 |
| E | 2.5 | 2.1 |
| aP | 9.2 | 10.4 |
| pC | 1.2 | 1.7 |

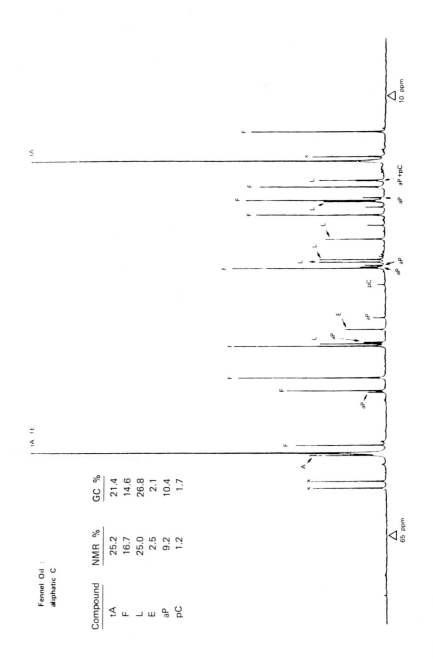

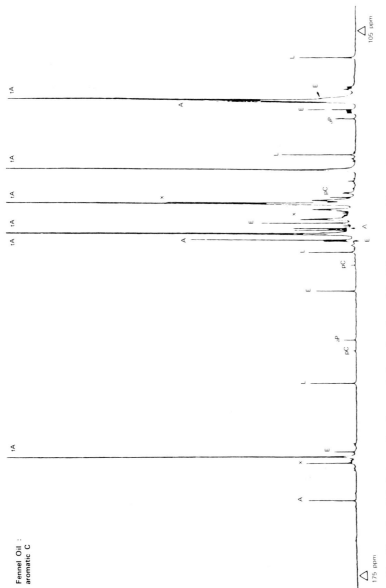

Fennel Oil : aromatic C

175 ppm

105 ppm

**Fig. 11.** The aliphatic (upper) and aromatic (lower) regions of the $^{13}$C spectrum of fennel oil, with comparative quantification of individual components from NMR and from GC. Key to abbreviations: tA = *trans*-anethole; F = fenchene; L = limonene; E = estragole; aP = α-pinene; A = anisaldehyde; pC = *para*-cymene. (Courtesy of V. Formacek, Bruker-Spectrospin, Zurich, Switzerland.)

termed bilayer membranes. The stability of these membranes, and hence of water – oil emulsions, is affected by several factors, such as the proportion of oil and water and the polar phospholipid content of the oil. Formation of a bilayer membrane is accompanied by an increased molecular order in the triglycerides: this leads to decreased mobility and shorter relaxation times, which can be measured by NMR. Thus, the influence of different factors on membrane and emulsion order and stability can be assessed. Although proton, carbon, and phosphorous spectra have been used for this type of study, deuterium NMR appears to give much more information. Stockton *et al.* (1974) used selectively deuterated phospholipids and fatty acids as probes to study the liquid crystalline phase of egg phosphatidylcholine in aqueous dispersion. Through observations of $T_1$ and $T_2$ values, and from order parameters, they were able to show that added cholesterol has no effect on the mobility of choline methyl groups but that the mobility of hydrocarbon chains was much reduced. Stockton and Smith (1976) later showed that cholesterol induces a higher degree of order in the fatty acid chains of phospholipids. Carbons down to position 10 in the chain were most affected, and they concluded that the condensing effect of cholesterol, which thickens the bilayer and lowers its permeability, came from interactions of acyl chains in the lipids with the rigid steroid nucleus and from solvation of the $3\beta$-hydroxyl group of cholesterol at the aqueous interface. Taylor *et al.* (1981) have studied the dynamic behavior of cholesterol itself in bilayer membranes using specifically deuterated cholesterol. Taylor and Smith (1981) have compared electron spin resonance (ESR) with deuterium and phosphorus NMR for the study of structures of lipids dispersed in water.

## 2. Identification of the Origin of Natural Products

The natural distribution of the minor isotopes in a given molecule depends on its origin: we see in Chapter 6, Section II,I, how deuterium – hydrogen ratios and $^{13}C/^{12}C$ ratios determined by MS give information on the geographical origin of natural products and on their mode of formation — chemical synthesis or biosynthesis. Deuterium NMR allows us to go one stage further. Effectively, MS measures only the *overall* molecular deuterium content, whereas deuterium NMR allows discrimination between the *relative deuterium contents of the different chemical sites* within a molecule. Approximately 1 molecule in 700 contains a deuterium atom: the probability of finding 2, 3, or more deuterium atoms in a given molecule is therefore 1 in $(700)^2$, $(700)^3$, etc. For this reason, natural abundance level deuterium spectra are very simple and reflect the quantities of individual monodeuterated species present. Following this reasoning, Martin and Martin (1981a) point out that the natural abundance spectrum of ethyl acetate shows three singlet signals that arise from the presence of compounds (**Va**), (**Vb**), and

(**Vc**), shown in Table 4. From signal intensities and integrals they were able to show that, relative to (**Va**), (**Vb**) is enriched and (**Vc**) depleted in deuterium compared to the expected values based on random statistical distribution. This behavior is repeated, for example in a series of vinyl derivatives (Martin and Martin, 1981a) and in ethyl derivatives (Martin and Martin, 1981b). A method based on such differences has been used by Martin *et al.* (1982a) to characterize anethole (**VI**) derived from different sources. They

$$CH_3-O-\underset{}{\bigcirc}-CH=CH-CH_3$$

**V I**

showed that NMR not only matched MS in distinguishing natural anetholes from samples prepared either synthetically or by isomerization of estragole but also was able to identify the natural source. For example, in a comparison of two natural anethole samples, they were able to report that fennel anethole is much depleted in deuterium in the terminal methyl group of the propenyl chain relative to its analog isolated from star anise.

Martin *et al.* (1982b) have shown how the same techniques can be used to reveal the origins of ethanols. In this case, the relative enrichment at the ethanol methylene group relative to that of the methyl group, along with an estimate of total deuterium content, distinguishes among alcohols derived from corn, sugarcane, wheat, apple, potato, and sugar beet and among synthetic ethanols produced in different countries, as well as differentiating between alcohols of natural and synthetic origin. Based on this, Martin and Martin (1982) have described a method to detect the addition of sugar to grape juice prior to fermentation, which results in the fortification of wines with nongrape alcohol.

More generally, Martin *et al.* (1982c) have suggested that these differences in deuterium distribution provide a natural label for investigation of chemical and biochemical reaction mechanisms avoiding problems arising from

**TABLE 4** Statistically Expected and Measured Ratios of Natural-Level Monodeuterated Ethyl Acetate Isomers

|  |  | Ratio | |
|---|---|---|---|
| | Compound | Statistical | Measured |
| (**Va**) | $CH_2D.CH_2.O.CO.CH_3$ | 3 | 3[a] |
| (**Vb**) | $CH_3.CHD.O.CO.CH_3$ | 2 | 2.4 |
| (**Vc**) | $CH_3.CH_2.O.CO.CH_2D$ | 3 | 2.3 |

[a] Compound (**Va**) is used as a reference relative to which (**Vb**) and (**Vc**) are measured.

the use of isotope-enriched probes. Isotope enrichment at a high level can markedly affect reaction rates and hence the quantitative distribution of final products, thus giving a distorted picture of reaction mechanisms.

In food legislation, the question of natural versus synthetic origin of food additives frequently arises. Although MS has provided a powerful means for detecting fraudulent practice in distinguishing natural from synthetic by D/H and $^{13}C/^{12}C$ ratios, it is not difficult to synthesize compounds that simulate the correct total natural isotope levels. Such "doctored" products would not survive scrutiny by NMR where each individual atomic site in the molecule can be examined either for deuterium or $^{13}C$ content. Although it is in theory possible to synthesize a compound that is totally nature-identical, this would be a mammoth task and certainly not commercially viable.

NMR is about five times faster than MS for these analyses. Its only disadvantage at present appears to be one of quantity of sample needed: gram quantities are required for NMR, whereas milligram quantities or less suffice for MS. In addition, for deuterium spectra, solids will pose a problem unless they are highly soluble in a suitable solvent. This situation, however, awaits only the development of more sensitive NMR methods or instrumentation.

## F. Kinetics, Equilibria, Complex Formation, and Reaction Mechanisms

The quantitative dependence of peak areas and the electronic environment dependence of peak positions can be exploited for ends other than simple structural elucidation or the determination of specific compounds in mixtures. Time-dependent changes in peak areas can be used to determine reaction rate constants and to establish kinetic profiles even for multicomponent mixtures. Changes in peak positions as a function of sample concentration, solvent, or temperature give information on the spatial organization of molecules in solution. The advantage of NMR in such studies is that observations reflect the instantaneous situation in solution at the time of measurement, and being noninvasive it does not perturb the sample.

### 1. Kinetics

The number of kinetic studies to report on foods is surprisingly limited, surprisingly, that is, in view of the fact that food materials spend all of their lives in a state of change. Low-resolution NMR has been used to study the rates of melting of fats and of solubilization of milk powders as described in Section II. For high-resolution NMR, only a study of the decomposition of trigonelline on roasting coffee (Viani and Horman, 1975) and a similar study of the thermal decomposition of chlorogenic acid (Horman and Viani, 1971) exist, and beyond these we must look to peripheral studies, such as the kinetics for the binding of coenzymes to lactate dehydrogenase isoenzymes from chicken (Lee, 1978). It is evident that rapid reactions cannot be followed by NMR, but reactions having half-lives of a minute or so and upwards can be studied with ease at different temperatures. For slow reactions, Le Botlan *et al.* (1980, 1982) have shown how extensive kinetic data can be recorded in a short time by the temperature programming of the sample during data acquisition. The examples described are not food oriented, but the variable-temperature NMR technique could certainly be used in model food systems to give not only rate information but also enthalpy and entropy values.

## 2. Equilibria

Equilibria between two interconverting species are easier to observe. The effect of temperature and concentration on the relative amounts of $\alpha$- and $\beta$-glucose in tautomeric equilibrium has been described by Hyvönen *et al.* (1977): they determined relative quantities by integrating the signals for the anomeric protons. Doddrell and Allerhand (1971) and Que and Gray (1974) have used $^{13}$C-NMR to study tautomeric equilibria in ketohexoses, such as fructose, where the anomeric proton is absent. In our own laboratories, we have seen (I. Horman, 1982, unpublished) that temperatures between 20 and 80°C have little or no influence on the $1:2$ equilibrium ratio of $\alpha$- and $\beta$-lactose ($\sim 1:2$), in close agreement with literature values determined by polarimetry. Proton NMR has been used to investigate the tautomeric ratios of furanose and pyranose products of the Amadori and Heyns rearrangements (Altena *et al.*, 1981). The conformation of macromolecules in solution is also a question of equilibrium; Cowburn *et al.* (1970) compared conformations of bovine $\alpha$-lactalbumin and lysosyme and were able to rationalize spectral differences in terms of specific sequence differences between the two proteins.

## 3. Complexes

A complex is a loosely and reversibly bound agglomeration of two or more molecules in an ordered fashion: such agglomerations are implicated in all sorts of food-related phenomena, such as color, aroma formation and retention, relative solubility of specific components, extractability, and synergic effects.

Complex formation is merely another kind of equilibrium. Relative to examples described immediately above, it differs only in the rate of interconversion of the interchanging forms. When exchange is slow, as in the case of $\alpha$-/$\beta$-tautomerism, separate signals are seen for the two forms: in complex equilibria, exchange is fast relative to the time scale of the NMR spectrometer, and the instrument can detect only one apparent form. Signals of partially complexed species are found at positions intermediate between those that would be observed in the two interconverting forms if they could be measured separately in a pure state. These positions displace linearly as a function of the relative proportions of the two forms present.

The basic theory behind the use of NMR to study association constants of bimolecular complexes such as AB forming reversibly from compounds A and B according to the relation A + B $\rightleftharpoons$ AB is summarized in a book by Foster (1969). From proton NMR, Horman and Viani (1972) described the naturally occurring complex between caffeine and the chlorogenate ion isolated from coffee as a hydrophobically bound $\Pi$-molecular complex in which the plane surface of caffeine sits parallel to and directly above the plane of the caffeoyl moiety of chlorogenate. Horman et al. (1980) showed the existence of a complex of a similar nature between chlorogenate and the vitamin thiamine, as part of a study refuting earlier reports of an antithiamine effect of coffee (see also the discussion of polyphenols in Section IV,A). Recently, Horman and Dreux (1983) have developed a new approach to the calculation of association constants and other complex parameters from concentration-dependent NMR peak displacements that appears to rationalize many anomalies reported in the literature to date.

Another form of complexation involves the use of lanthanide shift reagents to simplify spectra. The active site of such compounds is a polyvalent lanthanide cation such as europium or ytterbium chelated with organic ligands to render the substance soluble in nonpolar solvents. When a lanthanide cation is added to a solution of a compound bearing a polar group (hydroxyl, carbonyl, ester, etc.), a complex forms between that group and the lanthanide cation. The latter, being positively charged, changes markedly the electron density at the site of complexation and at all sites nearby, its effect falling off with increasing distance. The normal spectrum is thus spread out and better resolved, a fact that led to the description of shift reagents as the poor man's higher-resolution NMR. This approach can be useful in some specific problems; examples of applications pertaining to foods are the study of the configuration and conformation of nootkatone, an important grapefruit flavor constituent (Jones, 1974), and the simplification and interpretation of the $^{13}C$ spectra of saturated fatty acid methyl esters (Barton et al., 1978).

Peak broadening is another sign of complexation. Thus, Kainosho and Konishi (1976) reported $^{13}C$ spectra of the pericarps of star anise seeds that showed signals assignable to anethole (**VI**). Once the pericarps were crushed, peaks were broadened, and the authors suggested that, whereas in crushed material anethole spreads over the woody pericarp surface and is immobilized, in intact pericarps it probably exists as small droplets with a high motional state. Further biological applications of NMR are described by Feeney *et al.* (1977) in a book recording the proceedings of a symposium on NMR in biology, other articles of which may also be of interest to the food analyst.

## 4. Reaction Mechanisms

Reaction mechanisms are often amenable to NMR investigation. For example, it is interesting to observe changes in the peak positions shown in Fig. 12 as a sample of thiamine dichloride in $D_2O$ is taken through the pH range from 1 to 12 and to relate these changes to a series of chemical structures indicated in Scheme 1. The proton spectrum at pH 2 corresponds to structure (**VII**), in which peak assignments are shown. At pH 4, peak j disappears slowly from the spectrum without changing position, indicating deuterium exchange to give structure (**VIII**): the proton or deuteron at position j has thus become labile and behaves as a weak acid. At pH 6, the spectrum of a new species (**IX**) has appeared: this is evidently formed in an acid–base deprotonation mode, because between pH 4 and 6 peaks b and i displace steadily upfield until finally b has moved to higher field than a. At pH values between 9 and 10.5, a mixture of compounds (**IX, X, XI**) is observed. Compound (**X**), the "yellow form" of thiamine, is a short-lived intermediate on the way to the formation of (**XI**): the rate of disappearance of this intermediate can be measured by the decrease in surface with time of peak a. It can thus be concluded that this compound is more stable at pH 11 or higher than at pH 9.5, indicating that the transformation of (**X**) to (**XI**) is acid catalyzed. The displacement of peak d for structures (**X**) and (**XI**) indicates that the thiazole ring has opened and the positive charge on its N atom has been neutralized. The twin peaks observed for b and i at pH > 11 indicate *syn*- and *anti*-conformations of the *N*-formyl group (**XI** and **XII**). On lowering of the pH, compound (**XI**) is apparently directly transformed to compound (**IX**) and does not return via the "yellow form" as might perhaps be expected. Further lowering of the pH to 2 ultimately gives compound (**VII**) or (**VIII**), depending on whether the medium is $H_2O$ or $D_2O$. After a few hours at pH 14 an extremely complicated spectrum is obtained, indicating that extensive decomposition has taken place. Thus, with a few proton spectra that can be run on almost any NMR instrument, a detailed picture of the acid–base transformations of thiamine is almost totally revealed (I. Horman and E. Brambilla, 1982, unpublished). This subject, using various other forms of spectroscopy, has accounted in the past for several man-years of research endeavor [see Hoppmann (1982)]. Similar behavior can be observed in the $^{13}C$ spectra of thiamine.

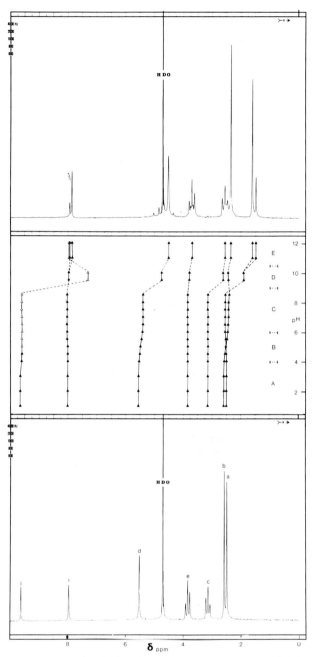

**Fig. 12.** The 80-MHz proton spectra of thiamine in D$_2$O at pH 1 (bottom) and at pH 12 (top), and the pH dependence of peak positions in the range pH 1–12 (middle). Five subranges of pH (A–E) are indicated in which the different compounds of Scheme 1 are stable as follows: range A, (**VII**); B, (**VIII, IX**); C, (**IX**); D, (**IX, X, XI, XII**); E, (**XI, XII**). Lettered peaks in the lower spectrum refer to structure (**VII**).

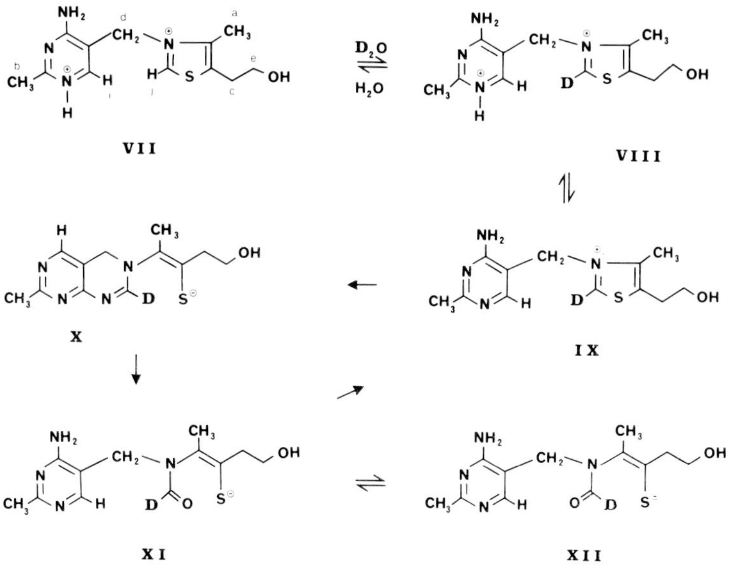

**Scheme 1**

Bisulfite cleaves thiamine into its pyrimidine and thiazole moieties at neutral pH, a reaction that can also be readily followed by NMR since the signal for the bridge methylene group d at $\delta 5.4$ is replaced by a peak at $\delta 4.4$ in the reaction product. Such mechanistic studies of the reaction of bisulfite with several biologically important pyrimidines are reported by Triplett (1977).

The potential of NMR in physicochemical studies of food systems has been little exploited to date. With modern ultrahigh-resolution instruments that give higher precision and sensitivity, and with the new techniques described, such experiments should now be easier than they have been in the past. It is hoped that the few examples cited here may stimulate further work on food applications in this interesting and exciting field.

## V. NEW APPLICATIONS

Reiterating the introduction, the past 3 years have seen an almost unbelievable development in new possibilities, not only through instrumental developments that have extended NMR to a much greater range of samples but also through a greater exploitation of the theory behind NMR. This has led to improvements in NMR sensitivity, the ability to handle solid samples, and 2D-NMR. Perhaps for the purists these techniques are not really new but merely recently available on a routine basis. There have been no reported applications of most of these techniques to date in foods or beverages, but they are worth describing here nonetheless to give an idea of potential uses. Examples have been created with the help of applications laboratory personnel of various NMR spectrometer companies.

In brief, these new possibilities are largely the result of an increased use of computers in pulsed NMR, both for control of data acquisition and for subsequent processing. Exciting pulses can be of long or short duration: they can be applied at a specific frequency or across a range of frequencies and can be made to excite either the nuclei to be observed or other nuclei in the molecule, or indeed both simultaneously. Once recorded, the data can be treated in several ways. Simple pulse sequences led first of all to various decoupling techniques, to solvent peak suppression, resolution or sensitivity enhancement, and automatic estimation of $T_1$ values. More evolved pulse sequences led then to the development of 2D-NMR, but before we go on to this, two other important improvements must be described.

### A. Low-Power Decoupling

It is frequently difficult to obtain quality proton-decoupled $^{13}C$ spectra of organic salts, or of any sample with a high inorganic salt content. The power fed into the sample through the proton decoupler leads to irregular local heating and to a marked deterioration of resolution and of signal-to-noise ratio. Levitt *et al.* (1982) have described a pulse sequence that permits total proton decoupling at about 0.5 W, less than one-tenth of the power require-ments for a conventional decoupler. The method, which they have named MLEV-16, can be implemented by hardware or software on almost any existing pulsed high-resolution spectrometer to give results such as are shown in Fig. 13, the $^{13}C$ spectrum of thiamine chloride hydrochloride. The MLEV-16 trace has a signal-to-noise ratio about five times higher, and peaks that in the regular decoupled spectrum may have been interpreted as closely spaced doublets are evidently sharply defined singlets.

### B. Solid-State Spectra

The major components of foods are solids, and many of them are of limited solubility. Their spectra must therefore be run in the solid state. In addition it is of interest for some samples to observe them in the solid state even if they are soluble. Cross-polarization magic angle spectroscopy (CPMAS) now makes this possible, overcoming problems of broad feature-less bands observed in the conventional spectra of solids. In favorable cases, CPMAS gives spectra with resolution comparable to that of normal solution

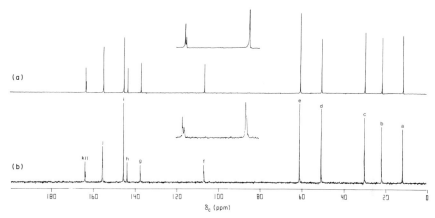

**Fig. 13.** The 50-MHz proton-decoupled $^{13}$C spectra of thiamine, (a) with the new low-energy MLEV-16 decoupler and (b) with the normal wide-band decoupler. Peaks j, k, and l are expanded in insert. Letters in (b) relate to structure (**VII**).

spectra. This is seen in Fig. 14, which shows the comparative $^{13}$C spectra of β-maltose (**XIII**) run in solution, as a solid with a normal probe, and as a solid with a magic angle probe.

**XIII**

Using spectra of solids, O'Donnell et al. (1981) have performed a $^{13}$C study of whole-seed protein and starch content for seeds of different leguminous origins, and Schaefer et al. (1979) have looked at the $^{13}$C spectrum of a $^{15}$N-labeled soybean, estimating the amino acid composition of the protein from resonances of carbons directly bound to nitrogen.

## C. Specific Recognition of Carbon Multiplicities

Proton-decoupled $^{13}$C spectra are simple but give no direct information on carbon multiplicities, that is, whether the different peaks come from $CH_3$, $CH_2$, CH, or quaternary carbons. Proton-coupled spectra can be used but are of low sensitivity and are difficult to interpret. Recently, techniques called APT, INEPT, and DEPT have been developed to overcome this problem. The latest of these, distortionless enhancement by polarization transfer (DEPT), was created by Doddrell et al. (1982), and further details

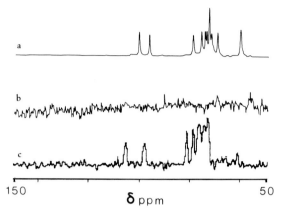

**Fig. 14.** $^{13}C$ spectra of maltose (**XIII**) run (a) in DMSO-$d_6$ solution, (b) as a solid in the regular liquid sample probe, and (c) as a solid in a magic angle probe.

are described by Bendall *et al.* (1982) and by Richarz *et al.* (1982a). The output of a DEPT experiment is presented in Fig. 15, run on a sample of rosemary oil. This contains $\alpha$-pinene, camphene, cineol, camphor, and bornyl alcohol as major constituents along with many more minor compounds. At high sensitivity, the total CH$_x$ spectrum reveals hundreds of peaks. However, as shown, these are readily separated into subsets corresponding to methyl, methylene, and methyne groups, which is a major aid in spectral interpretation.

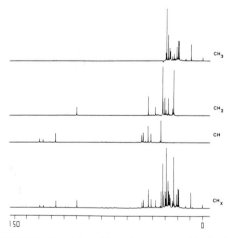

**Fig. 15.** Some DEPT $^{13}C$ subspectra of French rosemary oil in deuterobenzene. For each of the four DEPT spectra, 256 transients were accumulated in 20 min. (Reprinted by permission of Varian Associates, Zug, Switzerland.)

## D. Two-Dimensional NMR: J-Resolved Spectra

Although it is perhaps not apparent from the nature of the subspectra in Fig. 15, DEPT is, in fact, a two-dimensional experiment. However, for most two-dimensional work, we must become accustomed to a different kind of plot, such as those shown in Figs. 16 to 19. The principles of 2D-NMR, originally defined by Jeener (1971), are clear. Spectra, as conventionally represented, contain discretely independent forms of information. A proton spectrum, for example, contains information on chemical shifts and H–H coupling constants. When both types of information are plotted on the same axis, each hides the other. In two-dimensional spectroscopy, two such independent sets of data can be separated, and then peak intensities can be plotted as a function of the two variables, this time represented separately on two orthogonal axes. These axes now define a base plane for the spectrum, analogous to the baseline in a conventional one-dimensional representation. Thus, Fig. 16 is the so-called J-resolved $^{13}$C spectrum of lupane (**XIV**), J-resolved because C–H coupling constants and hence carbon multiplicities are represented on an axis perpendicular to the $^{13}$C chemical shift axis. The

**X I V**

information contained in this spectrum is the same as in a conventional proton-coupled spectrum, except that the multiplets for the different carbons (quartets for $CH_3$ groups, triplets for $CH_2$, doublets for CH, and singlets for quaternary carbons) are turned through an angle of 90° and appear along the $^{13}$C chemical shift axis in the positions that would be observed in a proton-decoupled spectrum. Whereas proton-coupled $^{13}$C spectra are invariably difficult or even impossible to interpret, interpretation of their two-dimensional equivalents is self-evident. Similar J-resolved plots can be used to simplify proton spectra.

## E. Carbon–Proton Heteronuclear Correlated Spectra

Figure 17 is a different type of plot and a different representation. In this case, we are looking perpendicularly down on the base plane of the spec-

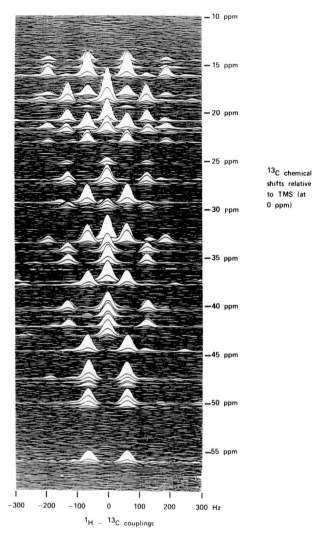

**Fig. 16.** The J-resolved $^{13}$C spectrum of lupane (**XIV**). (Courtesy of W. Ammann, Varian Associates, Zug, Switzerland.)

trum, and peak heights are represented in the form of contour maps exactly analogous to the geographer's representation of mountains, plains, and oceans. The spectrum is a $^1$H–$^{13}$C cross-correlated spectrum of coumarin (**XV**), with the normal proton spectrum spread along the horizontal axis and the proton-decoupled $^{13}$C spectrum on the vertical axis. Clearly, a signal on the two-dimensional map correlates each carbon with its attached protons,

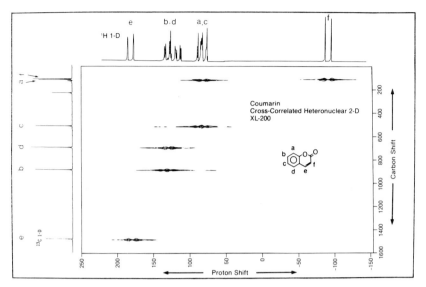

**Fig. 17.** Contour plot of the heteronuclear correlated two-dimensional data for coumarin (**XV**). The $^{13}$C separations among the carbons neatly distinguish proton signals that overlap extensively in the one-dimensional proton spectrum. (Reprinted by permission of Varian Associates, Zug, Switzerland.)

**X V**

and again a major simplification, this time of overlapping multiplets in the proton spectrum, is achieved. This type of plot clearly helps in carbon and proton spectral assignments.

## F. COSY Homonuclear Correlated Spectra

For sucrose (**XVI**) another sort of mapped plot is found in Fig. 18. This is known as a COSY-45 plot (COSY for correlated spectroscopy), or a $^1$H homonuclear correlated spectrum. The entire proton spectrum of sucrose, shown at the foot of the figure, is found in a narrow frequency spread of about 2 ppm. This segment becomes both the horizontal and the vertical axes, as shown. Correlated peak positions between the two spectra are inevitably located on the diagonal of the plot. But the signals of real interest are those that are symmetrically disposed on each side of the diagonal. These signals identify nuclei between which proton–proton coupling occurs, and

**X V I**

hence they give structural and conformational information. Thus, for example, proton Gl in (**XVI**) evidently not only couples with G2 as expected but also shows long-range coupling with Fl, G3, and G5. The square demonstrating the Gl–F5 coupling is drawn, as is another square indicating the expected coupling between F3 and F4.

Other homonuclear plots going under the names of SECSY and NOESY also yield structural and conformational information.

### G. INADEQUATE Carbon – Carbon Connectivity Plots (CCCP)

Probably the most intriguing two-dimensional experiment fulfills a dream common to all chemists who have ever engaged in structural elucidation. Which chemist has not at some time wished that he or she could take a photograph of a particularly difficult to identify molecule? To all intents and purposes, this can now be done using an experiment that goes under the surprisingly unassuming name of 2D-INADEQUATE, or CCCP. Using this method, we can perform what are known as carbon connectivity measurements on all of the C–C bonds in a molecule simultaneously. In other words, this implies the unambiguous detection of pairs of peaks in the regular proton-decoupled $^{13}$C spectrum coming from each pair of neighboring carbon atoms in the molecule. Because we can already determine from DEPT or J-resolved plots which signals in the normal spectrum correspond to $CH_3$, $CH_2$, CH, and quaternary carbons, we can immediately identify the respective multiplicities of neighboring carbons throughout the whole carbon skeleton of our molecule.

The example chosen to demonstrate this is again sucrose, and information can be presented either in a mapped plot form (Fig. 19a) or in profile spectra of adjacent pairs (Fig. 19b). The complete proton-decoupled $^{13}$C spectrum of sucrose is shown for reference under both output formats. It is immediately obvious that single peaks in the normal spectrum appear as

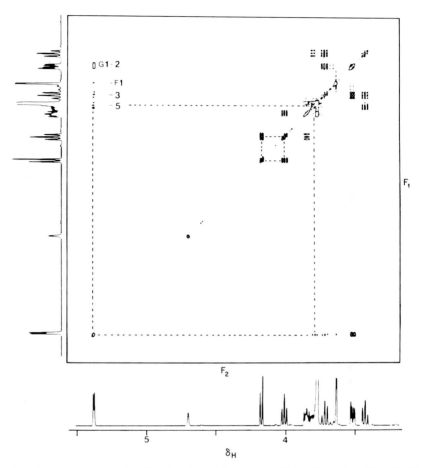

**Fig. 18.** The 500-MHz COSY-45 proton homonuclear shift correlated two-dimensional spectrum of sucrose (**XVI**) in $D_2O$ solution. Symmetrical off-diagonal signals reveal unexpected long-range couplings such as G1–F1, G1–G3, and G1–G5 in addition to the stronger couplings such as G1–G2 and F3–F4. Squares indicating G1–G5 and F3–F4 couplings are marked. (Reprinted by permission of Bruker-Spectrospin, Zurich, Switzerland.)

doublets in Fig. 19b and as twin dots in the mapped plot form. These doublets are in fact satellite peaks associated with each carbon and arising from $^{13}C-^{13}C$ one-bone coupling. They are present in the normal $^{13}C$ spectrum but have intensities less than one in $10^4$ relative to the major peaks. In the INADEQUATE experiment, the major peaks are eliminated, leaving only the satellites. Pairs of twin dots in the mapped plot, such as those circled, are symmetrically disposed in the direction of the horizontal axis about a 45° line passing through the center of the plot as drawn. The circled

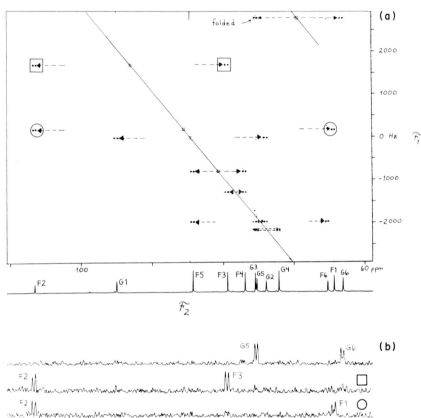

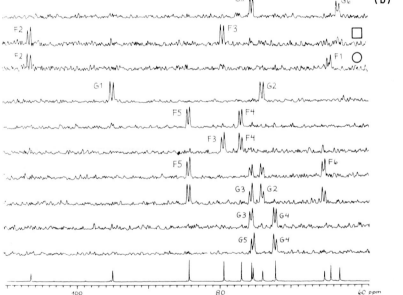

**Fig. 19.** The 100-MHz $^{13}$C INADEQUATE spectrum of sucrose (**XVI**) in D$_2$O solution, showing carbon–carbon connectivities, (a) in mapped plot form and (b) as individual rows of the matrix. Signals revealing the connectivity between F1 and F2 are shown in circles and those between F2 and F3 in squares. (Reprinted by permission of Bruker-Spectrospin, Zurich, Switzerland.)

dots indicate that the second peak from the right in the conventional spectrum, known to be a $CH_2$ peak, evidently has as partner the peak on the extreme left, which comes from a quaternary carbon. The latter peak has a second partner, as indicated by the pair of twin dots in the squares, this time a CH. This information is repeated in the profile spectra of Fig. 19b in the two rows also marked with circle and square. In fact, the three partners identified correspond to positions F1, F2, and F3 of the fructose moiety of sucrose. The process can be continued to identify the remaining elements of the carbon skeleton of both fructose and glucose.

The two-dimensional examples of coumarin and glucose come from a Varian applications report (Gray, 1982) and from a Bruker publication (Hull, 1982), respectively. An application of several two-dimensional techniques in the total structural and conformational analysis of lupane has been described by Ammann et al. (1982). These publications give many more references to two-dimensional studies.

## H. Further Developments

The performance of INADEQUATE in confirming the structure of sucrose perhaps loses some of its immediate impact because sucrose is a well-known compound, but the idea of applying this method to partially or totally unknown structures is more spectacular. However, before rushing off to buy a spectrometer with a view to identifying the few milligrams or less of a compound painstakingly isolated from a food material, it is worth noting that in the present-day state of the art at least 500 mg of compound is needed. This state of affairs may change, and in fact there is every reason to suppose it will. Already, newer approaches, such as the relayed coherence transfer method described by Bolton (1982), are being actively pursued. This method is similar to INADEQUATE expect that coupling information from neighboring sites is propagated first from proton to proton and then on to the attached carbons, which are ultimately observed. Richarz et al. (1982b) have used this approach to probe the structure of a trisaccharide, reporting it to be more sensitive than INADEQUATE and better resolved than homonuclear correlated proton spectra. This type of development may reasonably be expected to continue and can only serve to consolidate the position of NMR as the instrumental technique giving the most comprehensive structural information per milligram of substance examined.

## VI. CONCLUSIONS

In response to the inevitable question of whether these new techniques will be useful in the analysis of foods and beverages, the following points can be cited. First, the examples given here already relate directly or indirectly to food components. Second, it is not out of the question to apply these techniques to highly complex structures (or perhaps even to mixtures) as shown by recent elegant work on protein structure. Third, who could have foreseen at the time of Arnold's measurements of the ethanol proton spectrum in 1951 the number of NMR applications now possible in studies of food materials? It is surely reasonable to predict that the coming years will produce many interesting results that can only enrich our knowledge of the food we eat.

## REFERENCES

Albornoz, F., and Leon, V. (1980). *Acta Cien. Venez.* **31,** 20.
Altena, J. H., Van de Ouweland, G. A. M., Teunis, C. J., and Tjan, S. B. (1981). *Carbohydr. Res.* **92,** 43.
Ammann, W., Richarz, R., Wirthlin, T., and Wendisch, D. (1982). *Org. Magn. Reson.* **20,** 260.
Anders, U., Tittgemeier, F., and Hailer, G. (1976). *Z. Lebensm.-Unters.-Forsch.* **162,** 21.
Anonymous (1971). *Agric. Res. (Washington)* **20,** 7.
Arnold, J. T., Dharmatti, S. S., and Packard, M. E. (1951). *J. Chem. Phys.* **19,** 507.
Baianu, I. C. (1981). *J. Sci. Food Agric.* **32,** 309.
Barton, F. E., Himmelsbach, D. S., and Walters, D. B. (1978). *J. Am. Oil Chem. Soc.* **55,** 574.
Belleau, G., and Dadic, M. (1977). *J. Am. Soc. Brew. Chem.* **35,** 191.
Bendall, M. R., Doddrell, D. M., Pegg, D. T., and Hull, W. E. (1982). Bruker-Spectrospin Application Note "High Resolution Multipulse NMR Spectrum Editing and DEPT."
Bloch, F. (1946). *Phys. Rev.* **70,** 460.
Bloch, F., Hansen, W. W., and Packard, M. E. (1946). *Phys. Rev.* **69,** 127.
Blunt, J. W., Munro, M. H. G., and Swallow, W. H. (1979). *Aust. J. Chem.* **32,** 1339.
Bociek, S. M., Izzard, M. J., Morrison, A., and Welti, D. (1981). *Carbohydr. Res.* **93,** 279.
Bolton, P. (1982). *J. Magn. Reson.* **48,** 336.
Bouldoires, J. P. (1975). *In* "Application de la SM et de la RMN dans les Industries Alimentaires" (R. Ammon and J. Hollo, eds.), pp. 233–282. Commission Internationale des Industries Agricoles et Alimentaires, Paris.
Bovey, J. (1969). "Nuclear Magnetic Resonance Spectroscopy." Academic Press, New York.
Bradbury, J. H., and Collins, J. G. (1982). *Cereal Chem.* **59,** 159.
Brennan, M. (1982). *Technology (Ireland)* **13,** 39.
Brosio, E. (1981). *J. Food Technol.* **16,** 629.
Brosio, E. (1982). *J. Am. Oil Chem. Soc.* **59,** 1.
Bus, J., and Frost, D. J. (1974). *Recl. Trav. Chim. Pays-Bas* **93,** 2136.
Chapman, D., and Morrison, A. (1966). *J. Biol. Chem.* **241,** 5044.
Chapman, D., Richards, R. E., and Yorke, R. W. (1959). *Nature (London)* **183,** 44.
Chapman, D., Richards, R. E., and Yorke, R. W. (1960). *J. Am. Oil Chem. Soc.* **37,** 243.
Chen, S., Elofson, R. M., and MacTaggart, J. M. (1979). *J. Agric. Food Chem.* **27,** 435.
Coles, B. A. (1980). *J. Am. Oil Chem. Soc.* **57,** 202.
Corse, J., Lundin, R. E., Sondheimer, E., and Weiss, A. C., Jr. (1966). *Phytochemistry* **5,** 767.
Coveney, L. V. (1980). *Br. Food Manuf. Ind. Res. Assoc.* pp. 112–136.
Cowburn, D. A., Bradbury, E. M., Crane-Robinson, C., and Gratzler, W. B. (1970). *Eur. J. Biochem.* **14,** 83.
Cox, R. H., and Cole, R. J. (1977). *J. Org. Chem.* **42,** 112.

Coxon, B. (1980). *Dev. Food Carbohydr.* **2**, 351.
Currie, R. W., Jordan, R., and Wolfe, F. H. (1981). *J. Food Sci.* **46**, 822.
Dadic, M. (1977). *Brew. Dig.* **52**, 46.
Daun, P., and Biely, J. (1980). *J. Am. Oil Chem. Soc.* **57**, 380.
Desarzens, C., and Michel, F. (1982). *Rev. Fr. Corps Gras.* **29**, 419.
Desarzens, C., Besson, A., and Bouldoires, J. P. (1978). *Rev. Fr. Corps Gras.* **25**, 183.
Deyama, T., and Horiguchi, T. (1971). *Yakugaka Zasshi* **91**, 1383.
Doddrell, D., and Allerhand, A. (1971). *J. Am. Chem. Soc.* **93**, 2779.
Doddrell, D. M., Pegg, D. T., and Bendall, M. R. (1982). *J. Magn. Reson.* **48**, 323.
Ellison, R. A., and Kotsonis, F. N. (1976). *J. Org. Chem.* **41**, 576.
Endo, T., Misu, O., and Inaba, Y. (1969). *J. Oil Chem. Soc. (Tokyo)* **18**, 63.
Eppeley, R. M., Mazzola, E. P., Highget, R. J., and Bailey, W. J. (1977). *J. Org. Chem.* **42**, 240.
Feeney, J., Birdsall, B., Roberts, G. C. K., and Burgen, A. S. V. (1977). *In* "NMR in Biology" (R. A. Dwek, I. D. Campbell, R. E. Richards, and R. J. P. Williams, eds.), pp. 111–124. Academic Press, New York.
Ferren, W. P., and Seery, W. E. (1973). *Anal. Chem.* **45**, 2278.
Ferretti, A., Flanagan, V. P., Bondarovich, H. A., and Gianturco, M. A. (1968). *J. Agric. Food Chem.* **16**, 756.
Feucht, G., Hermann, K., and Heimann, W. (1971). *Z. Lebensm.-Unters.-Forsch.* **145**, 206.
Flament, I., Sonnay, P., and Ohloff, G. (1977). *Helv. Chim. Acta* **60**, 1872.
Formacek, V. (1979). Ph.D. thesis, Julius-Maximilians-Universitaet, Wuerzburg, Germany.
Formacek, V., and Kubeczka, K. H. (1979). *In* "Vorkommen und Analytik Aetherische Oele" (K. H. Kubeczka, ed.), pp. 130–138. Thieme, Stuttgart.
Formacek, V., and Kubeczka, K. H. (1982a). "Essential Oil Analysis by Capillary Gas Chromatography and Carbon-13 NMR Spectroscopy." Wiley, New York.
Formacek, V., and Kubeczka, K. H. (1982b). *In* "Aetherische Oele: Analytik, Physiologie, Zusammensetzung" (K. H. Kubeczka, ed.), pp. 42–53. Thieme, Stuttgart.
Foster, R. (1969). "Organic Charge Transfer Complexes." Academic Press, New York. (See Chapters 5, 6, 7.)
Glickson, J. D., Phillips, W. D., McDonald, C. C., and Poe, M. (1971). *Biochem. Biophys. Res. Commun.* **42**, 271.
Gray, G. A. (1982). Varian Instruments Application Report, "Two-Dimensional NMR on the XL-200," No. MAG2626.
Hirata, T., and Ishitani, T. (1978). *Agric. Biol. Chem.* **42**, 775.
Ho, C., Magnuson, J. A., Wilson, J. B., Magmuson, N. S., and Kurland R. J. (1969). *Biochemistry* **8**, 2074.
Holsinger, V. H., and Posati, L. P. (1974). *Abstr. Pap. Am. Chem. Soc.* **168**, AGFD 11.
Hoppmann, R. F. W. (1982). *N. Y. Acad. Sci.* **378**, 32–50.
Horike, S., Ohkuma, H., Tamai, H., and Akahoshi, R. (1982). *Nippon Nogei Kagaku Kaishi* **56**, 13.
Horman, I., and Brambilla, E. (1982). *Int. J. Vitam. Nutr. Res.* **52**, 134.
Horman, I., and Dreux, B. (1982). *Anal. Chem.* **55**, 1219.
Horman, I., and Viani, R. (1971). *Proc. 5th ASIC Colloq. Coffee, Lisbon, 1971* pp. 102–111.
Horman, I., and Viani, R. (1972). *J. Food Sci.* **37**, 925.
Horman, I., Carpenter, R. A., Brambilla, E., and Dreux, B. (1980). *Proc. 9th ASIC Colloq. Coffee, London, 1980* p. 99.
Horman, I., Brambilla, E., and Stalder, R. (1981). *Int. J. Vitam. Nutr. Res.* **51**, 385.
Hull, W. E. (1982). Bruker-Spectrospin Instruments "Two-Dimensional NMR, Aspect 2000."
Hyvönen, L., Varo, P., and Koivistoinen, P. (1977). *J. Food Sci.* **42**, 657.
Ishitani, T., Hirata, T., and Kato, K. (1978). *Agric. Biol. Chem.* **42**, 897.

Jackman, L. M. (1959). "Applications of NMR Spectroscopy in Organic Chemistry." Pergamon, Oxford.

Jeener, J. (1971). Ampere Summer School II, Basko Polje, Yugoslavia.

Johnson, L. F. (1971). *Anal. Chem.* **43**, 28A.

Johnson, L. F., and Shoolery, J. N. (1962). *Anal. Chem.* **34**, 1136.

Jones, R. A. (1974). *Flavour Ind.* **5**, 125.

Josefowicz, M. L., O'Neill, I. K., and Prosser, H. J. (1977). *Anal. Chem.* **49**, 1140.

Kainosho, M., and Konishi, H. (1976). *Tetrahedron Lett.* **51**, 4757.

Kim, K., Kurata, T., and Fujimaki, M. (1973). *Agric. Biol. Chem.* **38**, 53.

Kitamura, K., and Shibamoto, T. (1981). *J. Agric. Food Chem.* **29**, 188.

Kowaka, M., Kokubo, E., and Kuroiwa, Y. (1970). *Proc. 11th Brew. Convent.* pp. 35–45.

Lambelet, P. (1980). *Trav. Chim. Aliment. Hyg.* **71**, 119.

Leal, K. Z., Costa, V. E. U., Seidl, P. R., Campos, M. P. A., and Colnago, L. A. (1981). *Ciec. Cult. (Sao Paulo)* **33**, 1475.

Le Botlan, D., Berry, M., Mechin, B., and Martin, G. J. (1980). *J. Am. Chem. Soc.* **84**, 414.

Le Botlan, D., Bertrand, T., Mechin, B., and Martin, G. J., (1982). *Nouv. J. Chim.* **6**, 107.

Lechert, H., Maiwald, W., Kothe, R., and Basler, W. D. (1980). *J. Food. Proc. Preserv.* **3**, 275.

Lee, C. Y. (1978). *Biochem. Biophys. Acta* **527**, 289.

Levitt, M. H., Freeman, R., and Frenkiel, T. (1982). *J. Magn. Reson.* **47**, 328.

Lillford, P. J., Jones, D. V., and Rodger, G. W. (1980). *In* "Advances in Fish Science and Technology" (J. J. Connell, ed), pp. 495–497. Fishing News Books, Farnham.

Lowman, D. W., and Maciel, G. E. (1978). *Anal. Chem.* **51**, 85.

Maarse, H., and Van Os, F. H. L. (1973). *Flavour Ind.* **4**, 477.

Marmion, D. M. (1974). *J. Am. Oil Chem. Soc.* **57**, 495.

Martin, G. J., and Martin, M. L. (1981a). *C. R. Hebd. Seances Acad. Sci. Paris* **293**, 31.

Martin, G. J., and Martin, M. L. (1981b). *Tetrahedron Lett.* **22**, 3525.

Martin, G. J., and Martin, M. L. (1982). *J. Chem. Phys.* (in press).

Martin, G. J., Martin, M. L., Mabon, F., and Bricout, J. (1982a). *J. Am. Chem. Soc.* **104**, 2658.

Martin, G. J., Martin, M. L., Mabon, F., and Michon, M. J. (1982b). *Anal. Chem.* **54**, 2380.

Martin, G. J., Martin, M. L., Mabon, F., and Michon, M. J. (1982c). *Chem. Commun.* p. 616.

Martin, M. L., Delpuech, J. J., and Martin, G. J. (1980). "Practical NMR Spectroscopy." Heyden, London.

Mel'nikov, K. A. (1982). Maslo-Zhir. Prom-st. 44; CA 97, 22332q.

Merbach, A. E., Chastellain, F., and Hirsbrunner, P. (1975). *Mitt. Geb. Lebensmittelunters. Hyg.* **66**, 176.

Mills, B. L., and Van de Voort, F. R. (1981). *J. Am. Oil Chem. Soc.* **58**, 776.

Molyneux, R. J., and Wong, Y. (1975). *J. Agric. Food Chem.* **23**, 1201.

Müller-Stock, A., Joshi, R. K., and Buechi, J. (1973). *Helv. Chem. Acta* **56**, 76.

Murai, A., Kainosho, M., Takeuchi, Y., Kato, T., and Kimizuka, A. (1979). *Abstr. Pap. Am. Chem. Soc.* **177**, 1.

Nakel, G. M., and Haynes, L. V. (1972). *J. Agric. Food Chem.* **20**, 682.

Nielsen, L. V. (1976). *Milchwissenschaft* **31**, 598.

Noda, M., and Takahashi, T. (1979). *Yukagaku* **28**, 411; CA 92, 5688 q.

O'Donnell, D. J., Ackermann, J. J. H., and Maciel, G. E. (1981). *J. Agric. Food. Chem,* **29**, 514.

Ollis, W. D., Coxon, D. T., and Holmes, A. (1970). *Tetrahedron Lett.* p. 5241.

O'Neill, I. K., Trimble, M. L., and Casey, J. C. (1979). *Meat Sci.* **3**, 223.

O'Neill, I. K., Sargent, M., and Trimble, M. L. (1980). *Anal. Chem.* **52**, 1288.

Pawlowski, N. E., Nixon, J. E., and Sinnhuber, R. O. (1972). *J. Am. Oil Chem. Soc.* **49**, 387.

Pfeffer, P. E., Sampugna, J., Schwarz, D. P., and Shoolery, J. N. (1977). *Lipids* **12**, 869.

Piretti, M. V. (1975). *In* "Applications de la SM et de la RMN dans les Industries Alimentaires" (R. Ammon and J. Hollo, eds.), pp. 365–400. Commission Internationale des Industries Agricoles et Alimentaires, Paris.

Pople, J., Schneider, W., and Bernstein, H. (1959). "High Resolution Nuclear Magnetic Resonance." McGraw-Hill, New York.

Pratt, L., and Richards, R. E. (1953). *Trans. Faraday Soc.* **49**, 774.

Press, K., Sheeley, R. M., Hurst, W. J., and Martin, R. A., Jr. (1981). *J. Agric. Food Chem.* **29**, 1098.

Purcell, E. M., Torrey, H. C., and Pound, R. V. (1946). *Phys. Rev.* **69**, 37.

Pyysalo, H., and Enqvist, J. (1975). *Finn. Chem. Lett.* p. 136.

Pyysalo, H., Enqvist, J., Honkanen, E., and Pippuri, A. (1975). *Finn. Chem. Lett.* p. 129.

Que, L., Jr., and Gray, G. R. (1974). *Biochemistry* **13**, 146.

Ramirez, J. E., Cavanaugh, J. R., Schweizer, K. S., and Hoagland, P. D. (1975). *Anal. Biochem.* **63**, 130.

Richards, R, E., and Smith, J. A. S. (1951). *Trans. Faraday Soc.* **47**, 1261.

Richarz, R., Ammann, W., and Wirthlin, T. (1982a), Varian Application Note No. Z-15.

Richarz, R., Ammann, W., and Wirthlin, T. (1982b). Varian Application Note No. Z-17.

Richter, H., and Spiteller, G. (1979). *Chem. Ber.* **112**, 1088.

Richter, H., Obermann, H., and Spiteller, G. (1977). *Chem. Ber.* **110**, 1963.

Richter, I., and Lakshminarayana, G. (1979). *Chem. Phys. Lipids* **25**, 191.

Roberts, J. D. (1959). "Nuclear Magnetic Resonance." McGraw-Hill, New York.

Samuelsson, E. G., and Hueg, B. (1973). *Milchwissenschaft* **28**, 329.

Schaefer, J., and Stejskal, E. O. (1974). *J. Am. Oil Chem. Soc.* **51**, 210.

Schaefer, J., and Stejskal, E. O. (1975). *J. Am. Oil Chem. Soc.* **52**, 366.

Schaefer, J., Stejskal, E. O., and McKay, R. A. (1979). *Biochem. Biophys. Res. Commun.* **88**, 274.

Schweizer, T., and Horman, I. (1981). *Carbohydr. Res.* **95**, 61.

Scott, P. M. (1978). *J. Food Prot.* **41**, 385.

Shannon, P. V. R., Lloyd, R. O., and Cahill, D. M. (1969). *J. Inst. Brew.* **75**, 376.

Shoolery, J. N. (1973). Varian Application Note No. NMR-73-3.

Shoolery, J. N. (1975). Varian Application Note No. NMR-75-3.

Shoolery, J. N. (1982). Varian Application Note No. NMR-82-4.

Shoolery, J. N., and Jankowski, W. C. (1973). Varian Application Note No. NMR-73-4.

Shoolery, J. N., and Smithson, L. H. (1969). *J. Am. Oil Chem. Soc.* **47**, 153.

Sohde, K., Ogawa, T., and Matsushita, S. (1974). *J. Jpn. Oil Chem. Soc. Yukagaku* **23**, 228.

Spencer, J. F. T., Gorin, P. A. J., Hobbs, G. A., and Cooke, D. A. (1970). *Can. J. Microbiol.* **16**, 117.

Stagg, G. V., and Swaine. D. (1971). *Phytochemistry* **10**, 1671.

Stahl, H., and McNaught, R. P. (1970). *Cereal Chem.* **47**, 345.

Sterk, B. (1969). *Z. Lebensm.-Unters.-Forsch.* **140**, 154.

Stockton, G. W., and Smith, I. C. P. (1976). *Chem. Phys. Lipids* **17**, 251.

Stockton, G. W., Polnaszek, C. F., Leitch, L. C., Tulloch, A. P., and Smith, I. C. P. (1974). *Biochem. Biophys. Res. Commun.* **60**, 844.

Stoessl, A., and Stothers, J. B. (1978). *Can. J. Bot.* **56**, 2589.

Stoessl, A., Ward, E. W. B., and Stothers, J. B. (1976). *Tetrahedron Lett.* **37**, 3271.

Stoessl, A., Stothers, J. B., and Ward, E. W. B. (1978). *Can. J. Chem.* **56**, 645.

Stubblefield, R. D., Shotwell, O. L., Shannon, G. M., Weisleder, D., and Rohwedder, W. K. (1970). *J. Agric. Food Chem.* **18**, 391.

Takino, Y., Ferretti, A., Flanagan, V. P., Gianturco, M. A., and Vogel, M. (1967). *Can. J. Chem.* **45**, 1949.

Taylor, M. G., and Smith, I. C. P. (1981). *Chem. Phys. Lipids* **28**, 119.

Taylor, M. G., Akiyama, T., and Smith, I. C. P. (1981). *Chem. Phys. Lipids* **29**, 327.

Tisse, C., Artaud, J., Iatrides, M.-C., Zahra, J.-P., and Estienne, J. (1979). *Ann. Falsif. Expert. Chim.* **72**, 565.

Traitler, H., and Prévot, A. (1981). *HRC CC J. High Resolut. Chromatogr. Chromatogr. Commun.* **4,** 109

Triplett, J. W. (1977). *Diss. Abstr. Int.* **38**, 2195.

Trumbetas, J., Fiotiti, J. A., and Sims, R. J. (1978). *J. Am. Oil Chem. Soc.* **55,** 248.

Tulloch, A. P. (1982). *Lipids* **17,** 544.

Turczan, J. W., Goldwitz, B. A., and Medwick, T. (1977). *J. Agric. Food Chem.* **25,** 594.

Unterhalt, B. (1980). *Z. Lebensm.-Unters.-Forschung* **170,** 425.

Van Putte, K. (1975). *In* "Bruker MINISPEC Manual," pp. 73–75. Bruker, Karlsruehe.

Van Putte, K., and Van den Enden, J. J. (1973). *J. Sci. Instrum.* **6,** 910.

Van Putte, K., and Van den Enden, J. J. (1974). *J. Am. Oil Chem. Soc.* **51,** 316.

Viani, R., and Horman, I. (1975). *Proc. 7th ASIC Colloq. Coffee, Hamburg, 1975* pp. 273–278.

Viani, R., Bricout, J., Marion, J. P., Mueggler-Chavan, F., Reymond, D., and Egli, R. H. (1969). *Helv. Chim. Acta* **52,** 887.

Wehrli, F. W., and Wirthlin. T. (1976). "Interpretation of Carbon-13 NMR Spectra," pp. 247–264. Heyden, London.

Weinges, K. (1971). *Acta. Phys. Chim. Debrecina* **17,** 249.

Weisser, H. (1978). Bruker Spectrospin Report 78(1), 9.

Weisser, H. (1979). *In* "Food Process Engineering, 1979" (Finnish Ministry of Trade and Industry, eds.), pp. 326–336. Espoo, Finland.

Williams, D. H., and Fleming, I. (1966). "Spectroscopic Methods in Organic Chemistry." McGraw-Hill, New York.

# 8

# Minicomputers and Robotics

GERALD F. RUSSELL

*Department of Food Science and Technology*
*University of California*
*Davis, California*

ANALYSIS OF FOODS AND BEVERAGES
Copyright © 1984 by Academic Press, Inc.
All rights of reproduction in any form reserved.
ISBN 0-12-169160-8

## I. INTRODUCTION

Few, if any, developments in the multidisciplinary areas related to the analysis of foods and beverages have had so strong and ubiquitous an influence as the computer. Computers are no longer hidden away in sealed, protected chambers, presided over by high priests of computing. Rapid developments in various associated technologies have pushed them to levels of high capability, low cost, small size, and ease of use, where they are now beginning to be integrated into the routine world of most analytical laboratories. Virtually every analytical scientist is faced with decisions related to the purchase and use of analytical and control instrumentation that various vendors describe as being "computer controlled," "fully automated," "microprocessor controlled," and so on.

It is not the purpose of this chapter to review all the available instrumentation or methodologies that utilize computers. Specific instrumentation and protocols are covered in other chapters of this book. This chapter will outline some of the basic principles of operation around which computers function as measurement and control devices, with only limited reference to their use in data analysis programs. Selected references will be made to some applications that highlight particular applications related to the analysis of foods and beverages.

It is safe to say that the trend toward computer and microprocessor involvement in analytical and control applications is not likely to diminish in the future. With an ever-increasing and sometimes bewildering array of computerized products available from specific vendors, this newer technology can rapidly leave the scientist on very dangerous ground: specifically, the scientist may find himself or herself in the position of purchasing equipment controlled by the proverbial black box. It is the intent of this chapter to present an overview of the operating principles involved in new computers that will allow the reader to make knowledgeable decisions predicated upon what computers can, and cannot, do for their specific applications and needs.

Since the use of newer microcomputers in this area is still in its infancy, some items in this chapter may appear to the reader to be new terminology, but such terms are necessary for discussions with designers, vendors, and users of computer-controlled devices. Some knowledge of this field is assumed, but a small appendix on terminology at the end of the chapter is included to aid in the familiarization with some common terms.

## II. DEVELOPMENT OF COMPUTERS FOR LABORATORIES

This section is intended to give an overview of the development of computers suitable for analytical needs. Details of the functional attributes and uses will be presented in further sections.

### A. Mainframe Computers

The development of mini/micro–computers/processors has evolved from the traditional architecture of large mainframe computers. The large mainframe computers continue to occupy an important role in traditional university, government, and corporate computer centers. Their primary use is to provide very fast calculations of complex equations, and they have the ability to maintain and update large bases of information. They will continue to be important for scientific computations such as multivariate analyses and simulation programs. Mainframe computers are not well suited for real-time data acquisition and control of local laboratory or processing operations; these are more efficiently implemented on smaller, dedicated computers (see below). However, it is becoming increasingly recognized that an interface between local dedicated computers and larger mainframes through appropriate networks can be an efficient and optimal means of cost-effective laboratory analysis with concomitant storage and management of large data bases.

### B. Minicomputers

Kenneth Olson left the rarified atmosphere of MIT's Whirlwind computer project in 1957 with the goal of making smaller, more flexible, less expensive, transistorized computers. By 1960, Digital Equipment Corporation (DEC) marketed its first computer, the PDP-1. (The use of the phrase "Programmable Data Processor" in place of "computer" allowed many government and university laboratories to circumvent restrictions on purchasing "computers"). This first machine accommodated an astonishing range of peripherals, including one that revolutionized the way the world viewed computers — the cathode-ray tube (CRT) video screen.

Development of integrated circuit (IC) components and chips led to smaller component parts for the computers themselves. In 1965 the PDP-8 series was marketed as the first mass-produced minicomputer and brought computing power out of the cloistered machine room forever. This was followed in the 1970s with the highly successful 16-bit (Binary digIT) architecture of the PDP-11 series and the VAX-11 32-bit machines in the 1980s.

Digital Equipment Corporation was and probably still is the largest manufacturer of minicomputers. However, other vendors were competing in this new field, Data General and Hewlett Packard among others, each with their own architectures and special features to develop an aggressive and competitive market for laboratory and process-control minicomputers. These computers were designed around "board-sized" components, and each laboratory would typically configure a computerized system around optimal peripheral devices, often in a standard electronic 19-in. rack-mounted chassis. Hardware interfaces to instrumentation were frequently done by engineering teams at each location and vendor-supplied software for support of the computing systems was sparse, leaving the responsibility for system development largely with the user.

Newer minicomputer architectures have reduced the size of the hardware, lowered costs dramatically, and greatly expanded the availability of general-purpose software for the laboratory environment. It has become common to note that new generations of minicomputers have capabilities greater than previous generations of mainframe computers.

### C. Microcomputers

The development of large-scale integration (LSI) and very large scale integration (VLSI) chips allowed the evolution of further size and cost reductions for computer components. In many cases, an order-of-magnitude reduction in cost was possible while reliability and component functions increased. Often the cost of the associated peripheral equipment needed for laboratory use was the most important factor in cost-effectiveness analyses of computerization.

In addition to the laboratory-oriented microcomputers, there appeared in the late 1970s a dramatic increase in the availability of personal computers, with far-reaching effects. For example, two of the most successful personal computer series, the Apple and the TRS-80, brought the computer even further out of the computer-center concept and increased the computer literacy of a whole new generation of people, who rapidly developed programming and interfacing skills. This led to the use of many personal computers in laboratory and analytical applications where the power of microcomputers could be realized at relatively low costs. These microcomputers have user-friendly operating systems and support easily understood high-level computer languages such as BASIC. For example, Walling et al. (1982) described an Apple-II-Plus microcomputer interfaced to a gas chromatograph/mass spectrometer for data acquisition and manipulation of data.

## D. Microprocessors

The development of LSI and VLSI technologies has led to the incorporation of an entire central processing unit (CPU) on a single chip. The arithmetic and logic functions of a microprocessor can be built around very few component chips, and it requires only minimal space. Microprocessors are primarily used for dedicated tasks that can be handled with great speed. They do not usually have large memories, extensive peripherals, or operating systems and operate as stand-alone units driven by machine language instructions (see below). They are presently designed by engineering teams and implemented by vendors of specific instrumentation or devices, and they are particularly useful for dedicated tasks such as controlling laboratory instruments or robotic devices.

In summary, there exist levels of sophistication in computing power. There has been a continued decrease in cost and size of computers, with the amount of supported software and peripheral equipment determining whether a particular configuration is a mainframe computer, minicomputer, microcomputer, or microprocessor. In many cases the major distinction between a mainframe computer and a mini/microcomputer is not so much in the physical dimensions but rather in the size of the required support personnel and the bureaucracy associated with its use and maintenance.

## III. COMPUTER ARCHITECTURE

Computers all start off with a series of wires termed a bus. The various components of a computer system communicate with each other along the bus according to strict protocols. The heart of the system is the microprocessor or CPU. The addition of memory, and other peripherals, makes a computer. Each unit can communicate through one or more bus control paths such as a data bus, an address bus, or a control bus. Some architectures combine these functions in a single common bus.

## A. Central Processing Unit

The CPU and bus are analogous to the brain and spinal cord in the body. Data on the bus are transmitted and received using specified rates and protocols determined by control signals from the CPU. The CPU has two major functions: control and computation.

The control function establishes the synchronization and timing necessary for the computer system. For example, to execute a program instruction, the control unit will fetch the instruction from memory and decode the instruction into a form that can then be passed to the computational unit, which subsequently executes the decoded instruction.

The computation part of the CPU contains the arithmetic–logic unit (ALU). The ALU is tightly coupled with registers (accumulators) and status–control flags. The ALU executes arithmetic or logical operations decoded from the CPU control logic. Common arithmetic operations would include addition or subtraction of register contents, or comparison of register contents. The ALU also performs Boolean logical operations such as NOT, OR, and AND.

The registers or accumulators hold the results of arithmetic, comparison, or Boolean operations. In addition, a special register termed a program counter keeps track of the memory addresses of instructions that the CPU will follow. Some architectures include another register termed a stack pointer, which monitors stack-oriented functions within the CPU.

## B. Memory

The addition of memory accessible to the CPU along a bus allows rapid access to instructions and data. Each element of memory can be thought of as a series of cells. Each cell has its own address and also contains stored information. This information can be either data or instructions for the CPU to process.

Earlier computers used core memory, so named for the internal structure that was based on thousands of tiny ferrite rings or cores that could be magnetized in either of two directions equivalent to binary 0's and 1's. This was a permanent or nonvolatile memory such that, when power was removed, each memory cell retained its magnetic orientation, which could subsequently be read by the CPU. This was a read–write (R/W) type of memory, and the CPU could address each memory cell at random to yield the term *random access memory* (RAM).

Some newer types of memory are constructed from semiconductor methods (largely metal oxide semiconductor [MOS] technologies), which have lowered size and costs but may be "volatile"; that is, memory information may be lost if power is interrupted. Analytical scientists need to be aware of types of memory in their microcomputers and establish safety backup protocols for even momentary power interruptions. Early minicomputers commonly had only 4 or 8 kilobytes (KB) of memory because of its high cost. Subsequent cost reductions have made laboratory microcomputer systems available with megabytes (MB) of memory at reasonable costs. Often an initial purchase of additional memory will reduce software overhead and speed real-time data acquisition for laboratory routines.

Read only memory (ROM) is often supplied as a permanent form of storage for certain programming needs. The CPU can read instructions from an ROM but cannot write into or alter its contents. For dedicated microprocessors and minicomputers, programmable read only memory (PROM) chips allow users to "burn in" their own microcode. Some forms of erasable programmable read only memory (EPROM) can be programmed as with a PROM, but exposure to ultraviolet (UV) radiation will erase the memory, allowing subsequent reprogramming if necessary. Under electronic conditions slightly different from normal operating voltages, electronically alterable PROM (EAPROM) and electronically erasable PROM (EEPROM) chips have appeal as reprogrammable, but nonvolatile, memories for microcode. Such forms of permanent ROM are common in modern computers to initialize parameters, undertake self-diagnostic routines, and start up or "bootstrap" the system with sufficient programming ability to load larger operating system software from disks or other peripherals.

## C. Mass Storage

The internal memory of a microcomputer is limited in size and hence, along with the need for permanent and semipermanent storage of data, leads to the necessity for external data storage media. Previously used media such as punched paper tape and punched cards (and, to a lesser extent, magnetic tape reels) are being replaced in laboratory computers with hard and flexible disks, along with new types of high-density cassette tapes.

Flexible (or floppy) disks are a convenient storage medium where transport of data between computers or laboratories is desirable. They also provide a permanent or backup record of data with the ability to be overwritten and reused when necessary. Data access times on flexible disks are very much faster than sequential access tapes, with only fractions of a second required to find and read data into memory. Hard disks rotate at higher speeds and provides even faster random access to larger amounts of data.

At the present time, the predominant form of rapid, large-volume data storage is through the use of winchester hard disks. These are completely sealed disks of small size with very dense data storage but with rapid data transfer capabilities to the computer's memory and, vice versa, through direct memory access (DMA) protocols.

## D. Peripheral Equipment

The number of electronic and mechanical devices that can be interfaced to a computer and thus become peripherals is almost limitless. Some of the more common desirable pieces of equipment for laboratory computers will be described in this section.

Printers are needed to produce a permanent record of data or results and are available in a variety of forms. Common high-speed impact line printers for alphanumeric characters have been replaced by dot-matrix types for rapid routine printing of programs and draft-quality copies of data. Other, slower-impact printers of the daisywheel variety give better-quality printing for final reports, which can be easily assembled with word and data processors on laboratory computers.

A computer-programmable "clock" to time events is highly desirable, especially for software that is interrupt driven; requests for service from laboratory devices can come at specific time intervals under the control of such a programmable clock.

Among laboratory peripherals that have only recently been fully developed and whose need has only recently been recognized are high-quality graphics (Whitted, 1982). With ever-increasing amounts of data rapidly available, graphics presentations are frequently the only way to assess and analyze the data rapidly. Color graphics on CRT video screens may also need to be put into permanent hard-copy form through color printers or $X-Y$ plotters.

## IV. COMPUTER–LABORATORY INTERFACE

Computers have thus far been classified with some indication of different internal architectures. Personal computers and mainframes usually are not engineered to do much more than respond to commands from a terminal, process data, and display or print results. On the other hand, there is a need for analytical laboratory computers to communicate and interact directly with instruments in the laboratory. Computer engineers like to describe these two domains as the computer and the "real world."

The real world is not at all like the digital internal workings of a computer. Parameters to be measured and/or controlled, such as temperature, pressure, and flow rate, usually must be converted through transducers to electrical signals. These electrical signals are usually in analog form, and unlike the neat and tidy binary world of 0's and 1's they are often electrically noisy and not in appropriate voltage ranges for computers to convert to their domain. It is essential for the analytical scientist to know the basic concepts of how the computer – real world interface is made. Fortunately, much of the electrical engineering has been well developed and is available in chip and board form for many microcomputer products; hence the details of how these conversions are accomplished are not needed and are beyond the scope of this chapter.

## A. Digital-to-Analog Conversions

If a binary number in a computer is representative of a certain voltage or current value, it is possible to put a corresponding voltage on an output line through a digital-to-analog converter (DAC). For example, this could be done to make an analog meter read a certain value such as the level of a sampling chamber or its temperature. Two DACs could be used to set the $X$ and $Y$ voltage plates controlling the electron beam on an oscilloscope or a CRT with the computer changing the $X$ and $Y$ values to "draw" data on the screen. This is, in fact, a simple description of how graphics displays are generated. Other examples might be (1) to apply a DC voltage ramp to the opposing poles of a quadrupole mass spectrometer according to a computer-controlled timing ramp or protocol or (2) to jump to selected voltages to monitor only selected ions from the effluent of a gas chromatograph.

Digital-to-analog converters are commonly specified to produce analog voltage ranges such as $-10$ to $+10$ V, or 0 to $+5$ V, etc. The resolution of the DAC's analog output voltage depends on the number of binary input lines from the computer. Common resolutions are 8, 10, and 12 bits, with the cost increasing along with the precision. This translates into 1 part per $2^x$, where $x$ is the power of the binary input lines. For example, a 12-bit DAC with a 0- to $+10$-V range has a resolution of 10 V/4096 = 2.4 mV. The scientist deciding on the resolution required of a DAC should consider the precision needed for the experiments involved. There is excess precision if a 16-bit DAC (1 part in 65,536) is chosen for an experiment where pph precision would be the maximum final measurement possible.

## B. Analog-to-Digital Conversions

Most often an electrical signal from a transducer will be in analog form, and hence an analog-to-digital converter (ADC) is needed for a computer to assess the magnitude of the parameter being measured. The same considerations for precision of measurements discussed for DACs apply to ADCs; however, DACs are more complex electronic circuits and hence are more costly. If several analog signals are to be sampled, each is unlikely to have its own dedicated ADC as input to the computer. Rather, separate analog signals are connected to the ADC through an analog multiplexer (MUX), either sequentially or in an order determined by the computer's program. Each analog signal is then converted to binary form and strobed into the computer or its memory. There are many different types of ADCs available; two will be discussed here as representative of different configurations for specific applications.

A dual-slope integrator is a relatively inexpensive ADC that is well suited for applications with low sampling rates. This ADC has excellent immunity to random noise. It would typically be used for converting thermocouple temperature voltages where the sampling rate is low. Another common use is in gas–liquid chromatography (GLC) for GLC-signal ADC conversions where low sampling rates of 10 to 50 points/sec are adequate for GLC integration of peaks.

High-speed data conversions are typically done with successive-approximation ADCs. After selection of an analog channel by the MUX, conversions can be completed at typical rates of 10,000–100,000 samples/sec for a 12-bit ADC. Higher-resolution ADCs may have slower sampling rates but are seldom required for common laboratory instrumentation. This sampling rate is approximately the maximum that can currently be handled by a microcomputer CPU. Typically, the CPU must initiate a datum conversion, test appropriate flags for its completion, read the binary value into a register, and store the number in memory. Higher sampling rates are possible through a special DMA protocol on some computers. These protocols usually involve sampling conversions being triggered by an external clock; upon completion of the conversion, the hardware on the bus bypasses the CPU entirely and places the converted value directly in a specified memory location.

A lesser-known or less-appreciated feature of laboratory instruments is that many modern instruments with so-called digital readouts have an internal ADC that has already digitized an analog signal. This has been converted to binary coded decimal (BCD) to display the pH, temperature, absorbance, etc., and is seen in the digital display on the front of the instrument. Interfacing these intruments is often trivial, especially if the manufacturer has included a BCD outlet plug. It should be noted, however, that BCD is not a binary number and must be converted, either by software or a simple chip, to the corresponding binary number in order for the computer to handle it easily.

## C. Parallel Data Transmission

When high-speed data transmission is required, it can be most rapidly accomplished through parallel-line connections. This usually takes the form of an interface to the internal data bus, and on a 16-bit microcomputer 16 bits of data can be read simultaneously, as opposed to each bit being read independently. This works well for communication between digital devices located close to each other. However, some disadvantages limit the use of this method. There are no universal standards for communication protocols between devices, and long-distance communication becomes prohibitive because of the high cost of multiconductor wire cabling.

One standard for parallel interfacing that has become important for the laboratory is the IEEE-488 standard interface protocol. Many commercial instruments now contain, or offer as an option, an IEEE-488 interface connection port.

The IEEE-488 standard was largely developed by Hewlett Packard (who refer to it on their instruments as the HP-IB bus) and is designed to transmit a byte (8 bits) of information at a time. This method requires 3 additional control lines to implement the "handshake" between the transmitter and the receiver and 5 lines to coordinate the data transfer, yielding a total of 16 lines in connecting cables. Up to 15 devices each with its own unique address can be easily interconnected with the computer on a bus 20 m long (repeaters are required for longer distances) and can transmit 250 KB/sec (1 KB = 1024 bytes). This type of configuration is referred to as a multidrop system, with each device polled by the computer when control or information is requested.

## D. Serial Data Transmission

In many cases, parallel data transmission is not feasible. For example, if long distances are involved, cabling costs are prohibitive, or two computers may need to communicate over telephone lines. In such cases a different form of data transmission is required in order to send each information bit serially.

Since the information inside the computer is in "parallel" form, that is, in bytes or words of information, some form of interfacing is required. This is most often accomplished with a universal asynchronous receiver/transmitter (UART). The computer's UART may then communicate with another UART, for example, perhaps on a laboratory digital integrator for a gas chromatograph.

Timing and communication protocols are needed, and the most common of these is the Electronics Industry Association (EIA) RS-232C point-to-point protocol. This is a voltage conversion in which American Standard Code for Information Interchange (ASCII) characters and control signals are sent along a twisted pair of wires as a series of 1's ($-10$ to $-15$ V) and 0's ($+10$ to $+15$ V). Seven information bits allow 128 ASCII characters. An additional error-checking bit (a parity bit) is included to form an 8-bit (or 1-byte) string. This byte is encapsulated in a preceding start bit and a trailing stop bit. The RS-232C standard even specifies the shape of the computer–peripheral mating plugs required; therefore, the GLC integrator discussed above would be plug compatible with the computer if it had an RS-232C port. A typical rate of transmission is 9600 baud, approximately equal to 9600 bits/sec, which would equal about 1000 alphanumeric characters/sec. Slower transmission rates, such as 300 or 1200 bits/sec, may be required if communication is through modems on voice-grade telephone lines.

## V. SOFTWARE

Computers are clocked sequential machines that respond to a sequence of input instructions called a program. Programs in general are referred to as software; it is software that causes sequential machines to do what is desired. Generally, machines are supplied with some software; however, more software may have to be written to accommodate the requirements of a specific application. Recently, vendor-supplied software is becoming more "intelligent," and this is expanding the applications of microcomputers.

The task of software is to perform all necessary functions defined in a functional specification. Often a variety of hardware–software configurations is available, none of which exactly matches the requirements of the specification. Some compromises may be necessary. This is an important aspect of a new technology called software engineering.

Vendor-supplied software can be subdivided into the following categories: (1) operating systems, (2) utility programs, (3) general-purpose programming languages, and (4) vendor-supplied specific programs. From the end user's point of view these may all be "transparent," that is, not seen by the user. For example, a particular computer system may be arranged to power on, load its operating system, perform certain housekeeping utilities, and run a special-purpose program—all automatically, without the user being directly involved.

Operating systems allow certain very basic functions to be performed within the machine. Writing these programs usually requires a very detailed knowledge of the machine, and vendors normally supply operating systems to save all users from having to reinvent the wheel. Utility programs generally provide extensions to the basic operating system.

General-purpose programming languages such as FORTRAN, PL/1, PASCAL, and BASIC allow the user to write the programs in a language that is relatively easy to comprehend compared to the 0's and 1's that are the real food for the computer. The program text so written is referred to as the source program and is transformed into the object program by another program called a translator or compiler.

Many special-purpose programming languages have been developed. Some available graphics packages may be regarded as special-purpose programming languages. Vendor-supplied programs are often in object code, or machine and object form, so that they are more secure from the vendor's point of view (i.e., so that modification and distribution of vendor-supplied software is more difficult).

Tailored software has heretofore often been subcontracted to software houses by contractors or end users. There are now signs, however, that this situation is changing. Four factors are contributory.

1. As mentioned above, vendor-supplied software is becoming more "intelligent," expecially in the sense of being user friendly. Vendors are making it easier for end users of original equipment manufacturer (OEM) contractors to fulfill their own unique requirements within more general and powerful software packages.

2. Reducing the number of parties involved reduces time and confusion, making this route more economically attractive.

3. The end user is finding it more attractive to have software that he or she can modify independently.

4. More people are becoming available who can do this kind of work. (Fairly soon it will be common to find programmers with 10 years of programming experience.)

## A. Assembly–Machine Languages

At the most elementary programming level, all computers receive their instructions, manipulate data, make decisions, and communicate with the real world through binary numbers. The instruction set for a particular computer will be decoded from binary bytes or words, and these constitute the computer's machine language. For example, the DEC PDP-11 series machine language instruction "0110000010000001" tells the computer to add the contents of register 1 to the contents of register 2 and store the result in register 2. Strings of binary numbers are obviously difficult for humans to understand. Therefore, it is common to express them in octal or hexadecimal form. For example, the above binary number is 060201 in octal (base 8).

Some masochistic system programmers can work at the machine language level; most others work through an assembler that provides a mnemonic for each instruction. The assembler mnemonic for the example above would be "ADD R2,R1." A programmer writes instructions in assembler code, which is slightly more comprehensible; a computer program converts each assembler intruction into machine language code.

For real-time laboratory applications, it is often necessary to write programs at the assembler language level since it is the most efficient in execution speed. Often these programs will execute orders of magnitude more efficiently than the same instructions written in a high-level language such as BASIC. Also, many real-time applications require tasks that high-level languages cannot perform, such as controlling and reading data from an ADC.

## B. High-Level Languages

Most scientists find assembler languages cumbersome and painful. Hence other languages have been written that are more user friendly and more comprehensible by humans (or at least some of them). The instructions in high-level languages are converted to assembler – machine instructions, sometimes through machine independent p-code (pseudocode) intermediates, and subsequently assembled into the binary execution instructions the computer must have in order to function.

The development of high-level languages has evolved through several stages. Dessy and Starling (1980) discussed what some programmers have arbitrarily defined as four generations of computer languages.

1. First-generation languages include most of what was discussed above as assembler languages. They were cumbersome to use and difficult for other programmers to follow but still offered the best execution speed.

2. Second-generation languages were developed around more algebraic terms. FORTRAN was one of the first of this second generation of languages. FORTRAN programs are converted to machine language, and this object code is then loaded as the running program. Because of FORTRAN's and an operating system's overhead with continual need for error checking, bookkeeping, and similar functions, it executes more slowly than assembler language code.

BASIC was developed as an easily learned language for the beginning programmer. It usually operates through an "interpreter"; that is, as each BASIC instruction is encountered, it is converted to machine language code. This differs from a FORTRAN compiler in that, every time the same instruction is encountered (for example, in a loop doing the same steps thousands of times), each instruction must pass through the interpreter. For this reason, BASIC is among the slowest executing systems for laboratory computers, especially where real-time data acquisition is required.

3. Third-generation languages evolved that tried to be more structured and "top-down" organized. Examples include ALGOL, PASCAL, and MODULA-2. Structured programming languages are more easily read by other programmers since they are more modular; however, restrictions on programming style can decrease execution efficiency, especially for real-time applications. They presently are most widely used in statistical and

data analysis application programs, with limited use, to date, for real-time programming.

4. An example of a fourth-generation language is FORTH, developed by Charles Moore, formerly at the National Radiotelescope Laboratory. It is structured like third-generation languages but provides for rapid recompilation and redefinition of extensive "libraries" of smaller routines. Unlike other languages, which were developed for mainframe computers, FORTH was developed for small computers; however, it is probably more difficult to master than other languages. Some laboratory groups have employed it successfully [e.g., see Dessy and Starling (1980)], but its use as a widely accepted laboratory language remains to be seen. Aspects of the implementation of FORTH, along with various laboratory computer configurations, are discussed by Shoemaker (1983).

All the common programming languages are usually run through an operating system (OS). An OS is a overall software monitor that supervises the execution of programs and controls the computer system's resources. Rapid advances in OS design have aided the laboratory computer. The *de facto* standard for 8-bit computers is the CP/M OS. There presently exists no standard for the newer 16-bit computers, but UNIX, developed at Bell Laboratories (Kernighan and Morgan, 1982), is a strong contender.

Newer OS architectures allow foreground programs (such as a real-time data acquisition program for a laboratory instrument) to have priority, but since the needs of foreground programs usually leave idle time between demands for service, background programs can be run. This leads to OS designs by which a laboratory computer can execute foreground programs with high priority and also allow a multiuser environment where other users can analyze data, print results, compile new programs, etc. Frequently, the computer and OS can do these jobs so efficiently that even in the multiuser, multitasking environment each user is unaware of the other programs and operations being controlled and executed.

## C. Ergonomics in Design

Although each individual system will be the result of a certain amount of custom engineering to meet the specific requirements of the project, we can identify a number of desirable design qualities.

Perhaps one of the most important aspects of a computer system is the man–machine interface. Because a control operator spends up to 8 hours in front of one or more CRTs, fatigue can be a severe problem. Many CRTs in the past were block oriented; that is, they supported only alphanumeric characters on the display, in monochromatic form. Newer CRT terminals have their own "intelligence" and memory, which allows much finer resolution for graphics displays. Each pixel is individually defined as a single point, frequently with a large color selection (palette) available. The importance of high-resolution color graphics for operator interaction is just beginning to be recognized as newer bit-mapped graphics displays see increased usage in the laboratory.

The representation and configuration of CRT displays need careful consideration. The system operator should also have some control over the display of vital parameters in the case of an "upset" condition. The operator control and display console should be engineered in close cooperation with the end use. Operator convenience is of primary importance.

Since the size and operating conditions of computer systems are continually changing, they should be able to absorb changes and/or additions by simple addition or deletion of hardware. The software for such systems should be well documented and user friendly. It is desirable that the owner be able to carry out all required software maintenance.

It is a common practice to interface laboratory systems with mainframe corporate computers. Hardware provisions for this should be made in the initial design phase of the system. Software compatibility and communication protocol should also be considered.

A very high level of availability for the master station computer equipment is essential. System availability of 99.9% should be specified. This can be accomplished by having on-line "hot" spares for vital equipment or in most cases by ensuring the availability of a complete set of printed circuit boards. If the computer hardware is equipped with self-diagnostic capabilities, the average time for repair on the printed circuit board replacement level can be as low as 20 min. A 100% system availability is generally not required in laboratory environments, provided some means of local, manual control is possible.

## VI. LABORATORY APPLICATIONS

Details of specialized instrumental methods of analysis are covered in other chapters of this book; hence specific applications of computers to these methodologies will be only briefly mentioned, and no detailed coverage of commercially available, vendor-supplied turnkey systems will be provided.

### A. Data Acquisition and Control

Automation, from its beginning, has included three basic elements: sensors that permit measurement, actuators that provide control, and a decision maker that directs the control action. In human terms these elements correspond to the eyes (and other senses), hands, and brain. The advent of low-cost microcomputers and microprocessors has allowed sophisticated mechanical automation to be developed, although not as rapidly as might be possible with current technologies. A control engineer must design and install the control system, a programmer must write application programs, and it must be deemed appropriate to automate. The economic and social implications of these actions have been discussed by Ginzberg (1982).

A discussion of approaches to robotic automation of sample preparation for analytical laboratories was given by Hawk *et al.* (1982). They described features of a user-programmable, general-purpose laboratory robotic system and discussed sources of error in routine analytical methodologies. Previously supplied robotic devices for use in instrumentation such as chromatographic injection system [e.g., the WISP high-performance liquid chromatography (HPLC) system described by Garner (1982)] are sophisticated but have had limited abilities to be user reprogrammed, and their mechanical operations are limited. Although not directly related to analytical procedures, robotics applied to food packaging was discussed (Anonymous, 1982a) with a good overview of robotic options and vendors of. robotic systems available in America. A detailed report of the Japanese robotic industry and products was prepared by the Robotic Applications Center of Science Applications, Inc. (Anonymous, 1982b).

A robotic pipetting system was described that incorporated sophisticated software development on a microcomputer that was subsequently downloaded to an EPROM to provide a firmware-driven, microprocessor-based system suitable for handling highly radioactive and hazardous samples (Goerdinger and Klatt, 1982). Thompson (1982) described a system for accurate laboratory temperature control systems that emphasized the ease of interfacing a microcomputer (Apple-II) with instrumentation through the IEEE-488 bus.

An excellent review of automated and semiautomated analyses for the brewing industry was given by Buckee and Hickman (1975). Many of the procedures and methods outlined in their review are amenable to computer and/or microprocessor control. Of interest in this regard is the analytical and control system presented in the prestigious Cass lecture series by Hughes (1982), in which a distributed processing computer configuration was designed for a custom installation. A DEC PDP-11/60 minicomputer was a central supervising node in constant communication with small microcomputers dedicated to specific analytical and control functions. The supervisory minicomputer could then also be used in a background mode for report generation and data-base management with information constantly acquired from the satellite microcomputers, along with other laboratory data.

A paper given to introduce the subject of computers in the brewing industry with an overview of "beer and chips" was presented by Venn (1981). A twin CPU system for immediate failure backup in a monitoring and control system that centered on set-point-control operation was described by Kelsall (1982). In the sixth of the Cass lectures on brewing, Scopes (1982) discussed the advantages of computer modeling of brewing operations with input from analytical variables and appropriate previously assessed determinants to predict optimal brewing parameters, resource allocation, and assessment of proposed changes.

In summary, computerized data acquisition, control, and robotic systems are rapidly becoming accepted in modern analytical food and beverage applications. Even the most modern laboratories and industrial applications have need for robotic or automatic operation of what have been termed the 3-D jobs — dirty, dangerous, or difficult (not to mention dull or dehumanizing). Recently, Zenith Radio Corporation's Heath Company subsidiary announced delivery of Motorola 6808 microprocessor-based robots either assembled or in kit form (Erikson, 1982). Other vendors have also entered the small robotic equipment field for home, educational, and industrial use (Iverson, 1983). If the emergence of these small user-programmable microprocessor-controlled robots has the same effect as the introduction of the personal microcomputers a few years ago, there may be a similar dramatic increase in their acceptance and use.

## B. Spectroscopic Systems

Most commercial vendors of UV and light absorption spectrometers offer microprocessor control options with varying degrees of sophistication; unfortunately, most of these are simply turnkey systems and cannot be easily modified or reprogrammed by the user for special needs and applications. The popular Apple-II-Plus microcomputer has been interfaced to several different spectrophotometers to provide varying degrees of automation (Interactive Microware, 1981; Mitchell, 1981; Owen et al., 1981). These systems do provide users with the ability to customize the spectrophotometer automation for their own needs.

Stopped-flow spectrophotometric methods can be extremely useful for routine high-speed analytical assays that use either reaction-rate or equilibrium methods. Koupparis et al. (1980) described a stopped-flow analyzer controlled by a Rockwell AIM-65 microcomputer. Alcocer et al. (1983) described a simple interface of a DEC LSI-11/2 microcomputer to the commercial Durrum–Gibson stopped-flow spectrophotometer that allowed reaction-rate or equilibrium assays to be acquired in millisecond time ranges.

Applications of computers to nuclear magnetic resonance (NMR- spectroscopy were reviewed by Thibault and Cooper (1980), with the importance of the computer to Fourier transformation (FT) methodologies in NMR instrumentation stressed. Commercial electron spin resonance (ESR) spectrometers have been computerized by vendors; however, there is a continuing need for simulation programs to aid in the interpretation of spectra, especially those with complex hyperfine and superfine coupling. Recent ESR computer developments in this area have been reported by Daul et al. (1981), Moro and Freed (1980), and Misra and Vasilopoulos (1980).

## C. Chromatographic Methods

One of the areas of analytical methodology to which computers may contribute is with the use of Fourier transformation of physical domains such as time and frequency. In addition to the important use of this method for infrared (IR), NMR, and mass spectrometry (MS) discussed in other chapters, Fourier transformation of instrumental signals can often be used effectively to decrease noise in detector signals. Weber (1982) discussed FT methods to minimize noise in electrochemical detectors for liquid chromatography. Quantitative GC/FT/IR spectrometric methods of FT signal enhancement were presented by Sparks et al. (1982) and Lam et al. (1982).

Gas liquid chromatography occupies one of the most important roles in the analysis of foods and beverages. The use of microcomputers for laboratory automation of GLC was reviewed by Horn and Kummer (1980). There are many microprocessor-based GLC instruments with sophisticated analytical and control routines for calibration, integration, and reporting of analytical data. An excellent paper by Woerlee and Mol (1980) described a real-time data acquisition system for GLC analyses. Examples of microprocessor-controlled capillary column GLC with precise and flexible column oven temperature and carrier gas flow-rate controls were discussed by Nickel et al. (1981) and Rooney (1981).

A particularly important area of computer control of instrumentation important to the analysis of foods and beverages is the combination of GLC and mass spectrometry (GC/MS). Capillary GC/MS is now a mature, well-developed methodology with the GC/MS computer interface technologies among the most well developed in the analytical field; hence details are not reviewed here. Henneberg (1980) and Martinsen (1981) discussed various powerful computerized searching methods to match unknown spectra against libraries of standards. Halpern (1981) drew attention to the need for recognition of computer limitations and the need for skilled mass spectrometrists to provide ultimate interpretation of the data.

## D. Other Analytical Methods

One of the key areas that had an impact on the development of automated food analysis is the technique of flow injection. Details of this methodology are presented by Beecher and Stewart (Chapter 19, this volume), but it is important to note several inroads into the interfacing of microcomputers with laboratory instrumentation and simple robotic operations. Stewart *et al.* (1980) described a successful microprocessor-controlled system for automated multiple flow-injection analysis (AMFIA) using an IMSAI-8048 microprocessor; however, programming this microprocessor was somewhat difficult if frequent changes in analytical methods were required.

Methods involving AMFIA are attractive in many food analyses where large numbers of samples must be processed routinely. Brown *et al.* (1981a) described AMFIA automation with a more "friendly" Rockwell AIM-65 microcomputer, which supported BASIC as a high-level programming language. Through the use of a structured, layered software development they described a turnkey mode for routine assays, with built-in capability for fairly easy reprogramming for different assay protocols. This was further enhanced (Brown *et al.,* 1981b) with the use of a "learn" mode through which they demonstrated, with analyses of $B_6$ vitamers, how the microcomputer could monitor a manually performed assay protocol and subsequently use the learned time-base stream in a "run" mode to routinely repeat exactly the assay.

The above AMFIA papers and related works are particularly worth reviewing for scientists who wish to develop their own microcomputer-controlled experiments. These methods illustrate the use of distributed processing, with small low-cost microcomputer controllers operating independently, but with communication to larger minicomputers and mainframe computers for data-base management and report generation. They also highlight the rudimentary elements of laboratory robotic control and underscore the need for both hardware and software in such custom systems to be usable by nonprogrammers and technicians, who must ultimately utilize the systems.

With the rapid rise in the number of microcomputers and microprocessors in the laboratory environment, it is obvious that communication among them is highly desirable. Each independent processor has its own information to deliver, and communication to a larger minicomputer or mainframe allows concentration of systematically organized data bases for report generation and archiving of records.

As an example of desirable networking, consider a hypothetical but currently realistic example of a commercial food analysis laboratory asked to analyze a mixture suspected of containing a toxic substance. An analytical chemist works through a minicomputer that is connected through a local area network (LAN) to all instrumentation at the facility. He interrogates a data base about an HPLC analysis of the sample and finds a bioassay that revealed that a subsequently collected fraction number 37 from the HPLC effluent displayed toxic activity. He takes the sample tray and inserts it onto an autosampler on a GC/MS. Through the minicomputer and LAN, he instructs the autosampler to inject 8 $\mu$l onto a fused-silica GLC column with a split ratio of 50:1. The local GC/MS processor follows the minicomputer's instructions, obtains a total ion chromatogram, integrates each chromatographic peak, and records mass spectra every 5 sec, reporting all information back to the minicomputer. As each spectrum is received and determined to be different from its predecessor, it is searched against a local library of 30,000 spectra stored on a winchester disk. By the time the GC/MS run is complete, the minicomputer has reported the integrated areas of peaks from the gas chromatogram and assigned high probabilities of identification for all but one compound in the sample. None of the identified compounds is highly toxic.

The analyst then has the minicomputer establish network contact with Chemical Information Systems (CIS) and searches the unknown mass spectrum against their larger data base. Still no identification. He then networks to a service such as the third-generation data base and identification algorithms of STIRS MS at Cornell University Computing Center. Their program suggests a probable structure for the unknown spectrum. A network contact is made with an abstracting service such as DIALOG or ORBIT and a literature search through Chemical Abstracts is sent directly to the minicomputer's files and is printed locally. A search through the MEDLINE and TOXLINE data bases sends back toxicity data on this and related compounds. All this information is assembled through the minicomputer's word-processing programs and is immediately sent through electronic mail to terminals in the offices of all key company officials to decide what further analysis, confirmation, or action is needed.

The above example illustrates the desirability of computer–instrument networks; however, the networking protocols are not all uniform, and further development in this area is needed. Some indication of transmission protocols was presented in Section IV, but problems with high-speed networking protocols beyond the scope of this chapter still exist. With another human analogy, present transmission protocols have given computers "voice boxes" to communicate; however, one might speak German, another English, and another French. Aspects of computer networking were reviewed by Newell and Sproull (1982) and Dessy (1982), with current examples given by Miller et al. (1982).

The dream of a fully computerized analytical laboratory may never be realized, but the 1980s will see rapid developments in this area. There is an ever-increasing need for timely, accurate, and precise analyses of foods and beverages. The computer will aid the analytical scientist in achieving these objectives.

## APPENDIX: MICROCOMPUTER TERMINOLOGY

**Access time:** The time required to obtain data from main memory or from a storage device, such as a diskette or a winchester disk.

**Acoustic coupler:** A device for transmitting data over phone lines by converting electrical signals into audio signals and vice versa.

**Application:** A specific task, such as controlling a gas chromatograph, to which computer usage can be applied.

**Application program:** A computer program designed to meet specific user needs, such as a program that maintains chemical analysis data or monitors an automatic sampling robot.

**Architecture:** In the case of computers, the design or organization of the central processing unit (CPU) and its configuration on the bus.

**ASCII:** American Standard Code for Information Interchange, a binary code that has assigned a number to each alphanumeric character and several nonprinting characters used to control printers and communication devices.

**Asynchronous:** Referring to communications method in which data are sent as soon as they are ready, as opposed to methods in which data are sent at fixed intervals.

**Background processing:** The processing of one or more noninteractive services on a computer while the computer executes a higher-priority (foreground) service.

**Backup:** The copying of one or more files onto a storage medium for safekeeping, in case the original should be damaged or lost.

**BASIC:** Beginners' All-purpose Symbolic Instruction Code, a widely used interactive programming language that is especially well suited to microcomputers and beginning users. BASIC was developed at Dartmouth College.

**Baud:** A unit of data transmitting and receiving speed, roughly equal to 1 bit/sec. Common baud rates are 110, 300, 1200, 2400, 4800, and 9600.

**Bidirectional:** Having the ability to transfer data in either direction, especially on a bus.

**Binary:** (1) A number system with only two digits, 0 and 1, in which each symbol represents a decimal power of 2. (2) Any system that has only two possible states or levels, such as a switch that is either on or off. This is represented in computer circuitry by the presence of current (equivalent to 1) or the absence of current (equivalent to 0). All computer programs are executed in binary form.

**Bit:** A short term for Binary digIT, which can have only two possible values, 0 or 1. It is the smallest unit of data recognized by the computer. All data (letters, numerals, symbols) handled by a computer are digitized, that is, expressed entirely as a combination of 0's and 1's.

**Bit-map graphics:** A technology tht allows control of individual pixels on a display screen to produce arcs, circles, sine waves, or other images that block-addressing technology cannot accurately display.

**Board:** A plastic resin board containing electronic components such as chips and the electronic circuits needed to connect them. Also circuit board.

**Bus:** A group of parallel electrical connections that carry signals between computer components or devices.

**Byte:** The number of bits used to represent a character. For mini- and microcomputers, a byte is usually 8 bits.

**Cathode-ray tube (CRT):** A vacuum tube that generates and guides electrons onto a fluorescent screen to produce such images as characters or graphics displays on video display screens.

**Central processing unit (CPU):** Electronic components in a computer that control the transfer of data and perform arithmetic and logic calculations.

**Character:** A single printable letter (A–Z), numeral (0–9), or other symbol used to represent data. ASCII characters also include those that are not visible as characters, such as a space, a tab, a carriage return, or another control function.

**Chip:** A piece of semiconductor material containing microscopic integrated circuits.

**Circuit:** (1) A system of semiconductors and related electrical elements through which electrical current flows. (2) In data communications, the electrical path providing one-way or two-way communication between two points.

**COBOL:** COmmon Business-Oriented Language, a high-level programming language that is well suited to business applications involving complex data records (such as personnel files or customer accounts) and large amounts of printed output.

**Compatibility:** (1) The ability of an instruction, a program, or a component to be used on more than one computer. (2) The ability of computers to work with other computers that are not necessarily similar in design or capabilities.

**Computer network:** An interconnection of computer systems, terminals, and communications facilities.

**Core:** The older type of nonvolatile computer memory made of ferrite rings that represents binary data by switching the direction of polarity of magnetic cores.

**CP/M:** Control Program for Microprocessors, an operating system used by many microcomputers.

**Daisywheel:** A print head that forms full characters rather than characters formed of dots. It is shaped like a wheel with many spokes, with a letter, numeral, or symbol at the end of each spoke. The print method used is similar to that of a common typewriter.

**Database:** A large collection of organized data that is required to perform a task. Typical examples are laboratory inventory files or results of chemical analyses.

**Data communication:** The movement of coded data from a sender to a receiver by means of electrically transmitted signals.

**Data processing:** The application in which a computer works primarily with numerical data, as opposed to text.

**Dedicated computer:** A computer built for one special function, such as controlling a laboratory instrument.

**Diagnostic:** A program that checks the operation of a device, a board, or another component for malfunctions and errors and reports its findings.

**Direct memory access (DMA):** A method for transferring data to or from a computer's memory without CPU intervention.

**Disk:** A rigid, flat, circular plate with a magnetic coating for storing data.

**Diskette:** A flexible, flat, circular plate permanently housed in a black paper envelope with magnetic coating that stores data and software. Standard sizes are $5\frac{1}{4}$ and 8 in. in diameter.

**Distributed data processing:** A computing approach in which an organization uses computers in more than one location rather than using one large computer in a single location.

**Dot-matrix printer:** A printer that forms characters from a two-dimensional array of dots. More dots in a given space produce characters that are more legible.

**Draft-quality printer:** A printer, usually high speed, that produces characters that are easily readable but of less than typewriter quality. Typically used for internal documents for which type quality is not a major factor.

**Electronic Industries Association (EIA):** A standards organization specializing in the electrical and functional characteristics of interface equipment.

**Electronic mail:** A feature that allows short memos or messages to be sent to another computer.

**Ergonomics:** The science of human engineering that combines the study of human body mechanics and physical limitations, explaining the problem and indicating what to do next.

**File:** A collection of logically related records or data treated as a single item. A file is the means by which data are stored on a disk or diskette so they can be used at a later date.

**Foreground processing:** Top-priority processing. It has priority over background (lower-priority) processing.

**FORTRAN:** FORmula TRANslation, a widely used high-level programming language well suited to problems that can be expressed in terms of algebraic formulas. It is generally used in scientific applications.

**Graphics:** The use of lines and figures to display data, as opposed to the use of printed characters.

**Hard copy:** Output in a permanent form (usually on paper) rather than in temporary form, as on a CRT or visual display.

**Hard disk:** A disk that is not flexible, such as a winchester disk. It is more expensive than a diskette but is capable of storing much more data.

**Hardware:** The physical equipment that makes up a computer system.

**Hardwired:** Referring to a permanent, as opposed to a switched, physical connection between two points in an electrical circuit. Connections between two devices are typically hardwired, whereas all connections through a modem are switched, because they use telephone lines.

**Impact printer:** A printer that forms characters on paper by striking an inked ribbon with a character-forming element.

**Information services:** Publicly accessible computer repositories for data, such as Chemical Abstracts, MEDLINE, or STIRS MS.

**Instruction:** A command that tells the computer what operation to perform next.

**Integrated circuit (IC):** A complete electrical circuit on a single chip.

**Interface:** An electronic assembly that connects an external device, such as a printer, to a computer.

**K:** The symbol for the quantity $2^{10}$, or 1024. The K is uppercase to distinguish it from lowercase k, which is a standard international unit for kilo, or 1000.

**Kilobyte (KB):** 1024 bytes.

**Large-scale integration (LSI):** The combining of about 1000 to 10,000 circuits on single chip. Typical examples of LSI circuits are memory chips, microprocessors, calculator chips, and watch chips.

**Letter-quality printer:** The printer used to produce final copies of documents. It produces typing comparable in quality to that of a typewriter.

**Line printer:** A high-speed printer that prints an entire line of characters at a time.

**Magnetic tape (magtape):** Magnetic tape used as mass storage media and packaged on reels. It is often used as a backup device on larger computer systems.

**Mainframe:** A computer that is physically large and provides the capability to perform applications requiring large amounts of data.

**Mass storage:** A device like a disk or magtape that can store large amounts of data readily accessible to the CPU.

**Megabyte (MB):**   1,048,576 ($2^{20}$) bytes.

**Medium:**   The physical substance upon which data are recorded, for example, magnetic disks, magnetic tape, or punched cards.

**Memory:**   (1) The main high-speed storage area in a computer where instructions are temporarily kept for a program being run. (2) A device in which data can be stored and from which they can later be retrieved.

**Menu:**   A displayed list of options from which the user selects an action to be performed by typing a letter or by positioning the cursor.

**Menu driven:**   A computer system that primarily uses menus for its user interface rather than a command language.

**Microcomputer:**   A computer that is physically very small and that is based on (LSI) circuitry.

**Microprocessor:**   A single-chip CPU incorporating LSI technology.

**Minicomputer:**   A type of computer that is usually smaller than a mainframe in physical size. In general, its performance exceeds that of a microcomputer. Since minicomputers are more modular than mainframes, they can be configured to provide better price–performance systems.

**Mnemonic:**   A short, easy-to-remember name or abbreviation. Many commands in programming languages are mnemonic.

**Modem:**   MOdular/DEModulator, a device that converts computer signals (data) into high-frequency communications signals and vice versa. These high-frequency signals can be sent over telephone lines.

**Monitor (hardware):**   A CRT device that can be used as an output display.

**Monitor (software):**   A part of the operating system that allows the user to enter programs and data into memory and to run programs.

**MOS:**   Metal oxide semiconductor, the most common form of LSI technology.

**Multiprocessing:**   A scheduling technique that allows more than one job to be in an executable state at any one time. Thus, even with one CPU, more than one program can appear to be running at a time because the CPU is giving small slices of its time to each executable program.

**Multitasking:**   The execution of several tasks "at the same time" without one having to be completed before another is started. Although computers can perform only one task at a time, the speed at which a computer operates is so great that it appears as though several tasks are being performed simultaneously.

**Network:**   A group of computers that are connected to each other by communications lines to share information and resources.

**Nonvolatile memory:**   Memory that does not lose its contents when a processor's power supply is shut off or disrupted.

**On line:**   Referring to a process directly under control of the computer or to data that are introduced into the CPU immediately.

**Parallel transmission:**   The sending of more than one data bit at a time.

**Parity:**   A one-extra-bit code used to detect recording or transmission errors by making the total number of "1" bits in a unit of data, including the parity bit itself, odd or even.

**Peripheral:**   A device that is external to the CPU and main memory, such as a printer, modem, or terminal, but connected to it by appropriate electrical connections.

**Pixels:**   Picture elements, definable locations on a display screen that are used to form images on the screen. For graphics displays, screens with more pixels generally provide higher resolution.

**Port:**   A physical area for the connection of a communications line. This line can be between the CPU and anything external to it, such as a printer, another computer, a modem, or another communications line.

**Processor:** The functional part of the computer system that reads, interprets, and executes instructions.

**Program:** The complete sequence of instructions and routines needed to solve a problem or to execute directions in a computer. .

**Programming language:** The words, mnemonics, and/or symbols, along with the specific rules allowed in constructing computer programs. Some examples are BASIC, FORTRAN, and COBOL.

**RAM:** Random Access Memory, memory that can be both read and written into (i.e., be altered) during normal operation. RAM is the type of memory used in most computers to store the instructions of programs currently being run.

**Real-time:** Taking place during the actual occurrence of an event; referring to computer systems or programs that perform a computation during the actual time that a related physical process transpires.

**Remote:** Not hardwired, communicating via switched lines such as telephone lines. Usually applied to peripheral devices (e.g., printers, video terminals) that are located at a site away from the CPU.

**ROM:** Read Only Memory, memory containing fixed data or instructions that are permanently loaded during the manufacturing process. A computer can use the data in the ROM but cannot change it.

**Sequential or serial access:** Referring to media, such as magnetic tape, from which data or instructions can be retrieved only by passing through all locations between the one currently being accessed and the desired one.

**Serial transmission:** The sending of a single bit at a time.

**Single thread:** Referring to a simple operating system that executes any given task from beginning to end without interruption, as opposed to a multitasking system.

**Soft copy:** Alphanumeric or graphical data (or both) presented in nonpermanent form, such as on a video screen.

**Software:** The tasks or programs that make the computer perform a particular function.

**System:** A combination of software and hardware that performs specific processing operations.

**System board:** The main module (sometimes called a mother board). It contains the CPU, the memory, and the interface circuitry for the keyboard, a printer port, and a communications port.

**Tape:** A recording medium for data or computer programs. Tape can be in permanent form, such as perforated paper tape. Generally, tape is used as a mass-storage medium, in magnetic form, and has a far higher storage capacity than disk storage, but it takes much longer to write or recover data from tape than from a disk.

**Terminal:** An input–output (I/O) device used to enter data into a computer and record the output. Terminals are divided into two categories: hard copy (e.g., printers) and soft copy (e.g., video terminals).

**Time-sharing:** Providing service to many users by working on each one's task part of the time.

**Turnkey system:** A computer that is ready to be used without the addition of any hardware or software. It is complete as packaged for a particular application.

**Video terminal:** A terminal that displays data or instructions on a CRT.

**Volatile memory:** Memory that loses its contents when power is removed unless battery backup is available.

**Winchester disk:** A hard disk permanently sealed in a drive unit to prevent contaminants from affecting the read–write head, virtually eliminating the need for adjustment of the head by field-service personnel. The disk is capable of storing larger amounts of data than a diskette.

**Word:** The greatest number of bits a computer is capable of handling in any one operation. Usually subdivided into bytes.

## REFERENCES

Alcocer, J. A., Livingston, D. A., Russell, G. F., Shoemaker, C. F., and Brown, W. D. (1983). *J. Autom. Chem.* **5**, 83–88.

Anonymous (1982a). *Food Drug Packag.* (Jul), 18–21; (Aug), 16–19.

Anonymous (1982b). "Productivity, Automation, and Robotics in Japanese Manufacturing: Its Impact on U.S. and European Industry," Report H215. Frost and Sullivan, Inc., New York, New York.

Brown, J. F., Stewart, K. K., and Higgs, D. (1981a). *J. Autom. Chem.* **3**, 182–186.

Brown, J. F., Vanderslice, J. T., Maire, C., Brownlee, S. G., and Stewart, K. K. (1981b). *J. Autom. Chem.* **3**, 187–190.

Buckee, G. K., and Hickman, E. (1975). *J. Inst. Brew.* **81**, 399–407.

Daul, C., Schlaepfer, C. W., Mohos, B., Ammeter, J. H., and Gamp, E. (1981). *Comput. Phys. Commun.* **21**, 385–395.

Dessy, R. (1982). *Anal. Chem.* **54**, 1167A–1184A.

Dessy, R., and Starling, M. K. (1980). *Am. Lab. (Fairfield, Conn.)* **12**, 21–32.

Erikson, A. (1982). *Electronics* (Dec. 15), pp. 50.

Garner, J. K. (1982). *Am. Lab. (Fairfield, Conn.)* **14**, 130–138.

Ginzberg, E. (1982). *Sci. Am.* **247**, 67–75.

Goerdinger, D. E., and Klatt, L. N. (1982). *Anal. Chem.* **54**, 1902–1904.

Halpern, B. (1981). *CRC Crit. Rev. Anal. Chem.* **11**, 49–78.

Hawk, G. L., Little, J. N., and Zenie, F. H. (1982). *Am. Lab. (Fairfield, Conn.)* **14**, 96–104.

Henneberg, D. (1980). *Adv. Mass Spectrom.* **8B**, 1511–1531.

Horn, D., and Kummer, M. (1980). *In* "Measurement for Progress in Science and Technology: Proceedings of the 8th IMEKO Congress, May, 1979" (G. Striker, J. Solt, and T. Kemény, eds.), Vol. 2, pp. 513–522. North-Holland Publ., Amsterdam.

Hughes, C. J. (1982). *Brew. Guardian* **111**, 23–28.

Interactive Microware (1981). *Anal Chem.* **53**, 576A.

Iverson, W. R. (1983). *Electronics* (Jan. 27), pp. 79–81.

Kelsall, D. (1982). *Tech. Q. Master Brew. Assoc. Am.* **19**, 68–70.

Kernighan, B. W., and Morgan, S. P. (1982). *Science* **215**, 779–783.

Koupparis, M. A., Walczak, K. M., and Malmstadt, H. V. (1980). *J. Autom. Chem.* **2**, 66–75.

Lam, R. B., Sparks, D. T., and Isenhour, T. L. (1982). *Anal. Chem.* **54**, 1927–1931.

Martinsen, D. P. (1981). *Appl. Spectrosc.* **35**, 255–266.

Miller, R., Angelo, N. B., Sloan, T., Nunn, W. G., Koontz, J. P., Woodward, W. S., Hindler-Smith, J., Smith, D. F., Dessy, R., Chapple, I., Currie, J., Dueball, S., Thompson, M., Duchamp, D. J., and Olson, E. C. (1982). *Anal. Chem.* **54**, 1295A–1306A.

Misra, S. K., and Vasilopoulos, P. (1980). *J. Phys. C* **13**, 1083–1092.

Mitchell, N. (1981). *Varian Instrum. Appl.* **15**, 8–9.

Moro, G., and Freed, J. H. (1980). *J. Phys. Chem.* **84**, 2837–2840.

Newell, A., and Sproull, R. F. (1982). *Science* **215**, 843–852.

Nickel, W., Guidinger, D. C., Brown, A. C., Iwao, K. R., and Marshall, G. (1981). *Am. Lab. (Fairfield, Conn.)* **13**, 122–127.

Owen, G. S., Travis, D., and Green, T. (1981). *J. Chem. Educ.* **58**, 690–691.

Rooney, T. A. (1981). *Am. Lab. (Fairfield, Conn.)* **13**, 143–147.

Scopes, T. (1982). *Brew. Guardian* **111,** 20–22.

Shoemaker, C. F. (1983). *In* "Food Analysis: Principles and Techniques" (D. W. Gruenwedel and J. R. Whitaker, eds.). Dekker, New York.

Sparks, D. T., Lam, R. B., and Isenhour, T. L. (1982). *Anal. Chem.* **54,** 1922–1926.

Stewart, K. K., Brown, J. F., and Golden, B. M. (1980). *Anal. Chim. Acta* **114,** 119–127.

Thibault, C., and Cooper, J. W. (1980). *Magn. Reson. Rev.* **5,** 101–119.

Thompson, T. (1982). *Am. Lab. (Fairfield, Conn.)* **14,** 90–91.

Venn, E. J. (1981). *Brewer (London)* **67,** 275–278.

Walling, P. L., Gentille, T. E., and Gudat, A. E. (1982). *Am. Lab. (Fairfield, Conn.)* **14,** 22–32.

Weber, S. G. (1982). *Anal. Chem.* **54,** 2126–2127.

Whitted, T. (1982). *Science* **215,** 767–774.

Woerlee, E. F. G., and Mol, J. C. (1980). *J. Chromatogr. Sci.* **18,** 258–266.

# 9

# On-Line Methods

MANFRED MOLL

*Centre de Recherches TEPRAL*
*Champigneulles, France*

## I. INTRODUCTION

In this chapter, laboratory and production aspects of on-line analysis will be described. A great number of papers in both fields have been published since 1960.

The application of on-line methods has been easier in industrial laboratories than in production. It is clear that the introduction of a new instrument in the food industry is often regarded as complicating, disturbing, of little interest, time-consuming, etc. If management at a high level is not convinced of the utility of on-line methods there is no chance of introduction of this type of approach. On the other hand, technical management must fix the objectives in using on-line methods both in the laboratory and in production. On-line methods are very costly, using highly qualified techni-

ANALYSIS OF FOODS AND BEVERAGES

cians, but when they work correctly, what a change is seen in the industry!

It is impossible to describe in a few pages on-line methods in the food and beverage industry, and for this reason only specific experiences will be discussed.

## II. LABORATORY ON-LINE METHODS

The medical–pharmaceutical and chemical fields have made the greatest effort to introduce on-line and automated methods in laboratories. The food industry has mainly *adapted* from these special fields and applied these techniques.

Skeggs (1957) introduced in the clinical laboratory the air-segmented continuous-flow analysis by Technicon, which has been widely developed. Since 1957, new techniques have been introduced that will be discussed below. Reference is made to several publications: Furman (1976), Foreman and Stockwell (1975, 1979), Goldberg (1962), Devreux (1968), Buckee and Hickman (1975), Benard (1982), Ligot (1979), *Journal of Automatic Chemistry* (1978), Tunnell (1981), Devillers *et al.* (1982), and Siebert (1982).

Several methods are described that are or can be automated, such as gas chromatography (GC), column chromatography, high-performance liquid chromatography (HPLC), atomic adsorption spectroscopy, near-infrared (near-IR) reflectance spectroscopy, and many others.

### A. Physical Methods

#### 1. Automated Sample Preparation

An excellent review of automated sample preparation has been published recently by Burns (1981). Bartels *et al.* (1978) have described an automatic unit for solid–liquid extraction as a step in sample preparation. The unit consists of a mill and a centrifuge for separation of the phases. The controls allow automatic treatment of individual samples, without supervision, of many kinds of materials.

#### 2. Color

An automated color test apparatus was described by Curry *et al.* (1974). A spectrophotometric color method has been recommended by the American Society of Brewing Chemists (ASBC, 1976a) using absorbance of the liquid at 430 nm for a light path of 0.5 in. Correction for turbidity is made if the absorbance at 700 nm is $\leq 0.039$ times the absorbance at 430 nm. This method is easy to automate when an automatic sampler is used. Van Strien and Drost (1979) have proposed the use of a photometer equipped with an

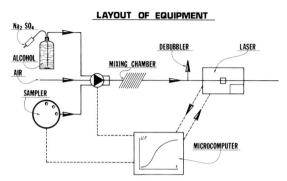

**Fig. 1.** General layout of equipment for the sampling and measuring of chill haze of liquids. [Reprinted with permission from Moll et al. (1981).]

interference filter (peak wavelength, 430 nm + 0.5/−0; band width, 3 nm ± 0.5; two cavities). This type of measurement has been adopted by the Analysis Committee of the European Brewery Convention (EBC) for industrial wort and for measuring beer color. This type of instrument can be automated.

### 3. Turbidity

Automatic equipment for the prediction of chill haze in liquids (patent pending) was developed by Moll et al. (1981). The layout of this equipment in represented in Fig. 1. This test allows measurement of haze and chill haze in beer in 15–20 min at −8°C.

### 4. Foam

Rasmussen (1981) developed the first apparatus for automated analysis of foam stability and measurement of the half-life of foam using the Blom (1957) method.

### 5. Viscosity

A new method for measuring wort and beer viscosity was described by Benard (1978b). The automatic oscillation viscometer gives good agreement with the Höppler viscometer ($r = 0.971$*** [*** = highly significant, $p = 0.001$], $n = 46$).

### B. Chemical and Biochemical Methods

### 1. Carbohydrates

**a. Chemical and Biochemical Methods Using the Autoanalyzer Technique.** An automatic method for estimating fermentable sugars was described by Cooper et al. (1961) with potassium ferricyanide as chromogen. The same method was also mentioned and improved by Gaines (1973) and by Vandercook et al. (1975).

Pinnegar and Whitear (1965) introduced the Autoanalyzer, using anthrone reagent, for the estimation of total carbohydrate content in wort and beer. An improved technique increasing the sensitivity of the anthrone method was discussed by Yadav et al. (1969) and by Jermyn (1975).

An automatic estimation of total carbohydrates using a reduction-oxidation electrode after hydrolysis was developed by Buckee (1972) following the technique of Porter and Sawyer (1972). A different method was described by Buckee (1973b) in which fermentable sugars of wort and beer are separated from unfermentable carbohydrates by desorption from a charcoal – kiesel – guhr column using 15% ethanol as the eluting agent. This method can be combined with the method previously described.

The determination of total reducing sugars in wort and beer by Autoanalyzer described by Sawyer and Dixon (1968) uses molybdophosphate as chromogen. Moll et al. (1975a) presented a technique based on studies by Brown (1961) and Bittner and McCleary (1963) in which the complex carbohydrates are hydrolyzed and the glucose formed is allowed to reduce the copper – neocuproin chelate, which is measured by colorimetry.

Hudson et al. (1976) mentioned an automatic procedure using p-hydroxybenzoic acid hydrazide as a color reagent for the detection of carbohydrates in solution. An automatic determination of glucose in liquids with hexokinase was described by Mor et al. (1973) and also by Moll et al. (1977a). Reducing sugars were determined by Robin and Tollier (1981) using tetrazolium blue.

**b. Column Chromatography.** In 1969 two teams described brewing carbohydrate analysis using column chromatography: Versuchs- und Lehranstalt für Brauerei (VLB) of Berlin [Dellweg et al. (1969), John et al. (1969a,b), Trenel and Emeis (1970), Trenel and John (1970), and Dellweg et al. (1970, 1971)], who used a polyacrylamide gel, Bio-Gel P2, 400 mesh, which made it possible to separate glucose polymers up to a degree of polymerization of 15 in 7 hr; and Otter et al. (1969, 1970) at Courage, London, who obtained similar results with a packed column using cellulose powder (Whatman C.F. 12).

Techniques using Bio-Gel P2 have been improved, as discussed by Enevoldsen (1970), Enevoldsen and Schmidt (1973, 1974), and Wight (1976); Schmidt and Enevoldsen (1976) used Bio-Gel P4, which they found more satisfactory.

**c. Gas Chromatography.** Gas chromatography is time-consuming because of the need to prepare the sample by trimethylsilylation; the injection into the gas chromatograph can be automated. The following authors have described this technique: Sweeley et al. (1963), Marinelli and Whitney (1966, 1967), Clapperton and Hollyday (1968) for worts and beers, Tuning

(1971) for worts after the technique of Beadle (1969), Moll *et al.* (1971) for diluent of proteolytic enzymes, John and Dellweg (1973) for malt wort, Devillers *et al.* (1974) for balance of a sugar factory, and Drawert and Leupold (1976) and Mouillet *et al.* (1977) for dairy products. A fast, one-step method for the silylation of sugars was described by Leblanc and Ball (1978); see also Jamieson (1976) for brewing materials.

   d. **High-Performance Liquid Chromatography.** Since the introduction of HPLC in the early 1970s a great number of publications has become available. The HPLC technique is easy to automate.

   Practical applications of HPLC for carbohydrates in food were mentioned by Brobst *et al.* (1973) and Palmer and Brandes (1974) for cation-exchange resins; Palmer (1975) for $\mu$ Bondapak/carbohydrate; Linden and Lawhead (1975) for $\mu$ Bondapak-AX/Corasil; Conrad and Palmer (1976) for $\mu$ Bondapak/carbohydrate; Schwarzenbach (1976) for Aminex; Scobell *et al.* (1977) for Aminex 50W-X4 ($Ca^{2+}$, $Ag^+$), Q 15 S ($Ca^{2+}$), and A-5 ($Ca^{2+}$); Richter and Woelk (1977) for Aminex, $\mu$ Bondapak, and Nucleosil; Helbert *et al.* (1977) for Aminex 50W-X4 and $\mu$ Bondapak; Timbie and Keeney (1977) for $\mu$ Bondapak; Gum and Brown (1977) for $\mu$ Bondapak and Whatman PXS-1025 PAC; Moll *et al.* (1978a) and Slotema (1978) for $\mu$ Bondapak; Noel *et al.* (1979) for Chromosorb LC9; Kato and Kinoshita (1980) for Aminex A-27, Lichrosorb $NH_3$, and $\mu$ Bondapak; Pomes (1980) for Aminex Q 15S; White *et al.* (1980) for Silica and Li Chroprep Si 60; Fitt *et al.* (1980) for Aminex 50W-X4; Fonknechten *et al.* (1980) for $\mu$ Bondapak; and Scobell and Brobst (1981) for Aminex HPX-42 ($Ca^{2+}$) and 50W-X4 ($Ag^+$). Bazard *et al.* (1981b) compared two techniques, $\mu$ Bondapak and Aminex HPX 87.

### 2. Acetoin and Diacetyl

   Brandon (1967) described for the first time a fully automatic procedure for the determination of diacetyl in beer using the hydroxylamine acetate technique. At a level of 0.11–0.12 ppm of diacetyl the standard deviation was ±4.8%. Brandon (1968) developed an automatic method for the determination of acetoin in wort and beer. A critical review of methods of determination of diacetyl was presented by Wainwright (1973).

   An automatic procedure for the determination of diacetyl and precursors was described by Bauer and Roth (1978) using the orthophenydiamine technique.

   A comparison between GC and an automated distillation method for the determination of diacetyl in beer was made by Buijten and Holm (1979). The correlation coefficient between the two methods is $r = 0.996***$, $n = 25$.

Inoue (1980) described an automatic stripping apparatus for determination of the vicinal diketones: the total analysis time is 30 min.

## 3. Sulfur Dioxide

An automatic method for the determination of sulfur dioxide in beer was described by Saletan and Scharoun (1965). The technique used is an adaptation of the method proposed by Stone and Laschiver (1957), namely, measurement of the color formed in the reaction of the sulfur dioxide from the beer with decolorized *p*-rosaniline and formaldehyde.

An automatic method for the determination of total sulfur dioxide in wine by distillation and iodometry was described by Sarris *et al.* (1970a).

## 4. Carbon Dioxide

Rees (1969) developed an automatic method for determination of carbon dioxide in beer by the Technicon Autoanalyzer. Carbon dioxide in the sample is fixed with 65% sodium hydroxide. The color of a weak sodium carbonate–sodium bicarbonate solution colored by addition of phenolphthalein is reduced.

Another method using a second generation Technicon Autoanalyzer has been described by Moll *et al.* (1975f). After fixation of the carbon dioxide with 0.1 $N$ sodium hydroxide, the carbon dioxide is liberated by acidification with 0.5 $N$ sulfuric acid. The liberated carbon dioxide is dialyzed through a special membrane and reacts with cresol red solution, which is decolorized and measured at 420 nm.

The correlation coefficient between the automatic method and the Cannizarro method described by De Clerck (1962) is $r = 0.97***$, $n = 38$.

## 5. Bitter Substances

A continuous-flow countercurrent extractor was developed by Cooper and Hudson (1961) to give continuous, automatic estimation of the amount of hop bitter substances in beer. Two other automatic methods having a production rate of 10 samples/hr were described by Jamieson and Mackinnon (1965) and by Pinnegar (1966b), who used a small-scale continuous extractor.

An automatic adaptation of the EBC (1975a) method for bitter substances in beer was developed by Moll *et al.* (1975b). The correlation coefficient between the two methods was $r = 0.994***$, $n = 45$. Several authors have introduced HPLC for the determination of individual bitter substances: Molyneux and Wong (1973), Siebert (1976), Schwarzenbach (1979), Verzele and Dewaele (1981a,b), Bodkin *et al.* (1980), Otter and Taylor (1978), Gill (1979), Verzele and De Porter (1978), Verzele *et al.* (1980), Lance *et al.* (1981), Whitt and Cuzner (1979), Dewaele and Verzele (1980), Bazard *et al.*

(1981a), Schulze *et al.* (1981), Gross and Schwiesow (1982), and Verzele *et al.* (1982).

Recently, near-infrared reflectance has been applied for the determination of alpha acids, moisture, beta acids, and hop storage index by Axcell *et al.* (1981).

## 6. Gravity

Henning and Hicke (1973) described an automated digital method used for wine that allows the calculation of the density or the apparent relative density from the resonance frequency of a flexural oscillator filled with the solution and excited to undamped oscillation by high frequency. Zaake and Otto (1975) described the Anton Paar DMA 50 apparatus; the differences in 47 extract determinations on malt infusions between pycnometer and DMA 50 were +0.02 to −0.03%. Benard (1975b, 1978a), Gauron and Conac (1978), and Conac and Rey (1981) also mentioned the use of the Anton Paar instrument with various liquids.

## 7. Alcohol

Ashurst (1963) developed an automatic colorimetric method for the estimation of the alcohol content of beer that involves treatment with hexanitratocerate after oxidation and dialysis to remove interfering carbohydrates.

Automatic distillation procedures have been applied by Sawyer and Dixon (1968), Sarris *et al.* (1969), Lidzey *et al.* (1970), and more recently by Moll *et al.* (1975d). The enzyme determination of ethanol (oxidation of ethanol to acetaldehyde in the presence of NAD and ADH, with the formed NADH then measured at 340 nm) was described by Mor *et al.* (1973) and by Moll *et al.* (1977a). The specificity of this method allows the determination of ethanol at very low concentration (i.e., in fruit juices, alcohol-free beer, etc.).

Automatic thermometric analysis (see Section II,B,8) has been used in spirits of 45° Gay–Lussac for the determination of ethanol. Dupont (1978) mentioned that this type of instrument allows 20 analyses to be made per hour with a precision of 0.1° Gay–Lussac of alcohol. Similar results were described by Tep *et al.* (1978) using the same technique with wine. The accuracy is better than 0.1% of alcohol in wine between 10 and 12% alcohol. High-performance liquid chromatography has been applied for the determination of ethanol, glycerol, and carbohydrate by Bazard *et al.* (1981b) and by Cieslak and Herwig (1982).

Recently, it was shown by Grandclerc *et al.* (1982) that the alcohol content of beer could be determined by near-infrared reflectance. The correlation coefficient was $r = 0.93$***, n $= 29$, between this method and the reference method.

## 8. Original Gravity

Sawyer and Dixon (1968) were the first to introduce automatic determination of original gravity of beer using the determination of alcohol and gravity lost. The agreement with the official method used in Great Britain is $\pm 2$ units of original gravity (distillation procedure). Several systems for the determination of original gravity have been proposed since the introduction of the Anton Paar Density Meter DMA 50 or 55, as described by Benard (1978a).

The SCABA instrument is equipped with an Anton Paar densitometer; the alcohol is measured by catalytic oxidation resulting in displacement of a Wheatstone bridge. By means of Tabarie's and Balling's formulas, original gravity, alcohol, real and apparent extract, and real attenuation can be calculated. Various tests with this instrument have been described by Schur et al. (1978) and by Korduner and Westelius (1981).

The Liquilyzer also uses the Anton Paar instrument, together with a refractometer and a calculator. A detailed description of this apparatus was published by Höhn (1977) and Rouiller (1977).

The Technicon Autoanalyzer uses the Anton Paar densitometer and a thermometric analyzer in which the sample is reacted with an alcoholic solution of perchlorate; the reaction is endothermic, and the variation of enthalpy, as measured by the thermometric detector, is proportional to the alcohol concentration. A calculator using the values of apparent extract and alcohol so obtained determines the original extract, real extract, and attenuation of the beer. Figure 2 shows an automated system for measuring the original gravity, alcohol content, and apparent extract of beer.

The Technicon instrument was tested by Lehuédé et al. (1979). With the reference method of the EBC (1975d), all three instruments, the Technicon, the Liquilyzer, and the SCABA, gave high correlations, as shown by Lehuédé et al. (1979) (see Table 1).

Original gravity of beer was determined by Grandclerc et al. (1982) by near-infrared reflectance. The correlation coefficient between the reference method and this new method was $r = 0.96$***, $n = 28$.

## 9. Nitrogen, Free Amino Acids, Amino Acids

**a. Total Nitrogen.** Information on several instruments for the automatic determination of nitrogen in grains has been summarized by Benard (1975a), Vincent and Shipe (1976), and Moll (1979a). A separate digestion with sulfuric acid and hydrogen peroxide is performed as a batch procedure, and the ammonium content of the acid digest is estimated automatically in a

**Fig. 2.** An automatic system for the measurement of original gravity of beer, alcohol content, and apparent extract.

continuous-flow system at a rate of 60 samples/hr. This type of procedure was described by Mitcheson and Stowell (1970), Concon and Soltess (1973), Bietz (1974), and Moll *et al.* (1975e). Buckee (1974) improved the method by including, after the digestion stage, an ammonia probe to replace the distillation–titration procedure.

Continuous digestion with a rotating glass helix was described by Garza (1965), Concon and Soltess (1973), Kramme *et al.* (1973), Bosset and Jenni (1973), Kubadinow (1974), Jackson *et al.* (1975), Wall and Gehrke (1975), Wall *et al.* (1975), who conducted a comparison with three other methods, and Elkei (1976).

**TABLE 1   Comparison with the Reference Method for Three Different Instruments**

| Parameters | SCABA | Liquilyzer[a] | Technicon |
|---|---|---|---|
| Number of trials | 54 | 35 | 49 |
| Correlation coefficient | $r = 0.990^{***}$ | $r = 0.995^{***}$ | $r = 0.996^{***}$ |
| Regression equation | $y = 0.99x + 0.52$ | $y = 0.994x + 0.12$ | $y = 1.00x - 0.06$ |

[a] Elimination of beers $> 14°$ Plato, which gave too high results.

Near-infrared reflectance spectroscopy was used to determine total nitrogen in cereals by Williams (1975), Watson et al. (1976, 1977), Moll et al. (1976), and Grandclerc et al. (1982).

**b. Free Amino Acids.** An automatic method for determination of free amino nitrogen in wort and beer using ninhydrin was described by Ashurst and MacWilliam (1963) and by Hampel (1973). For the same determination, Pinnegar (1966a), Skarsaune et al. (1973), and Mor et al. (1973) have employed the 2,4,6-trinitrobenzene-1-sulfonic acid (TNBS) reaction.

The EBC (1975b) method for determination of free amino nitrogen with ninhydrin has been automated by Moll et al. (1975c).

**c. Amino Acids.** The automatic column chromatography first described by Moore and Stein (1951, 1954), Spackman et al. (1958), and Moore et al. (1958) has been further developed by Jones and Pierce (1963), Moll et al. (1972), and Fujita et al. (1979). A general summary of these methods has been given by Moll (1979b).

High-performance liquid chromatography estimation of carbohydrates and free amino acids in the brewing industry was described by Helbert et al. (1977) at the Pittsburgh conference. Several techniques for the separation of amino acids by HPLC have been presented by Bayer et al. (1976), Molinar and Horvath (1977), Brown et al. (1978), Wilkinson (1978), Deelder et al. (1979), Lindroth and Mopper (1979), Schuster (1980), Bishop et al. (1980), Liebezeit and Dawson (1981), and Khayat et al. (1982). The important point in HPLC is careful preparation of the sample before injection to avoid erroneous results.

## 10. Mineral and Trace Elements

Today atomic absorption spectrophotometers with furnace oven and Zeeman effect are fully automated, with automatic samplers. This subject has been extensively reviewed by Charalambous (1981) and by Harvey (1981).

Skowzonski and Wedl (1982) described an automated coulometric titration method for the determination of chloride in wort and beer.

## 11. Phenols

Mathieu et al. (1965) have developed an automatic method of estimating phenolic substances that can yield colors with diazotized paranitraniline. The following compounds were estimated: phenol, 2-hydroxybenzoic acid, and 4-hydroxy-3-methoxymandelic acid.

Recently HPLC has improved the determination of phenols and polyphenols, as mentioned by Moll et al. (1978b), Hrazdina (1979), Jerumanis (1979), Outtrup (1981), and Mulkay (1982).

## 12. Total Acidity, Volatile Acidity

Sarris *et al.* (1970b) described an automatic method for volatile acidity and wine based on colorimetric determination after distillation of volatile acids. The indicator used is bromophenol blue.

## 13. Enzymic Activities in Foods and Food Raw Materials

**a. Proteolytic Activity.** An automated fluorometric method (using fluorescamine) for the determination of proteolytic activity in wheat extract is described by Preston (1974).

**b. Diastatic Power, α-Amylase.** Ashurst and MacWilliam (1963) estimated the reducing power of digested starch by measuring the reduction of color of an alkaline solution of potassium ferricyanide in a continuous-flow system. The glucose concentration values obtained and the corresponding value for diastatic activity were plotted graphically to estimate their relationship. An improved method, in which the starch digestion was carried out in a polyethylene coil immersed in a water bath at 20°C, was developed by Saletan and Scharoun (1965). The reducing power was determined as described above. Banasik (1971) developed an automated procedure for the determination of malt diastatic power and α-amylase activity. For the diastatic power the method of Saletan and Scharoun (1965) was slightly modified. For the α-amylase activity the β-amylase was inactivated by heating the malt infusion at 70°C in the presence of calcium ions.

Buckee (1973a) described a method for estimating automatically the diastatic activity in malt. A semiautomatic batch procedure was used for the dialysis stage, and the reducing power of the digested starch was measured automatically using a reduction–oxidation electrode in an Autoanalyzer system.

Rouiller (1980) has adapted both methods of Banasik (1971) on a Technicon Autoanalyzer (second generation). For the α-amylase activity three methods were compared: (1) inactivation of β-amylase by heat, (2) action of α-amylase on a substrate of limit dextrins followed by determination of the reducing sugars formed, and (3) iodine coloration. The last method gave the best correlation: $r = 0.87***$, $n = 36$, with the standard EBC (1975c) method.

## 14. Flavor

One of the most difficult problems in the food industry is to determine flavor components that are directly related to taste panel results. Today it is in many cases possible to determine negative flavor constituents in foods by various techniques.

The prediction of food quality from chemical analysis of raw materials is an objective still to be attained. Multidimensional analysis using chemical data are today able to classify *a posteriori* the flavor quality of food.

## C. Microbiological Methods

Harrison *et al.* (1974) described methods for the rapid detection of brewery spoilage microorganisms. Automation in microbiology and immunology and new approaches to the identification of microorganisms were both discussed in papers presented during the first International Symposium on Rapid Methods and Automation in Microbiology (Heden and Illeni, 1975). Papers presented at a second International Symposium on Rapid Methods and Automation in Microbiology have also been published (Johnston and Newsom, 1976).

Molzahn and Portno (1975) and Portno and Molzahn (1977) described a sterile, continuous, automatic microbiological sampler for liquids and gases with membrane filtration on a moving strip of membrane filter material. The method allows direct inoculation onto a nutrient agar surface and clearly shows variation in the level of airborne contamination over a period of time in any selected area. A mechanized plating instrument for viable counts of microbes in food was applied by Anagnostopoulos (1981). The French Microbiological Society (1980) provided a summary of papers concerning rapid and automatic methods in food microbiology.

Bourgeois and Mafart (1980) have listed several new techniques for the detection of microorganisms. The Bactometer, which was described by Hadley and Senyk (1975), detects and measures the metabolic activity of microorganisms by continually monitoring the impedance of the substrate. Lawless *et al.* (1974) studied beer spoilage bacteria and were able to detect six *Lactobacillus brevis* or *L. casei* cells per milliliter of beer in 2 days. Four or 5 days are necessary to detect *Pediococcus.*

In our laboratory we obtained the following results: detection of brewing yeasts, 1 cell in 41–47 hr, 16–32 cells in 30–38 hr; for *Lactobacillus brevis,* 3 cells in 36 hr, $10^4$ cells in 10.5 hr; for *L. bacillus plantarum,* $10^5$ cells in 12 hr. It is clear that these detection times are too long for the brewing industry, so that the method needs to be improved (D. Erbs and M. Aigle, private communication).

The results of rapid estimations of microorganisms in milk were published by Gnan and Luedecke (1982) for classification of 300 raw milk samples. Methods for quantifying the bacteria contents of cereal grains using impedance monitoring were described by Sorrells (1981). Harrison *et al.*

(1974) and Heysert *et al.* (1976) described a bioluminescence ATP assay method for rapid detection of brewery microorganisms.

Miller *et al.* (1978) used this method for the quantitation of variable yeasts for fermenter pitching in the brewery. Applications of this technique for determination of the bacteriological quality of milk were described by Bossuyt (1981) and by Waes and Bossuyt (1981).

Infrared (IR) analysis of headspace gas for the presence of $CO_2$ was evaluated by Threlkeld (1982) as a method of nondestructively sampling and segregating microbiologically contaminated citrus juices and fruit drinks. Comparison of IR analysis with the conventional plating method indicated that 89% of the samples determined to be positive by the conventional method were also detected by IR analysis.

## III. ON-LINE ANALYSIS

Many new and good instruments have been developed in research laboratories for the food industries, but in reality only very few have been installed. The reasons for this failure are:

1. Insufficient definition of the parameter to be automated.
2. Unrealistic expectations of the new automated instrument.
3. Insufficient preparation between research and development and industry for the transfer of the instrument.
4. Ineffective maintenance or absence of maintenance of the new instrument that immediately affects the accuracy of the measurements.

In this short overview only a few practical examples can be discussed. Several important books and articles should be mentioned: Association pour la Promotion Industrielle et Agricole (APRIA, 1980), Huskins (1977, 1981b, 1982), Hahn (1977a,b), Lloyd (1976), Hughes (1981), Dardenne (1981), Carr-Brion (1977), Houghton (1974), Wang *et al.* (1979), Boudier *et al.* (1977), Hasenböhler (1978), Hatch (1982), Ryu and Humphrey (1973), Schwartz and Cooney (1978), Forch (1976), and Enari *et al.* (1976).

### A. Sampling

The quality of results from on-line instruments depends on accurate sampling. An excellent chapter on sample handling is provided by Huskins (1981a). Sampling techniques for grains are mentioned by Fan *et al.* (1976) and Genzken (1979). Liquid-sampling devices are more complex: sterility and oxygen-free operations are often required in the beverage industry.

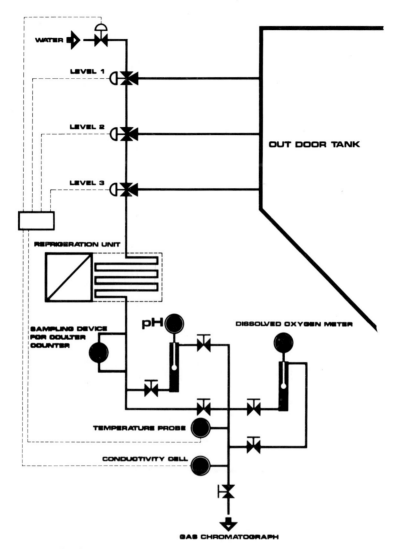

**Fig. 3.** Layout of an automatic sampling and analysis system.

Examples of automatic liquid samplers are described by Blachère *et al.* (1978) and Moll *et al.* (1979).

Figure 3 shows a process for the automatic representative taking of a beverage during fermentation. The sample is cooled to 0°C and passed to various instruments for measuring pH, dissolved oxygen, and temperature and to a Coulter counter and a gas chromatograph.

## B. Analysis of Solid Samples

### 1. Weight

Weighing equipment can be automated, as shown by Graham (1971), Pickering (1971), Nolte (1972), and Kalko and Crull (1975).

### 2. Moisture Content

Several techniques have been proposed for analysis of moisture contents but very few are used on line. Near-infrared reflectance applications for moisture content are described by Law (1977) and Moll et al. (1976). A microwave technique was proposed by Kraszewski (1974); Menu (1976) proposed an on-line apparatus for moisture content determination by hygrometer hyperfrequency. An IR adsorption technique was described by Verbiese (1981). During the drying of spent grains, a unique scheme based on feedback from steam-flow sensors was used to control levels of outlet moisture content as described by Myron and Shinskey (1975).

## C. Analysis of Liquid Samples

### 1. Primary Sensors in the Food Industry

Several automatic instruments are satisfactorily installed in the food industry. General articles in this field have been published by Soroko (1967), Hahn (1977a), Sullivan (1978), Desjardins and Sardet (1980), Boudier et al. (1977), and Prouveur (1975). A few examples of primary sensors are mentioned here. For temperature Michel (1971), Capteurs Français (1976b), and Kardos (1977), have published general discussions, and Mordt (1982) has treated temperature regulation in fermenters. For pressure Capteurs Français (1976a) summarizes absolute, absolute and differential, relative, and acceleration pressure gauges. For liquid level height a general article has been published by Clerc (1977), an acoustic device described by Hough et al. (1975), and a conductivity-operated level controller discussed by Moxey (1973). For flow-volume meters general articles have been published by Hofgrefe (1979), Boyer (1976), Lomas (1974), Stoten (1974), Walls (1977), Schumacker (1973), Kempf et al. (1976), Sanden and Hamm (1974), Sanden (1979), Hayward (1978), and Treiber (1978). For pH Bühler and Ingold (1976) have described the use of pH meters in fermenters.

### 2. Dissolved Oxygen

On-line oxygen determination in liquids is a delicate operation, and careful study of each situation in which it is to be used is required. Several hundreds of review articles have been published on this subject, among which the following references may be summarized.

Fatt (1976), Degh *et al.* (1976), Hitchman (1978), Lee and Tsao (1979), Dunn and Einsele (1975), Howard and Mawer (1977), Wackerbauer *et al.* (1975), Clack and Stevenson (1982), Bühler and Ingold (1976), Mukhopadhyay and Ghose (1976), Hahn and Hill (1980), Krebs (1975), Jandreau and Hahn (1978), and Moll *et al.* (1977c).

The continuous determination of dissolved oxygen in liquids presents several difficulties, such as:

1. Standardization of a physical method of determination of dissolved oxygen using a chemical calibration method is required. The iodometric Winkler method (1888) is adapted for the measurement and calibration of dissolved oxygen in water. The indigo–carmine method described by Jenkinson and Compton (1960) produces higher values than the Winkler method, as shown by the ASBC (1975, 1976b). The air calibration of oxygen analyzers is a delicate operation and is inaccurate for low levels of dissolved oxygen contents in beverages other than water in the range between 0 and 0.5 ppm.

2. The influence of $CO_2$ content, pH, temperature, and reducing or oxidizing substances in liquids is very significant, as partly illustrated by ASBC (1975, 1976), Hahn and Hill (1980), Howard and Mawer (1977), and Wisk *et al.* (1981).

3. The presence of bubbles affects the continuous determination of dissolved oxygen.

A means of avoiding bubbles on the polarograph cell was described by Moll *et al.* (1978c): a debubbling cell (see Fig. 4) in which the oxygen meter

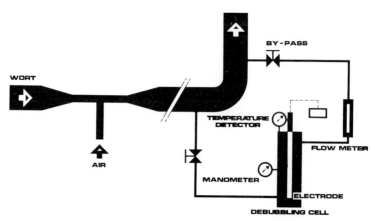

**Fig. 4.** Layout of equipment for the continuous measurement of dissolved oxygen in wort.

could be placed was set in a bypass pipe. The advantage of this arrangement was that it allowed the continuous measurement of dissolved oxygen in a wort or beer pipe. The calibration was done by the Winkler method or by air (the latter was less satisfactory).

### 3. Carbon Dioxide

Various methods for the automatic on-line measurement of dissolved carbon dioxide in beverages have been in use for several years. Pressure–temperature relationships based on Henry's law have been treated by Rohner and Tompkins (1970), De Brune *et al.* (1974), Steen (1973), and Fuellpack (1976). Infrared absorption has been discussed by Gamache (1968), Bedrossian (1981), Frant (1981), and Huben and Turner (1974). Kesson (1982) has treated carbon dioxide-selective membrane, leading to direct reading of carbon dioxide volume per volume of liquid.

### 4. Turbidity Measurement

Beverages such as water, wine, and beer (and also oil, etc.) are examined automatically on line for turbidity prior to packaging of the final product. The complexity of this type of measurement is described by Allen (1981), Kerker (1969), Groves (1978), Stockham and Fochthan (1978), Thorne and Beckley (1958), Thorne and Nannestad (1959), Thorne (1961), Thorne and Svendsen (1962), Claesson and Sandegren (1963, 1969), Gerster (1969), Clydesdale (1973), Sigrist (1975), and Moll (1984). The industrial objective is to find a determination of turbidity of the final product the results of which are correlated with the turbidity results after a period of storage. Most of the on-line turbidity instruments are based on measurement of light scattered at various angles.

Practical problems not detected by nephelometry may arise with variation in the chemical composition of haze. An example of this type was described by Leedham and Carpenter (1977), who reported that normal readings in a nephelometer had been obtained although the haze of the beer was noticeable to the eye. In this case the number of particles in the liquid of the size range 5–12 $\mu$m were determined by a Coulter counter, and their compositions were found to include degraded starch, calcium oxalate, protein, and silica.

### 5. Methods for Measurement of Microbial Growth

Several authors have described various techniques in this field, among them Wyatt (1973), Pringle and Mor (1975), and Cooney (1982). Many descriptions of applications in this field can be found in books and journals of biotechnology and fermentation. In the brewing industry the determination of yeast cells in suspension in wort and during fermentation presents difficulties because of colloids present in wort. These interfere with the measurement of yeast in suspension by light-scattering instruments. For this reason our choice was an automatic system of supply, developed by Moll *et al.* (1977b, 1978c), for the Coulter counter. The following measurements can be obtained by using this instrument: number of yeast cells in suspension, size distribution of the yeast cells, and average volume of the yeast cells (see Fig. 5).

Beyeler *et al.* (1981) described a new probe allowing the measurement of NAD(P)H-dependent culture fluorescence in a bioreactor. Metabolic alterations due to substrate or oxygen deficiency can easily be detected by fluorimetric measurements.

**Fig. 5.**   Interface for the continuous measurement of yeast cells in suspension.

## 6. On-Line Measurement of Density

An excellent and thorough review of density balance methods is given by Huskins (1982). The U-tube density meter was described by Dittrich (1964) and by Browne (1965). Similar instruments were used in the dairy industry by Depince (1981).

Feil and Zacharias (1971) used a sonic solution analyzer based on a modification of the sing-around principle. Another method for indirect density measurement in a fermenter using gas-bubble probes was described by Treiber (1975), Treiber and Höhn (1976), and Anthon (1977). Radiometric density measurement has been applied to starch solutions by Kempf et al. (1976), and Sommer (1980) applied a similar technique to the determination of density in wort.

The refractometer for density determination was mentioned by Asselmeyer (1965), Asselmeyer and Höhn (1965), Höhn (1965), and Asselmeyer et al. (1971).

A refractometer, Remat 20, for process control (lauter tun and beer filtration) was described by Matène (1979). A very simple device consisting of a balance-type principle in a bypass for the determination of density was developed by Krumtünger (1965).

Moller (1975) described the Platometer, which is based on the determination of differential pressure. It automatically compensates for temperature variations and provides an electric signal whose strength is directly proportional to the apparent extract of the fermenting wort.

The indirect measurement of density during fermentation was suggested by Holdom et al. (1976), Rischbieter (1977), Moll et al. (1978c), and Luckiewicz (1978), who determined the quantity of carbon dioxide produced. Mou and Cooney (1976), Monk (1978), Ruocco et al. (1980), and Bayer and Fuehrer (1982) measured the heat of fermentation.

## 7. Original Gravity of Beer

Very few industrial instruments are able to give satisfactory results for the on-line determination of original gravity of beer. Wagner and Lloyd (1977) and Lloyd (1979) described a microcomputer with a combination of specific-gravity sensor (Mettler Paar DMA 50) and a refractive index sensor (Anacon 31 S). The following measurements were obtained with this module: alcohol (by weight or volume), original gravity, and calories. It could also be used as a digital blender and batching system.

Forch (1977) described in his publication another technique for the on-line determination of original gravity of beer: a Steuma instrument working in a bypass using two refractive index sensors for the determination of alcohol and real extract combined with a microcomputer. Asselmeyer and Gschwind (1972) and Asselmeyer *et al.* (1977) developed an instrument for the determination of original gravity of beer using a radiometric density meter combined with a refractometer and a computer.

Graberg (1981) described a unit for the continuous control of beer strength during blending in which the alcohol content was determined by means of catalytic combustion of alcohol in an airstream. The density was measured using a tuning-fork densitometer.

## 8. Conductivity Measurement

On-line conductivity measurements to avoid liquid losses have been used in several applications. Differentiation of water and disinfection solutions was mentioned by Riemann (1974) and by Küster and Schlosser (1975). Brand differentiation was described by Paukner and Chaniotis (1956), Runkel (1971), and Riemann (1974). Differentiation between water and other solutions was used by Grafinger *et al.* (1965), Riemann (1974), Küster and Schlosser (1975), and Moll *et al.* (1979). Last runnings in mash filtration were proposed by Neugebauer (1970), and Küster (1977).

**REFERENCES**

Allen, T. (1981). "Particle Size Measurement," 3rd ed. Chapman and Hall, London.
Anagnostopoulos, G. D. (1981). *In* "The Quality of Food and Beverages (G. Charalambous and G. Inglett, eds.), Vol. 2, pp. 83–94. Academic Press, New York.
Anthon, F. (1977). *Monatsschr. Brau.* **30**, 408–409.
APRIA (1980). "Les Capteurs et les Industries Agro-Alimentaires." APRIA, Paris.
ASBC (1975). *J. Am. Soc. Brew. Chem.* **33**, 93–95.
ASBC (1976a) "Methods of Analysis," 7th ed. ASBC, St. Paul, Minnesota. (Beer 10 color.)
ASBC (1976b). *J. Am. Soc. Brew. Chem.* **34**, 115–118.
Ashurst, P. R. (1963). *J. Inst. Brew.* **69**, 457–459.
Ashurst, P. R., and MacWilliam, I. C. (1963). *J. Inst. Brew.* **69**, 394–397.
Asselmeyer, F. (1965). *Brauwelt* **105**, 41–42, 901–906.
Asselmeyer, F., and Gschwind, F. X. (1972). *Brauwissenschaft* **25**, 377–383.
Asselmeyer, F., and Höhn, K. (1965). *Brauwissenschaft* **18**, 165.
Asselmeyer, F., Hönh, K., and Klisch, W. (1971). *Brauwissenschaft* **24**, 73–80.
Asselmeyer, F., Feller, C. H., and Höhn, K. (1977). *Brauwissenschaft* **30**, 297–303.
Axcell, B., Tulej, R., and Murray, J. (1981). *Brew. Dig.* **56**, (6) 18, 19, 41; (11) 32–33.
Banasik, O. J. (1971). *Wallerstein Lab. Commun.* **34**, 45–52.
Bartels, H., Werder, R. D., Schurmann, W., and Arndt, R. W. (1978). *J. Autom. Chem.* **1**, 28–32.
Bauer, G., and Roth, R. (1978). French Patent 7803284.
Bayer, K., and Fuehrer, F. (1982). *Process Biochem.* **17** (7/8), 42–45.
Bayer, E., Grom, E., Kaltenegger, B., and Uhrmann, R. (1976). *Anal. Chem.* **48**, 1106–1109.
Bazard, D., Flayeux, R., and Moll, M. (1981a). *Ind. Aliment. Agric.* **98**, 55–61.

Bazard, D., Lipus, G., and Moll, M. (1981b). *Ind. Aliment. Agric.* **98**, 1033–1038.
Beadle, J. B. (1969). *J. Agric. Food Chem.* **17**, 904–906.
Bedrossian, J., Jr. (1981). *Tech. Q. Master Brew. Assoc. Am.* **17**, 87–91.
Benard, M. (1975a). *Bios (Nancy)* **6** (11), 390–393.
Benard, M. (1975b). *Bios (Nancy)* **6** (11), 394–395.
Benard, M. (1978a). *Ann. Nutr. Aliment.* **32**, 915–922.
Benard, M. (1978b). *Brauwissenschaft* **31**, 181–183.
Benard, M. (1982). *In* "Biotechnologie" (R. Scriban, ed.), pp. 463–476. Technique & Documentation, Lavoisier, Paris.
Beyeler, W., Einsele, A., and Fiechter, A. (1981). *Eur. J. Appl. Microbiol. Biotechnol.* **13**, 10–14.
Bietz, J. A. (1974). *Anal. Chem.* **46**, 1617–1618.
Bishop, C. A., Harding, D. R. K., Meyer, L. J., and Handcock, W. S. (1980). *J. Chromatogr.* **192**, 222–227.
Bittner, D., and McCleary, M. (1963). *Am. J. Clin. Pathol.* **11**, 423–424.
Blachère, H. T., Perringer, P., and Cheruy, A. (1978). *In* "Biotechnology," *Proc. 1st Eur. Congr. Biotechnol. Interlaken,* Part 3, pp. 65–68. Verlag Chemie, Weinheim.
Blom, J. (1957). *Eur. Brew. Conv. Proc. Congr., Copenhagen* pp. 51–56.
Bodkin, C. L., Clarke, B. J., Kavanagh, T. E., Moulder, P. M., Reitze, J. D., and Skinner, R. N. (1980). *J. Am. Soc. Brew. Chem.* **38**, 137–142.
Bosset, J. O., and Jenni, U. (1973). *Lab. Pract.* **22**, 578–580.
Bossuyt, R. (1981). *Milchwissenschaft* **36**, 257–260.
Boudier, J. F., Luquet, F. M., and Beringuer, R. (1977). *Ind. Aliment. Agric.* **94**, 7–26.
Bourgeois, C. M., and Mafart, P. (1980). *In* "Techniques d'analyse et de contrôle dans l'industrie agro-alimentaire" (C. M. Bourgeois and J. Y. Leveau, eds.), pp. 42–62, Technique et Documentation, Lavoisier, Paris.
Boyer, M. (1976). *Bios (Nancy)* **7** (3), 3–9.
Brandon, A. L. (1967). *Brew. Dig.* **42** (4), 96–101.
Brandon, A. L. (1968). *Proc. Am. Soc. Brew. Chem.* pp. 10–20.
Brobst, K. M., Scobell, H. D., and Steele, E. M. (1973). *Proc. Am. Soc. Brew. Chem.* pp. 43–46.
Brown, M. E. (1961). *Diabetes* **X**, 60–62.
Brown, A. S., Mole, J. E., Weissinger, A., and Bennett, J. C. (1978). *J. Chromatogr.* **148**, 532–535.
Browne, G. W. (1965). *Brew. Dig.* **40** (3), 84–87.
Bühler, H., and Ingold, W. (1976). *Process Biochem.* **11** (4), 19–24.
Buckee, G. K. (1972). *J. Inst. Brew.* **78**, 222–224.
Buckee, G. K. (1973a). *J. Inst. Brew.* **79**, 17–19.
Buckee, G. K. (1973b). *J. Inst. Brew.* **79**, 61–64.
Buckee, G. K. (1974). *J. Inst. Brew.* **80**, 291–294.
Buckee, G. K., and Hickman, E. (1975). *J. Inst. Brew.* **81**, 399–407.
Buijten, J. C., and Holm, B. (1979). *J. Autom. Chem.* **1**, 91–94.
Burns, D. A. (1981). *Anal. Chem.* **53**, 1403–1418.
Capteurs Français (1976a). "Commission Interministérielle des Appareils Electriques et Electroniques de Mesures," Vol. 1. Documentation Française, Paris.
Capteurs Français (1976b). "Commission Interministérielle des Appareils Electriques et Electroniques de Mesures," Vol. 2. Documentation Française, Paris.
Carr-Brion, K. G. (1977). *Meas. Control* **10**, 407–412.
Charalambous, G. (1981). *In* "Brewing Science" (J. R. A. Pollock, ed.), Vol. 2, pp. 177–180. Academic Press, New York.

Cieslak, M. E., and Herwig, W. C. (1982). *J. Am. Soc. Brew. Chem.* **40**, 43–46.

Clack, P. J., and Stevenson, R. G., Jr. (1982). *Brew. Dig.* **57** (5), 34–38.

Claesson, S., and Sandegren, E. (1963). *Proc. Eur. Brew. Conv. Congr., Bruxelles* pp. 221–232.

Claesson, S., and Sandegren, E. (1969). *Proc. Eur. Brew. Conv. Congr., Interlaken* pp. 339–347.

Clapperton, J. F., and Hollyday, A. G. (1968). *J. Inst. Brew.* **74**, 164–169.

Clerc. (1977). *Bios (Nancy)* **8** (7/8), 19–31.

Clydesdale, F. M. (1973). *Brew. Dig.* **48** (10), 46–53, 70.

Conac, M., and Rey, J. C. (1981). *Inf. Chim.* Nos. 216/217, pp. 169–179.

Concon, J. M., and Soltess, D. (1973). *Anal. Biochem.* **53**, 35–41.

Conrad, E. C., and Palmer, J. K. (1976). *Food Technol. (Chicago)* **30**(10), 84–92.

Cooney, C. L. (1982). *In* "Biotechnology" (H. J. Rehm and G. Reed, eds.), Vol. 1, pp. 73–114. Verlag Chemie, Weinheim.

Cooper, A. H., and Hudson, J. R. (1961). *J. Inst. Brew.* **67**, 436–438.

Cooper, A. H., Hudson, J. R., and MacWilliam, I. C. (1961). *J. Inst. Brew.* **67**, 432–435.

Curry, A. S., Gomm, P. J., Nicholson, D. J., and Patterson, D. A. (1974). *Lab. Pract.* **23**, 309–310, 315.

Dardenne, G. (1981). *Enjeux* **20**, 44–46.

De Brune, P., Cremer, J., Dorrenboom, J. J., and Witte, J. H. M. (1974). *Tech. Q. Master Brew. Assoc. Am.* **11**, 286–287.

De Clerck, J. (1962). "Cours de Brasserie" (Université de Louvain ed.), Vol. 2, 2nd ed., p. 709. Inst. Agronomique, Heverlee, Louvain.

Deelder, R. S., Linssen, H. A. J., Konijnendijk, A. P., and Van de Venne, J. L. M. (1979). *J. Chromatogr.* **185**, 241–257.

Degn, H. Balslev, I., and Brook, R. (1976). "Measurement of Oxygen." Elsevier Sci. Publ. Amsterdam.

Dellweg, H., Trenel, G., John, M., and Emeis, C. C. (1969). *Monatsschr. Brau.* **22**, 177–181.

Dellweg, H., John, M., and Trenel, G. (1970). *Proc. Am. Soc. Brew. Chem.* pp. 154–162.

Dellweg, H., John, M., and Trenel, G. (1971). *J. Chromatogr.* **57**, 89–97.

Depince, Y. (1981). *Inst. Syst.* No. 2, pp. 9–11.

Desjardins, M., and Sardet, D. (1980). *Ind. Aliment. Agric.* **97**, 1011–1014.

Devillers, P., Cornet, C., and Detavernier, R. (1974). *Ind. Aliment. Agric.* **91**, 833–839.

Devillers, P., Detavernier, R., and Leroux, M. (1982). *Ind. Aliment. Agric.* **99**, 497–504.

Devreux, A. (1968). *Ind. Chim. Belge, Nr. Spéc.* **33**, 199–203.

Dewaele, C., and Verzele, M. (1980). *J. Chromatogr.* **197**, 189–197.

Dittrich, G. (1964). *Brauwelt* **104**, 902–908.

Drawert, F., and Leupold, G. (1976). *Chromatographia* **9**, 447–453.

Dunn, I. J., and Einsele, A. (1975). *J. Appl. Chem. Biotechnol.* **25**, 707–720.

Dupont, P. (1978). *Ann. Nutr. Aliment.* **32**, 905–914.

EBC, Analytica III (1975a). *Schweiz. Brau. Rundsch.* p. E 60.

EBC, Analytica III (1975b). *Schweiz. Brau. Rundsch.* p. E 61.

EBC, Analytica III (1975c). *Schweiz. Brau. Rundsch.* p. E 34.

EBC, Analytica III (1975d). *Schweiz. Brau. Rundsch.* p. E 55.

Elkei, O. (1976). *Anal. Chim. Acta* **86**, 63–68.

Enari, T. M., Home, S., and Pajunen, E. (1976). *Eur. Brew. Conv., Comput. Process Contr. Symp., Monogr. III, Strasbourg* pp. 361–373.

Enevoldsen, B. S. (1970). *J. Inst. Brew.* **76**, 546–552.

Enevoldsen, B. S., and Schmidt, F. (1973). *Eur. Brew. Conv. Proc. Congr., Salzburg* pp. 135–148.

Enevoldsen, B. S., and Schmidt, F. (1974). *J. Inst. Brew.* **80**, 520–533.

Fan, L. T., Lai, F. S., and Wang, R. H. (1976). *Adv. Cereal Sci. Technol.* **1**, 77–111.
Fatt, I. (1976). "Polarographic Oxygen Sensors." CRC Press, Cleveland, Ohio.
Feil, M. F., and Zacharias, E. M. (1971). *Brew. Dig.* **46** (11), 76–80.
Fitt, L. E., Hassler, W., and Just, D. E. (1980). *J. Chromatogr.* **187**, 381–389.
Fonknechten, G., Bazard, D., Flayeux, R., and Moll, M. (1980). *Bios (Nancy)* **11** (6), 60–63.
Forch, M. (1976). *Eur. Brew. Conv., Comput. Process Contr. Symp., Monogr. III, Strasbourg* pp. 328–360.
Forch, M. (1977). *Monatsschr. Brau.* **30**, 487–493.
Foreman, J. K., and Stockwell, P. B. (1975). "Automatic Chemical Analysis." Horwood, Chichester.
Foreman, J. K., and Stockwell, P. B. (1979). "Topics in Automatic Chemical Analysis." Horwood, Chichester.
Frant, M. S. (1981). *In* "The Quality of Foods and Beverages" (G. Charalambous and G. Inglett, eds.), Vol. 2, pp. 241–252. Academic Press, New York.
French Microbiology Society (1980). "Méthodes rapides et automatiques en microbiologie alimentaire." Institut Pasteur Lille, France.
Fuellpack (1976). *Brauwelt* **116**, 833.
Fujita, K., Takeuchi, S., and Ganno, S. (1979). *In* "Liquid Chromatographic Analysis of Food and Beverages" (G. Charalambous, ed.), Vol. 1, pp. 81–97. Academic Press. New York.
Furman, W. B. (1976). Continuous Flow Analysis: Theory and Practice. Dekker, New York.
Gaines, T. P. (1973). *J. Assoc. Off. Anal. Chem.* **56**, 1419–1424.
Gamache, L. D. (1968). *Proc. Am. Soc. Brew. Chem.* 120–124.
Garza, A. C. (1965). *Autom. Anal. Chem., Technicon Symp.* pp. 61–65.
Gauron, C., and Conac, M. (1978). *Bios (Nancy)* **9** (10), 26–32.
Genzken, K. (1979). *Getreide, Mehl Brot* **32**, 323–326.
Gerster, R. A. (1969). *Tech. Q. Master Brew. Assoc. Am.* **6**, 218–220.
Gill, R. (1979). *J. Inst. Brew.* **85**, 15–20.
Gnan, S., and Luedecke, L. O. (1982). *J. Food Prot.* **45**, 4–7.
Goldberg, R. D. (1962). *Proc. Am. Soc. Brew. Chem.* pp. 140–146.
Graberg, B. (1981). *Brew. Distilling Int.* **11**, 29.
Grafinger, L., Neugebauer, K., and Schaefer, P. (1965). *Brauwelt* **105**, 1209–1214.
Graham, C. S. (1971). *Brew. Dig.* **46** (11), 72–75, 80.
Grandclerc, J., Lebordais, B., Carnielo, M., and Moll, M. (1982). *Brau. Rundsch.* **93**, 177–179.
Gross, R. W., and Schwiesow, M. H. (1982). *J. Am. Soc. Brew. Chem.* **40**, 116–117.
Groves, M. J., ed. (1978). "Particle Size Analysis," Heyden, Philadelphia.
Gum, E. K., Jr., and Brown, R. D., Jr. (1977). *Anal. Biochem.* **82**, 372–375.
Hadley, W. K., and Senyk, G. (1975). *In* "Microbiology, 1975" (D. Schlessinger, ed.), pp. 12–21. Am. Soc. Microbiol. Washington, D.C.
Hahn, C. W. (1977a). *Tech. Q. Master Brew. Assoc. Am.* **14**, 59–69, 87–93.
Hahn, C. W. (1977b). "The Practical Brewer" (H. M. Broderick, ed.), 2nd ed., pp. 395–398. Master Brew. Assoc. Am., Madison, Wisconsin.
Hahn, C. W., and Hill, J. C. (1980). *J. Am. Soc. Brew. Chem.* **38**, 53–60.
Holdom, R. S., Spivey, M. J., and Larsen, V. F. (1976). *J. Appl. Bacteriol.* **41**, 255–257.
Hampel, W. (1973). *Eur. Brew. Conv. Proc. Congr., Salzburg* pp. 25–32.
Harrison, J., Webb, T. J. B., and Martin, P. A. (1974). *J. Inst. Brew.* **80**, 390–398.
Harvey, J. V. (1981). *In* "Brewing Science" (J. R. A. Pollock, ed.), Vol. 2, pp. 277–290. Academic Press, New York.
Hasenböhler, A. (1978). Ph.D. thesis, University Tübingen, Tübingen.
Hatch, R. T. (1982). *Annu. Rep. Ferment. Processes* **5**, 291–311.
Hayward, A. T. J. (1978). *Mes., Reg., Autom.* **43** 59–74.

Heden, G. G., and Illeni, T., eds. (1975). "New Approaches to the Identification of Microorganisms and Automation in Microbiology and Immunology." Wiley, New York.

Helbert, J. R., Herwig, W. C., Wagener, R. E., Cieslak, M. E., and Chicoye, E. (1977). Pittsburgh Conference, Cleveland.

Henning, G., and Hicke, E. Z. (1973). *Anal. Chem.* **265**, 97–104.

Hitchman, M. L. (1978). "Measurement of Dissolved Oxygen." Wiley, New York.

Hofgrefe, W. (1979). *Brauindustrie* **64**, 161–171.

Höhn, K. (1965). *Brauwelt* **105**, 907–910.

Höhn, K. (1977). *Bios (Nancy)* **8** (9), 24–31.

Hough, J. S., Wadeson, A., and Welsby, V. G. (1975). *Tech. Q. Master Brew. Assoc. Am.* **12**, 9–14.

Houghton, D. R. (1974). *Brewer* **60**, 63–69.

Howard, G. A., and Mawer, J. D. R. (1977). *J. Inst. Brew.* **83**, 144–152.

Hrazdina, G. (1979). *In* "Liquid Chromatographic Analysis of Food and Beverages" (G. Charalambous ed.), Vol. 1, pp. 141–159. Academic Press, New York.

Huben, W. S., and Turner, G. S. (1974). *Brew. Guardian* **103**(7), 23–25.

Hudson, G. J., John, P. M. V., Bailey, B. S., and Southgate, D. A. T. (1976). *J. Sci. Food Agric.* **27**, 681–687.

Hughes, C. J. (1981). *Brew. Guardian* **110**(11), 41–46.

Huskins, D. J. (1977). "Gas Chromatographs as Industrial Process Analysers." Hilger, Bristol.

Huskins, D. J. (1981a). "General Handbook of On-Line Process Analysers," pp. 67–149. Horwood, Chichester.

Huskins, D. J. (1981b). "General Handbook of On-Line Process Analysers." Horwood, Chichester.

Huskins, D. J. (1982). "Quality Measuring Instruments in On-Line Process Analysis." Horwood, Chichester.

Hysert, D. W., Kovecses, F., and Morrison, N. M. (1976). *J. Am. Soc. Brew. Chem.* **34**, 145–150.

Inoue, T. (1980). *J. Am. Soc. Brew. Chem.* **38**, 8–12.

Jackson, C. J., Morley, F., and Porter, D. G. (1975). *Lab. Pract.* **24**, 23–25.

Jamieson, A. M. (1976). *J. Am. Soc. Brew. Chem.* **34**, 44–48.

Jamieson, A. M., and Mackinnon, C. G. (1965). *Am. Brew.* **98**(10), 37–39.

Jandreau, G. L., and Hahn, C. W. (1978). *J. Am. Soc. Brew. Chem.* **36**, 44–48.

*J. Autom. Chem.* (1978). **1**.

Jenkinson, P., and Compton, J. (1960). *Proc. Am. Soc. Brew. Chem.* pp. 73–77.

Jermyn, M. A. (1975). *Anal. Biochem.* **68**, 332–335.

Jerumanis, J. (1979). *Eur. Brew. Conv. Proc. Congr., Berlin-West* pp. 309–319.

John, M., and Dellweg, H. (1973). *Monatsschr. Brau.* **26**, 145–152.

John, M., Skrabei, H., and Dellweg, H. (1969a). *FEBS Lett.* **5**, 185–186.

John, M., Trenel, G., and Dellweg, H. (1969b). *J. Chromatogr.* **42**, 476–484.

Johnston, H. H., and Newsom, S. W. B., eds. (1976). "International Symposium on Rapid Methods and Automation in Microbiology, 2nd" Learned Information (Europe) Ltd., Oxford and New York.

Jones, M., and Pierce, J. S. (1963). *Eur. Brew. Conv. Proc. Congr., Bruxelles* pp. 101–134.

Kalko, J., and Crull, W. (1975). *Monatsschr. Brau.* **28**, 145–150.

Kardos, P. W. (1977). *Chem. Eng.* **84**, 79–83.

Kato, T., and Kinoshita, T. (1980). *Anal. Biochem.* **106**, 238–243.

Kempf, W., Friedrich, I., Hoepke, C. H., and Veldkamp, J. (1976). *Starch: Chem. Technol.* **28**, 54–68.

Kerker, M. (1969). The Scattering of Light and Other Electromagnetic Radiation." Academic Press, New York.
Kesson, J. (1982). *Brew. Guardian* **111** (10), 18–20.
Khayat, A., Redenz, P. K., and Gorman, L. A. (1982). *Food Technol. (Chicago)* **36**, 46–50.
Korduner, H., and Westelius, R. (1981). *Eur. Brew. Conv. Proc. Congr., Copenhagen* pp. 615–622.
Kramme, D. G., Griffen, R. H., Hartford, C. G., and Corrado, J. A. (1973). *Anal. Chem.* **45**, 405–408.
Kraszewski, A. (1974). *Ind. Aliment. Agric.* **91**, 1403–1405.
Krebs, W. M. (1975). *Tech. Q. Master Brew. Assoc. Am.* **12**, 176–185.
Krumtünger, A. (1965). *Brauwelt* **105**, 378–381.
Kubadinow, N. (1974). *Z. Zuckerind.* **24**, 65–67.
Küster, J. (1977). *Brauwelt* **117**, 1068–1077.
Küster, J., and Schlosser, H. (1975). *Brauwelt* **115**, 75–81.
Lance, D. G., Kavanagh, T. E., and Clarke, B. J. (1981). *J. Inst. Brew.* **87**, 225–228.
Law, D. P. (1977). *Cereal Chem.* **54**, 874–881.
Lawless, P., Shaw, J., and Kraeger, S. J. (1974). Presented at the *Annu. Meet. Am. Soc. Microbiol.*
Leblanc, D. J., and Ball, A. J. S. (1978). *Anal. Biochem.* **84**, 574–578.
Lee, Y. H., and Tsao, G. T. (1979). *Adv. Biochem. Eng.* **12**, 35–38.
Leedham, P. A., and Carpenter, P. M. (1977). *Proc. Eur. Brew. Conv. Congr., Amsterdam* pp. 729–744.
Lehuédé, J. M., Flayeux, R., and Moll, M. (1979). *9th Technicon Symp., Paris* pp. 1–15.
Lidzey, R. G., Sawyer, R., and Trockwell, P. B. (1970). *Lab. Pract.* **20**, 213–219.
Liebezeit, G., and Dawson, R. (1981). *HRC CC, J. High Resolut. Chromatogr. Chromatogr. Commun.* **4**, 354–356.
Ligot, J. (1979). *Bios (Nancy)* **10** (11/12), 20–27.
Linden, J. C., and Lawhead, C. L. (1975). *J. Chromatogr.* **105**, 125–133.
Lindroth, P., and Mopper, K. (1979). *Anal. Chem.* **51**, 1667–1674.
Lloyd, A. K. (1976). *Brew. Guardian* **105**, 27–29, 33.
Lloyd, M. (1979). *Tech. Q. Master Brew. Assoc. Am.* **16**, 182–185.
Lomas, D. (1974). *Brewer* **60**, 447–451.
Luckiewicz, E. T. (1978). *Tech. Q. Master Brew. Assoc. Am.* **15**, 190–197.
Marinelli, L., and Whitney, D. (1966). *J. Inst. Brew.* **72**, 252–256.
Marinelli, L., and Whitney, D. (1967). *J. Inst. Brew.* **73**, 35–39.
Matène, H. (1979). *Brauwissenschaft* **32**, 29–33.
Mathieu, P., Treuil, J. J., and Revol, L. (1965). *Ann. Biol. Clin.* **23**, 671–677.
Menu, J. (1976). *Mes., Regul., Autom.* **41**, 61–63.
Michel, E. (1971). *Brauwelt* **111**, 695–702.
Miller, L. F., Mabee, M. S., Gress, H. S., and Jangaard, N. O. (1978). *J. Am. Soc. Brew. Chem.* **36**, 59–62.
Mitcheson, R. C., and Stowell, K. G. (1970). *J. Inst. Brew.* **76**, 335–339.
Molinar, I., and Horvath, C. (1977). *J. Chromatogr.* **142**, 623–640.
Moll, M. (1979a). *In* "Brewing Science" (J. R. A. Pollock, ed.), Vol. 1, pp. 12, 14. Academic Press, New York.
Moll, M. (1979b). *In* "Brewing Science" (J. R. A. Pollock, ed.), Vol. 1, pp. 91–94. Academic Press, New York.
Moll, M. (1984). *In* "Brewing Science" (J. R. A. Pollock, ed.), Vol. 3. Academic Press, New York.

Moll, M., Vinh, T., Flayeux, R., and Bastin, M. (1971). *Bios (Nancy)* **2** (3), 3–11.

Moll, M., Vinh, T., and Noel, J. P. (1972). *Bios (Nancy)* **3**, 293–310.

Moll, M., Flayeux, R., and Lehuédé, J. M. (1975a). *Bios (Nancy)* **6** (3), 94–97.

Moll, M., Flayeux, R., and Lehuédé, J. M. (1975b). *Brauwissenschaft* **28**, 134–136.

Moll, M., Flayeux, R., and Lehuédé, J. M. (1975c). *Ind. Aliment. Agric.* **92** (6), 635–637.

Moll, M., Flayeux, R., and Lehuédé, J. M. (1975d). *Bios (Nancy)* **6** (3), 90–93.

Moll, M., Flayeux, R., and Lehuédé, J. M. (1975e). *Ind. Aliment. Agric.* **92** (6), 631–633.

Moll, M., Flayeux, R., and Lehuédé, J. M. (1975f). *Bios (Nancy)* **6** (3), 85–89.

Moll, M., Flayeux, R., and Lehuédé, J. M. (1976). *Bios (Nancy)* **7** (11), 3–6.

Moll, M., Duteurtre, B., Lehuédé, J. M., and Faivre, J. (1977a). *Schweiz. Brau. Rundsch.* **88**, 263–265.

Moll, M., Weber, J. C., and Delorme, J. J. (1977b). French Patent 2, 402, 202.

Moll, M., d'Hardemare, C., and Midoux, N. (1977c). *Tech. Q. Master Brew. Assoc. Am.* **14**, 194–196.

Moll, M., Bazard, D., and Flayeux, R., (1978a). *Bios (Nancy)* **9** (9), 23–26.

Moll, M., Flayeux, R., Bazard, D., and Mouet, A. (1978b). *Groupe Polyphénols, Bull. Liaison* **8**, 364–368.

Moll, M., Duteurtre, B., Scion, G., and Lehuédé, J. M. (1978c). *Tech. Q. Master Brew. Assoc. Am.* **15**, 26–29.

Moll, M., Midoux, N., and d'Hardemare, C. (1978d). U. S. Patent 4, 129, 029.

Moll, M., Delorme, J. J., and Weber, J. C. (1979). U. S. Patent 4, 165, 643.

Moll, M., Vinh, T., Bazard, D., Vincent, L. M., and André, J. C. (1981) *J. Am. Soc. Brew. Chem.* **39**, 15–19.

Moller, N. C. (1975). *Tech. Q. Master Brew. Assoc. Am.* **12**, 41–45; see also *Brygmesteren,* 1975, **32**, 155–167.

Molyneux, R. J., and Wong, Y. I. (1973). *Proc. Am. Soc. Brew. Chem.* pp. 71–74.

Molzahn, S. W., and Portno, A. D. (1975). *Eur. Brew. Conv. Proc. Congr., Nice* pp. 479–487.

Monk, P. R. (1978). *Process Biochem.* **13** (12), 4–5, 8.

Moore, S., and Stein, W. H. (1951). *J. Biol. Chem.* **192**, 663–681.

Moore, S., and Stein, W. H. (1954). *J. Biol. Chem.* **211**, 907–913.

Moore, S., Spackman, D. H., and Stein, W. H. (1958). *Anal. Chem.* **30**, 1185–1190.

Mor, J. R., Zimmerli, A., and Fiechter, A. (1973). *Anal. Biochem.* **52**, 614–624.

Mordt, H. E. (1982). *Brygmesteren* **39**, 305–321.

Mou, D. G., and Cooney, C. L. (1976). *Biotechnol. Bioeng.* **18**, 1371–1392.

Moxey, P. (1973). *Brew. Distilling Int.* **3** (3), 27–28.

Mouillet, L., Luquet, F. M., and Boudier, J. F. (1977). *Ann. Falsif. Expert. Chim.* **70**, 145–155.

Mukhopadhyay, S. N., and Ghose, T. K. (1976). *Process Biochem.* **11** (6), 19–27, 40.

Mulkay, P. (1982). Thesis, University of Louvain, Louvain, Belgium.

Myron, T. J., and Shinskey, F. G. (1975). *Tech. Q. Master Brew. Assoc. Am.* **12**, 235–242.

Neugebauer, K., Weber, H., and Kupprion, R. (1970). *Brauwelt* **110**, 701–704.

Noel, D., Hanai, T., and D'Amboise, M. (1979). *J. Liq. Chromatogr.* **2** (9), 1325–1336.

Nolte, C. (1972). *Brew. Dig.* **47**, 74–77.

Otter, G. E., and Taylor, L. (1978). *J. Inst. Brew.* **84**, 160–164.

Otter, G. E., Popplewell, J. A., and Taylor, L. (1969). *Eur. Brew. Conv. Proc. Congr., Interlaken* pp. 481–495.

Otter, G. E., Popplewell, J. A., and Taylor, L. T. (1970). *J. Chromatogr.* **49**, 462–468.

Outtrup, H. (1981). *Eur. Brew. Conv. Proc. Congr., Copenhagen* pp. 323–333.

Palmer, J. K. (1975). *Anal. Lett.* **8** (3), 215–224.

Palmer, J. K., and Brandes, W. B. (1974). *J. Agric. Food Chem.* **22**, 709–712.

Paukner, E., and Chaniotis, N. (1956). *Brauwelt* **96**, 1587–1590.

Pickering, J. W. (1971). *Tech. Q. Master Brew. Assoc. Am.* **8**, 41–44.
Pinnegar, M. A. (1966a). *J. Inst. Brew.* **72**, 62–64.
Pinnegar, M. A. (1966b). *J. Inst. Brew.* **72**, 366–368.
Pinnegar, M. A., and Whitear, A. L. (1965). *J. Inst. Brew.* **71**, 398–400.
Pomes, A. F. (1980). *J. Am. Soc. Brew. Chem.* **38**, 67–70.
Porter, D. G., and Sawyer, R. (1972). *Analyst* **97**, 569–575.
Portno, A. D., and Molzahn, S. W. (1977). *Brew. Dig.* **52** (3), 44–48, 50.
Preston, K. R. (1974). *Cereal Chem.* **52**, 451–458.
Pringle, J. R., and Mor, J. R. (1975). *In* "Methods in Cell Biology. XI. Yeast Cells" (D. M. Prescott, ed.), pp. 131–168. Academic Press, New York.
Prouveur, P. (1975). *Bios (Nancy)* **6** (4), 126–133.
Rasmussen, J. N. (1981). *Carlsberg Res. Commun.* **46**, 25–36.
Rees, T. C. (1969). *J. Inst. Brew.* **75**, 465–468.
Richter, K., and Woelk, H. U. (1977). *Starch: Chem. Technol.* **29**, 273–277.
Riemann, J. (1974). *Brauwelt* **114**, 1219–1227.
Rischbieter, S. (1977). *Brauwelt* **117**, 719–720, 1010–1014, 1634–1636.
Robin, J. P., and Tollier, M. T. (1981). *Sci. Aliments* **1**, 233–246.
Rohner, R. L., and Tompkins, J. R. (1970). *Proc. Am. Soc. Brew. Chem.* pp. 111–117.
Rouiller, M. (1977). *Bios (Nancy)* **8** (9), 38–41.
Rouiller, M. (1980). *Bios (Nancy)* **11** (2), 17–26.
Runkel, U. D. (1971). *Brauwelt* **111**, 1389–1395.
Ruocco, J. J., Coe, R. W., and Hahn, C. W. (1980). *Tech. Q. Master Brew. Assoc. Am.* **17**, 69–76.
Ryu, D. D. Y., and Humphrey, A. E. (1973). *J. Appl. Chem. Biotechnol.* **23**, 283–295.
Saletan, L. T., and Scharoun, J. (1965). *Proc. Am. Soc. Brew. Chem.* pp. 198–207; see also *Wallerstein Lab. Commun.*, 1965, **28**, 191–207.
Sanden, U. (1979). *Brauwelt* **119**, 460–462.
Sanden, U., and Hamm, H. (1974). *Brauwelt* **114**, 23–25.
Sarris, J., Morfaux, J. N., Dupuy, P., and Hertzog, D. (1969). *Ind. Aliment. Agric.* **86**, 1241–1246.
Sarris, J., Morfaux, J. N., and Darvin, L. (1970a). *Connaissance Vigne Vin* **4**, 431–438.
Sarris, J., Morfaux, J. N., Dupuy, P., and Hertzog, D. (1970b). *Ind. Aliment. Agric.* **87**, 115–121.
Sawyer, R., and Dixon, E. J. (1968). *Analyst* **93**, 669–679, 680–687.
Schmidt, F., and Enevoldsen, B. S. (1976). *Carlsberg Res. Commun.* **41**, 91–110.
Schulze, W. G., Tiwg, P. L., Henckel, L. A., and Goldstein, H. (1981). *J. Am. Soc. Brew. Chem.* **39**, 12–15.
Schumacher, R. (1973). *Monatsschr. Brau.* **26**, 172–178, 225.
Schwartz, J. R., and Cooney, C. L. (1978). *Process Biochem.* **13**, (2), 3–4, 6–7, 24.
Schur, F., Anderegg, P., and Pfenninger, H. (1978). *Brau. Rundsch.* **89**, 229–234.
Schuster, R. (1980). *Anal. Chem.* **52**, 617–620.
Schwarzenbach, R. (1976). *J. Chromatogr.* **117**, 206–210.
Schwarzenbach, R. (1979). *J. Am. Soc. Brew. Chem.* **37**, 180–184.
Scobell, H. D., Brobst, K. M., Steele, E. M. (1977). *Cereal Chem.* **54**, 905–917.
Scobell, H. D., and Brobst, K. M. (1981). *J. Chromatogr.* **212**, 51–64.
Siebert, K. J. (1976). *J. Am. Soc. Brew. Chem.* **34**, 79–90.
Siebert, K. J. (1982). *J. Am. Soc. Brew. Chem.* **40**, 9–14.
Sigrist, W. (1975). *Wasser, Luft Betr.* **19** (2), 3–7.
Skarsaune, S. K., Judd, C. L., and Banasik, O. J. (1973). *Brew. Dig.* **48** (7), 48–54, 77.
Skeggs, L. T. (1957). *Am. J. Clin. Pathol.* **28**, 311–322.

Skowronski, J. S., and Wedl, D. J. (1982). *J. Am. Soc. Brew. Chem.* **40**, 75–77.
Slotema, F. P. (1978). *Brew. Dig.* **53** (8), 48–52.
Sommer, G. (1980). *Monatsschr. Brau.* **33**, 405–406.
Soroko, O. (1967). *Wallerstein Lab. Commun.* **30** (102/103), 123–134.
Sorrells, K. M. (1981). *J. Food Prot.* **44**, 832–834.
Spackman, D. M., Stein, W. H., and Moore, S. (1958). *Anal. Chem.* **30**, 1190–1206.
Steen, P. W. (1973). *Brew. Dig.* **48**, (10), 54–55.
Stockham, J. D., and Fochtman, E. G., eds. (1978). "Particle Size Analysis." Ann Arbor Sci.
  Publ., Ann Arbor, Michigan.
Stone, I., and Laschiver, C. (1957). *Proc. Am. Soc. Brew. Chem.* pp. 46–55; see also *Wallerstein
  Lab. Commun.,* 1957, **20** (71), 361–375.
Stoten, T. A. (1974). *Brewer* **60**, 452–454.
Sullivan, K. (1978). *Brewer* **64**, 396–400.
Sweeley, C. C., Bentley, R., Makita, M., and Wells, W. W. (1963). *J. Am. Chem. Soc.* **95**,
  2497–2507.
Tep, Y., Brun, S., Leboeuf, J. P., and Bard, M. (1978). *Ann. Nutr. Aliment.* **32** (5), 899–904.
Thorne, R. S. W. (1961). *J. Inst. Brew.* **67**, 191–199.
Thorne, R. S. W., and Beckley, R. E. (1958). *J. Inst. Brew.* **64**, 38–46.
Thorne, R. S. W., and Nannestad, I. (1959). *J. Inst. Brew.* **65**, 175–188.
Thorne, R. S. W., and Svendsen, K. (1962). *J. Inst. Brew.* **68**, 257–270.
Threlkeld, C. H. (1982). *J. Food Sci.* **47**, 1222–1225.
Timbie, D. J., and Keeney, P. G. (1977). *J. Food Sci.* **42**, 1590–1591, 1599.
Treiber, K. (1975). Ph. D. thesis, University of Munich, Munich.
Treiber, K. (1978). *Brauwissenschaft* **31**, 279–287.
Treiber, K., and Höhn, K. (1976). *Brauwissenschaft* **29**, 221–226.
Trenel, G., and Emeis, C. C. (1970). *Starch: Chem. Technol.* **22**, 188–191.
Trenel, G., and John, M. (1970). *Monatsschr. Brau.* **23**, 6–10.
Tuning, B. (1971). *Eur. Brew. Conv. Proc. Congr., Estoril* pp. 191–196.
Tunnell, D. A. (1981). "Technicon Near Infrared Bibliography." Technicon, Tarrytown, New
  York.
Vandercook, C. E., Price, R. L., and Harrington, C. A. (1975). *J. Assoc. Off. Anal. Chem.* **58**,
  482–487.
Van Strien, J., and Drost, B. W. (1979). *J. Am. Soc. Brew. Chem.* **37**, 84–88.
Verbiese, Y. (1981). *Ind. Alim. Agric.* **98**, 759–768.
Verzele, M., and De Porter, M. (1978). *J. Chromatogr.* **166**, 320–326.
Verzele, M., and Dewaele, C. (1981a). *J. Inst. Brew.* **87**, 232–233.
Verzele, M., and Dewaele, C. (1981b). *J. Am. Soc. Brew. Chem.* **39**, 67–69.
Verzele, M., Van Dyck, J., and Claus, H. (1980). *J. Inst. Brew.* **86**, 9–14.
Verzele, M., Dewaele, C., and Van Kerrebroeck, M. (1982). *J. Chromatogr.* **244**, 321–326.
Vincent, K. R., and Shipe, W. F. (1976). *J. Food Sci.* **41**, 157–162.
Wackerbauer, K., Teske, G., Tödt, F., and Graff, M. (1975). *Proc. Eur. Brew. Conv. Congr.,
  Nice* pp. 757–790.; see also *Monatschr. Brau.,* 1977, **30**, 44–57.
Waes, G., and Bossuyt, R. (1981). *Milchwissenschaft* **36**, 548–552.
Wagner, B., and Lloyd, M. (1977). *Monatschr. Brau.* **30**, 122–124.
Walls, W. R. (1977). *Brewer* **63**, 377–379.
Wainwright, T. (1973). *J. Inst. Brew.* **79**, 451–470.
Wall, L. L., and Gehrke, C. W. (1975). *J. Assoc. Off. Anal. Chem.* **58**, 1221–1226.
Wall, L. L., Gehrke, C. W., Neuner, T. E., Cathey, R. D., and Rexroad, P. R. (1975). *J. Assoc.
  Off. Anal. Chem.* **58**, 811–817.

Wang, D. I. C., Cooney, C. L., Demain, A. L., Dunnill, P., Humphrey, A. E., and Lilly, M. D. (1979). "Fermentation and Enzyme Technology," pp. 212–237. Wiley, New York.

Watson, C. A., Carville, D., Dikeman, E., Daigger, G., and Booth, G. D. (1976). *Cereal Chem.* **53,** 214–222.

Watson, C. A., Shuey, W. C., Banasik, O. J., and Dick, J. W. (1977). *Cereal Chem.* **54,** 1264–1269.

White, C. A., Corran, P. H., and Kennedy, J. F. (1980). *Carbohydr. Res.* **87,** 165–173.

Whitt, J. T., and Cuzner, J. (1979). *J. Am. Soc. Brew. Chem.* **37,** 41–46.

Wight, A. W. (1976). *Starch: Chem. Technol.* **28,** 311–315.

Wilkinson, J. M. (1978). *J. Chromatogr. Sci.* **16,** 547–552.

Williams, P. C. (1975). *Cereal Chem.* **52,** 561–576.

Winkler, L. W. (1888). *Ber. Dtsch. Chem. Ges.* **21,** 2834–2854.

Wisk, T. J., Weiner, J. T., and Siebert, K. J. (1981). *J. Am. Soc. Brew. Chem.* **39,** 147–153.

Wyatt, P. J. (1973). *In* "Methods in Microbiology" (J. R. Norris and D. W. Ribbons, eds.), Vol. 8, pp. 183–263. Academic Press, New York.

Yadav, K., Weissler, H., Garza, A., and Gurley, J. (1969). *Proc. Am. Soc. Brew. Chem.* pp. 59–69.

Zaake, S., and Otto, G. (1975). *Monatsschr. Brau.* **28,** 183.

# 10

# Molecular Structure – Activity Analyses of Artificial Flavorings

A. J. HOPFINGER
R. H. MAZUR

*Searle Research and Development*
*Skokie, Illinois*


H. JABLONER

*Technical Center*
*Hercules Incorporated*
*Wilmington, Delaware*

ANALYSIS OF FOODS AND BEVERAGES

## I. INTRODUCTION AND BACKGROUND

The assumption that biological activity is a function of chemical structure constitutes the fundamental postulate of the biochemical sciences. Success in the pharmaceutical industry depends upon identifying structure–activity relationships (SARs) and exploiting them in the development of new therapeutic agents. In view of this, it is not too surprising that the pharmaceutical sciences have led in the identification and implementation of new methods, procedures, and technologies that aid in defining SARs.

One new approach is the application of computational chemistry techniques and/or statistical methods to establish SARs. This new means of identifying SARs is, by its nature, computer based and has correspondingly been called computer-assisted drug design. However, the *reliable* application of the computational techniques requires periodic experimental physicochemical characterization of key compounds to verify the accuracy of the calculations. This aspect of computer-assisted drug design is often forgotten and/or neglected. Nevertheless, the overall goal of computer-assisted drug design is to transform SARs into quantitative structure–activity relationships (QSARs).

The realization of a particular taste and/or olfactory response involves the same general biochemical components as are at play in drug action. Consequently, the procedures used to establish QSARs in pharmaceutical studies should also be applicable in the design of artificial flavorings. In fact, one can argue that QSAR methods stand a better chance of being successful in artificial flavoring applications than in drug design for the following reasons.

1. General sites of action are often known.
2. The compounds are usually direct acting, so metabolic activation is not a consideration.
3. The desired biological end point is well defined (although not necessarily easy to measure quantitatively).

The major limitation in evaluating new chemical entities as flavoring agents is that in-human testing generally constitutes the only means of deciding the worth of a compound. There is little in the way of primary, *in vitro* screening for projected flavoring agents. Hence, extensive toxicity testing needs to be done on all compounds being considered. This is in contrast to drug development, where only compounds that pass primary and often secondary (animal) tests are subjected to toxicity studies.

An associated limitation to in-human testing of potential artificial flavoring agents, which significantly impacts on QSAR methods, is the quality of the biological measurements. A QSAR is only as good as the biological activity from which it is constructed. Sensory responses, and their measurements, are very much subjective and individualistic. This limitation can be minimized, as is usually the case, by enlarging the sample size. Ultimately the quality of the measured sensory response represents a practical compromise based upon time and cost.

QSAR methods are beginning to be applied to the design of artificial flavoring agents. These procedures have been employed most extensively in the design of artificial sweeteners. As such, development of sweetener QSARs constitutes a focal point in this review. However, the overall objectives of this paper are first to present and discuss current QSAR methods and then to show how they are applied in a design mode.

## II. QUANTITATIVE STRUCTURE–ACTIVITY RELATIONSHIPS

An individual attempting to construct a QSAR generally has little or no control over the representation of the biological activity. The activity measurements must be treated as a set of fixed quantities, and it is the physicochemical properties, and their measures, that can be selected in development of a QSAR. Out of this situation arises the question of what physicochemical properties should be considered and how they should be measured or calculated. The purpose of this section is to describe physicochemical properties that can be computed. Other chapters in this book deal with physical and chemical measurements.

### A. Additive Property Models

The simplest model of a molecule, with respect to computing molecular properties, is to assume the molecular property to be the sum of the property values of the individual constituent atoms or groups of atoms. Extensive tables (Rekker, 1977; Hansch and Leo, 1979) of atomic and group (fragment) property values have been compiled to facilitate implementation of this model. The prevalent physicochemical properties employed in QSARs using an additive property model are:

1. $\log P$, the water–octanol partition coefficient (Hansch and Clayton, 1973).
2. $\sigma$, the Hammett constant (Charton, 1974, 1975).
3. MR, the molecular refractivity index (Hansch et al., 1973).
4. $pK_a$, the ionization constant (Seiler, 1974).
5. $E_s$, the Taft steric constant (Verloop, 1972).

This last descriptor, $E_s$, represents an attempt to enhance the additive model for a nonadditive contribution: spatial steric interactions. Other corrections, based upon the chemical bonding topology, have been employed in the additive model. These include proximity, bond type, ring, and group shape correction features (Bowden and Woolridge, 1973; Leo et al., 1975; Verloop et al., 1976).

Molecular connectivity (Kier and Hall, 1976), which is based upon graph theory (Wilson, 1972), represents an empirical alternative to an additive model employing physicochemical properties. Indices derived from the molecular connection table (chemical bonding topology) using mathematical functions are evaluated as potential activity correlates in molecular connectivity theory. As such, this approach is completely mathematical and has no direct physicochemical basis. Its strength is that correlation indices can always be generated (provided that one knows how to assign intrinsic relative weights to individual atom types).

## B. Hansch Analysis

The most successful, and the most often used, method of constructing a QSAR is that of Hansch et al. (1965). This method employs multidimensional regression analysis to correlate structure to activity in a chemically congeneric set of compounds. The structural features have been traditionally derived from additive property models. However, recent applications of Hansch analysis recognize any and all molecular descriptors as potential correlates to activity (Martin et al., 1973; Silipo and Hansch, 1975). Prominent in this line of thinking is Hansch himself, who now freely uses indicator variables as correlates (Hansch et al., 1975). An indicator variable has a value of 1 if some user-defined property is present in a compound and a value of 0 if the property is absent. It is important to point out that Hansch analysis is based upon a biological action model. By deriving the general QSAR equation associated with the action model, it is possible to justify conceptually the general usage of any molecular descriptor in a correlation analysis.

Hansch began by assuming that, for any congeneric series of compounds, one particular "reaction" could be critical and rate determining. If $K_x$ is an equilibrium, or rate, constant for this rate-determining reaction (which is possibly, but not necessarily, at the site of action of the drug), $C$ is the applied concentration, that is, dose, and $A$ represents the probability of a drug molecule reaching this critical site in a given time interval, then the expression for the rate of biologic response is

$$\frac{d(\text{response})}{dt} = ACK_x \qquad (10\text{-}1)$$

Hansch was aware of the work of Collander (1954) and determined to relate $A$ to $\log P$ and changes in $A$, as one compared molecules in a congeneric series, to changes in $\log P$. These changes can be expressed in the form of substituent constants using an additive model, which Hansch and co-workers (1963) have termed $\pi$ constants, and which are rigorously defined as

$$\pi = \log(P_x/P_h) \qquad (10\text{-}2)$$

$$\pi = \log P_x - \log P_h \qquad (10\text{-}3)$$

where $P_x$ and $P_h$ are the partition coefficients of substituted and parent molecules, respectively.

For a reference standard, partition, and consequently $\pi$, coefficients have all been measured in the 1-octanol/water system. Clearly this system has been chosen as a model for biologic lipid and aqueous phases. A negative $\pi$ value thus indicates a change toward greater affinity for the aqueous phase, and a positive value indicates greater affinity for the lipid phase. In essence, $\pi$ expresses the relative free-energy change on moving a derivative from one phase to another. Thus, another name given the additive model when only thermodynamic properties are being considered is *linear free-energy model.*

Hansch next chose to assume as a working hypothesis that the probability $A$ is related to $\log P$ for the complete molecule, or to $\pi$ for changes in congeneric series, through a Gaussian distribution function

$$A = a \exp[-(\pi - \pi_0)^2/b] \qquad (10\text{-}4)$$

where $a$ and $b$ are constants and $\pi_0$ is the $\pi$ value corresponding to the maximum in the distribution. The choice for this functional relationship is based on the fact that, in many series of compounds tested in biologic systems, as the relative lipophilicity was increased, activity rose to a maximum, fell off, and eventually reached 0. Indeed, *in vivo* it is generally to be expected that, for highly water-soluble drugs (low or negative $\log P$), the probability of reaching some distant receptor site will be low because of rapid excretion. Additional descriptors could be used to define the probability $A$. For example, electronic indices, such as charge density, might be used in Eq. (10-4) if metabolic activation is suspected to affect $A$. Overall, Eq. (10-4) represents an entry point for a menu of possible different correlation descriptors in constructing the QSAR.

Hansch pointed out that approximately linear relationships between biologic activity with $\log P$ can be expected when compounds within a limited range of $\log P$ values, either all higher or all lower than the optimum, are chosen for study.

The substitution of the expression for $A$ given in Eq. (10-4) into the fundamental rate equation, Eq. (10-1), gives

$$\frac{d(\text{response})}{dt} = a \exp[-(\pi - \pi_0)^2/b]CK_x \qquad (10\text{-}5)$$

If it is stipulated that $C$ is measured as the compound concentration necessary to produce a particular constant response ($LD_{50}$, $LD_{95}$, $ED_{50}$, percentage inhibition, etc.) in a fixed time interval, then $d(\text{response})/dt$ can be considered to be constant, and Eq. (10-5) becomes

$$\log(1/C) = -k_1\pi^2 + k_2\pi\pi_0 - k_3\pi_0^2 + \log k_x + k_4 \qquad (10\text{-}6)$$

where $\pi_0$ is a constant, being the value of $\pi$ that yields a log $P$ that, in turn, yields an optimum activity. Thus Eq. (10-6) is an expression relating biologic activity as measured through $\log(1/C)$ to a measurable free-energy difference parameter $\pi$ and to the unknown rate constant $K_x$ for the critical reaction.

The final major assumption inherent to the complete derivation of the structure–function equation is, as first suggested by Hansen (1962), that $K_x$ is a function of the electron release or withdrawal of the substituents, the hydrophobicity of the drug, and the stereochemical geometry. Thus log $K_x$ can be expressed as

$$\log K_x = k_5\pi + k_6\sigma + k_7E_s \qquad (10\text{-}7)$$

Substitution of Eq. (10-7) into Eq. (10-6), noting that $(1/C)$ is a measure of biologic response, BR, leads to the working equation

$$\log \text{BR} = -k_1\pi^2 + k_2\pi + k_3\sigma + k_4E_s + k_5 \qquad (10\text{-}8)$$

where $k_2 = C_1(\pi_0 + C_2)$ and $k_5 = C_3 - C_4\pi_0^2$.

Clearly any number of thermodynamic descriptors could be used with, or in place of, $\pi$ in Eq. (10-7). The same holds for electronic and spatial descriptors regarding $\sigma$ and $E_s$, respectively. Thus, Eq. (10-7) provides another location in the derivation of the general QSAR equation where a variety of user-selected molecular features can be included as potential activity correlates in Hansch analysis.

## C. Conformational Analysis

The major limitation to additive property models of molecular structure is that three-dimensional features cannot be determined. However, it is abundantly clear that conformation, or, more generally, molecular shape, is an important factor in dictating biological activity. Thus, there is a necessity to have a computational means to analyze molecular conformation.

There are two principal methods of performing computational conformational analyses: molecular mechanics (Boyd and Lipkowitz, 1982) and quantum mechanics (Segal, 1977). Molecular mechanics considers a molecule as a set of balls (the atoms), coated with sticking paste, connected by a set of springs (the bonds) according to a prescribed set of valence angles. Quantum mechanics views the molecule as a set of nuclei in space with electrons moving about the nuclei.

In both cases the goal of the calculation is to minimize the energy of the molecule as a function of its geometry. The energy minimization can involve different types and numbers of geometric degrees of freedom, which include bond lengths, bond angles, and torsional bond rotation angles. An often employed approximation of this calculation, especially for large molecules, is achieved by holding the valence geometry constant and minimizing the energy as a function of only torsional bond rotations.

The choice of the representation of the molecular energy, both in molecular mechanics and quantum mechanics, is critical to both computational time and accuracy of a calculation. Generally, for "typical" organic molecules considered in chemical research, molecular orbital theory approximations (Pople and Beveridge, 1970) are used in the quantum mechanical investigations.

The unproven, but reasonable, assumption implicit to SAR-directed conformational studies, both experimental and theoretical, is that one of the stable intramolecular conformers is the "active" conformation. A difficulty in applying conformational data in a quantitative design mode is selection of conformational features for QSAR development. Moreover, molecular shape properties are preferable features to have available in design studies. Conformation is a component of shape in that conformation defines the location of atoms in space. The properties of the atoms, most notably their "sizes," constitute an additional set of factors needed to specify molecular shape.

Some structure – activity studies have employed conformational features. These include: (1) an interatomic distance within a molecule (Kier, 1968; Ham, 1974), (2) a set of interatomic distances within a molecule (Kier, 1970; Rohrer et al., 1979), (3) a set of atomic coordinates within a molecule (Weinstein et al., 1973; Weintraub and Hopfinger, 1973), and (4) a set of critical intermolecular binding distances (Crippen, 1979). Various shape descriptors include: (1) molecular volume (Yamamoto, 1974), (2) molecular surface area (D. Pensak, unpublished work), and (3) spatial potential surfaces of a molecule with respect to a test species (Bartlett and Weinstein, 1975; Weinstein, 1975), which have also been considered in establishing structure – activity relationships.

A very powerful tool for visualizing three-dimensional molecular properties, including potential surfaces, is computer graphics (Max *et al.*, 1981). Computer graphics is particularly useful in the qualitative comparison of two or more molecules.

Hopfinger has developed a general theory of quantitatively comparing molecular shapes using common overlap steric volume (Hopfinger, 1980, 1981; Battershell *et al.*, 1981; Hopfinger and Potenzone, 1982) and, more recently, descriptors derived from superimposed molecular potential energy fields of pairs of molecules (Hopfinger, 1983). This theory has allowed a "marriage" between Hansch analysis and conformational analysis.

## D. Electronic Structure Calculations

Molecular orbital theory provides electronic, as well as conformational, data for inclusion in QSAR development. Electronic properties of a molecule can be expected to control biological activity when chemical reactions, as opposed to only physical interactions, are part of the mechanism of biological action. Electronic indices should also be used in modeling physical interactions as part of the means for estimating binding energies.

The electronic indices most often considered in molecular design studies include: atomic charge densities; bond, group, and/or molecular dipole moments; orbital energy levels, especially the highest occupied molecular orbital (HOMO), the lowest unoccupied molecular orbital (LUMO), and their difference; and orbital wave function coefficients.

Several different molecular orbital methods have been used in SAR investigations. These include simple Hückel theory, or HT (Streitweiser, 1961), extended Hückel theory, or EHT (Hoffman, 1963), CNDO (Pople *et al.*, 1965) and NDDO (Kikuchi, 1977) approximations, MINDO/3 (Bingham *et al.*, 1975), and PCILO (Coubeils *et al.*, 1971).

## E. Statistical Methods

Multidimensional linear regression analysis is far and away the most often employed statistical method used to construct QSARs. Its popularity is coupled to the acceptance of the Hansch approach to drug design described earlier. The techniques and pitfalls of regression analysis have been well described (Topliss and Costello, 1972; Goodford, 1973).

Other, less used statistical methods in quantitative molecular design investigations include discriminate analysis, cluster analysis, multiple-factor analysis, and pattern recognition procedures (Kowalski and Bender, 1973; Perrin, 1973; Matthews, 1975). Pattern recognition may prove particularly useful when the design objective is a complicated profile of biological activities, as opposed to being only a maximized potency and minimized toxicity.

## III. ANALYSIS AND DESIGN OF ARTIFICIAL SWEETENERS

As mentioned earlier, the majority of effort in the design and synthesis of artificial flavoring agents has been directed toward eliciting a sweet response. The adverse health effects of high levels of sugar intake have been a driving force in the development of artificial sweeteners. However, the deleterious side effects (toxicity) realized in several sweetening agents have led to their commercial demise. Thus, when we speak of designing chemical entities that elicit a sweet response, it is essential to qualify this with the word *safety*.

### A. Classifications of Sweeteners

Four chemical classes have members that exhibit a sweet response: proteins, dipeptides, synthetic organic products, and natural organic products.

Three proteins, all from fruits of West African plants, have been identified as sweeteners. Pertinent facts regarding these proteins are summarized here:

1. Miraculin, a glycoprotein, has a protein component with a molecular weight of $42,000 \pm 3000$ (aproximately 373 amino acids). The protein has not been sequenced, although the composition is known. The size and properties of the carbohydrate remain largely unknown. It is not a true sweetener, but it does make sour substances seem sweet (Bartoshuk *et al.*, (1974).

2. Monellin, a protein of 94 amino acids (MW = 11,069), is composed of two separate polypeptide chains, designated A and B. The A chain contains 44 amino acids; the B chain, 50. The independent chains are not sweet by themselves, while the intact protein has a relative sweetness of 1500–2000 times that of sucrose for a 7% solution. Thermal and pH denaturation destroys sweetness (Higginbotham, 1979).

3. Two unique single-chain proteins, termed Thaumatin I and Thaumatin II (each with MW $\approx 21,000 \pm 600$ and 1600 times sweeter than sucrose for a 7% solution) have been identified. Thaumatin 0 is a mixture of I and II. A freeze-dried aluminum complex form of Thaumatin 0 is Talin, a sweetener being developed by Tate & Lyle.* The high molecular weight of Thaumatin delays the onset of a sweet response as compared to sucrose. Thaumatin exhibits a sweet, licorice-like aftertaste for up to 1 hr that can be diminished with increasing dilution (van der Wel *et al.*, 1972).

* London, England.

Mazur *et al.* (1969) reported that L-aspartyl-L-phenylalanine methyl ester (**I**) is 100–200 times sweeter than sucrose on a weight basis.

(I)

This discovery has led to intense research on the identification and development of small peptide sweeteners. Dipeptide-based sweeteners that are 5000–20,000 times as sweet as sucrose have now been reported (Fujino *et al.,* 1976).

Synthetic organic sweeteners exhibit diverse chemical structures. Prominent among these compounds are saccharin (**II**), cyclamate (**III**), and the alkoxynitroaniline P-4000 (**IV**). Each of these three sweeteners has been identified as possessing possible carcinogenic properties.

(II)          (III)          (IV)

The natural organic sweeteners include steviol, stevioside, rebaudioside, glycyrrhizin, phyllodulcin, perillartine, and the dehydrochalcones. Many of these natural materials have been synthesized and derivatized with the goal of optimizing the sweet nature of the molecule.

Sweeteners can also be characterized in terms of sensory response evaluation. This is a most important consideration in a QSAR analysis of a chemically congeneric data base of potential sweeteners. Sweetness potency (SP), relative to a sucrose reference (SR), is the normal biological measure. Sweetness potency is, however, inversely a function of the concentration of SR employed in testing. Thus, meaningful comparisons of SP require a common base concentration of SR.

The quality of the sweet taste is also an important factor, especially with respect to commercialization. The obvious difficulty in characterizing the quality of sweet taste is the inability to design a physicochemical experiment that measures this sensory entity. Taste quality is a subjective judgment. However, taste can be described as a combination of the four so-called primary tastes: sweet, sour, salty, and bitter. Thus, taste quality can be represented in terms of a sweet-to-nonsweet taste component ratio (S/NS). Protocols and corresponding applications for measuring SP and S/NS have been reported (Acton *et al.,* 1970; Acton and Stone, 1976; Swartz and Furia, 1977; DuBois *et al.,* 1981a,b).

Lastly, sensory responses need to be characterized also in terms of their temporal profiles. Thus, the onset time (OT) of perceiving the sweet sensation and the corresponding extinction time (ET) should be part of the set of biological descriptors used to define the sweet response.

Consequently, a total sweet sensory evaluation would include the measures SR, SP, S/NS, OT, and ET.

Further, SP should be expressed on a molar concentration base and not the customary weight base. That is, SP should be made by comparing relative numbers of molecules, not the molecular weights of numbers of molecules. As will be evident in the QSAR case studies, only SR, on a weight-base scale, is normally available (i.e., measured).

## B. Physicochemical Models and Theories of Sweeteners

The sweet taste response is thought to be the result of a complementary interaction of the sweetener with a receptor protein embedded on the taste cell surface. Binding of the sweetener to the receptor protein induces taste cell depolarization, which, in turn, initiates the sweet signal into the nervous system.

Formulation of a general theory as to how molecular structure and sweetness are interrelated has been thwarted by the diversity of chemical structure observed in sweetening agents. Current thinking is that there are multiple sweet receptor sites (Beets, 1978; Jakinovich, 1981; Tanimura and Shimada, 1981). A model of the sweet taste receptor, in which many dissimilar binding sites are proposed to exist in the hydrophobic $\alpha$-helical region of a cell membrane bound protein, has been designed to account for available experimental information (Crosby and DuBois, 1980). An alternate hypothesis to multiple sweet receptor sites is multiple (different) spatial interactions (bindings) of sweet molecules to a single receptor site. This line of thinking has received little attention.

The most widely accepted molecular requirement for a sweetener is the presence of an A–H (proton donor) and a B (proton acceptor) hydrogen-bonding moiety (Shallenberger and Acree, 1967). The H and B atoms are 2.5–4.0 Å apart. This theory was expanded to account for high-potency sweeteners by including a third binding site (X) some 3.5–5.5 Å from both the A and B atoms (Kier, 1972). The generality of this A–H/B/X model consigns it to being of limited value in sweetener design.

In a series of molecular orbital calculations on diverse sweeteners, using the CNDO approximation, a correlation between taste response and the pseudohydrogen bond – electrostatic intramolecular energy $\Delta E$ of the H and B atoms was observed (Hopfinger and Jabloner, 1981). Sweetness potency exhibits a parabolic dependence on $\Delta E$ such that SP is maximum for $\Delta E = -1.9$ to $-2.0$ kcal/mole. It was speculated that $\Delta E$ is an indirect, and inverse, measure of how strongly the A – H/B groups can form intermolecular hydrogen bonds (presumably at the receptor site). It was also noted that a very large number of A – H/B systems can adopt molecular geometries in which the A – H can hydrogen bond to a carbonyl oxygen and the B can hydrogen bond to an amide proton of two adjacent amino acids located in the center of a $\beta$-turn conformation (Venkatachalam, 1968).

## C. QSAR Analyses of Sweeteners

### 1. Perillartine and Analogs

Iwamura has carried out QSAR analyses of the structure – taste relationship for 49 perillartines (see Chart 1) and 20 nitro- and cyanoaniline derivatives, (V) and (VI), respectively (Iwamura, 1980).

(V)                    (VI)

The activity of the perillartines, log $A$, was expressed by the logarithm of the taste potency (Acton and Stone, 1976) irrespective of the taste quality, that is, irrespective of sweetness-to-bitterness ratio. The sweet intensity for the nitro- and cyanoanilines is the logarithm of the molar sweetening potency and calculated from the original weight ratio.

The STERIMOL parameters (Verloop et al., 1976) were used to evaluate the steric dimensions of compounds. It must be stressed that this approach is not an explicit three-dimensional conformational analysis. The $L$ parameter expresses the length of the substituent $R_1$ along the bond axis that connects $R_1$ and the oxime carbon in perillartines (see Chart 1); $R_1$ is the rest of the molecule from which the common oxime end is subtracted. In the aniline derivatives, $L$ is the length of the whole substituted benzene moiety along the bond axis between $C_5$ and the N atom of the nitro or the C atom of the cyano group. The $W_1$, $W_r$, $W_u$, and $W_d$ are the molecular widths in the

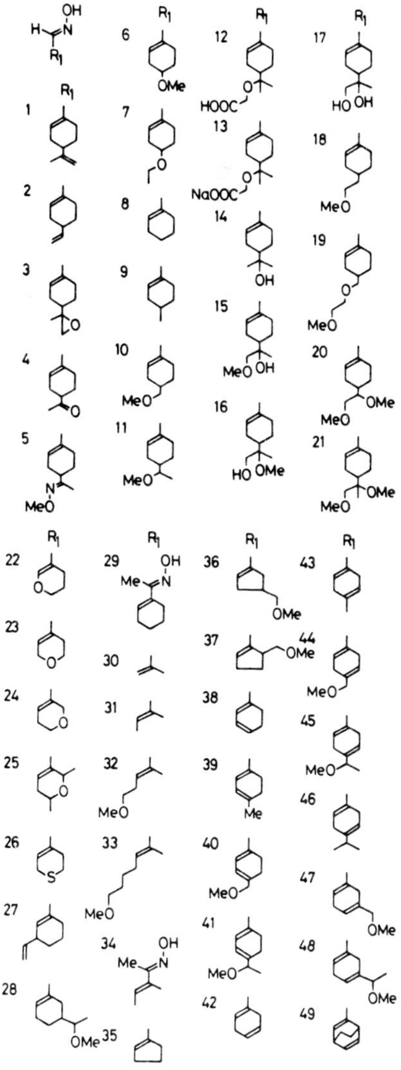

**Chart 1**

directions perpendicular to the $L$ axis and rectangular to each other; $W_1$ in the perillartine series is taken as the width in direction to which the 4-substituent projects in the fully extended (staggered) conformation, and the $C_1–C_2$ double bond is assumed to adopt this direction. In the nitro- and cyanoaniline series, it is directed to the same side where the amino group exists, and the 2-substituents are assumed to extend in this direction also.

The width in the direction opposite to $W_1$ is expressed by $W_r$; $W_u$ and $W_d$ are the widths upward and downward, respectively, of the molecule when one views it from the basic end along the $L$ axis, locating the $W_1$ to the left. These $W$ parameters correspond to the original $B_n$ ($n = 1$ to 4) parameters reported by Verloop et al. (1976) and are shown schematically in Fig. 1.

The hydrophobicity of the compounds, log $P$, where $P$ is the 1-octanol/ water partition coefficient of a whole molecule, was estimated using the additive–constitutive model. For the perillartine series, the procedure of Hansch and Leo (1979) was employed. The log $P$ values for the aniline derivatives were evaluated according to the method developed by Fujita (1983).

Of the various combinations of the steric and other substituent parameters as independent variables, the equation

$$\log A = 0.63 \log P \,(\pm 0.31) + 0.19L \,(\pm 0.10)$$
$$- 0.48W_1 \,(\pm 0.18) - 0.62W_u \,(\pm 0.24) + 2.87 \,(\pm 0.78)$$
$$n = 38, r = 0.91, s = 0.32 \tag{10-9}$$

gave the best correlation for the perillartines. In Eq. (10-9) and the following equations, $n$ is the number of compounds included in the analysis, $r$ is the multiple correlation coefficient, and $s$ is the standard deviation. The figures in parentheses are the 95% confidence limits.

The results indicate that taste potency, whether sweet and/or bitter, increases as log $P$ becomes larger, as the "length" ($L$) of the molecule increases, and as the "width" ($W_1$) of the molecule increases. However, the activity decreases as $W_u$, a measure of molecular thickness, increases. Thus

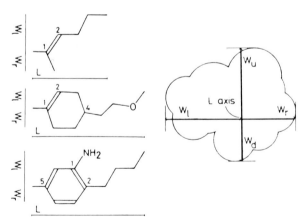

**Fig. 1.** Schematic representation of the steric parameters $L$, $W_1$, $W_r$, $W_u$, and $W_d$. [Reprinted with permission from Iwamura (1980).]

Eq. (10-9) states that sweet perillartines should be lipophilic, long, wide, and thin molecules. It is important to note that 11 compounds were deleted from consideration in formulating Eq. (10-9). These compounds are major outliers.

The optimum regression equations generated for, respectively, the nitroanilines and cyanoanilines are:

$$\log A = 0.50L \ (\pm 0.17) - 1.42W_1 \ (\pm 1.22) + 4.17 \ (\pm 4.37)$$
$$n = 14, \ r = 0.90, \ s = 0.32 \qquad (10\text{-}10)$$

$$\log A = 0.53L \ (\pm 0.27) - 2.14 \ (\pm 2.24)$$
$$n = 6, \ r = 0.94, \ s = 0.29 \qquad (10\text{-}11)$$

The results indicate that the longer 2-substituents enhance sweetness in both series of compounds. The insignificance of $W_1$ in Eq. (10-11) may be due to a narrow range of variation in the structure of the cyanoaniline derivatives in this direction. Since the coefficients of the $L$ term in Eqs. (10-10) and (10-11) overlap within the 95% confidence interval, these two sets of compounds were combined to give:

$$\log A = 0.52L \ (\pm 0.14) - 1.34W_1 \ (\pm 1.08) + 3.71 \ (\pm 3.49)$$
$$n = 20, \ r = 0.90, \ s = 0.32 \qquad (10\text{-}12)$$

The addition of a log $P$ and a $\sigma$, or $\sigma^+$, term singly or in combination did not significantly enhance the results. These findings are at odds with those of Deutsch and Hansch (1966), who have analyzed the relative sweetness (RS) of nine substituted nitroanilines and generated the following QSAR:

$$\log(\text{RS}) = 1.610 \log P - 1.831\sigma + 1.729$$
$$n = 9, \ r = 0.936, \ s = 0.383 \qquad (10\text{-}13)$$

where the $\sigma$ value of the 2-substituent is directed to the nitro group at the 5-position. Equation (10-13) can be improved by replacing the electronic term $\sigma$ with $\sigma^+$. From these equations, Deutsch and Hansch suggest a dependency of RS on hydrophobic bonding and basicity of the compound. The activity data used by these authors is a ratio of weights of sucrose and of the compound tested, which gave the same intensity. Thus, two different activity scales were used in the construction of Eqs. (10-10) and (10-13).

Iwamura (1980) concludes that the log $P$ term in Eq. (10-13) is a consequence of a high correlation between $L$ and log $P$ for the compounds used by Deutsch and Hansch (1966). Thus, hydrophobicity (log $P$) is not important in specifying the SP in 5-nitro- and 5-cyanoanilines; rather it is the steric dimensions of the substituent that are important.

Figure 2 is a schematic illustration of the considerations on binding of the perillartines to the receptor. In Fig. 2a the maximal contour viewed from the

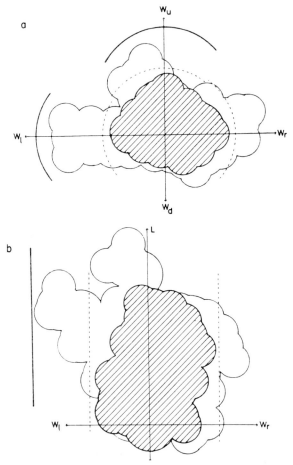

**Fig. 2.** Perillartine-receptor binding model. The maximum contour of the sweet perillartine derivative is striped. Unfigured is that of the bitter analogs. Solid lines represent the spatial walls for taste potency, and dashed lines the bitter barriers. (a) The view from the $C_1$ atom to the $L$ direction. (b) The view from the up side. [Reprinted with permission from Iwamura (1980).]

C-1 atom to the $L$ direction of the sweet perillartine molecules is overlapped with those of the bitter analogs. Fig. 2b is the view from the up side.

## 2. L-Aspartyldipeptide Analogs

Iwamura (1981) has also investigated the structure–sweetness relationship in four classes of L-aspartyldipeptides using the same procedures as employed in the perillartine work. The four classes of compounds are as follows ($N$ = the number of compounds in each class):

1. L-aspartic acid amides, $N = 66$.

$$\text{L-Asp-NHC}^1\text{H(R}_1)\text{C}^2\text{H(R}_2')\text{R}_2$$
(VII)

2. L-aspartylaminoethyl esters, $N = 61$.

$$\text{L-Asp-NHC}^1\text{H(R}_1)\text{C}^2\text{H(R}_2')\text{OCOR}_2$$
(VIII)

3. L-aspartylamino propionates, $N = 31$.

$$\text{L-Asp-NHC}^1\text{H(R}_1)\text{C}^2\text{H(R}_2')\text{COOR}_2$$
(IX)

4. L-aspartylamino acetates, $N = 59$.

$$\text{L-Asp-NHC}^1\text{H(R}_1)\text{COOR}_2$$
(X)

QSARs were generated for each of the four data bases. A representative multidimensional linear regression equation is that developed for the L-aspartylaminoethyl esters:

$$\log(\text{SP}) = 0.67\sigma^* (\pm 0.48) + 3.36 L_2 (\pm 1.03) - 0.29 L_2^2 (\pm 0.08)$$
$$+ 4.18(W_u)_1 (\pm 0.88) - 0.85(W_u)_1^2 (\pm 0.18)$$
$$- 0.53 L_1 (\pm 0.18) - 11.33$$
$$n = 51, r = 0.88, s = 0.27 \qquad (10\text{-}14)$$

In such equations the $L_i$ parameter expresses the length of substituent $R_i$ to the rest of the molecule; $(W_u)_i$ is the width upward of $R_i$ when one views it from the connecting end along the bond axis defining $L_i$. The electronic parameter, $\sigma^*$, was estimated for the structure substituted on the common aspartylamino moiety, so that the electronic effect is directed to the peptide bond. Ten compounds of the original 61 do not fit Eq. (10-14); hence, $n = 51$.

The analysis, in composite over the four classes of L-aspartyldipeptides, suggests that the electron-withdrawing effect of substituents directed to the peptide bond, and the steric dimensions of the molecules, are important in eliciting the sweet taste. The values of the regression coefficients of the asterisked term in the QSAR equations for L-aspartic acid amides, L-aspartylaminoethyl esters, and L-aspartylamino propionates are all approximately 0.7, suggesting that these three classes of dipeptide sweeteners interact in a common manner at the receptor. However, the $\sigma^*$ regression coefficient for the QSAR of the L-aspartylamino acetates is approximately 1.5. This value, along with an examination of the optimum steric parameter

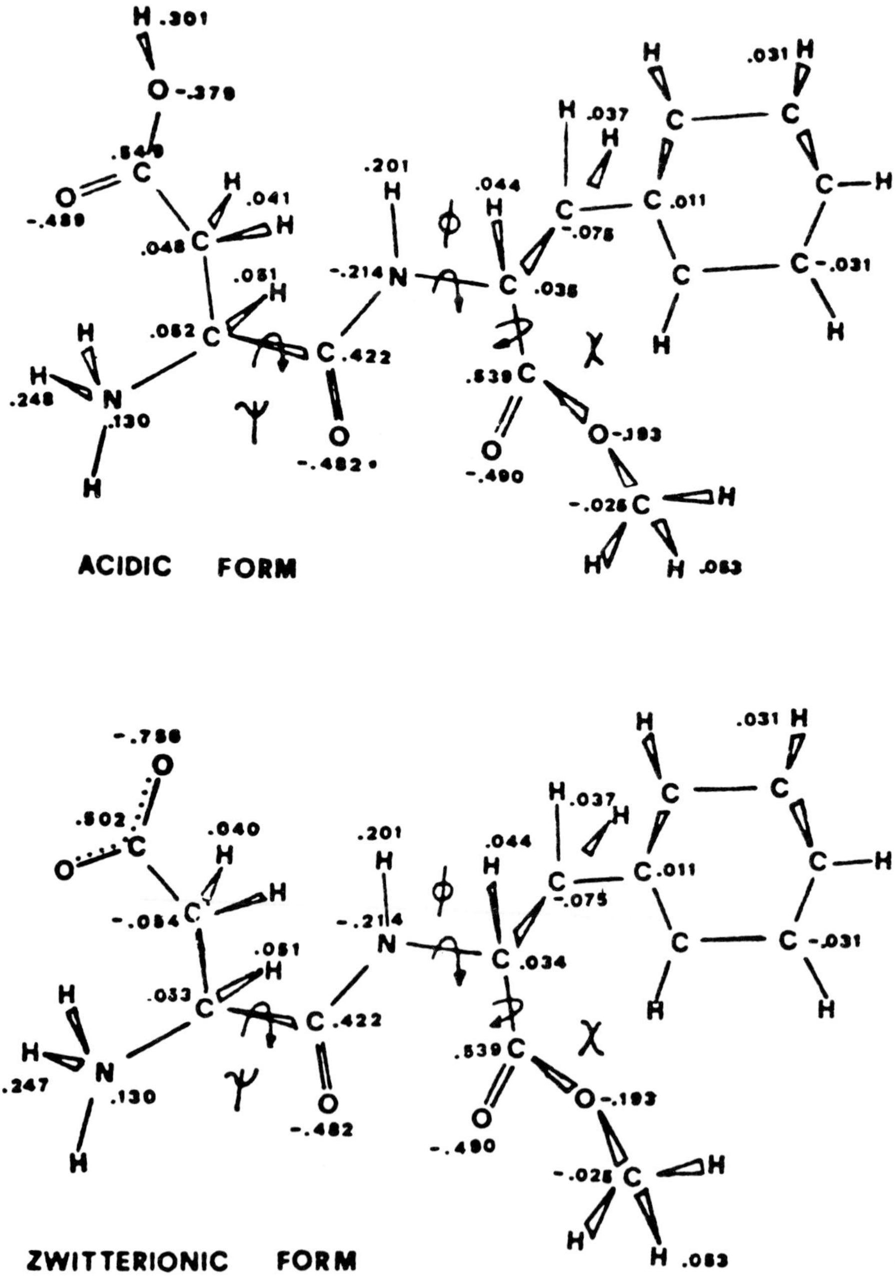

**Fig. 3.** Schematic molecular models of α-APM with fractional electronic charges. [Reprinted with permission from Lelj *et al.* (1976).]

values in the QSARs, suggested to Iwamura (1981) that the L-aspartylamino acetates can fit better to the receptor than the other three classes of compounds.

Lelj and co-workers (1976, 1980) have carried out a combination of experimental and theoretical studies to elucidate the "active" three-dimensional structure of $\alpha$-L-aspartyl-L-phenylalanine methyl ester ($\alpha$-APM). The pH dependence of molecular conformation in aqueous solution was examined by nuclear magnetic resonance (NMR) spectroscopy. Molecular mechanics-based conformational energy calculations were made for the three torsional angles ($\psi$, $\phi$, $\chi$) defined in Fig. 3. The numbers next to the atoms in Fig. 3 indicate the electronic charge densities at the atom sites.

A total of nine conformers have been identified in these conformational analyses that are common to the acidic, zwitterionic, and basic forms of $\alpha$-APM. No conformer state is populated more than 30% for any ionic form, indicating that $\alpha$-APM is a quite flexible molecule over a wide range of solution pH. Lelj et al. (1976, 1980) have deduced their choice for the "sweet" conformer by insisting that the A–H/B hydrogen-bonding requirement be met by the aspartic acid —NH$_3^+$ and COO$^-$, respectively. Three of the nine conformers meet these constraints. Two of these three structures are dismissed because the A–H/B grouping is partially shielded from participating in intermolecular hydrogen bonding by other portions of $\alpha$-APM. Thus, the conformation of $\alpha$-APM, shown embedded in a postulated receptor site in Fig. 4, is the choice of Lelj et al. (1976, 1980) to be the "sweet" conformation.

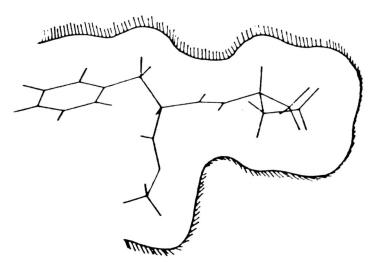

**Fig. 4.** Possible inner section of the receptor site as limited by the van der Waals radii of the postulated "sweet" conformation of $\alpha$-APM. [Reprinted with permission from Lelj et al. (1976).]

## IV. APPLICATION OF STRUCTURE–ACTIVITY METHODOLOGY TO OTHER SENSORY AGENTS

The application of quantitative molecular design methods to develop sensory agents other than sweeteners has been sparse with the exception of local anesthetics. Anesthetics have been given particular attention because they also fall into the drug category. Part of the reason for the lack of QSAR input into the development of sensory-active compounds reflects the biological activities and corresponding measurements. As already noted for sweeteners, in addition to potency there is the issue of quality. Thus, the intrinsic problem of quantitating relative potency based upon subjective evaluation is further complicated by trying to account quantitatively for perceived differences in the quality of induced sensory response.

### A. Olfactory Compounds

Amoore has pioneered the QSAR approach to the field of olfaction. He originally proposed that the size and shape of a molecule are the key physicochemical parameters for stimulation in olfaction (Amoore, 1952). A simple index of molecular size and shape for five classes of odors (ethereal, camphoraceous, musky, floral, and minty) was devised by Amoore and Venstrom (1967) to test this hypothesis. Meaningful correlations were found between odor quality and the molecular shape–size index. Amoore (1971) extended this work and was able to establish a correlation of 0.90 between odor quality and an extended molecular shape index when 25 compounds were compared with benzaldehyde.

It has also been proposed that the stimulation of olfactory receptors occurs through a mechanism involving the low-energy molecular vibrations of the odor agent (Wright, 1966). Wright and Robson (1969) have correlated a pattern of far-infrared frequencies with the bitter almond odor. A relationship between pheromone activity and molecular vibration in insects has also been reported (Wright and Brand, 1972).

Four user-defined physicochemical properties, apolarity, proton receptor capacity, proton donor capacity, and electron inductivity, have been used to establish both qualitative and quantitative discrimination in human beings (Dravnieks and Laffort, 1972). Advanced statistical methods and procedures have been applied to quantitatively differentiate olfactory quality using physicochemical parameters (Wright and Michaels, 1964; Guttman, 1968; Schiffman, 1974). This approach offers an opportunity to numerically analyze biological potency measures that are dependent upon differential perception quality. Wright and Michaels (1964) considered a data base of 50 odorants and 9 odorant standards. A standard versus test correlation was carried out, and the resulting $50 \times 50$ correlation matrix was factor analyzed with an eight-factor space emerging as significant.

Doving and Schiffman (1974) used multidimensional scaling techniques to further analyze the data base studied by Wright and Michaels (1964). Schiffman (1974) was able to demonstrate that the data base can be represented as a two-dimensional solution for 91% of the compounds, as is shown in Fig. 5. The solution divides the stimuli into approximately two groups: a larger subset containing predominantly "pleasant" compounds and another containing "adverse" odorants.

Schiffman (1974) next attempted to establish a pseudo structure–activity relationship by weighting a set of "physicochemical" variables relative to one another so as to reproduce the two-dimensional solution found via multidimensional scaling. The purpose of this procedure is to maximize the configurational similarity of the psychologically determined space (olfactory perceptions) with the space generated by physicochemical parameters. The physicochemical variables (along with means and variances) used in the fitting analysis and their corresponding weights are reported in Table 1. The regenerated space using the weighted physicochemical variables is shown in Fig. 6. Again, Fig. 6 is the best attempt, using physicochemical variables, to reproduce Fig. 5 generated from psychological (olfaction) variables. The overall results indicate that Wright's molecular vibration model of olfaction (Wright, 1966) better explains the olfaction data than does the size–shape model of Amoore (1952, 1971). However, the choice of physicochemical variables selected in the analysis does, in part, prejudice the resulting fit.

## B. Local Anesthetics

There have been several QSAR analyses of activity of local anesthetics. The examples reported here have been chosen to demonstrate the application of both molecular connectivity (Kier and Hall, 1976) and electronic structure indices as independent variables in separate QSARs based on regression equations. Agin *et al.* (1965) have determined the minimum nerve-blocking concentrations (MBCs) for 30 widely varying chemical structures. They observed a good correlation between log(MBC) and an approximation of the London or dispersion-interaction energy $E_L$ between a molecule and a conducting surface. This energy was deduced from the approximation $E_L = \alpha I / 8r^3$, where $\alpha$ is the electronic polarizability of the molecule, $I$ the ionization potential, and $r$ the interacting distance. The

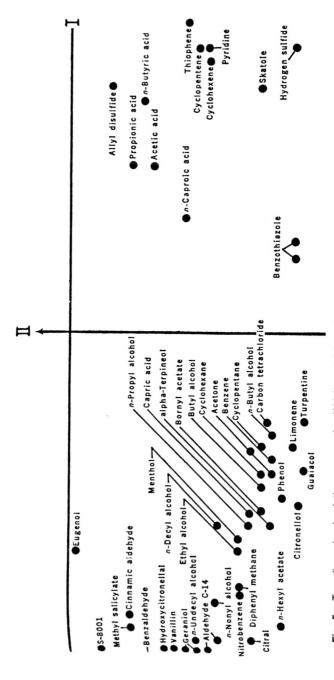

**Fig. 5.** Two-dimensional solution representing Wright and Michaels' psychophysical olfactory data for 50 stimuli. [Reprinted with permission from Schiffman (1974).]

**Fig. 6.** Space regenerated from weighting the physicochemical variables shown in Table 1 in an attempt to reproduce the psychological space illustrated in Fig. 5. [Reprinted with permission from Schiffman (1974).]

values for $I$ were eclectic values from the literature, simple Hückel molecular orbital theory approximations, and extrapolated estimates. The conclusion drawn from the correlation was that anesthetic potency is a result of an interaction between some portion of the nervous tissue and the anesthetic agent that is dominated by a dispersion, as opposed to an electrostatic, potential component.

This same set of molecules and six additional compounds were subsequently analyzed using molecular connectivity to describe their structures (Kier et al., 1975). The simple $^1\chi$ term (Kier and Hall, 1976) was found to correlate closely with log(MBC) via the relationship

$$\log(\text{MBC}) = 3.60 - 0.779 \ ^1\chi \qquad n = 36, r = 0.982, s = 0.409 \quad (10\text{-}15)$$

The data based and observed and predicted values of log(MBC) are reported in Table 2.

At first glance it is difficult to understand how both $^1\chi$ and $E_L$ could, independently, correlate with log(MBC). However, the polarizability, $\alpha$, of a

**TABLE 1  Weights Applied to Standard Scores for Physicochemical Variables to Achieve the Regenerated Space in Fig. 6.**[a]

| Physicochemical variable | Mean | Variance | Weight |
|---|---|---|---|
| Molecular weight | 116.57 | 1788.64 | 6.24 |
| Number of double bonds | 0.74 | 0.55 | 0.51 |
| Phenol | 0.13 | 0.11 | 2.33 |
| Aldehyde | 0.10 | 0.09 | 3.21 |
| Ester | 0.05 | 0.05 | 0.24 |
| Alcohol | 0.26 | 0.19 | 2.54 |
| Carboxylic acid | 0.13 | 0.11 | 5.05 |
| Sulfur | 0.08 | 0.07 | 3.44 |
| Nitrogen | 0.08 | 0.07 | 3.15 |
| Benzene | 0.33 | 0.27 | −0.14 |
| Halogen | 0.03 | 0.02 | −0.34 |
| Ketone | 0.03 | 0.02 | −0.19 |
| Cyclic | 0.31 | 0.21 | 4.56 |
| Mean Raman intensity | | | |
|    Below 175/cm | 0.51 | 3.14 | 0.01 |
|    176–250/cm | 2.36 | 9.30 | 3.57 |
|    251–325/cm | 1.65 | 7.10 | −0.75 |
|    326–400/cm | 1.56 | 5.74 | 3.81 |
|    401–475/cm | 2.10 | 7.23 | 1.65 |
|    476–550/cm | 1.54 | 5.22 | −3.63 |
|    551–625/cm | 2.07 | 7.09 | −0.69 |
|    626–700/cm | 1.07 | 5.14 | −1.16 |
|    701–775/cm | 2.36 | 11.01 | 0.07 |
|    776–850/cm | 4.36 | 13.84 | 3.04 |
|    851–925/cm | 3.44 | 15.77 | 0.24 |
|    926–1000/cm | 2.06 | 8.29 | 0.36 |

[a] From Schiffman (1974).

molecule is a strong function of its chemical topology. That is, $\alpha$ is a function of molecular connectivity. In this particular data base a regression fit between $\alpha$ and $^1\chi$ yields

$$\alpha = 9.503 \; {}^1\chi + 8.50 \qquad n = 36, \, r = 0.993, \, s = 3.108 \qquad (10\text{-}16)$$

Of course, the difficulty in employing molecular connectivity indices in QSAR studies is the inability to infer an action mechanism or model from the resulting correlation equation.

## C. Alternate Flavor Chemistry Applications

Relatively little QSAR work has been reported in the literature of flavor chemistry that has not already been mentioned. Marmor (1982) is using

**TABLE 2**   Local Anesthetic Activity and Polarizability[a]

| | $\alpha$ | | log MBC | |
|---|---|---|---|---|
| Anesthetic | Observed | Calculated | Observed | Calculated |
| Methanol | 8.2 | 10.9 | 3.09 | 2.79 |
| Ethanol | 12.9 | 14.7 | 2.75 | 2.47 |
| Acetone | 16.2 | 17.7 | 2.6 | 2.23 |
| 2-Propanol | 17.6 | 17.7 | 2.55 | 2.23 |
| Propanol | 17.5 | 19.3 | 2.4 | 2.09 |
| Urethane | 23.2 | 27.3 | 2.0 | 1.44 |
| Ether | 22.5 | 24.0 | 1.93 | 1.71 |
| Butanol | 22.1 | 24.0 | 1.78 | 1.71 |
| Pyridine | 24.1 | 24.8 | 1.77 | 1.65 |
| Hydroquinone | 29.4 | 32.1 | 1.4 | 1.05 |
| Aniline | 31.6 | 28.4 | 1.3 | 1.35 |
| Benzyl alcohol | 32.5 | 33.4 | 1.3 | 0.935 |
| Pentanol | 26.8 | 28.6 | 1.2 | 1.33 |
| Phenol | 17.8 | 28.4 | 1.0 | 1.35 |
| Toluene | 31.1 | 28.4 | 1.0 | 1.35 |
| Benzimidazole | 40.2 | 33.7 | 0.81 | 0.901 |
| Hexanol | 31.4 | 33.2 | 0.56 | 0.949 |
| Nitrobenzene | 32.5 | 36.8 | 0.47 | 0.651 |
| Quinoline | 42.1 | 38.3 | 0.3 | 0.528 |
| 8-Hydroxyquinoline | 44.7 | 42.6 | 0.3 | 0.174 |
| Heptanol | 36.0 | 37.9 | 0.2 | 0.567 |
| 2-Naphthol | 45.4 | 42.0 | 0.0 | 0.228 |
| Methyl anthranilate | 48.9 | 45.6 | 0.0 | −0.072 |
| Octanol | 40.6 | 42.5 | −0.16 | 0.186 |
| Thymol | 47.3 | 44.3 | −0.52 | 0.052 |
| o-Phenanthroline | 57.8 | 52.1 | −0.8 | −0.602 |
| Ephedrine | 50.2 | 50.3 | −0.8 | −0.453 |
| Procaine | 67.0 | 72.6 | −1.67 | −2.29 |
| Lidocaine | 72.5 | 72.8 | −1.96 | −2.31 |
| Diphenhydramine | 79.5 | 78.2 | −2.8 | −2.75 |
| Tetracaine | 79.7 | 81.5 | −2.9 | −3.03 |
| Phenyltoloxamine | 79.9 | 78.1 | −3.2 | −2.74 |
| Quinine | 93.8 | 91.5 | −3.6 | −3.85 |
| Physostigmine | 82.4 | 75.4 | −3.66 | −2.52 |
| Caramiphen | 87.0 | 87.0 | −4.0 | −3.48 |
| Dibucaine | 103.6 | 105.2 | −4.2 | −4.97 |

[a] From Kier and Hall (1976).

molecular shape features, including common and difference volume and surface measures, to seek out structure – organoleptic relationships that may exist among tobacco flavorants. To this end he has uncovered a correlative trend between the shape of a sizable region of the cedrol molecule and that of a number of cedarwood flavorants of diverse chemical structure.

Several private organizations having research programs in artificial flavoring have recently hired scientists trained in QSAR and/or molecular modeling techniques. A few of these companies have, in fact, embarked upon software development programs to provide computer-assisted chemical design capabilities. Presumably, there is a growing amount of proprietary QSAR information on flavoring agents that will one day be available in the literature.

## V. SUMMARY AND PROSPECTS

The pharmaceutical industry has pioneered in the application of computer-assisted drug design methods in ongoing research programs. To a significant degree this is a consequence of the straightforward role of computational chemical techniques in aiding lead drug identification and subsequent optimization. Moreover, the quality of the biological activity measures, coupled with the high purity of the chemical entities, has provided reliable data to use in pharmaceutical QSAR studies.

For the flavor industry to take advantage of the benefits of quantitative molecular design, more stringent measures of the biological responses are absolutely necessary. The artificial sweetener area represents an example in flavor chemistry where high-quality measures of SP have made it possible for QSAR techniques to be successfully applied. Alternate artificial flavoring areas could be well served by following the sweetener example.

A great amount of methodology in molecular design is available for artificial flavoring QSAR applications. Much of the methodology is formatted in easy-to-use software packages. Thus, computer-assisted molecular design can be construed as an analytical tool ready for use by flavor chemists.

## REFERENCES

Acton, E. M., and Stone, H. (1976). *Science* **193**, 584.
Acton, E. M., Leaffer, M. A., Oliver, S. M., and Stone, H. (1970). *J. Agric Food Chem.* **18**, 1061.
Agin, D., Hersh, L., and Holtzman, D. (1965). *Proc. Natl. Acad. Sci.* U.S.A. **53**, 952.
Amoore, J. E. (1952). *Perfum. Essent. Oil Rec.* **43**, 321.

Amoore, J. E. (1971). *Nature* (*London*) **233**, 270.

Amoore, J. E., and Venstrom, D. (1967). *In* "Olfaction and Taste" (T. Hayashi, ed.), Vol. 2, p. 357. Pergamon, Oxford.

Barlett, J., and Weinstein, H. (1975). *Chem. Phys. Lett.* **30**, 441.

Bartoshuk, L. M., Gentile, R. L., Moskowitz, H. R., and Meiselman, H. L. (1974). *Physiol. Behav.* **12**, 449.

Battershell, C., Malhotra, D., and Hopfinger, A. J. (1981). *J. Med. Chem.* **24**, 812.

Beets, M. G. J. (1978). "Structure-Activity Relationships in Human Chemoreception," Chap. 3. Applied Science, London.

Binghan, R. C., Dewar, M. J. S., and Lo, D. H. (1975). *J. Am. Chem. Soc.* **97**, 1302.

Bowden, K., and Woolridge, R. H. (1973). *Biochem. Pharmacol.* **22**, 1015.

Boyd, D. B., and Lipkowitz, K. B. (1982). *J. Chem. Educ.* **59**, 269.

Charton, M. (1974). *CHEMTECH* No. 502.

Charton, M. (1975). *CHEMTECH* No. 245.

Collander, R. (1954). *Physiol. Plant.* **7**, 420.

Crippen, G. M. (1979). *J. Med. Chem.* **22**, 988.

Crosby, G. A., and DuBois, G. E. (1980). *Trends Pharmacol. Sci.* p. 372.

Coubeils, J. L., Courriere, Ph., and Pullman, B. (1971). *Acad. Sci.* (*Paris*) **272**, 1813.

Deutsch, E. W., and Hansch, C. (1966). *Nature* (*London*) **211**, 75.

Dravnieks, A., and Laffort, P. (1972). *In* "Olfaction and Taste" (D. Schneider, ed.), Vol. 4. Wissenschaftliche Verlagsgesellschaft, Stuttgart.

DuBois, G. E., Crosby, G. A., and Stephenson, R. A. (1981a). *J. Med. Chem.* **24**, 408.

DuBois, G. E., Dietrich, P. S., Lee, J. F., McGarraugh, G. V., and Stephenson, R. A. (1981b). *J. Med. Chem.* **24**, 1269.

Fujino, M., Wakimasa, M., Tanaka, K., Aoki, H., and Nakajima, N. (1976). *Chem. Pharm. Bull.,* **24**, 2112.

Fujita, T., (1983). *Prog. Phys. Org. Chem.* (in press).

Goodford P. J. (1973). *In* "Advances in Pharmacology and Chemotherapeutics," (M. Daniels, ed.), p. 52. Academic Press, New York.

Guttman, L. (1968). *Psychometrika* **33**, 469.

Ham, N. S. (1974). *In* "Molecular and Quantum Pharmacology" (E. D. Bergmann and B. Pullman, eds.), p. 261. Reidel-Dordrecht, Holland.

Hansch, C., and Clayton, J. M. (1973). *J. Pharm. Sci.* **62**, 1.

Hansch, C., and Leo, A. (1979). "Substituent Constants for Correlation Analysis in Chemistry and Biology." Wiley (Interscience), New York.

Hansch, C., Muir, R. M., Fujita, T., Maloney, P. P., Geiger, F., and Streich, M. (1963). *J. Am. Chem. Soc.* **84**, 2817.

Hansch, C., Steward, A. R., Iwasa, J., and Deutsch, E. W. (1965). *Mol. Pharmacol.* **1**, 205.

Hansch, C., Unger, S. H., and Forsythe, A. B. (1973). *J. Med. Chem.* **16**, 1217.

Hansch, C., Silipo, C., and Steller, E. E. (1975). *J. Pharm. Sci.* 1186.

Hansen, O. R. (1962). *Acta Chem. Scand.* **16**, 1593.

Higginbotham, J. D. (1979). *In* "Development in Sweeteners—1" (C. A. M. Hough, K. J. Parker, and A. J. Vlitos, eds.), p. 87. Applied Science, London.

Hoffman, R. (1963). *J. Chem. Phys.* **39**, 1397.

Hopfinger, A. J. (1980). *J. Am. Chem. Soc.* **120**, 7196.

Hopfinger, A. J. (1981). *J. Med. Chem.* **24**, 818.

Hopfinger, A. J. (1983). *J. Med. Chem.* **26**, 990.

Hopfinger, A. J., and Jabloner, H. (1981). *In* "The Quality of Foods and Beverages" (G. Charalambous and G. Inglett, eds.), p. 83. Academic Press, New York.

Hopfinger, A. J., and Potenzone, R., Jr. (1982). *Mol. Pharmacol.* **21**, 187.

Iwamura, H. (1980). *J. Med. Chem.* **23**, 308.

Iwamura, H. (1981). *J. Med. Chem.* **24**, 572.

Jakinovich, W., Jr. (1981). *Brain Res.* **210**, 69.

Kier, L. B. (1968). *J. Pharmacol. Exp. Ther.* **164**, 75.

Kier, L. B. (1970). *In* "Fundamental Concepts in Drug–Receptor Interactions" (J. Danielli, J. Moran, and D. Triggle, eds.), p. 15. Academic Press, New York.

Kier, L. B. (1972). *J. Pharm. Sci.* **61**, 1394.

Kier, L. B., and Hall, L. H. (1976). "Molecular Connectivity in Chemistry and Drug Research." Academic Press, New York.

Kier, L. B., Hall, L., Murray, W. J., and Randic, M. (1975). *J. Pharm. Sci.* **64**, 1971.

Kikuchi, O. (1977). *Bull. Chem. Soc. Jpn.* **50**, 593.

Kowalski, B. R., and Bender, C. F. (1973). *J. Am. Chem. Soc.* **95**, 586.

Lelj, F., Tancredi, T., Temussi, P. A., and Toniolo, C. (1976). *J. Am. Chem. Soc.* **98**, 6669.

Lelj, F., Tancredi, T., Temussi, P. A., and Toniolo, C. (1980). *Farmaco, Ed. Sci.* **35**, 988.

Leo, A., Jow, P. Y. C., Silipo, C., and Hansch, C. (1975). *J. Med. Chem.* **18**, 865.

Marmor, R. S. (1982). Presented at the *Symp. Chem. Tob. Tob. Smoke, AGFD, 184th Am. Chem. Soc. Mtg., Kansas City,* Sept. 12–17.

Martin, Y. C., Bustard, T. M., and Lynn, K. R. (1973). *J. Med. Chem.* **16**, 1089.

Matthews, R. J. (1975). *J. Am. Chem. Soc.* **97**, 935.

Max, N. L., Malhotra, D., and Hopfinger, A. J. (1981). *Comput. Chem.* **5**, 19.

Mazur, R. H., Schlatter, J. M., and Goldkamp, A. H. (1969). *J. Am. Chem. Soc.* **91**, 2684.

Perrin, C. L. (1973). *Science* **183**, 551.

Pople, J. A., and Beveridge, D. C. (1970). "Approximate Molecular Orbital Theory." McGraw-Hill, New York.

Pople, J. A., Santry, D. P., and Segal, G. A. (1965). *J. Chem. Phys.* **43**, 5129.

Rekker, R. F. (1977). "The Hydrophobic Fragment Constant," Pharmacochemistry Library, Vol. 1. Elsevier, New York.

Rohrer, D. C., Fullerton, D. S., From, A. H. L., and Ahmed, K. (1979). *ACS Monogr. Ser. No.* 112, p. 259.

Schiffman, S. S. (1974). *Science* **185**, 112.

Segal, G. A., ed. (1977). "Methods of Electronic Structure Calculations, Parts A and B." Plenum, New York.

Seiler, P. (1974). *Eur. J. Med. Chem.* **9**, 663.

Shallenberger, R. S., and Acree, T. C. (1967). *Nature (London)* **216**, 480.

Silipo, C., and Hansch, C. (1975). *J. Am. Chem. Soc.* **97**, 6849.

Streitweiser, A. (1961). "Molecular Orbital Theory for Organic Chemists." Wiley, New York.

Swartz, M. L., and Furia, T. E. (1977). *Food Technol. (Chicago)* **11**, 51.

Tanimura, T., and Shimada, I. (1981). *J. Comp. Physiol.* **141**, 265.

Topliss, J. G., and Costello, R. J. (1972). *J. Med. Chem.* **15**, 1066.

van der Wel, H., Van Soest, T. C., and Royers, E. C. (1972). *Eur. J. Biochem.* **31**, 221.

Venkatachalam, C. M. (1968). *Biopolymers* **6**, 1425.

Verloop, A. J. (1972). *In* "Drug Design" (E. J. Ariens, ed.), Vol. 3, p. 133. Academic Press, New York.

Verloop, A., Hoogenstraaten, W., and Tipker, J. (1976). *In* "Drug Design" (E. J. Ariens, ed.), Vol. 7, p. 165. Academic Press, New York.

Weinstein, H. (1975). *Int. J. Quantum Chem.* **2**, 59.

Weinstein, H., Apfelder, B. Z., Cohen, S., Maayani, S., and Sokolovsky, M. (1973). *In* "Conformation of Biological Molecules and Polymers" (E. D. Bergmann and B. Pullman, eds.), p. 531. Academic Press, New York.

Weintraub, H. J. R., and Hopfinger, A. J. (1973). *J. Theor. Biol.* **41**, 53.

Wilson, R. (1972). "Introduction to Graph Theory." Academic Press, New York.
Wright, R. H., (1966). *Nature* (*London*) **209**, 571.
Wright, R. H., and Brand, J. M. (1972). *Nature* (*London*) **239**, 225.
Wright, R. H., and Michaels, K. (1964). *Ann. N.Y. Acad. Sci.* **116**, 535.
Wright, R. H., and Robson, A. (1969). *Nature* (*London*) **222**, 290.
Yamamoto, K. (1974). *J. Biochem.* **76**, 385.

# 11

# Bioassays

## CHANDRALEKHA DUTTAGUPTA

Department of Gynecology/Obstetrics
Albert Einstein College of Medicine
Yeshiva University
Bronx, New York


## ELI SEIFTER

Departments of Biochemistry and Surgery
Albert Einstein College of Medicine
Yeshiva University
Bronx, New York

ANALYSIS OF FOODS AND BEVERAGES
Copyright © 1984 by Academic Press, Inc.
All rights of reproduction in any form reserved.
ISBN 0-12-169160-8

## I. INTRODUCTION AND AIMS

This volume combines the efforts of experts in the variety of chemical and physical techniques that are used in the quantitative analysis of foods and beverages. The relevance of this chapter, devoted to biological assays, to the other chapters requires discussion. Additionally, the aims of bioassays and the factors that influence the results also require consideration.

Bioassay (biological assay) is of two types. Originally, bioassays involved the use of whole organisms to measure the *activity* of pharmacologic or nutritional agents. For example, laboratory animals were used to assay the potency of antisera to diphtheria toxin, the insulin activity of pancreatic extracts, the curariform activity of plant extracts, the glucogenic potential of amino acids, the growth-promoting effects of food extracts on rats raised on defined or semisynthetic diets (vitamin A), or the cure of a deficiency disease in animals raised on defined diets (vitamin D). Not only were laboratory animals used to bioassay materials, but plants were used to assay the growth effects of gaseous, solid, or mineral metabolites of animal origin. More recently, bacteria, fungi, and animal tissues or cells in culture have been used to assay the biotic or antibiotic activity of extracts of other living systems.

Even people have been used to bioassay activities such as digitalis or lipotropic activity; in these cases, volunteers (frequently medical students or prisoners) and even nonvoluntary persons (mainly prisoners) were used for bioassay. Substances were assayed both in healthy individuals and in patients with comparable stages of a known disease. The first type of bioassay, then, involved whole organisms to measure the response or activity to an impure agent that was usually of biologic origin.

The first bioassay to receive acceptance by official international bodies was developed by Paul Ehrlich, founder of laboratory medicine. His bioassay for diphtheria antitoxin defined an *amount* of dried antiserum that protected *standardized* guinea pigs against *described signs* produced by the injection of specified amounts of a *reference or standard* toxin preparation. The assay was introduced into medicine in the 1890s and was adopted as an international standard by the League of Nations in 1921. Ehrlich pointed out that one could report the results of tests as either the amount of a preparation required to elicit a given response or the quantitative response elicited by a given amount of preparation. He suggested that the strength of a preparation be judged relative to a standard amount of a standard preparation that produces a defined response. This concept is still in use, and it expresses the idea that it is easier (and more accurate) to quantitate the amount of preparation that produces a predetermined response than it is to quantitate the response.

Another type of bioassay widely used in assessing the pharmacologic or toxic activities of dietary ingredients or foods employs excised tissues or organ preparations that are studied *in vitro*. The criteria used for quantitating the *in vitro* activities of the extracts are similar to those used in Michaelis–Menten equations that describe the influence of substrates and inhibitors on enzyme kinetics; that is, in place of reaction velocity a biological parameter such as muscle tension, secretory activity of an isolated exocrine gland, or heartbeats per minute is used to measure the acetylcholine-like activity that is present in a food or tissue extract. Similarly, extracts can be tested *in vitro* for their ability to potentiate or inhibit the effects of standard amounts of added acetylcholine. The examples of food ingredients that contain intrinsic or pesticidal residues influencing acetylcholine metabolism have been chosen for discussion because they represent types of problems that are of importance in chemical and biological assessment of food toxicities.

It is not the aim of this chapter to provide an atlas or workbook of methods; rather, its aim is to familiarize chemists with some biological techniques that are useful in evaluating the biological activities and functions of foods and their constituents and to indicate circumstances in which bioassay may be useful.

Bioassay is a different type of analytic approach that may be used to supplement the information obtained by chemical analysis. *Chemical* analysis may provide knowledge of the identity of a given compound present in a food of plant origin, let us say vitamin $K_1$ and the amount of vitamin $K_1$ present. *Bioassay* of the food tells us how active that vitamin $K_1$ is, for example, in prothrombin synthesis. Bioassay can tell us whether substances exist in food samples that diminish the activity (e.g., dicumarol) or hasten its biodegradation (e.g., pesticides that induce hepatic enzymes that degrade vitamin $K_1$). Bioassay can also tell us when the biological activity of a substance is increased or potentiated by other dietary ingredients (e.g., by compounds that increase the uptake or alter the distribution of the test chemical).

The proof of the pudding is in the eating, and the overall nutritional effect of the pudding is based only in part on the individual ingredients. Some of the effects are due to the interactions of the ingredients in the process of food preparation; most of the effect is due to the metabolic interactions of the ingredients and, further, to their interactions with substances already present in the body. Discovering these effects is the essence of bioassay.

In the hierarchy of chemists (soapers being at the bottom), analytical chemists rate fairly high. They tell us what and how much is cooking. Even higher on the scale of chemists are the physical chemists. They know from the analytic chemists what and how much is cooking, but they tell us two additional things: why it is cooking (i.e., what forces are responsible for the reaction) and what is the chemical activity (i.e., what portion $\alpha$ of the molecules in a given reaction concentration is actually participating in a given reaction). Fortunately, biologists do not have to be rated on a chemical scale. Nonetheless, we feel that we would be up there with the physical chemists. We are both concerned with the *activities* of reactants.

## II. PRINCIPLES OF BIOASSAY

### A. Acetylcholine

Toxic dietary factors influencing acetylcholine metabolism and response may be examined in relation to acetylcholine synthesis and breakdown and the receptor activity (Griffith and Dyer, 1968; Nutrition Society Symposium, 1971; Michelson and Danilov, 1971; Mayer, 1980; Taylor, 1980).

### 1. Dietary and Other Factors Influencing Acetylcholine Synthesis

The synthesis of acetylcholine in animals, including humans, is dependent upon the presence of adequate amounts of choline. Choline levels themselves are the result of the amount of choline ingested (mostly as lecithin) and the amount of choline synthesized *in vivo*. Choline synthesis requires, at various stages, pyridoxal and nutrients (methionine, betaine) that can supply a ready amount of reactive methyl groups as *S*-adenosylmethionine. It is clear that diets deficient in pyridoxine or methionine are also choline deficient; however, choline-deficient diets are generally deficient in methionine rather than pyridoxine. Several other methods are known to produce choline deficiency in nervous tissue (and therefore acetylcholine deficiency) even though methionine intake appears to be adequate. Mammals ingest some compounds (mainly nicotinic acid derivatives) that are "detoxified" by N-methylation prior to their excretion. These compounds, so-called methyl traps, decrease the amount of methionine available for choline and phospholipid synthesis. Because the liver is a major site of choline biosynthesis and export (as part of lipoproteins), liver diseases such as cirrhosis (including choline-deficiency cirrhosis) intensify choline deficiency by inhibiting hepatic choline synthesis from methionine and diminishing the export of even choline that is synthesized. It is possible that some of the neurologic disease secondary to cirrhosis reflects decreased levels of acetylcholine. Some of the neuromuscular disorders of cirrhosis may represent diminished myelin synthesis secondary to diminished choline biosynthesis.

## 2. Acetylcholine Receptors and Activities

**a. Muscarinic Actions of Acetylcholine.** The main functions of acetylcholine have been characterized by the mimicry of these responses by extracts of plant toxins. These functions are muscarinic and nicotinic. Acetylcholine acts directly on appropriate exocrine glands to stimulate secretion of fluids such as saliva, tears, HCl, $HCO_3^-$, etc. This activity is also elicited by the ingestion of muscarine derived from poisonous mushrooms, by neurine (vinyl trimethylammonium salts) derived from the bacterial decomposition of choline during food spoilage, and by the alkaloid pilocarpine. Associated with the secretory activity of these compounds is stimulation of smooth muscle contraction (in the gastrointestinal tract, bronchi, gallbladder, urinary bladder, etc.). These effects are called the muscarinic actions of acetylcholine. Whereas acetylcholine has effects in addition to these, muscarine's action is limited to this described scope of acetylcholine's action. A characteristic of muscarinic activity is the binding of the compound (at sites that normally bind acetylcholine) and induction of the same electrical (depolarization) phenomena induced by acetylcholine. A second feature of the muscarinic activities (including depolarization) is the competitive inhibitory action of some solanaceous alkaloids such as atropine. It is clear that the toxicity of atropine (which inhibits muscarine activity) is influenced by dietary choline intake, choline and acetylcholine synthesis, and intake of noncholine (but muscarinic) constituents of food. Similarly, *in vitro* assays for smooth muscle contraction due to muscarinic extracts of food will be influenced by the presence of agents that block the binding of acetylcholine or of other agents that prolong the activity of endogenous acetylcholine.

**b. Nicotinic Actions of Acetylcholine.** Acetylcholine also has effects—low-dose stimulation of autonomic ganglion cells and high-dose blockade of their functions—that are mimicked by nicotine. This is the main nicotinic action of acetylcholine. Additionally, nicotine, like acetylcholine, stimulates the nerve muscle junction to produce muscle contraction. This nicotinic action is blocked by curare and succinylcholine-like compounds. It is clear that *in vitro* bioassays for nicotinic action (nerve muscle preparations) using plant extracts as agonists or antagonists must take into account (1) the presence of interfering agents in the extract and (2) the choline status of the animal used as the source of the muscle. Standardization of the tissue preparation usually requires the preparations to be made from standardized animals.

**c. Combined Muscarinic and Nicotinic Actions.** Some materials consumed because of their presence in masticatory mixtures (seeds of *Areca catecher*) contain alkaloids such as arecoline that have both nicotinic and muscarinic activities. Additionally, the same seed contains some alkaloids that have primarily nicotinic activity. Ingestion or chewing of the nut itself or of pure arecoline causes both nicotinic and muscarinic actions.

Bioassay in the intact animal or even certain tissue preparations may not be reflective of toxicity that would occur in humans. For example, the activity of arecoline is strongly influenced by the other cholinergic agents, such as nicotine (either by inhalation or chewing), or by anticholinergic agents (atropinic agents used to control gastrointestinal motility or acid secretion) or antihistaminic agents ($H_1$ antagonists) with atropinic side effects. These modifications of toxic responses by other dietary drugs or nutrients constitute the basis of a burgeoning area of bioassay called drug–nutrient interactions. A more complete discussion of these bioassays is beyond the scope of the present chapter.

## B. Unsaturated Fatty Acids

Unsaturated fatty acids are important nutrients (Alfin-Slater and Aftergood, 1975), because they are constituents of membranes (Green and Tzagoloff, 1966) and precursors of important hormones, prostaglandins, prostacyclins, and thromboxanes (Van Drop et al., 1964; Bergström et al., 1964). Unsaturated fatty acids (linoleic, linolenic, and arachidonic acids) are present as part of triglycerides of adipose tissues and phospholipids of membranes of cells and organelles. Unsaturated fatty acid deficiency can occur even if the storage fat is rich in these fatty acids, if lipase activity is diminished because of hormonal control. For example, unsaturated fatty acid deficiency has been observed in patients who receive continuous intravenous nutrition with glucose and amino acid solutions, even though these individuals have large amounts of stored unsaturated fatty acids. Apparently, continous infusion of glucose causes the release and maintenance of high levels of circulating insulin; and it is the circulating insulin that inhibits the mobilization of stored essential fatty acids by suppressing lipase activity.

Excess intake of essential fatty acids (greater than 4% of caloric intake) results in storage of these fatty acids, which can then undergo enzymatic and nonenzymatic oxidation, yielding pigmented compounds (ceroids) that may deplete the body's glutathione, ascorbic acid, or vitamin E stores. Excess intake of unsaturated fatty acids is considered a risk factor in carcinogenesis; its mechanism of action may be due to the depletion of protective reducing agents. Excess intake of unsaturated fatty acids may pose problems for the homeostasis of immunity; that is, they tend to favor the synthesis of the immune-inhibiting prostaglandins.

Plant oils that are rich in unsaturated fatty acids may contain traces of toxic materials of plant origin such as saponins and other sterols. Of more importance is the presence of pesticide residues such as lipid-soluble organic insecticides and lipid-miscible organometallics used as fungicides.

Both chemical and biological assays of unsaturated fats fall short of measuring their toxicities, because the tests (assays) are run on samples treated very differently from the way fats are treated in food processing.

Fats used in cooking (boiling) are influenced mainly by the aqueous reactions; that is, the reactions are occurring at or below the boiling point of water. However, in frying and broiling, temperatures are always above the boiling point and nonaqueous reactions are favored (Sugimura, 1981). Under these conditions a variety of addition reactions occur, and one is concerned about the present tendency to use the very light (polyunsaturated) oils for frying or roasting. We suggest that analytic data of fully *processed* foods are required to evaluate potential toxicity due to unsaturated fatty acids.

## C. Amino Acids

Adverse effects due to ingestion of disproportionate amounts of amino acids have been attributed to amino acid toxicities, antagonisms, or imbalances (Elvehjem, 1956; Harper, 1956; Harper *et al.*, 1970). In addition to amino acids consumed from food proteins or as food supplements, some amino acids are used as therapeutic–pharmacologic agents and also as flavoring and sweetening substances. The use for these purposes can result in consumption of greater quantities of individual amino acids than are ordinarily consumed in foods and may produce toxic effects (Milne, 1968; Harper *et al.*, 1970; Harper, 1973). Another consideration is the level of certain amino acids or related aminonitriles, which because of their presence in food items that are used as staples constitute a potential danger to health (Fowden *et al.*, 1967; Liener, 1969, 1980; Thompson *et al.*, 1969; Sarma and Padmanavan, 1969; Bell, 1973; Padmanavan, 1980).

Toxicity due to ingestion of excessive amounts of free amino acids is a much-studied subject and too complex for consideration in the present chapter. Here, we consider several amino acid toxicities produced by foods that are ingested as natural food rather than as supplements or additives.

A number of plants that are usual food for animals or humans contain toxic amino acids or some of their metabolites.

## 1. Lathyrogens

Most aminonitriles present in foods and feeds produce their toxicities only when the offending crop is used as a staple. Unfortunately, large populations (e.g., in India, China) consume crops containing significant amounts of lathyrogens. Seeds from some species of *Lathyrus*, commonly *L. sativus* and less frequently *L. cicera* and *L. clymenum,* are toxic when ingested by humans and domestic animals (Sarma and Padmanavan, 1969; Padmanavan, 1980).

Classical lathyrism is a neurological disease of humans, producing muscular weakness, irreversible paralysis of the legs, and, in extreme cases, death. Selye (1957) proposed that the term *neurolathyrism* be used for the classic disease and *osteolathyrism* for the experimental disease produced by ingestion of *Lathyrus odoratus* (sweet pea) seeds. Neurolathyrism appears to be related to specific failure of pyridoxal-requiring reactions, for example, the synthesis of some sphingolipids and other myelin components. Osteolathyrism produces structural weaknesses in tendons and ligaments due to inhibition of the maturation of their structural proteins.

Most lathyrogens present in legumes exert their toxicity by interfering with pyridoxine-dependent enzymes. Lathyrism causes connective tissue disease characterized by a loss of tensile strength of collagen and a loss of elasticity of elastin. In both cases, posttranslational modification of proteins normally results in the oxidation of $\epsilon$-amino groups of specific lysyl residues to aldehydes. These aldehyde moieties then cross-link to specific amino groups (or other functional groups) and cause a marked increase in the molecular weights and physical properties of the proteins. The enzymes that catalyze the posttranslational changes are called lysyl oxidases. They are pyridoxal-containing enzymes, related to the diamine oxidases.

*Lathyrus sativus* seeds contain two unusual amino acids, L-homoarginine (Rao *et al.*, 1963) and $\beta$-$N$-oxalyl-L-$\alpha$, $\beta$-diaminopropionoic acid (Rao *et al.*, 1964; Murti *et al.*, 1964). The latter compound produces the neurological disease (Bell, 1973).

*Lathyrus sativus* seeds contain a second neurotoxin, $N$-$\beta$-D-glucopyranosyl-$N$-$\alpha$-L-arabinosyl-$\alpha$, $\beta$-diaminopropionitrile (Rukmini, 1968, 1969), found to be highly toxic for chicks. Whether it is one of the factors contributing to human lathyrism is not clear.

Another field legume, *Vicia sativa* (common vetch) produces at least two neurotoxins, $\beta$-cyanoalanine and $\gamma$-glutamyl-$\beta$-cyanoalanine (Ressler, 1962). The toxicity of $\beta$-cyanoalanine to rats can be largely blocked by injection of pyridoxal (or by feeding pyridoxine). Also, $\beta$-cyanoalanine-treated rats excrete unusually large amounts of cystathionine, probably because the conversion of cystathionine to cysteine is a pyridoxine-dependent reaction and is inhibited by the cyanoalanine.

Although the aminonitriles bind and inactivate pyridoxal enzymes, the binding is selective; that is, not all pyridoxal enzymes bind the aminonitriles, and there appears to be recognition of the toxic agent by the enzyme.

A goal of bioassays and nutritional studies should be to determine whether supplemental pyridoxine can prevent all the toxicities of lathyrism, in which case lathyritic crops may continue to be used as feeds and foods. If all signs are not prevented by pyridoxal, safe limits of lathyrogen consumption will have to be established.

## 2. Canavanine

The toxicity of some legumes and their sprouts (alfalfa, soy, etc.), resulting from the presence of the amino acids canaline (an ornithine analog) and canavanine (an arginine analog), is the subject of current investigation (Bell, 1973). These amino acids are part of a cyclic mechanism used by some legumes for storing and releasing ammonia and urea nitrogen. The finding that ingestion of large amounts of canavanine-containing foods causes specific toxicity because of interference with arginine metabolism suggests that bioasssay of test materials in systems (whole animal, cell cultures, etc.) that respond to arginine deficiency or supplementation will be useful in assessing the toxicity inherent in specific crops.

Canavanine toxicity is associated with inhibition of arginine incorporation into proteins; that is, it can be activated by arginine-activating enzymes. In most cases activated canavanine is not incorporated into proteins in the presence of arginine. Nonetheless, some of the analog *is* incorporated into arginine-rich nucleoproteins. In monkeys (and human volunteers) the arginine analog is incorporated into nuclear proteins of leukocytes and other cells with a rapid turnover. These altered proteins are detected as foreign by the immune system, and antibodies are developed against them. The presence of autoantibodies against the nuclei of such cells causes a disease related to lupus erythematosus. The role of bioassay in studying this phenomenon is important, because the toxicity of canavanine is dependent on the arginine and lysine content of the diet as well as on factors not yet elucidated.

## 3. L-3,4-Dihydroxyphenylalanine (L-Dopa)

Pods of *Vicia faba* (broad beans, or fava beans) are rich in L-dihydroxyphenylalanine (L-dopa). Consumption of fava beans is associated with a human disease called favism. Persons genetically deficient in glucose-6-phosphate dehydrogenase (G6PD) are found to be at risk for favism. Kosower and Kosower (1965) showed that reduced glutathione (GSH) levels were diminished when G6PD-deficient erythrocytes were incubated with L-dopa. Decreased GSH levels favor hemolysis, one of the signs of favism. Foods containing L-dopa may produce undesirable neurotoxic side effects, such as nausea, vomiting, and dizziness (Van Woert and Bowers, 1970).

Bioassays involving the presence of G6PD defects in red cells appear to be more important than measuring the levels of dopa in the blood in order to determine the safety of a food, because materials other than L-dopa frequently influence GSH stability.

## 4. Djenkolic Acid

Djenkolic acid is a sulfur-containing amino acid present in free state in djenkol beans, the seeds of *Pithecolobium lobatum,* a leguminous tree (Van Veen, 1966). The bean is widely consumed by people in Far Eastern islands, particularly Java and Sumatra. The bean is not frequently toxic; however, it can produce serious kidney disease as a result of crystallization of the amino acid in renal tubules. Because of its relative stability, djenkolic acid escapes metabolic degradation and crystallizes out in the urine (Liener, 1969, 1980). Neither bioassay nor chemical assay can reliably predict djenkolic acid toxicity. This is more dependent on host factors, such as preexisting renal disease, pH of urine produced, and other dietary and metabolic factors.

## 5. Seleno Amino Acids

Some western soils (e.g., in Idaho and Oklahoma) contain selenium (selenite) in appreciable amounts. Crops such as wheat have a moderate ability to synthesize seleno amino acids, whereas other plants may accumulate inorganic selenites. Foods rich in selenium produce toxicity in both humans and animals. Rosenfeld and Beath (1964) showed that chronic selenium poisoning (alkali disease) can be produced in pigs fed on selenium-rich corn-based diets. In humans symptoms of selenium poisoning include dermatitis, fatigue, dizziness, and loss of hair.

*Lecythis ollaria* (monkey's coconut) and certain species of *Astragalus* have high ability to make selenium compounds, particularly seleno amino acids, such as selenocystathionine (Kerdal-Vegas *et al.*, 1965) and selenomethylcysteine (Trelease *et al.*, 1960).

Although crops containing either seleno amino acids or inorganic selenites are toxic, those containing organic selenium are far more toxic and are unacceptable for human consumption. It has been pointed out that a serious illegal practice has occasionally been employed in this country whereby grain known to be high in selenium content has been diluted with grain from nonseleniferous soils during milling.

To determine the levels of selenium in flour and grains, chemical analysis is preferable to bioassay. Bioassay may play a role in the development of varieties of grains that have weak abilities to synthesize seleno amino acids.

## D. Vitamins

This section concerns (1) the occurrence of excess vitamin intake and the possible resultant toxic effects and (2) the antivitamin effects of some foodstuffs that cause described toxicity in humans and other animals.

### 1. Toxicity of the Vitamins

Vitamin toxicity is presently a subject of public concern because vitamins are freely available and, when consumed in excess, can cause toxicity. Because some qualify as toxic substances, consideration of their presence in a balanced diet is required.

**a. Fat-Soluble Vitamins.** *i. Vitamin A and β-Carotene.* In humans, vitamin A is obtained from both plant and animal sources. In food of plant origin the main source is carotenoid pigments, particularly β-carotene. β-Carotene is cleaved in the intestinal mucosa to vitamin A, which is stored in the liver and fat as palmitate ester. Dietary vitamin A is also obtained from milk, meat, and eggs.

Vitamin A toxicity due to ingestion of natural foodstuffs is rare. However, there are reports of toxicity produced by consumption of polar bear liver or of livers of large fishes such as shark, halibut, and cod that are capable of storing high concentrations of vitamin A (Nater and Doeglas, 1970). Diets producing levels of serum vitamin A greater than 120 µg% may be suspected of containing abnormally high amounts of vitamin A or agents (surfactants) that interfere with vitamin A metabolism.

Bioassays of toxic amounts of vitamin A are based on findings of weight losses in rats and bone fracture and increased rates of bone turnover in animals. In culture, toxicity of vitamin A is seen by excessive mucoid differentiation and secretion by tissues that normally keratinize. There is no account of vitamin A toxicity produced by ingestion of carotenoids. However, children may demonstrate xanthosis when they consume baby food rich in carotenoids, and adults may exhibit other signs of carotenemia. Pigmentation disappears when the dietary souce of pigment is removed (Josephs, 1944).

*ii. Vitamin D.* Unlike other vitamins, vitamin D, a sterol vitamin, can be made in the body. The biological reaction takes place in the skin by the conversion of 7-dehydrocholesterol to cholecalciferol (vitamin $D_3$) under the influence of sunlight. Conversion of vitamin $D_3$ to the active hormones occurs in the liver and kidneys. Vitamin D is plentiful in fish liver oils. In the northern latitudes, especially in cities, there is a necessity for an exogenous source of this vitamin; this is usually supplied commercially by irradiation of yeast ergosterol to ergocalciferol (vitamin $D_2$). Vitamin $D_3$ can be made by irradiation of foods of animal origin; these are then added to milk.

There is no report of occurrence of vitamin D toxicity from overexposure of sunlight. The source of vitamin D affects both its potency and toxicity; Harris et al. (1939) reported that ergocalciferol ($D_2$) has more cardiovascular and renal toxicity than $D_2$ from tuna fish oil. In pigs (Burgisser et al., 1964) and New World monkeys (Hunt et al., 1969), $D_3$ was found to produce more toxicity than $D_2$. In humans no such relative toxic effects were reported. The vehicle of administration is also an important variable in vitamin D potency and toxicity; Lewis (1935) reported that vitamin D is more potent in milk than when it is provided in corn oil. Vitamin D toxicity produced by cod-liver oil is also variable. This may be due to the vitamin A content of milk and fish liver oil, which influences the vitamin D toxicity. During the past two decades several cases of an idiopathic hypercalcemia of infants consuming a formula diet have been reported. This incidence of the syndrome reached a peak in England when milk was fortified with approximately 1600 IU/quart, and the problem subsided when fortification was reduced to 400 IU/quart (Hays and Hegsted, 1973). Vitamin D toxicity is commonly detected by measuring serum calcium; toxicity can be suspected with a level more than 12 mg/dl (Hays and Hegsted, 1973). Among other clinical findings, increase of blood citrate and cholesterol and decreased alkaline phosphatase (bone) activity are notable.

*iii. Vitamin E.* Vitamin E, in its most active form, $\alpha$-tocopherol, occurs in the lipid fraction of plants. Hypervitaminosis E is not common; however supplemental intake of vitamin E or vitamin E-rich materials (wheat germ and wheat germ oil) is suspected of causing some toxicities. In one human study, of subjects given a dose of 300 mg/day, several reported headache, nausea, fatigue, and blurred vision (King, 1949). Also, according to other reports gonadal functions were disrupted (Beckman, 1955). Other nonspecific degenerative changes were reported in animal systems with toxic doses of vitamin E (Ostwald and Briggs, 1966).

**b. Water-Soluble Vitamins.** For most water-soluble vitamins toxicity from food and diet is not known. However, external administration of large doses can give rise to toxic effects. Because the chemistry of the water-soluble vitamins is well known, they are best analyzed by chemical means.

## 2. Antivitamin Effects in Foodstuffs

**a. Antivitamin A Effects.** *i. Lipoxidase.* The enzyme lipoxidase is present in many foodstuffs. Its presence in raw soybean is important because of the great consumption of such beans by humans. This enzyme oxidizes and destroys carotene (Sumner and Douce, 1939). Shaw et al. (1951) showed that 30% or more ground soybean in the diet of dairy calves reduced the blood levels of both vitamin A and carotene.

*ii. Citral.* Citral is a common constituent of orange oil. When administered subcutaneously or by mouth to rabbits and monkeys, it was reported to cause damage to vascular endothelia, which could be prevented by vitamin A (Leach and Lloyd, 1956). Citral is present in foods like marmalade and fruit juices flavored with orange oil. Liener (1969, 1980) suggested that excessive consumption of such food products capable of damaging blood vessels may contribute to cardiovascular disease.

**b. Antivitamin D Effects.** The first antivitamin effect of any foodstuff was reported by Mellanby (1926), who observed the antagonistic action of certain cereals to vitamin D. Later, this action was found to be due to phytic acid (Harrison and Mellanby, 1939). Bruce and Callow (1934) had earlier described the inhibition of calcium absorption by phytic acid. Since a main effect of vitamin D is to promote calcium absorption by stimulating the biosynthesis of calcium-binding proteins, agents that interfere with calcium uptake or synthesis of certain intestinal proteins will have antivitamin D activity.

A number of naturally occurring substances have such antivitamin D effects. The appearance of rickets and growth inhibition in New Zealand lambs fed green oats and other green feeds during winter months was reported in 1944 by Fitch and Ewer. These experiments were confirmed in lambs and guinea pigs (Ewer and Bartrum, 1948; Ewer, 1950). Vitamin D administration cured the sickness. Grant (1951) suggested that the $\beta$-carotene present in green oats was probably responsible for this antivitamin D effect. Another antivitamin D compound was isolated from the stems and leaves of certain fresh vegetables by Raoul *et al.*, (1957). The antivitamin D property of pig liver was also demonstrated (Coates and Harrison, 1957). A rachitogenic effect of unheated soybean meal or soy protein was observed by Carlson *et al.* (1964a,b). Autoclaving the meal decreased its rachitogenic property. Increased vitamin D supplementation prevented the rachitogenic action of the unheated product.

Although the activity of crystalline vitamin $D_3$ is usually determined by bioassay in rats fed a rachitogenic diet, vitamin D toxicity is not readily assessed by bioassay. Whereas vitamin D is assayed in rachitogenic animals on a defined calcium- and phosphorus-containing diet, the toxicity of vitamin D is even more strongly dependent on the dietary calcium and phosphorus contents. Thus, when excess vitamin D is provided in corn oil, it is not necessarily occurring together with excess calcium intake. However, if excess vitamin D intake is due to excess intake of milk, it is almost certain to be occurring in the presence of excess intake of calcium.

**c. Antivitamin E Effects.** Vitamin E prevents encephalomalacia, muscular dystrophy, liver necrosis, dialuric acid-induced hemolysis, and fetal resorption. Among the dietary constituents that inhibit these actions of vitamin E the following are noteworthy:

1. High content of legumes such as beans, peas, and alfalfa.
2. High amounts of polyenic acids.
3. Low levels of dietary sulfur amino acids and selenium.
4. Very high amounts of vitamin A.
5. High levels of nitrites, nitroso derivatives, and some oxidant in the diet.

The behavior of the antivitamin E factor in kidney beans *(Phaseolus vulgaris)* causes an increased incidence of liver necrosis in rats and muscular dystrophy in chicks and lambs (Hogue *et al.*, 1962; Hintz and Hogue, 1964, 1970; Desai, 1966). The incidence of muscular dystrophy was rectified by vitamin E supplementation. *Pisum sativum* was shown to contain a factor that interfered with the effectiveness of vitamin E in prevention of fetal resorption in rats (Sanyal, 1953). In encephalomalacia of chicks the utilization of vitamin E from alfalfa was shown to be interfered with (Singsen *et al.*, 1955). Alfalfa *(Medicago sativa)* contains an ethanol-soluble fraction that may interfere with the absorption of vitamin E (Pudelkiewicz and Matterson, 1960).

**d. Vitamin A and Vitamin E Interaction.** The possible sparing action of vitamin E on body stores of vitamin A has been under discussion for a long time. In the early observation of Davies and Moore (1941), depletion of large hepatic vitamin A stores was accelerated in the absence of vitamin E. After 222 days vitamin E-deficient rats also became vitamin A deficient, whereas hepatic vitamin A reserves were still present in vitamin E-sufficient animals. These early experiment were confirmed later on by Cawthorne *et al.* (1968). Jenkins and Mitchell (1975) also showed the sparing actions of vitamin E using a large excess of tocopherol together with massive doses of vitamin A. A study on humans was done in India (Kusin *et al.*, 1974) in which massive doses of vitamin A were given to children at 6-month intervals. Supplemental vitamin E (500 mg) was also given to one group along with vitamin A. Supplemental vitamin E increased the absorption of vitamin A from 67 to 82%. Similar results were obtained by Arnrich and Arthur (1980) in animal systems.

**e. Antivitamin K Effects.** The best-known antivitamin K substance is dicumarol, which was isolated by Stahmman *et al.* (1941) from spoiled sweet clover hay. Farm animals that had eaten sweet clover had developed a fatal hemorrhagic condition known as sweet clover disease (Kingsbury, 1964). Dicumarol, administered orally to humans or animals, lowers the prothrombin levels of blood, which leads to the dysfunction of the clotting mechanism. Bioassays for dicumarol-related substances have usually been based on "prothrombin time" (Quick, 1957) and have been very reproducible. Newer methods, based on interference with $CO_2$ uptake into the $\gamma$-carboxy group of some peptide-bound glutamate residues are accurate and rapid (Friedman et al., 1977).

**f. Antithiamine Effects.** Green (1937) first demonstrated that Chastek paralysis among silver foxes on a raw fish diet was associated with vitamin $B_1$ deficiency. Although antithiamine factors occur predominantly in viscera of carp, they are also found in mny other freshwater and marine fish (Somogyi, 1949). The antithiamine compounds were also found to be present in various plants, as has been reviewed by Liener (1969, 1980).

The antithiamine factors of bacterial origin were identified by a group of Japanese workers and were reviewed by Somogyi (1973). Antithiamine compounds present in foods are of several types: enzymes, naturally occurring reductants, and food additives. Fish muscle and certain mollusks contain heat-labile factors, thiaminases. Thiaminase I catalyzes an exchange of the hydroxyethyl methyl thiazole ring for other tertiary amines (trimethylamine) to yield substituted thiamine molecules that are inactive or are antagonistic to thiamine. Thiaminase II catalyzes the hydrolysis of thiamine into its methyl pyrimidine and thiazole portions. The presence of such activities can be determined only by bioassay. The organisms preferred are fungi (rhizopus and mucor) that can synthesize both thiazole and pyrimidine moieties of thiamine but cannot couple them. The toxic principles can be destroyed by boiling.

Among the heat-stable antithiamine factors are some polyphenols of plant origin, such as the tannins and caffeic acid. Similarly, some hemes of animal origin also inactivate thiamine. Some other preservatives or bleaches added to food (e.g., bisulfite) can split the thiamine, much as do the heat-labile thiaminases.

The antithiamine factor in foods consumed in northern provinces of Thailand has been recently reported by Vimokesant et al. (1982). Fermented tea leaves and betel nuts (tannins?) were shown to cause destruction of appreciable amounts of dietary thiamine. Thiamine deficiency was corrected by *removing* suspected antithiamine foods from the diet.

**g. Antiniacin Effects.** A characteristic symptom of niacin deficiency in dogs was observed when they were fed a *Sorghum vulgare* (called jowar in India) (Belavady and Gopalan, 1965). This is of significance for human nutrition, because pellagra is endemic in parts of India where people consume jowar. Jowar contains sufficient free nicotinic acid and also has a high leucine content. It is probable that leucine interferes with niacin metabolism because conversion of nicotinic acid to nucleotide forms is inhibited by leucine. In corn, niacin exists in a bound form, making itself unavailable to humans and animals. Treatment with alkaline solutions frees the bound niacin and renders it usable (Cravioto *et al.*, 1952; Kodicek and Wilson, 1960). The main antiniacin effect seen (through bioassay) in animals eating crops related to corn or sorghum is due to low levels of tryptophan (a precurser of niacin) coupled with an excess (imbalance) of leucine.

**h. Antibiotin Effects.** Raw egg white contains a protein, avidin, that is known to form a stable complex with biotin in the intestine and renders the vitamin biologically inactive (Eakin *et al.*, 1940). Gilgen and Leuthardt (1962) isolated a protein fraction from chicken liver, which also binds added biotin quickly and irreversibly. Creation of biotin deficiency in rats by the addition of avidin can be used as a bioassay for biotin. Animals maintained in such a marginally biotin-deficient state can be used to assay compounds with antibiotin activity. There is evidence that some pharmacologic agents act as antibiotins.

**i. Antivitamin B₆ Effects.** In plant and animal tissues, most vitamin $B_6$ activity is associated with pyridoxal and pyridoxamine and their phosphorylated forms. Dry heat used to sterilize some milk products and even extended autoclaving result in the destruction of the pyridoxal (it becomes bound to amino residues of proteins) and of the pyridoxamine (it becomes bound to the reducing ends of polysaccharides). In some cases, reactions beyond binding occur and antagonists may be formed from the Schiff bases. The antivitamin $B_6$ activity of some naturally occurring amino acids and their precursors is discussed elsewhere in this chapter (see Section II,C).

The growth of chickens was retarded when they were fed flax or linseed *(Linum usitatissimum)* meals (Kratzer, 1946); supplemental pyridoxine reversed or prevented the growth retardation (Kratzer and Williams, 1948). Klosterman *et al.* (1967) isolated a vitamin $B_6$ antagonist from flaxseed that was identified as 1-amino-D-proline (linatine). Sasaoka *et al.* (1976) reported marked changes in amino acid metabolism when rats were given 1-amino-D-proline. Amino proline reacts with pyridoxal to form a hydrazone, in which pyridoxal is tightly bound. In a recent report the bioavailability of vitamin $B_6$ from commonly consumed foods was discussed (Gregory and Kirk, 1981). Dietary fiber seems to have a weak inhibitory effect on the utilization of the vitamin.

## E. Hormones

Some food and feed crops contain sufficient amounts of hormonally active materials to influence or interfere with normal metabolism in individuals ingesting the food. This is especially true for crops used as staples. Our concern with these substances derives from the fact that their activity is best detected and measured by bioassay. The elucidation and bioassay of all food compounds affecting hormonal activity is a long-range aim, but already bioassay of foods or extracts containing these materials can provide much necessary information on this special class of toxic agents.

A large variety of pro- or antihormone activities have been described in foodstuffs. Perhaps those of most importance in nutrition–endocrine relations are (1) those that mimic some of the hormonal actions of ovarian-derived hormones, the estrogens, and (2) those inhibiting thyroid hormone biosynthesis, the goitrogens.

## 1. Estrogens

**a. Chemistry.** Estrogens are compounds of animal origin and are defined in terms both of their chemical structure and of their biological activity. In addition to the features common to all steroids (see the steroid skeleton below), important structural features of estrogens (estradiol, estrone) are the presence of an aromatic ring A (with a C-3 OH group) and hydroxl or keto functions in ring D. Whereas progesterone, glucocorticoids, and mineralo-

corticoids, all $C_{21}$ steroids, and the androgens, $C_{19}$ steroids, have a C-19 methyl group, estrogens are $C_{18}$ compounds lacking this functional group. Some $C_{21}$ and $C_{19}$ compounds demonstrate estrogenic activity *in vivo,* because they may be converted to estrogens.

**b. Biology.** Estrogens are required for the development and growth of female secondary sex characteristics, primarily the uterus and some associated structures as well as the mammary glands and structures associated with them in the breasts. These properties of estrogens lend themselves to ready quantitation, and they serve as the basis of bioassays performed under highly standardized conditions, using either immature or ovariectomized animals. Estrogens have other functions: they can regulate gonadotrophic secretion in either sex and can induce the synthesis of hormone-specific proteins in females. These properties are not generally exploited for bioassay purposes. Estrogens also have important functions in males, some of which are synergistic with those of the male hormones and others that are antagonistic to them. At high levels of estrogen intake, all functions are antagonistic. Although estrogen intake by males is a very serious health problem, bioassays based on male animals have generally not been used, because the response to estrogens is strongly dependent on endogenous androgen levels. Standardizing animals in this respect is difficult.

**c. Food Sources.** Most of the estrogenic compounds of food are derived from plant sources (Bradbury and White, 1954; Liener, 1969, 1980; Stob, 1973; Labov, 1977). Among the plants commonly used for human food, the following have been reported to contain substances with estrogenic activity in experimental or farm animals: carrots (Ferando *et al.*, 1961), soybeans (Carter *et al.*, 1955; Naim *et al.*, 1973; Lockhart *et al.*, 1978; Drane *et al.*, 1980), wheat, rice, oat, barley, potatoes, apples, cherries, plums (Bradbury and White, 1954), wheat bran, wheat germ, rice bran, and rice polish (Booth *et al.*, 1960). Estrogenic activity has also been reported in such vegetable oils as cottonseed, safflower, wheat germ, corn, linseed, peanut, olive, soybean coconut, and refined or crude rice bran (Booth *et al.*, 1960).

Estrogens of purely animal origin may be present to some extent in cow's milk, especially in colostrum; however, estrogens from this souce do not enter food products.

Bovine fat is normally nonestrogenic, but in animals, including the human female, treatment with certain estrogens produces storage of estrogen in the body fat. There is little danger that the consumption of milk or dairy foods provides estrogens in sufficient quantity to produce physiological responses in humans.

Estrogens of plant origin contribute to infertility to animals, especially livestocks that graze on certain clovers (Moule *et al.*, 1963), but for humans their occurrence in plants in most cases is only of curious interest. The plants containing such compounds are diluted out in a balanced or mixed diet. The quantity of estrogenic compounds in foods is too low to exert physiologic effects. However, in recent years concern has been expressed because of the increased usage of soybean for human consumption, especially since the discovery of various estrogen compounds in soybeans (Drane *et al.*, 1980).

**d. Form of Estrogenic Compounds in Foods: Isoflavones and Coumestans.** Most estrogenic compounds in food are chemically identified as isoflavones and coumestans and occur naturally as glycosides. These estrogenic compounds warrant scrutiny, not only because of their reported occurrence in food used for human consumption but also because of their physiologic effect on animals.

The isoflavones and coumestans are biogenetically similar; that is, they are synthesized by similar biosynthetic pathways. The isoflavones usually occur in plants in a bound form, probably glycosidic. The sugar moiety is attached to one or more of the hydroxyl groups located at various positions of the isoflavone nucleus (Liener, 1969).

Isoflavones have been isolated from soybeans (Carter *et al.*, 1955; Nilson,

1962; Naim *et al.*, 1974). In addition to estrogenic activity, soybean isoflavones also inhibit lipoxygenase activity and exert and antihemolytic effect on erythrocytes subjected to peroxidation (Naim *et al.*, 1976). This antioxidant effect is shared by other plant phenols (flavones, anthocyanins) not having estrogenic activity. Because antioxidants protect against rancidity, carotene decomposition, etc., it is important that *both* the estrogenic and antioxidant properties of soybean be assayed.

*i. Genistein.* Genistein has been isolated from the soybean (Walz, 1931). Genistein is heat stable and weakly estrogenic in nature. When administered orally, it is much less potent than either diethylstilbestrol or estrone in causing uterine hypertrophy in immature and intact mice (Stob, 1973; Bickoff *et al.*, 1962). It stimulates protein synthesis in the uterus of ovariectomized rats (Noteboom and Gorski, 1963) and can displace estradiol from receptor sites of uterine tissues (Shutt, 1967). It has been reported that, in sheep, genistein circulates as a glucuronide in the blood and is rapidly excreted (Shutt *et al.*, 1967). The presence of genistein and its estrogenic activity has not been studied extensively in humans. Although soybeans are widely used in human nutrition and have estrogenic action due to genistein and other compounds, only two studies have explored the problem. Wada and Fukushima (1963) reported the estrogenic activity of soybeans as 6 µg of diethylstilbestrol equivalent per kilogram, and Walter (1941) reported that hexane-extracted soybeans contain 0.1% glucoside genistein.

A newly described isoflavone, glycitein, has also been reported in soybean (Naim *et al.*, 1973).

*ii. Biochanin A.* Biochanin A, an isoflavone derivative, has been isolated from Bengal gram *(Cicer arietinum)* and red clover *(Trifolium pratense).* Its activity is heat labile (Siddiqui, 1945); perhaps this explains the observation that drying clover causes a loss in its estrogenic activity (Alexander and Watson, 1951).

*iii. Coumestrol.* Coumestrol, an estrogenic compound, was first isolated from alfalfa (Bickoff *et al.*, 1962). It was also found in soybeans and soybean sprouts (Wada and Yuhara, 1964). The reported levels of coumestrol in soybean are 0.05 – 30 µg/g (Lockhart *et al.*, 1978). Coumestrol is heat stable. Bickoff *et al.* (1969) have summarized the chemical and biological properties of coumestrol. It is more potent than genistein in stimulating the uterine hypertrophy in animals yet much less potent than diethylstilbestrol or estrone. It can inhibit hypophyseal gonadotrophin function (Leavitt and Wright, 1965) and can also stimulate protein synthesis in the uterus of oophorectomized rats (Noteboom and Gorski, 1963).

*iv. Zearalenone.* Zearalenone, a weak estrogen, is a $\beta$-resorcylic acid lactone produced particularly by *Fusarium roseum* acting on corn, wheat, barley, oats, etc. (Mirocha *et al.*, 1977). Zearalenone, like estradiol and diethylstilbestrol, induces biphasic alterations in serum gonadotrophin levels in oophorectomized monkeys. It also promotes estrus in adult mice. Zearalenone has importance because its estrogenic activity is associated with the presence of a lactone that may induce cell proliferation. Bioassays for both activities need to be done, especially on stored grains.

## 2. Goitrogens

**a. Chemistry.** The goitrogens are of several types and have different chemical activities. Therefore, several different assays may be required to differentiate the various goitrogenic activities. For example, many food plants contain either thiocyanate anions, thiocyanate esters, or thiocyanate glycosides. After ingestion, these compounds become part of the thiocyanate anion pool of the body (Montgomery, 1969, 1980; Van Eatten, 1969).

**b. Biology.** Thiocyanate ($SCN^-$) inhibits thyroid hormone biosynthesis, because it competes with iodide ion ($I^-$) for the iodine-concentrating system in the thyroid gland. An additional source of thiocyanate is dietary cyanide intake. Such cyanide is usually in the form of cyanogenic glycosides, of which laetrile has become the most familiar. Cyanide rising from the metabolism of dietary precursors is converted to thiocyanate, which enters the thiocyanate pool. Bioassays employing radioactive iodide anion can determine whether food ingredients inhibit iodide uptake by the thyroid gland. Bioassay can also determine whether radioactive iodine once taken up by the gland is released after the test animal ingests the dietary ingredient being studied. Bioassay can determine whether the radioactivity is present as iodide or as organic iodine.

Other classes of antithyroid agents are present in foods, and these activities can be detected by specific biological and chemical assays. Some goitrogens inhibit or reverse the second step in thyroid hormone biosynthesis, as summarized following:

Step 1:        Concentration of $I^-$ by thyroid (competitively inhibited by $SCN^-$)

Step 2:                      $I^- + H_2O_2 \xrightarrow{\text{peroxidase}}$ enzyme-bound $I^0$, $OI^-$

Goitrogens of the thiopyrimidine, thiourea, or thiooxazolidone type found in some cruciferous plants inhibit the forward reaction by consuming the peroxide, as follows:

$$\underset{\substack{|\\O(N)\\R'}}{HS-C=NR} + H_2O_2 \longrightarrow \left[\underset{\substack{|\\O(N)\\R'}}{\cdot S-C=NR}\right]_2 + 2\,H_2O$$

where R and R′ are parts of the carbon chain. Goitrogens also decrease the half-lives of the enzyme-bound oxidized products by chemically reducing them:

$$\text{Enzyme-bound OI}^- + 2\text{ RSH} \rightarrow (\text{RS})_2 + H_2O + I^-$$

Some goitrogens inhibit the deiodination of thyroxine (T4) to the more active triiodothyronine (T3), whereas others do not. Therefore, the body fluids of animals suspected of being fed goitrogens need to be analyzed for relative amounts of T4 and T3. Additionally, other bioassays (heat production) and the thyroid biochemical assays previously described are required in order to determine the mechanism of action of dietary goitrogens involved in thyroid hormone synthesis or function.

**c. Food Sources.** Some goitrogens present in pasture crops and weeds, or present in some seed meal extracts fed as supplements to dairy animals, are present in the milk of these animals and therefore are present in some dairy products.

Goitrogenic activity is clearly a toxicity of some foodstuffs. Ways of preventing or counteracting the toxicity depend importantly on understanding the modes of their actions. If a goitrogen is determined by bioassay to inhibit only the first step in thyroid hormone synthesis, it might be possible to counteract its toxicity simply by increasing the iodide supplement for both livestock and humans. If bioassay determines that goitrogenic activity of a food is due to inhibition of later stages of T3 synthesis, animals might receive T3 supplements; however, such a solution would be unacceptable for humans. In such a case, limits of goitrogen present in food would have to be established, even if this required special processing to remove goitrogens from constituents destined for human consumption.

## F. Natural Toxins

This section concerns problems related to the analysis of some microbial and seafood toxins that are closely associated with human foods and animal feeds. The toxicity problems are reviewed in detail (Riemann and Bryan, 1969, 1979; National Research Council, 1973; Schantz, 1974; National Academy of Sciences, 1975; U.S. Department of Health and Human Services, 1982).

### 1. Microbial Toxins

Microbial toxins are of bacterial and fungal origins. The amount of toxin produced depends on temperature, pH, moisture, and incubation time.

**a. Bacterial Toxins.** There are three types of food-borne bacterial diseases that result from bacterial growth or the production of toxins: (1) those that are in foods before they are eaten (e.g., staphylcoccus and botulinum), (2) those that grow in the intestine and other sites of the infected host (e.g., salmonella and shigella), and (3) those that are due to a dual mechanism (e.g., *Clostridium perfringens, Bacillus cereus,* and *Vibrio parahaemolyticus.*)

Staphylococci are recognized by staining characteristics (gram positive), morphology of dividing cultures (grapelike clusters), and culture characteristics (nonmotile and nonsporulating). Three species of *Staphylococcus,* namely, *S. aureus, S. epidermidis,* and *S. saprophyticus,* are differentiated on the basis of coagulase and endonuclease production, aerobic and anaerobic mannitol fermentation, and cell wall composition (Baird-Parker, 1974).

Illnesses resulting from staphylococcus enterotoxins are common and widespread; however, the true incidence of staphylococcal food poisoning is unknown, because staphylococcal enteritis is not a reportable disease in the United States and other countries (Bergdoll, 1979). Cooked protein-rich foods such as ham, poultry, beef, fish and shellfish, and eggs and milk are implicated in outbreaks of staphylococcal enteritis (de Figueiredo, 1970). However, organisms present before processing of foods are involved in outbreaks of food poisoning only if the cooking process is inadequate. Postprocessing contamination and proliferation is more often the causative factor (Bergdoll, 1979).

Enterotoxins are proteins (Bergdoll *et al.,* 1959), and their antigenic properties are used in their detection and analysis (Surgalla *et al.,* 1954). Five enterotoxins have been found on the basis of their reaction toward specific antibodies A through E, which are well characterized (Bergdoll, 1979). Staphylococcus enterotoxins are much more stable than most proteins. The active site is relatively resistant to proteolytic enzymes (such as trypsin). They have relative heat stability.

The detection and quantitation of enterotoxins are currently accomplished by immunologic techniques based on their reactions toward specific antibodies (Silverman *et al.,* 1969; Casman and Bennett, 1965; Hall *et al.,* 1965; Genigeorgos and Sadler, 1966). The most frequent method for the detection of enterotoxins in unknown materials is by Ouchterlony plate technique (Bergdoll *et al.,* 1965). A single-gel diffusion tube test of Oudin has also been adapted (Hall *et al.,* 1965; Read *et al.,* 1965). The most sensitive of the double-gel techniques employs microslides. It can be used to detect the presence of the toxin and also to estimate the toxicity of a particular bacterial strain (Casman and Bennett, 1965; Hall *et al.,* 1965). Solid-phase radioimmunoassay is sometimes used for the detection and assay of enterotoxins in culture supernatants and in food extracts (Collins *et al.,* 1972, 1973; Johnson *et al.,* 1971, 1973).

Botulism is a rare yet highly fatal food-borne disease. The causative agent is a gram-positive anaerobic bacillus, *Clostridium botulinum*. This organism is one of a group of heat-resistant endospore-forming bacteria that germinate to form rod-shaped structures during their vegetative stage. The vegetative structures proliferate and produce exotoxin, utilizing food sources (such as canned food) as a substrate. Many foods (canned fruits and vegetables, cheese spreads, smoked and canned fish, pot pies, etc.) provide substrates for the production of botulin, the botulinum toxin. Low pH inhibits germination of spores and vegetative growth; therefore, acidity regulates toxin production.

Although several types of botulism bacteria are recognized, almost all of them produce a specific neurotoxin that causes typical botulism. Immunologic properties differentiate one bacterium from the other (Wilson and Hayes, 1973; Sakaguchi, 1979). Neurotoxicity produced by the organism is due to the inhibition of acetylcholine exocytosis, and this can be used to bioassay the activity of suspected toxins.

Identification of bolutin is usually dependent upon the very sensitive mouse test. Clear supernatants of food extracts are injected intraperitoneally into mice at doses of 0.5 ml. Neurologic signs may appear in 1 to 24 hr, depending upon the toxin concentration (Sakaguchi, 1979). Recently, immunologic techniques, including the reversed passive hemagglutination techinique, have been introduced (Evancho *et al.*, 1973). Except for reversed passive hemagglutination, none of the immunologic techniques is as sensitive as the mouse assay.

**b. Fungal Toxins (Mycotoxins).** Since the discovery of aflatoxin in 1960, there has been increased concern over the health hazards resulting from the toxigenic molds. These compounds are generally classified under the name mycotoxins. Mold spores are ubiquitously distributed, and foods are readily contaminated with them. Spore germination and mold growth are determined by various factors such as relative humidity, moisture content, and temperature (Busby and Wogan, 1979). A major mycotoxin is aflatoxin, produced by *Aspergillus flavus* (Sargeant *et al.*, 1961; Schoental, 1967; Busby and Wogan, 1979). Foods usually contaminated are peanuts, legumes, rice, corn, barley, cottonseed meal, cereal foods, fruits, etc. (Golumbic and Kulik, 1969).

Physicochemical (Wilson and Hayes, 1973) and biological assays (Wogan *et al.*, 1971; Busby and Wogan, 1979) have been developed for determination of aflatoxins.

Although aflatoxins produce a variety of serious diseases, including liver tumors in humans and in wildlife and experimental animals that eat aspergillus-rotted peanuts, beans, etc., chemical identification and analysis are more useful than bioassay, because the latter requires long periods of feeding in animals. Aflatoxins are lactones that acylate host tissues. The significance of this property is that aflatoxins can be hydrolyzed to relatively nontoxic products by moist heat in the presence of mild base. Also, aflatoxin activity can be detected by transforming and promoting activity against cells in culture. This may form the basis of a bioassay in the near future.

A number of other molds have been found to produce toxic metabolites (e.g., ochratoxins and sterigmatocystins) (Busby and Wogan, 1979). Like *Aspergillus, Penicillium* has various species that are toxin-producing food contaminants. Cyclochlorotin, luteoskyrin citrinin, citromycetin, and citreovididin are important mold-produced toxins (Miller, 1973; Wilson and Hayes, 1973; Sakabe *et al.*, 1964; Busby and Wogan, 1979).

An important toxic metabolite (a furan derivative) has been reported to be produced by mold-damaged sweet potatoes (Wilson *et al.*, 1970). Identification and assay of mycotoxins affecting this crop are urgently needed. The sweet potato is an economical and highly nutritious crop. Fungal infection of the stored crop itself and of its dehydrated and processed forms limits the utility of this potentially major agricultural resource.

Among human mycotoxicoses, ergotism probably was the first one described. The syndrome has its origin in the ingestion of infected rye or other cereals or flour made from these and of forage grasses (Van Rensburg and Altenkirk, 1974). The characteristics of ergot poisoning can be classified into two distinct types, gangrenous and convulsive. Ergot alkaloids are smooth muscle stimulants and therefore cause vasoconstriction. This property is responsible for the gangrenous aspects of ergotism. Convulsive and hallucinatory aspects of ergotism are caused by neurohormonal effects of serotonin antagonism and inhibition of both the stimulatory and inhibitory actions of epinephrine by ingested ergotamines.

Although the food-related problem of ergotism has largely been eliminated for humans, contamination of certain types of grasses still causes problems for domestic animals.

Ergot derivatives are of importance in medicolegal problems relating to the use of these compounds for hallucinogenic or abortifacient purposes. The vehicle of administration of these compounds is frequently deliberately contaminated sugar or contaminated cookies. Chromatographic separation followed by chemical identification, though difficult, is superior to bioassay.

## 2. Seafood Toxins

In addition to botulism, another serious problem existing in canned fish is that of seafood toxins. These toxins are produced either by algae or bacteria, which are then transmitted through the food chain, or intrinsically by fish and mollusks.

**a. Toxins Produced by Algae or Bacteria.** Marine food toxicants produced by microorganisms such as algae are difficult to control, because conditions that cause sporadic algae "blooms" are not fully understood.

*i. Shellfish Poisoning.* Consumption of shellfish from some coastal areas and brackish waters can result in paralysis and death in humans and animals. Saxitoxin, the agent responsible for most shellfish poisoning, orginates in marine dinoflagellates, *Gonyaulax catenella* (Sommer and Meyer, 1937; Sommer *et al.*, 1937; Schantz, 1974), and other dinoflagellates (Taylor and Scliger, 1979). Shellfish (mainly clams and mussels) acquire the toxin by ingestion of the dinoflagellates. Mussels bind the poison without themselves becoming toxic; however, the toxin is readily released after the mussel is consumed by humans.

The poison was isolated from California mussels and Alaska clams (Schantz *et al.*, 1957; Mold *et al.*, 1957). The clam–mussel poison is a substituted tetrahydropurine (Wong, 1971) that functions as a powerful neurotoxin. Saxitoxin blocks propagation of nerve impulses and skeletal muscle response by interference with sodium permeability.

The United States Food and Drug Administration (FDA) has set standards of acceptance for the paralytic poison of not more than 400 mouse units (MU) or approximately 80 $\mu$g per 100 g of shellfish meat. The official method of assay was developed by the Association of Official Agricultural Chemists (Schantz, 1974). The poison is extracted from the shellfish and serial dilutions are injected intraperitoneally into white mice weighing 19–21 g. From the death times the amount of poison in 100 g of meat is calculated.

Other types of shellfish poisons have been reported, and these have been reviewed (Schantz, 1974; Taylor and Seliger, 1979).

*ii. Cigutera Poisoning.* Cigutera poisoning is the most prevalent public health problem involving seafood. Almost all fishes involved in this type of poisoning are palatable shore species. The poison is found in the liver and other viscera and in the muscle. Although 15–20 species of fish are known to cause toxicity (sea bass, snapper, barracuda, grouper, eel), as many as 300 species may actually be involved. Detection and assay techniques are similar to those employed for the other shellfish poisons.

**b. Intrinsic Fish Toxins.** Some species of fish are intrinsically poisonous, and their toxins are associated with their defense mechanisms. In many cases the toxic principles have been isolated, characterized, and assayed.

Puffer fish poisoning has long been recognized in China and Japan. The most poisonous species belong to the family Tetraodontidae, and the poison is known as tetrodotoxin. The toxin has received much notoriety because of its use as a secret weapon by intelligence agencies. Because of this special use of the toxin, very much is known of its chemistry and pharmacology. Tetrodotoxin is an aminoperhydroquinazoline (Tsuda *et al.*, 1964; Woodward, 1964). Its physiological effects and the bioassay procedure are similar to those of saxitoxin (Schantz, 1974).

## III. METHODS OF BIOASSAY

### A. Selection of a Test System

The selection of a test system is based upon the type of information desired. If it is desired to determine the amount of a specific substance (e.g., thiamine) present in a sample, the answer can be obtained with certainty only by chemical analysis. We could approximate these analyses using bioassays, for example, by measuring the amount of labeled cocarboxylase formed from $^{32}$P-ATP in the presence of the appropriate kinase, or we could measure the amount of growth of a thiamine-requiring mutant microorganism. However, the absolute amount of thiamine present may not be the appropriate question to ask. We may, on occasion, be more interested in the thiamine activity present in a food sample. If the sample contains thiamine monophosphate, thiamine diphosphate, and other phosphorylated derivatives (all of which have vitamin $B_1$ activity), chemical analysis might not give us the type of information we desire. The enzymatic assays would also underestimate the amount of vitamin $B_1$ activity present. Bioassays using either animals or mutant microoganisms would yield information appropriate to the question.

From the problems related to thiamine analysis, another type of relationship between chemical and bioassays is evident. Just as chemical analysis for thiamine in a food might underestimate the total vitamin $B_1$, there are frequent occasions when analysis of thiamine might grossly overestimate its activity. If a food contained an antagonist of thiamine, then the amount of the vitamin found by chemical analysis would overestimate its biological activity. In addition to the antithiamine discussed earlier, foods might be contaminated with pesticides that increase the thiamine requirement, and feeds may contain added antibacterial agents (related structurally to thiamine) that interfere with thiamine analysis. The foregoing discussion demonstrates that the type of information desired determines the choice of test system and that roles exist for both chemical and biological analysis.

In selecting among bioassays for use in determining a suspected substance in a food, consideration should be given to that test system that best describes the biological activity of the agent. For example, if a grain is suspected to be contaminated with a naturally occurring coumarin or to be contaminated with a coumarin-containing pesticide, bioassays should be done in addition to chemical assays. A primary bioassay would measure the influence of food extracts on clotting mechanisms in rodents maintained on defined diets and fed graded doses of vitamin K. The clotting times of animals fed the grain as the sole foodstuff would also be carried out. Of somewhat less importance would be enzymatic bioassays for inhibition of carboxylation of glutamyl residues; still less significant would be bioassays to measure inhibition of oxidative phosphorylation, although all of these assays measure antivitamin K activity.

## B. Microbiological Analysis

Microbiological analysis is employed when a microbial response to a food constituent or its metabolite is more sensitive or specific than physical and/or chemical determinations. In microbiological assays, the selected microorganism is used as a reagent to measure the activity of the test material. An ideal test organism should have four primary characteristics: (1) it should be sensitive to the substance being assayed; (2) it should be easily cultivated; (3) it should have some metabolic function or response that is measurable; and (4) it must not be readily susceptible to variation in either sensitivity or phase. In addition, it should have specificity and, preferably, be a nonpathogen (Gavin, 1956).

If a culture medium (the basal medium) supports no growth in the absence of a substance and full growth with it, the plot of growth versus concentration of the substance should be linear over an appreciable range of concentration. Concentration of an unknown substance is obtained as the ratio of the slopes of the linear part of the unknown and the known curves (Hutner et al., 1958).

Since the discovery of auxotrophic mutants of *Neurospora* (Beadle and Tatum, 1941; Beadle, 1945), mutants of fungi, bacteria, and protozoa have been used to elucidate the biosynthetic pathways of amino acids, vitamins, and other metabolites. Lactobacilli are widely used for vitamins and amino acid bioassays, because their nutritional requirements are well known. They grow well in synthetic and semisynthetic media and produce good amounts of lactic acid, which can be determined by titration or chemical analysis. In aerobic and microaerophilic conditions, they grow well and are nonpathogens. The validity of some lactobacillus assays have been established by use of other assay organisms, by repeat assay, or by independent chemical assays. *Saccharomyces carlsbergenesis* is used for the determination of total vitamin $B_6$ activity, because it responds to pyridoxine, pyridoxal, and pyridoxamine (Freed and Association of Vitamin Chemists, Inc., 1966).

The principles and basic requirements of microbiological assay are described in several articles (Gavin, 1956; Hutner *et al.*, 1958; Freed and Association of Vitamin Chemists, Inc., 1966). The general techniques need the following considerations.

## 1. Apparatus

**a. Glassware and Pipettes.** Screw cap glass tubes are recommended for maintenance of stock cultures to prevent media from drying out. Use of a glass heating tape for unfreezing joints is recommended (Moos and Chen, 1957). Amber glass containers should be used for media containing light-sensitive compounds. For aerobic organisms, flasks are preferred over the glass tubes.

For microbiological experiments, cotton-plugged pipettes are used. These and other glassware are sterilized by dry heat or are purchased already sterile.

**b. Sterilization.** Autoclaves are usually used for the purpose of sterilizing media, although filtration of labile ingredients is frequently carried out. The time and pressure may vary for different purposes. For example, sterilization for media is usually performed at a pressure of 15 lb/in.$^2$ for 2–15 min. The shorter period is preferred for preventing carmelization. Media prepared for rapidly growing organisms are usually sterilized at atmospheric pressure for 10–15 min.

## 2. Procedure

**a. Induction and Selection of Mutants.** Nutritional mutant auxotrophs are extremely important as analytical reagents. The efficiency of selection of mutants has been improved manifold with the Davis–Lederberg penicillin technique. In a penicillin-sensitive population, penicillin kills dividing cells. In a bacterial population, 99.9% are killed by a dose of ultraviolet irradiation. The survivors are then allowed to grow in a basal or minimal medium with penicillin. Penicillin is then removed, the survivors are seeded heavily in miminal agar, and the substance for testing is applied to the agar. A good portion of the cells that require this compound will grow.

**b. Microbial Culture.** Most bacteria can be preserved well under lyophilized conditions (Weiss, 1957). Lyophilized cultures can be obtained from American Type Culture Collection.* This technique is useful when microbiological assays are performed only periodically, because it eliminates the necessity of frequent transfers of cultures.

Lactic acid bacteria are usually maintained in stab cultures in yeast–glu-

* 12301 Parklawn Drive, Rockville, Maryland 20852.

cose – agar media. Stabs are incubated at 30 – 37°C for 18 – 48 hr or longer to obtain a visible line of growth.

Yeasts and molds are grown on agar slants. Inocula are prepared 6 – 18 hr prior to the start of an assay by transferring a stab culture to an inoculum broth.

**c. Culture Media.** In preparing an assay medium from natural material, it is necessary to remove quantitatively the substance to be assayed for. For example, in a classical riboflavin assay medium, riboflavin is removed using light and alkali, which destroy the vitamin. However, for assaying unknown compounds, this procedure is not always practical and therefore semisynthetic assay media are used. The natural (nonsynthetic) part of the medium, such as casein hydrolysate, is stripped (prior to hydrolysis) of vitamins, purines, salts, amino acids, etc.; except for the factor to be assayed, these factors are then added back.

The ingredients of the assay medium are dissolved in water and adjusted for pH, and the solution is diluted so that the concentration is twice that desired in the assay tubes.

**d. Measurement of Growth.** Generally, microbial procedures that are used for analysis depend upon the growth response of the microorganism to the environment. The growth may be measured by numerical counts, by optical density, by weight, or by area. The response may be graded in proportion to the concentration of the test material or by a definite end point.

*i. Turbidimetric Method.* The turbidimetric method has come into favor because it is rapid and accurate. It is also sensitive in comparison to the other available methods, such as agar-plate technique, where there is a hindered diffusion of nutrient in the solid media. Usually the turbidity is measured colorimetrically or nephelometrically.

*ii. Paper Disk Method.* The paper disk method is used for various purposes, such as for vitamin assays (Jukes and Williams, 1954; Simpson, 1956). Standard solutions or test extracts are pipetted on paper disks, which are placed on solid assay medium. Measuring the zones produced by known amounts of standard permits quantitative determination of unknowns. This method has the advantage of being rapid and allowing analysis of a large number of samples.

*iii. Metabolic Response.* In response to dietary ingredients, certain microorganisms produce measurable metabolites or undergo changes in some metabolic function that can be measured. Carbon dioxide production, oxygen uptake, nitrate reduction, and luminescent and antiluminescent activity have been used for bioassay.

**e. Calculations.** *i. Classical Technique.* A standard curve is drawn by plotting concentrations of the unknown substance against the optical density (O.D.) of the culture, using uninoculated and unsupplemented tubes as blanks. *Titer values* are determined by interpolation of the standard curve using the observed O.D. as the reference.

*ii. Linearization of Vitamin Standard Curves.* Kavanagh (1977) observed that nonlinear standard curves are much more frequent than straight lines in microbiological turbidimetric assays for growth-promoting substances. Therefore, he adopted new equations (analogous to log dose–response curves used in pharmacology) that yield superior results to those obtained using the older formulation.

### 3. Radiometric Microbiological Analysis for Biologically Active Compounds

While it is essential that the test organism require the unknown substance that is to be assayed, this of itself may not make it suitable for practical work; there are some other limitations of a microbiological assay. Some organisms can synthesize certain vitamins from derivatives or breakdown products, whereas humans may not be able to do so. For example, a processed food may contain some forms of thiamine (dihydrothiamine, thiamine disulfide, split products of sulfitolysis) that may be utilized by some mutant fungi for thiamine synthesis but cannot be used by humans. In such a case the bioassay would *overestimate* the actual thiamine content of the food. In a related situation a bioassay may *underestimate* the nutrient content of a dietary extract.

To alleviate some of the problems inherent in bioassays involving growth of microorganisms, there have been developed new methods using radioactive substances to measure the biologically active test materials. Some examples are listed below.

**a. Vitamin B$_6$.** A radiometric microbiological assay of vitamin B$_6$ has recently been reported (Guilarte and McIntyre, 1981) and extended to determine the contents of total vitamin B$_6$ or its specific forms in various food items (cereals) using high-performance liquid chromatographic (HPLC) techniques (Guilarte *et al.*, 1981).

The radiometric analysis for vitamin B$_6$ activity is based upon an increase in the activity of B$_6$-containing enzymes in a culture of B$_6$ auxotrophs, usually yeasts. The test culture system contains [$^{14}$C]carboxyl-labeled valine and graded doses of the food extracts. As B$_6$ activity of the extracts is increased, growth increases, as does the activity of B$_6$-containing amino acid decarboxylases and transaminases. Decarboxylation of valine or $\alpha$-ketovaline releases $^{14}$CO$_2$, which is trapped and counted. The system is an improvement over some other methods, but it also has some drawbacks. It may be influenced by the amounts of other branched-chain amino acids and requires control of cultural conditions, mainly pH. If the system increases in alkalinity, the organism preferentially deaminates the added radioamino acids. Under these conditions, $CO_2$ will tend to be fixed. If the medium is acidic, the organisms will tend to decarboxylate amino and keto acids at a greater rate than normal. Radiometric assays based upon the use of apo-$B_6$ enzymes may become useful in refining these assays.

    **b. Vitamin $B_{12}$.** Vitamin $B_{12}$ ($\alpha$-5,6-dimethylbenzimidazolylcobamide) is synthesized only by microorganisms. In nature vitamin $B_{12}$ enters the food chain because it is synthesized by intestinal bacteria and euglenoids that are ingested by larger organisms. Animals that are vegetarians obtain vitamin $B_{12}$ via one of two mechanisms: (1) they have rumen (compartmentalized stomachs with selective regurgitation) in which bacterial fermentations occur, or (2) they have enlarged ceca in which similar fermentations occur. In these cases, absorption of the bacterially produced vitamin $B_{12}$ can occur *in situ.* Monogastric vegetarian animals, such as most rodents, obtain vitamin $B_{12}$ by coprophagy or the reingestion of the products of the intestinal flora.

Humans obtain vitamin $B_{12}$ largely from the ingestion of animal products, including the small amounts present in milk as well as the relatively larger amounts present in eggs. A very small amount of vitamin $B_{12}$ may enter the diet via some types of bacterial fermentation products. A significant amount of vitamin $B_{12}$ enters the Western diet as a food supplement ($\alpha$-5,6-dimethylbenzimidazolylcobamide).

Microbiological assay for vitamin $B_{12}$ employs lactobacilli (Donner factor) and euglenoids, including *Ochromonas* (Baker and Frank, 1967; Skeggs, 1967). These methods are very useful in assaying vitamin $B_{12}$ present in food supplements or in animal tissues, because (1) media used for the growth assay are highly standardized and (2) the vitamin $B_{12}$ present is almost entirely true vitamin $B_{12}$ (i.e., $\alpha$-5,6-dimethylbenzimidazolylcobamide). However, some of the vitamin $B_{12}$ used for animal feed supplements is prepared from sewage or from concentrates of industrial fermentations. Based upon the presence of foreign nitrogen bases in these waste materials or perhaps as a result of genetic influences, some microorganisms synthesize vitamin $B_{12}$ analogs containing a different base in place of 5,6-dimethylbenzimidazole. For example, lactobacilli can synthesize a benzotriazole or nitrobenzimidazole (pseudo-vitamin $B_{12}$'s) when supplied with these nitrogen bases. Pseudo-vitamin $B_{12}$'s have variable activity in microbial assays, and they may be inert in the nutrition of monogastric mammals, somewhat less active than true vitamin $B_{12}$, or moderately antagonistic to vitamin $B_{12}$. To alleviate these problems, and to develop a bioassay that yields results that have relevance to nutrition, radioassays for vitamin $B_{12}$ have been developed (Lau *et al.*, 1965). They are based on methodology used to determine vitamin $B_{12}$ content of human serum and are modified for the R proteins (nonspecific binders) of vitamin $B_{12}$ analogs (Kolhouse *et al.*, 1978).

In the radioassay method, the test sample is deproteinized by heat and centrifuged, and the supernatant containing $B_{12}$ is converted to cyanocobalamin and used for further analysis. The assay reagents contain an intrinsic factor from which R protein has been removed, and that therefore does not bind pseudo-vitamin $B_{12}$. The intrinsic factor is bound covalently to polymer beads. The beads are mixed with graded amounts of $^{57}Co$-cyanocobalamin, and incubation of the mixtures occurs. After 2 hr the mixtures are centrifuged and the radioactivity in the pellet is measured.

## C. Tissue Culture Analysis

Tissue culture is a major and unique bioassay system that gives information, not available from whole-animal experiments, regarding the direct action of a reagent on cells or tissues. An important advantage of culture experiments over the whole-animal system is in their relative costs: the former are less expensive and require less time. The major limitation of culture experiments in contrast to the whole-animal system is their inability to yield information on systemic effects of the test materials.

The *in vitro* culture of animal cells poses problems not seen in the culture of most microorganisms. Some of these problems exist because tissue culture is a developing field and much methodologic research is still required. Some of the problems and limitations are intrinsic to the method and relate to the behavior of cells in culture. For example, bacteria are grown in pure culture, uncontaminated by other varieties or species of organisms. Animal cells are often contaminated with microorganisms that have intrinsically faster growth rates than do the animal cells themselves. Moreover, animal cells have more rigid nutritional and environmental requirements than do bacteria. The bacterial contamination of mammalian cell cultures is therefore a problem. However, the problem is prevented in part by the use of antibiotics in the culture medium.

Tissue culture deals with the *in vitro* culture of cells, tissues, or organs under controlled conditions. For growing cells, an environment approximating *in vitro* conditions is desirable; these have been reviewed recently (Jacoby and Paston, 1979). In addition to environmental and nutritional requirements (temperature, osmotic pressure, pH, $Pco_2$, $Po_2$, energy source, amino acids, etc.), cells in culture are influenced importantly by factors of biological origin. Two of these factors, hormones (Sato and Reid, 1978) and tissue matrices (Reid and Rojkind, 1979), have regulatory effects on cells in culture; however, more work is required to elucidate the role of such agents. The role of such agents is beyond the scope of the present work; our concern is with techniques that have well-established use.

## 1. Culture Medium

The medium is the single most important factor in tissue culture. The formulation and composition of culture media have been investigated since the first medium (frog lymph) was used by Arnold (1887) and by Harrison (1907). The medium provides the *physical* conditions (such as a specified pH and osmotic pressure) and the *chemical* substances required by the tissue culture. Media consist of two parts: a defined synthetic part and a biological fluid (e.g., serum, plasma clots, or tissue extracts, such as embryo extracts). Osmolality, ionic balance, pH, gas phase, etc., are very important cultural variables. Media are often buffered with phosphates and bicarbonates; the latter must be equilibrated with a $CO_2$-containing atmosphere.

**a. Chemically Defined Media.** Defined essential nutrients of culture media are shown as follows (Waymouth, 1972; Paul, 1975; Jacoby and Paston, 1979):

1. Major ions, such as $Na^+$, $K^+$, $Ca^{2+}$, $Mg^{2+}$, $Cl^-$, $PO_4^{3-}$, and $HCO_3^-$, etc., are added as a balanced salt solution (BSS).

2. Trace metals, such as Fe, Cu, Co, Mn, Zn, and Mo, are added; Se (McKeehan *et al.*, 1976) is also essential for several cell types in some culture media.

3. The major energy source is usually D-glucose. Occasionally other energy sources (keto acids, short chain fatty acids, and nucleosides) are used.

4. Amino acids, vitamins, and hormones are added to the medium according to the needs of the specific cells or tissues. Cells in culture have amino acid requirements (arginine, glutamine, cysteine, asparagine, etc.) in addition to those of the intact animal. Normally these amino acids are synthesized by an organ (liver, kidneys) and released into the circulation and made available to the other tissues and cells. The same may be true for some altered forms of some vitamins. Defined media commonly used with serum or serum protein are Eagle's media (minimum essential medium, MEM; basal medium, BME), Delbecco's medium (DME), McCoy's medium, and Ham's medium (Ham and McKeehan, 1979); these are commercially available.

**b. Biological Fluid.** The biological (nondefined) portion of the culture medium has several described functions; additionally, it probably has other functions. The viscosity contributed to the medium by the protein portion has effects related to matrix-influenced growth patterns, and the decrease in surface tension due to the proteins influences the chemical (thermodynamic) activities of a number of ions, nutrients, and gases involved in cellular metabolism. Similarly, the biological portion contributes to extracellular oncotic (colloid osmotic) pressure and thereby influences water and ionic balances within the cells. Some of the biological fluid constituents may also affect cell – cell interactions and influence the pattern of cultural growth; related glycosaminoglycans may have mitogenic (or antimitogenic) activity as well.

The biological portion may influence the nutrition of cells in culture in several ways. Some cell types are able to absorb (by endocytic processes) the protein constituents and, after hydrolysis, utilize the amino acids. The biological fluid contains a large number of transport proteins for nutrients such as copper, iron, and vitamin A. These transport proteins enhance the absorption of the nutrients by cells. Not infrequently, defined media are deficient in trace nutrients and growth is dependent upon the undefined biological portion to provide not only carrier proteins but the nutrients (Cu, Fe, Zn) as well.

The biological additives also provide growth-regulator substances, for example, steroid hormones and polypeptide hormones and their transport proteins. It is clear that much work is required to define the composition and activities of the biological fluids used in tissue culture.

## 2. Preparation of Cells or Tissues for Culture

In tissue and organ culture, small pieces of tissue are placed in a medium and allowed to develop. Special precautions to prevent the tissue from becoming disorganized in organ culture include gauze supports for the tissue. In cell culture, individual cells (usually obtained by trypsin digestion of tissues) are placed into culture. Cultures of this kind are called primary cultures. After cells divide repeatedly, they often can be "passaged." For this purpose, cells are usually separated by trypsin and used to inoculate new vessels containing fresh media.

Several criteria are used to classify cells, such as multiplication potential, anchorage dependence, density-dependent inhibition, karyotype, etc.

## 3. Preparation of Samples from Foods

No established methodology for food analysis by tissue culture exists; therefore what follows is our suggestion to establish such a method. Foods can be extracted with ethanol, water, or dimethyl sulfoxide (DMSO) (Bouck and DiMayorca, 1979). Dimethyl sulfoxide and alcoholic extracts will most likely require concentrating, because limits for the safe use of alcohol and DMSO suggest that they be used at less than 1% in the final culture medium. Blanks containing only the solvent should be run with test extracts.

Sterilization by filtration (cellulose acetate or nitrate, $0.45\mu m$) is usually recommended to prevent viral or mycoplasma contamination (Telling and Radlett, 1970). Positive pressure (rather than suction) is applied (Paul, 1975). Filtration of solutions high in DMSO or other organic solvents requires special filter beds, because the solvents disrupt the cellulose acetate. In these cases, solutions of the test material are added to the culture media after the latter have been filtered.

## 4. Measurements of Biological Activity

A variety of measurements is used to assess the activity of extracts. Most important among these are survival or plating efficiency of cells put into culture, their doubling times, their growth curves, and their ability to synthesize specific metabolites. Additionally, cultures should be examined microscopically to determine whether the treatment has altered cell morphology or prevented alterations in morphology and growth habit that occur in culture.

## D. Enzymatic Analysis

Enzymatic methods of analysis have been used previously to analyze food for nutrient content. For example, glucose oxidase has been used to quantitate glucose in dried egg products. $\beta$-Galactosidase has been used to estimate the lactose content of milk products, and sucrase has been used to evaluate sucrose as an additive to alcoholic and nonalcoholic beverages. Similarly, collagenase and trypsin have been used to evaluate the collagen, gelatin, or other gel-forming proteins in food, and lipoxygenase has been used to assess the amount of unsaturated fat present. More recently, enzymes have been used to assay for components that influence more subtle aspects of food quality, such as color, clarity, aroma, and consistency. The methods are rapid and allow a reliable assessment of various components of foods and beverages. These techniques also bypass problems that are very difficult to resolve by the conventional methods. To obtain a successful enzymatic analysis a working knowledge of the reagent changes and experience in handling enzymes is necessary.

Enzymes are highly specialized proteins that catalyze specific biochemical reactions. During the course of reactions some enzymes undergo conformational changes, giving rise to protein moieties with varying tendencies of denaturation. Not only is enzymatic action strongly dependent on pH, temperature, ionic strength, and other environmental factors, but denaturation and loss of enzymatic activity are also influenced by these variables.

A recent and important development has been the use of immobilized enzymes (Wingard *et al.*, 1976; Chibata, 1978). To immobilize enzymes, agents that covalently bind the enzymes to a fixed bed are used. Although fixed enzymes have slower turnover numbers than do soluble enzymes, they undergo very little thermal denaturation and can be handled much more easily than the free enzymes.

## 1. Handling of Reagents

Test systems for a number of different assays have been described for enzymatic analysis of foods and beverages (Bergmeyer, 1978; Anonymous, 1979). The substance to be analyzed is most often a natural product usually present together with metabolic precursors and with metabolic products, which may have some similar chemical and physical properties. It is important that handling of the specimen or mixing with reagents not convert one related chemical into another. For example, in preparation of lyophilized material, care must be taken not to form air bubbles or foam. Reagents (buffers, inorganic ions, enzymes, coenzymes, cofactors, and substrates) also are prepared and stored under conditions appropriate to the specific assay.

## 2. Enzymes

The criterion most important in choosing between a chemical assay and an enzymatic assay is the need for specificity. Suppose it is desired to determine whether a given sample of olives contains natural L(+) as opposed to racemic (synthetic) DL-lactic acid or mixtures thereof. This is most readily accomplished by the use of enzymes specific for the natural isomer used by lactic dehydrogenase, because other optically active ingredients would interfere with polarimetric determinations. Similarly, D-amino acid oxidase may be used to determine whether a feed has been supplemented with a natural or synthetic amino acid. The most important attribute of an enzyme is its specificity.

## 3. Preparation of Samples

Sample preparations are described in Anonymous (1979). Filtration is essential for a turbid liquid, and if the filtrate is strongly colored, decolorization by activated charcoal, polyamide, or polyvinylpolypyrrolidone is recommended.

Solid or semisolid samples are mixed throughly, extracted with water, and deproteinized. The deproteinized, neutralized supernatant is usually used directly for analysis; however, sometimes concentration by chromatography or drying may be required.

## 4. Methods Used to Analyze Enzymatic Reactions

Several methods are available for following enzyme reactions quantitatively (Dixon et al., 1980).

**a. Spectrophotometric Method.** Products of many enzyme reactions absorb light in either the visible or the ultraviolet range. Substrates and cofactors (i.e., reactants) have different absorption spectra from their products; therefore the progress or extent of the reaction can be followed by the decrease in the optical density associated with the reactant and the increase in the optical density of the spectrum associated with the product. Coenzymes NAD and NADP have an absorption band at 340 nm in the reduced state but not in the oxidized state. This makes a large number of dehydrogenases useful for nicotinamide-dependent reactions. For example, lactic acid in a sample can be determined by the stoichiometric reduction of $NAD^+$ (as measured by increase in O.D. at 340 nm) it causes in the presence of lactic dehydrogenase. Similarly, ethanol content of samples can be measured by the reduction of $NAD^+$ in the presence of alcohol dehydrogenase. This method can be used to determine many food additives (malic acid, sorbitol, glutamic acid, etc.). Dehydrogenases that employ flavin coenzymes can be used similarly to measure some food ingredients (succinic acid, D-amino acids). In this case, reduction of the flavin by the test material causes a drop in absorption at the wavelength (450 nm) used. In some systems the amount of reduced coenzyme formed is measured by its reoxidation (increase in O.D.) by an exogenous electron acceptor.

**b. Fluorescence.** Fluorescence is widely used to assay NAD-dependent oxidations, because the sensitivity of the fluorometric measurements of NADH is many times greater than the optical density measurements. To measure fluorescence, light of a given wavelength (365 nm to measure NADH) is used as the incident or existing source; NADH or other fluorescent materials dissipate some of the light energy and increase the wavelength of some of the absorbed light (i.e., it is emitted as a higher wavelength, 465 nm). Whereas this method can detect NADH at 1% of the limits of NADH detection by spectrophotometry, it is generally less accurate (there is more interference) and more difficult to carry out than spectrophotometry.

**c. Kinetic Studies.** NADH and other enzyme products are sometimes determined by continuous sampling of the reaction mixture, as opposed to permitting the equilibrium concentrations to form. This type of study lends itself to analysis with automated equipment, particularly that of the Technicon, since many of these analyses measure the influence of substrate concentration (test material) on the initial rate of the reaction.

## 5. Radioenzymatic Analysis

The analysis of carnitine (3-carboxy-2-hydroxy-$N,N,N$-trimethyl-1-propanaminium hydroxide) illustrates this method. The method is used to analyze *important* constituents such as hormones, vitamins, or other metabolites present in concentrations so low that chemical assays or most bioassays fail to analyze the test material adequately.

As its name suggests, carnitine is present in meat and animal tissues. It is also present in foods of plant origin. Carnitine facilitates fatty acid transport through the inner mitochondrial membrane. Although most mammals can synthesize carnitine, under some conditions (choline deficiency, scurvy, diabetes, some renal diseases) endogenous carnitine may not be adequate to meet the requirements for fatty acid oxidation, and the amount of dietary carnitine may become important. The most sensitive and specific method for the measurement of carnitine employs [1-$^{14}$C]acetyl coenzyme A (McGarry and Foster, 1976; Pace *et al.*, 1978).

A protein-free sample containing carnitine is incubated with radioactive acetyl coenzyme A and the enzyme carnitine acetyl transferase (CAT).

$$\text{L-Carnitine} + [1\text{-}^{14}\text{C}]\text{acetyl-CoA} \xrightleftharpoons{\text{CAT}} O\text{-acetyl-L-carnitine} + \text{CoA}$$

Present in the reaction mixture is sodium tetrathionate, which is not reactive with acetyl coenzyme A but which is reactive with the SH group of coenzyme A. As the reaction proceeds, the labeled acetyl moiety is transferred from coenzyme A to carnitine, forming radioactive acetyl carnitine and liberating free coenzyme A (SH) which is then inactivated by the tetrathionate. Therefore the reaction goes to completion. The formed radiolabeled acetyl-L-carnitine is separated from acetyl-CoA by passing the mixture through a Dowex anion-exchange resin and determining the radioactivity of the effluent liquid.

Radioenzymatic analysis is applicable to the determination of a variety of trace substances in food, including catecholamines and phenols of plant origin.

## IV. CONCLUSIONS

There are two main conclusions that may be drawn from the work reviewed. First, bioassay, at present a useful adjunct to conventional food analytic schemes, has not developed to the point of being equally important with the physical and chemical analytic schemes. Second, the field of bioassay is developing rapidly and will, in the future, assume a more important role in food analysis.

Bioassay tells us the amount of a specific ingredient present in a food sample; of equal or more importance, it also tells us the biologic response produced by that ingredient when it is present in combination with other ingredients being ingested simultaneously. For example, if it were desired to determine inorganic magnesium in a sample containing hydroxypolycarboxylic acids such as citric acid, atomic absorption spectrometry of aqueous extracts could yield accurate results of the magnesium content. However, this would not indicate the availability of the magnesium for uptake from the intestinal tract. As is well known, citrate, operating by several mechanisms, reduces the uptake of magnesium, accounting for some of the cathartic action of magnesium citrate. Even chemical analyses for citric acid would not yield a strong indication of the biological acitivy of the sample. Bioassay based upon $Mg^{2+}$-dependent (not chelated or deionized) enzymes such as the phosphokinases would yield relevant data. Similarly, bioassays based upon citrate and magnesium inhibition of some $Ca^{2+}$-dependent enzymes, such as $Ca^{2+}$-specific ATPases, would be good measures of the biological activity of magnesium in the presence of citrate.

Besides providing an accurate chemical analysis of a material and a description of the biological properties contained in the food, bioassay and chemical analysis have an important research function. In part, our fields have grown through interactions of the following type. An important biological variable (disease) is observed, and food extracts involved in producing or ameliorating the disease are prepared. The active factors are isolated, their activity being bioassayed at each stage of purification. The identity of the material is determined, and a pure material is made available. Other biological activities of the pure materials are then observed in isolated (usually enzyme) systems. Synthesis of the substance and its reactive intermediates is accomplished. The reactive intermediates are asssayed and reveal new biological differences. For research purposes, the bioassays should include both gross (whole-organism) assays and molecular (enzyme) assays, since both of these approaches are necessary to advance the field.

NOTE ADDED IN PROOF

After the proofs of this paper were received by the authors, a paper entitled "Dietary carcinogens and anticarcinogens" (Ames, 1983) appeared. It emphasizes the role played by the *whole* diet in modifying the biological activity of single dietary components (in this case carcinogenicity). This view is similar to the main theme of our review.

**REFERENCES**

Alexander, G., and Watson, R. H. (1951). *Aust. J. Agric. Res.* **2**, 480.
Alfin-Slater, R. B., and Aftergood, L. (1975). *In* "Modern Nutrition in Health and Disease, Dietotherapy" (R. S. Goodhart and M. E. Shils, eds.), p. 117. Lea & Febiger, Philadelphia, Pennsylvania.
Ames, B. N. (1983). *Science* **221**, 1256.
Anonymous (1979). "Methods of Enzymatic Food Analysis." Boehringer, Mannheim, West Germany.
Arnold, J. (1887). *Arch. Mikrosk. Anat.* **30**, 205.
Arnrich, L., and Arthur, V. A. (1980). *Ann. N.Y. Acad. Sci.* **355**, 109.
Baird-Parker, A. C. (1974). *Ann. N.Y. Acad. Sci.* **236**, 7.
Baker, H., and Frank, O. (1967). *In* "Vitamins: Chemistry, Physiology, Pathology, Methods" (P. György, and W. N. Pearson, eds.), Vol. VII, pp. 293. Academic Press, New York.
Beadle, G. W. (1945). *Physiol. Rev.* **25**, 643.
Beadle, G. W., and Tatum, E. L. (1941). *Proc. Natl. Acad. Sci. U.S.A.* **27**, 499.
Beckman, R. (1955). *Z. Vitam. Horm. Fermentforsch.* **7**, 153, 281.
Belavady, B., and Gopalan, C. (1965). *Lancet* **2**, 1220.

Bell, E. A. (1973). *In* "Toxicants Occurring Naturally in Foods," Natl. Res. Counc., 2nd ed., p. 153. Natl. Acad. Sci., Washington, D.C.

Bergdoll, M. S. (1979). *In* "Food-Borne Infections and Intoxications" (H. Riemann, and F. L. Bryan, eds.), 2nd ed., p. 443. Academic Press, New York.

Bergdoll, M. S., Sugiyama, H., and Dack, G. M. (1959). *Arch. Biochem. Biophys.* **85,** 62.

Bergdoll, M. S., Borja, C. R., and Avena, R. M. (1965). *J. Bacteriol.* **90,** 1481.

Bergmeyer, H. U., ed. (1978). "Principles of Enzymatic Analysis." Verlag Chemie, Weinheim.

Bergström, S., Danielsson, H., and Samuelsson, B. (1964). *Biochim. Biophys. Acta* **90,** 207.

Bickoff, E. M., Livingstone, A. L., Hendrickson, A. P., and Booth, A. N. (1962). *J. Agric. Food Chem.* **10,** 410.

Bickoff, E. M., Spencer, R. R., Witt, S. C., and Kunckles, B. E. (1969). *U.S., Dep. Agric., Tech. Bull.* No. 1408.

Booth, A. N., Bickoff, E. M., and Kohler, G. O. (1960). *Science* **131,**

Bouck, N., and DiMayorca, G. (1979). *In* "Methods in Enzymology, Vol. 58: Cell Culture" (W. B. Jacoby, and I. H. Pastan, eds.), p. 296. Academic Press, New York.

Bradbury, R. B., and White, D. E. (1954). *Vitam. Horm.* **12,** 207.

Bruce, H. M. and Callow, R. K. (1934). *Biochem. J.* **28,** 517.

Burgisser, H., Jacquier, C., and Leuenberger, M. (1964). *Schweiz. Arch. Tierheilkd.* **106,** 714.

Busby, W. F., Jr., and Wogan, G. N. (1979). *In* "Food-Borne Infections and Intoxications" (H. Riemann and F. L. Bryan, eds.), 2nd ed., p. 519. Academic Press, New York.

Carlson, C. W., McGinnis, J., and Jensen, L. S. (1964a). *J. Nutr.* **82,** 366.

Carlson, C. W., Saxena, H. C., Jensen, L. C., and McGinnis, J. (1964b). *J. Nutr.* **82,** 507.

Carter, M. W., Matrone, G., and Smart, W. G., Jr. (1955). *J. Nutr.* **55,** 639.

Casman, E. P., and Bennett, R. W. (1965). *Appl. Microbiol.* **13,** 181.

Cawthorne, M. A., Bunyan, J., Diplock, A. T., Murrell, E. A., and Green, J. (1968). *Br. J. Nutr.* **22,** 133.

Chibata, I. ed. (1978). "Immobilized Enzymes: Research and Development." Wiley, New York.

Coates, M. E., and Harrison, G. E. (1957). *Proc. Nutr. Soc.* **16,** 21.

Collins, W. S., Metzger, J. F., and Johnson, A. D. (1972). *J. Immunol.* **180,** 852.

Collins, W. S., Johnson, A. D., Metzger, J. F., and Bennett, R. W. *Appl. Microbiol.* **25,** 774.

Cravioto, R. O., Massieu, G. H., Cravioto, O. Y., and Figueroa, F. (1952). *J. Nutr.* **48,** 453.

Davies, A. W., and Moore, T. (1941). *Nature (London)* **147,** 794.

de Figueiredo, M. P. (1970). *N.Y. Agric. Exp. Stn., Geneva Res. Cir.* No. 23.

Desai, I. D. (1966). *Nature (London)* **209,** 810.

Dixon, M., Webb. E. C., Thorne, C. J. R., and Tipton, K. F., eds. (1980). *In* "Enzymes: Third Edition." Academic Press, New York.

Drane, H. M., Patterson, D. S. P., Roberts, B. A., and Saba, N. (1980). *Food Cosmet. Toxicol.* **18,** 425.

Eakin, R. E., Snell, E. E., and Williams, R. J. (1940). *J. Biol. Chem.* **136,** 801.

Elvehjem, C. A. (1956). *Fed. Proc., Fed. Am. Soc. Exp. Biol.* **15,** 965.

Evancho, G. M., Ashton, D. H., and Briskey, E. J. (1973). *J. Food Sci.* **38,** 764.

Ewer, T. K. (1950). *Nature (London)* **166,** 732.

Ewer, T. K., and Bartrum, P. (1948). *Aust. Vet. J.* **24,** 73.

Ferando, R., Guilleus, M. M., and Guirrilott-Vinet, A. (1961). *Nature (London)* **192,** 205.

Fitch, L. W. N., and Ewer, T. K. (1944). *Aust. Vet. J.* **20,** 220.

Fowden, L., Lewis, D., and Tristram, H. (1967). *Adv. Enzymol.* **29,** 89.

Freed, M., and Assoc. of Vitam. Chem., Inc., eds. (1966). "Methods of Vitamin Assay," 3rd ed., p. 37. Wiley (Interscience), New York.

Friedman, P. A., Rosenberg, R. D., Hauschka, P. V., and Fitz-James, A. (1977). *Biochim. Biophys. Acta* **494,** 271.

Gavin, J. J. (1956). *Appl. Microbiol.* **4**, 323.
Genigeorgis, C., and Sadler, W. W. (1966). *J. Food Sci.* **31**, 605.
Gilgen, A., and Leuthardt, F. (1962). *Helv. Chim. Acta* **45**, 1833.
Grant, A. B. (1951). *Nature (London)* **168**, 789.
Green, D. E., and Tzagloff, A. (1966). *J. Lipid. Res.* **7**, 587.
Green, R. G. (1937). *Minn. Wildlife, Dis. Invest.* **3**, 83.
Gregory, J. F., III, and Kirk, J. R. (1981). *Nutr. Rev.* **39**, 1.
Griffith, W. H., and Dyer, H. M. (1968). *Nutr. Rev.* **26**, 1.
Guilarte, T. S., and McIntyre, P. A. J. (1981). *J. Nutr.* **111**, 1861.
Guilarte, T. S., Shane, B., and McIntyre, P. A. (1981). *J. Nutr.* **111**, 1869.
Gulambic, C., and Kulik, M. (1969). *In* "Aflatoxin" (L. A. Goldblat, ed.), p. 319. Academic Press, New York.
Hall, H. E., Angelotti, R., and Lewis, K. H. (1965). *Health Lab. Sci.* **2**, 179.
Ham, R. G., and McKeehan, W. L. (1979). *In* "Methods in Enzymology, Vol. 58: Cell Culture" (W. B. Jacoby, and I. H. Pastan, eds.), p. 44. Academic Press, New York.
Harper, A. E. (1956). *Nutr. Rev.* **14**, 225.
Harper, A. E., Benevenga, N. J., and Wohlhueter, R. M. (1970). *Physiol. Rev.* **50**, 428.
Harper, A. E. (1973). *In* "Toxicants Occurring Naturally in Foods," Natl. Res. Counc., 2nd ed., p. 130. Natl. Acad. Sci., Washington, D.C.
Harris, R. S., Ross, B. D., and Bunker, J. W. M. (1939). *Am. J. Dig. Dis.* **6**, 81.
Harrison, D. C., and Mellanby, E. (1939). *Biochem. J.* **33**, 1660.
Harrison, R. G. (1907). *Proc. Soc. Exp. Biol. Med.* **4**, 140.
Hays, K. C., and Hegsted, D. M. (1973). *In* "Toxicants Occurring Naturally in Foods," Natl. Res. Counc., 2nd ed., p. 235. Natl. Acad. Sci., Washington, D.C.
Hintz, H. F., and Hogue, E. E. (1964). *J. Nutr.* **84**, 283.
Hintz, H. F., and Hogue, D. E. (1970). *Proc. Soc. Exp. Biol. Med.* **131**, 447.
Hogue, D. E., Pector, J. F., Warner, R. G., and Loosli, J. K. (1962). *J. Anim. Sci.* **21**, 25.
Hunt, R. D., Garcia, F. G., and Hegsted, D. W. (1969). *Am. J. Clin. Nutr.* **22**, 358.
Hutner, S. H., Cury, A., and Baker, H. (1958). *Anal. Chem.* **30**, 849.
Jacoby, W. B., and Pastan, I. H., eds. (1979). "Methods in Enzymology, Vol. 58: Cell Culture." Academic Press, New York.
Jenkins, M. Y., and Mitchell, G. V. (1975). *J. Nutr.* **105**, 1600.
Johnson, H. M., Bukovic, J. A., Kauffmann, P. E., and Peeler, J. T. (1971). *Appl. Microbiol.* **22**, 837.
Johnson, H. M., Bukovic, J. A., and Kauffmann, P. E. (1973). *Appl. Microbiol.* **26**, 309.
Josephs, H. W. (1944). *Am. J. Dis. Child* **67**, 33.
Jukes, T. H., and Williams, W. L. (1954). *In* "The Vitamins: Chemistry, Physiology, Pathology, Methods" (W. H. Sebrell, and R. S. Harris, eds.), Vol. I, p. 468. Academic Press, New York.
Kavanagh, F. (1977). *J. Pharm. Sci.* **66**, 1520.
Kerdel-Vegas, F., Wagner, F., Russell, P. B., Grant, N. H., Alburn, H. E., Clark, D. E., and Miller, J. A. (1965). *Nature (London)* **205**, 1186.
King, R. A. (1949). *J. Bone Jt. Surg.* **31B**, 443.
Kingsbury, J. M. (1964). *In* "Poisonous Plants of the United States and Canada." Prentice-Hall, Englewood Cliffs, New Jersey.
Klosterman, H. J., Lamoureux, G. L., and Parsons, J. L. (1967). *Biochemistry* **6**, 170.
Kodicek, E., and Wilson, P. W. (1960). *Biochem. J.* **76**, 27P.
Kolhouse, J. F., Kondo, H., Allen, N. C., Podell, E., and Allen, R. H. (1978). *N. Engl. J. Med.* **299**, 785.
Kosower, N. S., and Kosower, E. M. (1965). *Nature (London)* **215**, 285.
Kratzer, F. H. (1946). *Poult. Sci.* **25**, 541.

Kratzer, F. H., and Williams, D. E. (1948). *J. Nutr.* **36,** 297.

Kusin, J. A., Reddy, V., and Shivakumar, B. (1974). *Am. J. Clin. Nutr.* **27,** 774.

Labov, J. B. (1977). *Comp. Biochem. Physiol. A* **57,** 3.

Lau, K. S., Gottlieb, C., Wasserman, L. R., and Herbert, V. (1965). *Blood* **26,** 202.

Leach, E. H., and Lloyd, J. P. F. (1956). *Proc. Nutr. Soc.* **15,** xv.

Leavitt, W. W., and Wright, P. A. (1965). *J. Exp. Zool.* **160,** 319.

Lewis, J. J. (1935). *J. Pediatr.* **6,** 326.

Liener, I. E. (1969). *In* "Toxic Constituents of Plant Foodstuffs" (I. E. Liener, ed.), p. 409. Academic Press, New York.

Liener, I. E. (1980). *In* "Toxic Constituents of Plant Foodstuffs" (I. E. Liener, ed.), 2nd ed., p. 429. Academic Press, New York.

Lockhart, G. L., Jones, B. L., and Finney, K. F. (1978). *Cereal Chem.* **55,** 967.

McGarry, J. D., and Foster, D. W. (1976). *J. Lipid Res.* **17,** 277.

McKeehan, W., Hamilton, W. G., and Ham, R. G. (1976). *Proc. Natl. Acad. Sci. U.S.A.* **73,** 2023.

Mayer, S. E. (1980). *In* "The Pharmacologic Basis of Therapeutics" (L. S. Goodman and A. Gilman, eds.), p. 56. Macmillan, New York.

Mellanby, E. (1926). *J. Physiol. (London)* **61,** 24.

Michelson, N. J., and Danilov, A. F. (1971). *In* "Fundamentals of Biochemcial Pharmacology" (M. Bacq, ed.), p. 221. Pergamon, Oxford.

Miller, J. A. (1973). *In* "Toxicants Occurring Naturally in Foods," Natl. Res. Counc., p. 508. Natl. Acad. Sci., Washington, D.C.

Milne, M. D. (1968). *Clin. Pharmacol. Ther.* **9,** 484.

Mirocha, C. J., Pathre, S. V., and Christiansen, C. M. (1977). *In* "Mycotoxins in Human and Animal Health" (J. V. Rodricks, C. W. Hesseltine, and M. A. Mehlman, eds.), p. 345. Pathotox Publ. Illinois.

Mold, J. D., Bowden, J. P., Stanger, D. W., Maurer, J. E., Lynch, J. M., Wyler, R. S., Schantz, E. J., and Riegel, B. (1957). *J. Amer. Chem. Soc.* **79,** 5235.

Montgomery, R. D. (1969). *In* "Toxic Constituents of Plant Foodstuffs" (I. E. Liener ed.), p. 143. Academic Press, New York.

Montgomery, R. D. (1980). *In* "Toxic Constituents of Plant Foodstuffs" (I. E. Liener, ed.), 2nd ed., p. 143. Academic Press, New York.

Moos, G. E., and Chen, J. P. (1957). *Chemist-Analyst* **46,** 72.

Moule, G. R., Braden, A. W. H., and Lamond, D. R. (1963). *Anim. Breed.* **31,** 139. (Abstr.)

Murti, V. V. S., Seshadri, T. R., and Venkitasubramanian, T. A. (1964). *Phytochemistry* **3,** 73.

Naim, M., Gestetner, B., Kirson, I., Burk, Y., and Bondi, A. (1973). *Phytochemistry* **12,** 169.

Naim, M., Gestetner, B., Zilkah, S., Birk, Y., and Bondi, A. (1974). *J. Agric. Food Chem.* **22,** 806.

Naim, M., Gestetner, B., Bondi, A., and Birk, Y. (1976). *J. Agric. Food Chem.* **24,** 1174.

Nater, J. P., and Doeglas, H. M. G. (1970). *Acta Derm. Venereal.* **50,** 109.

Natl. Acad. Sci. (1975). "Prevention of Microbial and Parasitic Hazards Associated with Processed Foods." Natl. Acad. Sci., Washington, D.C.

Natl. Res. Counc. (1973). "Toxicants Occurring Naturally in Foods." Natl. Acad. Sci., Washington, D.C.

Nilson, R. (1962). *Acta Physiol. Scand.* **56,** 230.

Noteboom, W. D., and Gorski, J. (1963). *Endocrinology (Baltimore)* **73,** 736.

Nut. Soc. Symp. (1971). *Fed. Prod., Fed. Am. Soc. Exp. Biol.* **30,** 130.

Ostwald, R., and Briggs, G. M. (1966). *In* "Toxicants Occurring Naturally in Foods," Natl. Res. Counc., 1st ed., p. 183. Natl. Acad. Sci., Washington, D.C.

Pace, J. A., Wannemacher, R. W., and Neufeld, H. A. (1978). *Clin. Chem.* **24,** 32.

Padmanavan, G. (1980). *In* "Toxic Constituents of Plant Foodstuffs" (I. E. Liener, ed.), p. 239. Academic Press, New York.

Paul, J. (1975). *In* "Cell and Tissue Culture," 5th ed. Churchill, London.

Pudelkiewicz, W. J., and Matterson, L. E. (1960). *J. Nutr.* **71,** 143.

Quick, A. J. (1957). *In* "Hemorrhagic Diseases." Lea and Febiger, Philadelphia, Pennsylvania.

Rao, S. L. N., Ramachandran, L. K., and Adiga, P. R. (1963). *Biochemistry* **2,** 298.

Rao, S. L. N., Adiga, P. R., and Sarma, P. S. (1964). *Biochemistry* **3,** 432.

Raoul, Y., Marnay, C., leBoulch, N., Prelot, M., Guerillot-Vinet, A., Bazier, R., and Baron, C. (1957). *C. R. Acad. Sci. (Paris)* **244,** 954.

Read, R. B., Jr., Bradshaw, J., Pritchard, W. L., and Black, L. A. (1965). *J. Dairy Sci.* **48,** 420.

Reid, L. M., and Rojkind, M. (1979). *In* "Methods in Enzymology, Vol. 58: Cell Culture" (W. B. Jacoby, and I. H. Pastan, eds.), p. 263. Academic Press, New York.

Ressler, C. (1962). *J. Biol. Chem.* **237,** 733.

Riemann, H., and Bryan, F. L., eds. (1969). "Food-Borne Infection and Intoxications." Academic Press, New York.

Riemann, H., and Bryan, F. L., eds. (1979). "Food-Borne Infection and Intoxications: Second Edition." Academic Press, New York.

Rosenfeld, I., and Beath, O. A. (1964). "Selenium: Geobotany, Biochemistry, Toxicity and Nutrition." Academic Press, New York.

Rukmini, C. (1968). *Indian J. Biochem.* **5,** 182.

Rukmini, C. (1969). *Indian J. Biochem.* **7,** 1062.

Sakabe, N., Goto, T., and Hirata, Y. (1964). *Tetrahedron Lett.* **27–28,** 1825.

Sakaguchi, G. (1979). *In* "Food-Borne Infections and Intoxications" (H. Riemann and F. L. Bryan, eds.), 2nd ed., p. 389. Academic Press, New York.

Sanyal, S. N. (1953). *Calcutta Med. J.* **50,** 409.

Sargent, K., Sheridan, A., O'Kelly, J., and Carnaghan, R. B. A. (1961). *Nature (London)* **192,** 1096.

Sarma, P. S., and Padmanavan, G. (1969). *In* "Toxic Constituents of Plant Foodstuffs" (I. E. Liener, ed.), p. 267. Academic Press, New York.

Sasaoka, K., Ogawa, T., Moritoki, K., and Komoto, M. (1976). *Biochim. Biophys. Acta* **428,** 396.

Sato, G., and Reid, L. (1978). *Biochem. Mode Action Horm. II* p. 219.

Schantz, E. J. (1974). *In* "Toxic Constituents of Animal Foodstuffs (I. E. Liener, ed.), p. 425. Academic Press, New York.

Schantz, E. J., Mold, J. D., Stanger, D. W., Shavel, J., Riel, F. J., Bowden, J. P., Lynch, J. M., Wyler, R. S., Riegel, B., and Sommer, H. (1957). *J. Am. Chem. Soc.* **79,** 5230.

Schoental, R. (1967). *Annu. Rev. Pharmacol.* **7,** 343.

Seyle, H. (1957). *Rev. Can. Biol.* **16,** 3.

Shaw, J. C., Moore, L. A., and Sykes, J. F. (1951). *J. Dairy Sci.* **34,** 176.

Siddiqui, S. (1945). *J. Sci. Ind. Res.* **4,** 68.

Silverman, S. J., Espeseth, D. A., and Schantz, E. J. (1969). *J. Bacteriol.* **98,** 437.

Simpson, J. S. (1956). *Med. Lab. Technol.* **13,** 474.

Singsen, E. P., Potter, L. M. Bunnell, R. H., Matterson, L. D., Stinson, L., Amato, S. V., and Jungherr, E. L. (1955). *Poult. Sci.* **34,** 1234.

Shutt, D. A. (1967). *J. Endocrinol.* **37,** 231.

Shutt, D. A., Axelson, A., and Linder, H. R. (1967). *Aust J. Agric. Res.* **18,** 647.

Skeggs, H. R. (1967). *In* "Vitamins: Chemistry, Physiology, Pathology, Methods" (P. Gyorgy and W. N. Pearson, eds.), Vol. 7, p. 277. Academic Press, New York.

Sommer, H., and Meyer, K. F. (1937). *Arch. Pathol.* **24,** 560.

Sommer, H., Whedon, W. F., Kofoid, C. A., and Stohler, R. (1937). *Arch. Pathol.* **24,** 537.

Somogyi, J. C. (1949). *Helv. Physiol. Pharmacol. Acta* **7, 24C.**

Somogyi, J. C. (1973). *In* "Toxicants Occurring Naturally in Foods," Natl. Res. Counc., 2nd ed., p. 254. Natl. Acad. Sci., Washington, D.C.

Stahmman, M. A. Huebner, C. F., and Link, K. P. (1941). *J. Biol. Chem.* **138,** 513.

Stob, M. (1973). *In* "Toxicants Occurring Naturally in Foods," Natl. Res. Counc., 2nd ed., p. 550. Natl. Acad. Sci., Washington, D.C.

Sugimura, T. (1981). *In* "Accomplishment in Cancer Research" (J. G. Fortner and J. E. Rhoads, eds.), p. 61. Lippincott, Philadelphia, Pennsylvania.

Sumner, J. B., and Dounce, A. L. *Enzymology* **7,** 130. (1939).

Surgalla, M. J., Bergdoll, M. S., and Dack, G. M. (1954). *J. Immunol.* **72,** 398.

Taylor, D. L., and Seliger, H. H., eds. (1979). "Toxic Dinoflagellate Blooms." Elsevier/North-Holland, New York.

Taylor, P. (1980). *In* "The Pharmacologic Basis of Therapeutics" (L. S. Goodman and A. Gilman, eds.), p. 91. Macmillan, New York.

Telling, R. C., and Radlett, P. J. (1970). *Adv. Appl. Microbiol.* **13,** 91.

Thompson, J. F., Morris, C. J., and Smith, I. K. (1969). *Annu. Rev. Biochem.* **38,** 137.

Trelease, S. F., DiSomma, A. A., and Jacobs, H. L. (1960). *Science* **132,** 618.

Tsuda, K., Tachikawa, R., Sakai, K., Tamura, C., Amakasu, O., Kawamura, M., and Ikuma, S. (1964). *Chem. Pharm. Bull.* **12,** 642.

U.S. Dept. Health and Hum. Serv. (1982). "Disease Transmitted by Food: A Classification and Summary," 2nd ed. U.S. Dept. Health and Human Services, Washington, D.C.

Van Drop, D. A., Beerthuis, R. K., Nugteren, D. H., and Vokeman, H. (1964). *Biochim. Biophys. Acta* **90,** 204.

Van Eatten, C. H. (1969). *In* "Toxic Constituents of Plant Foodstuffs" (I. E. Liener, ed.), p. 103. Academic Press, New York.

Van Rensberg, S. J., and Altenkirk, B. (1974). *In* "Mycotoxins" (I. F. H. Purchase, ed.), p. 69 Elsevier, New York.

Van Veen, A. G. (1966). *In* "Toxicants Occurring Naturally in Foods," Natl. Res. Counc. 1st ed., p. 176. Natl. Acad. Sci., Washington, D.C.

Van Woert, M. H., and Bowers, M. B. (1970). *Experientia* **26,** 161.

Vimokesant, S., Kunjara, S., Rungruangsak, K., Nakornchi, S., and Panijpan, B. (1982). *Ann. N.Y. Acad. Sci.* **378,** 123.

Wada, H., and Fukushima, S. (1963). *Jpn. J. Zootech. Sci.* **34,** 243.

Wada, H., and Yuhara, M. (1964). *Jpn. J. Zootech. Sci.* **35,** 87.

Walter, E. D. (1941). *J. Amer. Chem. Soc.* **63,** 3273.

Walz, E. 1931. *Leibigs. Ann.* **489,** 118.

Waymouth, C. (1972). *In* "Growth, Nutrition and Metabolism in Cells in Culture" (G. H. Rothblat and V. J. Cristofalo, eds.), Vol. 1, p. 11. Academic Press, New York.

Weiss, F. A. (1957). *In* "Manual of Microbiological Methods" (H. J. Coon and M. J. Peklczar, Jr., eds.), p. 99. McGraw-Hill, New York.

Wilson, B. J., and Hayes, A. W. (1973). *In* "Toxicants Occurring Naturally in Foods," Natl. Res. Counc., p. 372. Natl. Acad. Sci., Washington, D.C.

Wilson, B. J., Yang, D. T. C., and Boyd, M. R. (1970). *Nature (London)* **227,** 521.

Wingard, L. B., Katchalski-Katzir, E., and Goldstein, L., eds. (1976). "Immobilized Enzyme Principles" Academic Press, New York.

Wogan, G. N., Edwards, G. S., and Newbern, P. M. (1971). *Cancer Res.* **31,** 1936.

Wong, J. L. (1971). *J. Am. Chem. Soc.* **93,** 7344.

Woodward, R. B. (1964). *Pure Appl. Chem.* **9,** 49.

# 12

# X-Ray Analysis

GENE S. HALL

*Department of Chemistry*
*Rutgers, The State University of New Jersey*
*New Brunswick, New Jersey*

ANALYSIS OF FOODS AND BEVERAGES

## I. INTRODUCTION

When humans first inhabited this earth, their diet was simple and relatively free of toxic metals. Today, however, the environment has undergone drastic changes with the influx of many toxic metals that are accumulating in the food chain. In addition to increased knowledge concerning toxic metals, humans have become conscious of their diet, making sure to obtain the proper balance of essential elements while at the same time eliminating or reducing the levels of toxic elements in their food. In order to monitor the concentrations of essential and toxic elements in foods and beverages, an analytical technique that is accurate, sensitive, multielemental, rapid, and reliable is required.

Today there are several analytical techniques that can accomplish these requirements for quantitative analysis of foods and beverages. One of the techniques applicable to trace element analysis is the X-ray technique, which can be divided into two major areas: proton-induced X-ray emission (PIXE) and X-ray fluorescence (XRF). In this chapter, the fundamentals of X-ray analysis will be given. However, every detail and aspect of X-ray analysis cannot be presented here, and the reader is referred to several review articles for an in-depth discussion of PIXE (Johansson and Johansson, 1976; Khan and Crumpton, 1981a,b) and XRF (Woldseth, 1973; Jenkins and de Vries, 1975).

## II. FUNDAMENTALS OF X RAYS

The production of X rays in target atoms can be brought about by exposing the target to either charged particles (electrons, protons, alphas, or heavy ions) or primary X rays. The primary X rays can be produced by an X-ray tube, radioisotopes, or, more recently, by a synchrotron (Sparks, 1980).

### A. Basic Concepts

X rays were discovered in 1895 by Röntgen (1898). They are electromagnetic waves that cover the spectrum between 0.01 and 10 nm (124–0.124 keV) being enclosed by gamma rays on the short-wavelength (high-energy) side and by the vacuum ultraviolet (UV) region on the long-wavelength (low-energy) side. However, the region between 0.02 and 0.2 nm (62–0.62 keV) is the region most useful for quantitative–qualitative analysis of the elements fluorine through uranium.

In order to understand the origins of X rays, one must return to the simple

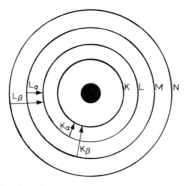

**Fig. 1.** Bohr model, showing electron transitions between energy levels ($E$) giving X rays: $K_\alpha = E_K - E_L$; $K_\beta = E_K - E_M$; $L_\alpha = E_L - E_M$; $L_\beta = E_L - E_N$.

model of the Bohr atom (Fig. 1). There are two mechanisms responsible for the production of X rays in target atoms. They are (1) direct ionization of inner-shell electrons, usually a K- or an L-shell electron, by bombardment with charged particles or (2) absorption of X rays by the photoelectric effect. Both mechanisms create inner-shell vacancies. The target atom is thus in an excited state and the atom must revert to its original state to regain stable energy. This is accomplished by a transition of an electron from a higher shell to the vacant electron site, and there is a loss in energy due to this transition. Since energy is neither created nor destroyed but is instead conserved, the loss of energy is accompanied by the emission of electromagnetic radiation. The emitted radiation carries an amount of energy equivalent to the energy difference between the two shells of the electron transition.

## B. Characteristic X-Ray Energies

Since each kind of atom has a unique electronic configuration, each element will have a unique characteristic X-ray spectrum. Atom bombardment with primary X rays or charged particles will result in the production of at least two ($K_\alpha$ and $K_\beta$ or $L_\alpha$ and $L_\beta$) or more characteristic X rays, depending on the atom. A list of some characteristic X rays is presented in Table 1. Figure 2 shows a typical PIXE X-ray spectrum from a hair sample; it illustrates the elements represented by the characteristic X rays.

## C. Production of X Rays by Charged Particles

The most popular method for producing characteristic X rays by charged particles is that of bombarding target atoms with protons. This method is

**TABLE 1   Characteristic X Rays**

| Z | Element | X-Ray | Energy (keV)[a] | Wavelength (nm) |
|----|---------|-------|-----------------|-----------------|
| 13 | Al | $K_\alpha$ | 1.49 | 0.832 |
| 20 | Ca | $K_\alpha$ | 3.69 | 0.336 |
| 25 | Mn | $K_\alpha$ | 5.89 | 0.211 |
| 26 | Fe | $K_\alpha$ | 6.39 | 0.194 |
| 29 | Cu | $K_\alpha$ | 8.03 | 0.154 |
| 30 | Zn | $K_\alpha$ | 8.62 | 0.144 |
| 33 | As[b] | $K_\alpha$ | 10.51 | 0.118 |
| 37 | Rb | $K_\alpha$ | 11.88 | 0.104 |
| 47 | Ag | $K_\alpha$ | 21.99 | 0.056 |
| 48 | Cd | $K_\alpha$ | 22.98 | 0.054 |
| 48 | Cd | $L_\alpha$ | 3.13 | 0.396 |
| 80 | Hg | $L_\alpha$ | 9.99 | 0.120 |
| 82 | Pb[b] | $L_\alpha$ | 10.55 | 0.118 |
| 92 | U | $L_\alpha$ | 13.61 | 0.091 |

[a] Data from Ortec (1976).
[b] The arsenic K X ray occurs at the same energy as the lead L X ray, and the two peaks cannot be resolved even in the wavelength dispersive mode.

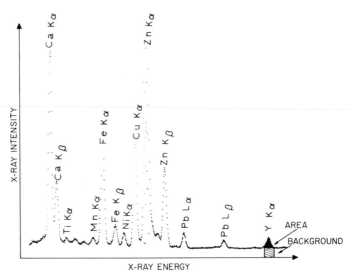

**Fig. 2.** Typical PIXE X-ray spectrum of ashed hair showing trace elements that are identified by their energy.

known as PIXE. Proton-induced X-ray emission was first demonstrated in 1970 by Johansson and Johansson. Using a proton energy of 3 MeV, they were able to obtain minimum detection limits (MDL) of $10^{-11}$ to $10^{-13}$ g for the elements sodium through uranium. Other charged particles such as alpha particles (Uemura *et al.,* 1978) and $^{16}O$ (Chaturverdi *et al.,* 1975) have been used in some laboratories to produce X rays. Since the energy of the bombarding protons are in the range 1 – 5 MeV, they have great penetrating power. The high proton fluxes generated with accelerators produce an abundance of X rays in the target, which allows shorter analysis times (less than 5 min in some cases). However, high proton fluxes cause high count rates, which can create large dead times, detector saturation, and poorer resolution for the detected X rays and hence a decrease in the MDL and sensitivity.

## D. Production of X Rays by Radioactive Sources

Radioactive sources that emit X rays by electron capture or by alpha particle (i.e., $^{241}Am$) emission can be used to excite target atoms. As mentioned in Section II,A, higher-energy X rays (primary X rays) can cause ionization of inner-shell electrons by the photoelectric effect.

There are certain requirements that must be met when radioactive sources are used to produce secondary (fluorescent) X rays in the target atom. A table of photoelectric excitation cross sections (ionization probabilities) as a function of energy for different elements must be consulted in order to choose the proper excitation source. This is because the primary X-ray energy must be greater than the binding energy of the ejected electron so that ionization of the atom can occur. However, the photoelectric excitation cross sections for inner-shell ionization decrease continuously as binding energies decrease. Therefore, a limited range of elements can be analyzed with one source, and a different source has to be used for other elements. To cover the analysis of several elements in a sample, an array of four, six, or eight standard radioisotopes in an annular source assembly has to be employed. The most commonly used monochromatic X-ray sources include $^{55}Fe$, $^{109}Cd$, $^{153}Gd$, $^{238}Pu$, and $^{241}Am$. Table 2 lists the elemental ranges for which these sources can be used. However, because of Compton scattering from the target, it is best to have the excitation X rays slightly higher in energy than the binding energy at the maximum photoelectric cross section. For example, the maximum photoelectric cross section for the K shell of titanium occurs at 4.962 keV. Therefore, $^{55}Fe$ would be the best source for titanium analysis. A general rule of thumb is to select a source that has an excitation photon that is at least 2 keV higher in energy than the $K_\alpha$ peak from the heaviest element in the sample to be analyzed.

**TABLE 2   Radioisotopes in X-Ray Fluorescence**[a]

| Nuclide | Element X-rays excited usefully |
|---------|-------------------------------|
| ³H/Ti | Na(K)–Cu(K) |
| ⁵⁵Fe | Al(K)–Cr(K) |
| ²³⁸Pu | Ca(K)–Br(K) |
|  | W(L)–Pb(L) |
| ¹⁰⁹Cd | Ca(K)–Mo(K) |
|  | W(L)–U(L) |
| ¹⁵³Gd | Mo(K)–Ce(K) |
| ²⁴¹Am | Elements up to Tm |

[a] Data from Rhodes (1977).

The main advantages of using these radioactive sources for X-ray analysis are that (1) they give the best signal-to-background ratio for quantitative analysis and (2) a portable X-ray analysis system can be easily assembled and carried to remote places for operation. This latter point has been demonstrated with the use of an XRF system using ⁵⁵ Fe and ²³⁸Pu as the excitation sources for trace element analysis on Venus (Surkov et al., 1982). The main disadvantage is the low counting rate of the secondary X rays, which is due to the low intensity of the primary X rays produced by the source.

## E. Production of X Rays by X-Ray Tubes

X-ray tubes are the most popular means for the production of primary X rays in XRF analysis.

The operation of an X-ray tube is based on the bombardment of a target material (anode) with a beam of electrons generated by applying a high current to the cathode. The electrons produced are accelerated along the anode as a result of a large potential difference applied between the anode and cathode. The X-ray continuum is then produced as a result of rapid deceleration of the bombarding electrons due to multiple interactions with the electrons in the anode as the bombarding electrons pass through the target material. The energy lost as a result of slowing down will be converted into a continuum of X-radiation (bremsstrahlung), and there will be a sharp maximum energy $E_{max}$ corresponding to the maximum energy of the electrons. This maximum energy is proportional to the applied voltage and can be calculated by

$$E_{max} \text{ (keV)} = 0.001 * V$$

where $V$ is the accelerating voltage applied across the X-ray tube. The total intensity of the primary X rays over the entire spectrum can be expressed by (Olsen, 1975)

$$I = 1.4 * 10^{-9} * i * Z * V^2$$

where $i$, $Z$, and $V$ represent tube current, atomic number of the anode, and tube voltage, respectively. This expression is important in selecting the appropriate X-ray tube. Commercially available XRF systems contain several X-ray tubes necessary to analyze the elements fluorine through uranium. The anode is made of chromium, molybdenum, rhodium, silver, tungsten, or gold. Improvements in detection sensitivity are being made using pulsed and polarized X rays.

## III. INSTRUMENTATION IN X-RAY ANALYSIS

There are two instrumental setups in X-ray analysis. They are an XRF system and a PIXE system. An XRF system uses an X-ray tube or radioisotopes to produce characteristic X rays, and a PIXE system uses an accelerator to produce characteristic X rays.

### A. X-Ray Fluorescence

In XRF, a beam of primary X rays is directed onto a sample being analyzed, causing it to emit secondary (fluorescent) X rays. These X rays are characteristic of each element in the sample and the total number of X rays is directly proportional to the concentration of the element from which they were produced.

There are two modes in which the secondary X rays, produced in XRF, can be detected. They are the wavelength dispersive mode (WDXRF), in which the X rays are analyzed according to wavelength, and the energy dispersive mode (EDXRF), in which the X rays are analyzed according to energy.

### B. Accelerators

In PIXE, energetic protons used for sample bombardment are mainly produced by a Van de Graaff accelerator. To produce accelerated protons, one must begin with $H_2$ gas in an ion source. The ion source, such as a duoplasmatron, produces hydride ions ($H_2^-$), which are then injected into the low-energy end of the accelerator. These ions can now be accelerated because they are attracted to the positive high voltage of several mega-electron-volts in the center of the machine. As the rapidly moving particles cross a thin carbon foil, both electrons are stripped, and the resulting protons are repulsed from the terminal. In a tandem accelerator, the total acceleration equals twice the terminal voltage.

## C. Target Chamber and Beam Line

The bombardment of samples takes place in a target chamber. There are two types of design, one employing an internal beam for analysis and one employing an external beam for analysis. For internal beam analysis, the chamber is designed to hold a high vacuum ($> 10^{-5}$ atm), and the X rays are detected external to the chamber. Irradiations that take place in this chamber result in the loss of volatile elements during the analysis, and the beam current must be kept low to prevent deterioration of the targets. In addition, because of the time involved in evacuating the chamber and then obtaining a good beam on target, one must have a multiple-sample positioner.

For external beam analysis, the chamber need not be operated under vacuum. Many analyses can be carried out at atmospheric pressure or in a helium environment. Analyses carried out using an external beam will eliminate or diminish problems due to the loss of volatile elements, because the system is not under vacuum. In addition, the helium cools the targets so that higher beam currents can be used, and it reduces the background due to charging effects. A chamber should have a means to measure the proton current, either a Faraday cup or a surface barrier detector to measure Rutherford backscattered protons. In Fig. 3, a typical external beam arrangement is illustrated. This system is much less expensive to construct than an internal beam system, because no vaccum is required in the chamber. However, since the detector is often placed inside the chamber, it should be protected from the helium environment, because diffusion of helium through the beryllium window can result in detector failure.

The construction of a PIXE external beam system is subject to much variety, depending on its intended use and the type of accelerator available. With the external beam system illustrated in Fig. 3, the accelerated beam is focused in the quadrupole magnet and passes through a tantalum collimator backed by a gold diffuser foil. This foil, located 60 cm before the target, defocuses the beam by multiple scattering to make it uniform. The energy lost as the beam passes through this foil is approximately 30 keV. Next the particles pass through three graphite collimators separated by 6-cm intervals. Each collimator has a 1-cm aperture that serves to refocus the beam

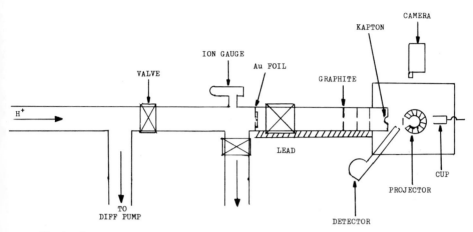

**Fig. 3.** Experimental setup for an external beam PIXE system. The length of the beam line from the first valve to the Faraday cup is 120 cm. The drawing is not to scale.

and select only the homogeneous center. The beam cross section on target should be scanned to ensure that it is homogeneous.

Approximately 60 cm downstream from the gold foil, the scattered collimated beam emerges from the vacuum of $10^{-6}$ atm through a 25-$\mu$m Kapton window and into a helium-filled target chamber. The Kapton window, located 7 cm from the target, is chosen because it is strong, pure, heat stable, and resistant to radiation damage. Also, no interferences due to the interaction of the protons with the window will appear in the X-ray spectrum. In the example cited, the beam emerges from the accelerator with an energy of 4 MeV. The total energy loss of the beam after traveling through the gold, the Kapton window, and the helium is approximately 0.4 MeV. Therefore, the final energy on target is 3.6 MeV. Because of the energy and current loss due to multiple scattering of the proton beam in the diffuser foil, the Kapton exit window, and the helium, an accelerator capable of producing a proton beam of at least 3 MeV and a current of at least 200 nA must be available.

The samples are introduced to the beam by a slightly modified Kodak Carousel projector, which can hold up to 140 samples. A stream of helium cools the targets as they are irradiated. In addition, the external beam can allow a detector to be placed within 1 cm of the target so that high count rates can be obtained; this allows a shorter analysis time (in some cases less than 1 min). The beam is finally dumped into a graphite Faraday cup, from which the charge is integrated. The entrance of the Faraday cup is sealed with Kimfol to prevent escape of charge.

With the development of the external beam, several laboratories have developed the proton microprobe (see the Proceedings of the International Conference on PIXE, 2nd, 1981) with beam diameters of several microns. These beams are useful for lateral distribution studies.

## D. X-Ray Detection Systems

There are two modes in which X rays can be detected, energy dispersive mode and wavelength dispersive mode.

### 1. Energy Dispersive Mode

In the energy dispersive mode, X rays are detected and analyzed according to their energies. The X rays are detected using a solid-state detector (semiconductor or scintillator) in which the energy of the X ray is deposited in the solid-state crystal. This mode can detect X rays of different energies simultaneously and produces real-time qualitative information on the composition of the sample. The experimental setup for an energy dispersive system is illustrated in Fig. 4. Commercial EDXRF systems were introduced in the early 1970s.

### 2. Wavelength Dispersive Mode

In the wavelength dispersive mode, a system of slits or collimators and a monochromator are used to isolate a selected wavelength for X-ray analysis. The X rays are analyzed sequentially by a goniometer. A collimator directs a parallel beam of the secondary X rays onto a monochromator (a crystal of LiF, Ge, polyethylene, or ADP), which separates the different wavelengths and diffracts them into a proportional gas flow or NaI(Tl) detector. At

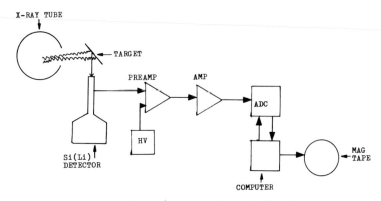

**Fig. 4.**  Experimental setup for energy dispersive X-ray fluorescence.

certain angles of the goniometer, corresponding to each different wavelength present, a peak of radiation is obtained. From the angles at which the goniometer obtained peaks, each element in the sample can be identified, and its concentration is proportional to the intensity of the measured radiation. The choice of the crystal is determined by which elements are being detected. The commerical instruments are equipped with several crystals to cover the analysis of the elements fluorine through uranium. The experimental setup for a wavelength dispersive system is illustrated in Fig. 5. Commercial WDXRF was first introduced in the early 1950s.

### 3. Detectors

There are two types of detectors commonly used in modern-day X-ray analysis systems. They are the solid-state detector and the proportional counter.

Solid-state detectors can be divided into two groups: the semiconductor and the scintillation detectors. The semiconductor detector can be a lithium-drifted silicon crystal [Si(Li)], a lithium-drifted germanium crystal [Ge(Li)], or an intrinsic germanium crystal (pure germanium). These detectors have to operate at liquid nitrogen temperatures to reduce thermal noise, and with lithium-drifted detectors the cooling is needed to prevent diffusion of the lithium through the crystal. The lithium produces an intrinsic zone within the silicon or germanium crystal that can be regarded as an ionization

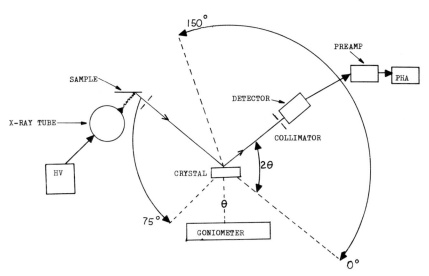

**Fig. 5.** Experimental arrangement for wavelength dispersive X-ray fluorescence. The analyzing crystal and detector move in their respective planes.

chamber. When an X ray is absorbed in the crystal, an electron-hole pair is created. The pair is collected by applying a high voltage (0.7 – 1.5 kV) across the intrinsic zone. Applying the high voltage (bias) produces a charge pulse at the detector terminal.

The choice of a semiconductor detector depends on which elements the experimenter would like to analyze. If the emphasis is on heavy elements, that is, elements whose characteristic X rays are greater than 10 keV, then an intrinsic germanium detector is recommended. If the emphasis is mainly on lower-energy X rays, then an Si(Li) detector is recommended. However, both detectors can perform well in both energy regions.

The second group of solid-state detectors is the scintillation detector. A popular scintillation detector used in WDXRF analysis is the thallium-activated sodium iodide crystal [NaI(Tl)] used as the phosphor coupled to a photomultiplier tube (PMT). In this detector, X rays excite iodine valence bond electrons into a conduction band, thereby leaving a positive hole behind. When the electrons combine with the positive holes, an excited state is formed and the atom rapidly decays to its normal ground state with the emission of radiation (410 nm). This radiation is then directed onto an antimony – cesium (Sb/Cs) photocathode, thereby producing a burst of electrons, which are then accelerated by a potential of 100 V or more to a series of 10 dynodes. Each dynode produces more electrons, which are accelerated to the next dynode, etc. Once the final dynode has been reached, the original photoelectron will have been multiplied by a factor of $10^6$. When the multiplied electrons reach the anode of the PMT, a small current is produced that is then amplified. This current is proportional to the incident X-ray energy. The scintillation detectors have a much poorer resolution than the semiconductor detectors.

The scintillation detector in a WDXRF system is used to measure the intensity of the radiation and not its energy. The detector is used for measuring the intensity of X rays between 0.02 nm and 0.14 nm. For $K_\alpha$ X rays these include the elements tungsten to gallium, and for $L_\alpha$ X rays the elements tungsten to uranium.

The second type of detector used in X-ray analysis is the proportional counter. A popular type of proportional counter used in WDXRF systems is the gas flow counter. The detector consists of a gas-filled hollow cylinder (grounded) with a thin wire (anode) along its radial axis. The gas is a mixture of 95% argon and 5% methane. One end of the cylinder is fitted with a thin window (1 – 6 $\mu$m of mylar coated with 0.1 nm of aluminum) through which X rays may travel. Under normal conditions, the gas mixture between the electrodes is a good insulator. Because of this, a high voltage can be applied to the electrodes to form an electric field in the gas mixture. Primary X rays interacting with the gas mixture cause ionization and the production of ion pairs (electron-positive ions). The energy required to produce these pairs in the gas mixture is 30 eV. As a result, the electrons are accelerated toward the anode, producing more and more ion pairs on the way (this is the avalanche phenomenon). This results in an increase in the ionization current, which can be measured with the appropriate electronics. The number of ion pairs produced by the X ray is proportional to the energy of the X ray and thereby proportional to the measured current.

Depending on the design of the proportional counter, resolutions approaching that of a solid-state detector ($<150$ eV) can be achieved.

Typically it requires 296 eV to produce an ion pair in germanium at $77°K$ (liquid nitrogen temperature) as compared to 30 eV in a proportional counter. This large difference in energy accounts for the higher resolution and detection efficiency (100%) of semiconductor detectors. The resolution of these detectors is not as good as that obtained with a wavelength dispersive system, but the detector can detect different X-ray energies simultaneously. A filter is often placed in front of the detector to reduce the counting rate and to attenuate low-energy X rays. Again, in WDXRF systems, the detector is used to measure the intensity of the selected X ray and is used for measuring $K_\alpha$ and $L_\alpha$ X rays with wavelengths between 1.8 and 0.13 nm. This range includes the $K_\alpha$ X rays for the elements fluorine to germanium and the $L_\alpha$ X rays for the elements molybdenum to tungsten.

The proportional counter has a short dead time and can be used in high count rate measurements. Because of this, the dynamic range of the WDXRF spectrometer is increased (i.e., one set of instrument parameters can handle wider concentration variations).

## 4. Electronics

In the energy dispersive system, a small charge is produced at the detector terminal. This small charge is then reshaped, integrated, amplified, and converted into a voltage pulse by the preamp. This pulse is proportional to the deposited X ray. The output of the preamp is then sent to the main amplifier for pulse shaping. The main amplifier also contains circuitry for pileup rejection. The pileup rejector is needed because the pulses can become distorted due to leading edge, trailing edge, and sum peak effects. Pulse distortion can also be diminished by using an on-demand beam system. The rejector maintains linearity between the energy of the X ray and the output pulse. When the rejector is activated, signals from the detector cannot be processed, because the system is disabled. The linear (analog) output signals (0 to $+10$ V) from the main amplifier are then sent to an analog-to-digital converter (ADC). The computer then transfers the signals from the ADC directly into memory. The X-ray spectrum can now be viewed on a cathode-ray tube (CRT) in real time, and the experimenter can observe the X-ray spectrum grow. The electronic setup used in a PIXE and an EDXRF system are identical. The setup is illustrated in Fig. 4.

In the wavelength dispersive system, the charge produced at the terminal of the detector is reshaped and amplified by a preamplifier, and the signal is then sent to a pulse-height analyzer (PHA).

## 5. Resolution

Resolution is very important in X-ray analysis, because of the small differences in energy of the various characteristic X rays. It is defined in terms of the ability of the detector to differentiate between two X rays of similar energy and is expressed as the full width of the X-ray peak at half its maximum (FWHM).

Some of the factors that contribute to the resolution of the X-ray analysis system include detector voltage, shaping time of the main amplifier, counting rate, energy of the detected X ray, energy to produce an electron-hole pair, and system noise. Manufacturers can produce Si(Li) and intrinsic germanium detectors with resolutions less than 150 eV at the manganese K line. It should be noted that the resolution varies with energy, and it is generally agreed that in the intercomparison of different detector systems the resolution be quoted in reference to the manganese $K_\alpha$ X ray. The best resolution (2 eV) is obtained in the wavelength dispersive mode.

## IV. MULTIELEMENTAL ANALYSIS

### A. Sample Collection

One of the most important aspects of X-ray analysis is sample collection. All samples should be collected in some type of acid-washed conventional polyethylene (CPE) or borosilicate glass containers. Lindstrom and Moody (1977) determined the concentrations of different elements in different plastic containers and recommend CPE for sample storage and collection.

When collecting the sample, care should be taken to be sure the sample is representative of the bulk. The sample size will depend on the type of information required by the experimental plan. If the standard deviation of the individual sample is known in advance or can be reasonably estimated, then a statistical approach to sampling can be applied. According to Walpole and Myers (1972), statistical sampling can be described by the expression

$$n = (Z\sigma/E_{\mathrm{m}})$$

where $n$ is number of samples, $Z$ is a constant (standard normal) from tables (Natrella, 1963), $\sigma$ is the standard deviation of individual samples, and $E_m$ is the tolerable error in the estimation of the mean. Often the data needed to determine the minimum number of samples ($n$) are not available at the time of sampling. Another approach suggested by Harris and Cummings (1964) is to use the *n-n-n* rule. The rule states that equal numbers ($n$) of field samples, spiked blanks, and field blanks should be analyzed along with the calibration standards and controls.

## B. Sample Preparation

Neither PIXE nor XRF requires any sample preparation if the concentration of the elements is above 0.2 ppm. For liquids, a sample size of 5–40 $\mu$l can be deposited on a thin backing material attached to a slide frame, vacuum dried, and irradiated. For solids, pellets of the material can be made, or the solid can be converted to a liquid by fusion with common fluxes, and the sample can then be irradiated.

The rate-determining step for obtaining quantitative results is sample preparation. It should be avoided whenever possible because of introducing contaminants into the sample and obtaining less than 100% recovery of the elements. However, to obtain a homogeneous sample and to increase the sensitivity by preconcentration when the elemental concentrations are less than 0.2 ppm, sample preparation is required. Sample preparation should be carried out in a Class 100 clean hood. All plasticware and glassware should be acid washed, and reagents should be of ultrapure quality.

There are four popular methods for sample preparation. They are freeze-drying (lyophilization), wet and dry ashing, solvent evaporation, and chelation.

Freeze-drying is often used in conjunction with other methods of sample preparation (such as wet and dry ashing). For foods, if the material contains some moisture, freeze-drying is recommended to reduce the sample mass. Using freeze-drying to reduce the mass for biological materials, Iyenger *et al.* (1980) obtained a recovery rate of better than 98% for many elements.

Ashing techniques are often used for preparation of biological materials. Foods are largely composed of an organic matrix consisting of carbon, oxygen, hydrogen, and nitrogen. The advantage of removing this matrix is twofold. First, the detection limits of elements with higher atomic numbers are increased since the specimen now contains more of those elements in a given quantity of the sample. Second, the background radiation due to bremsstrahlung is reduced.

In deciding what technique to employ to remove the organic matrix, the benefits of wet ashing versus dry ashing should be considered. Wet ashing involves digestion of the matrix with high-purity, strong, oxidizing acids in a Teflon digestion bomb. Dry ashing can be accomplished by heating the sample in a muffle furnace ($500°C$) or by passing a gaseous plasma over the sample at low temperatures ($50-100°C$). Gorsuch (1970), in his monograph on the destruction of organic materials, recommends wet digestion because of the high recovery rates and excellent reproducibility using an internal standard. However, according to Mangelson et al. (1979), $HClO_4$ and $H_2SO_4$ do not dry sufficiently on backing materials and use of HCl, HF, and $HNO_3$ produces a nonhomogeneous solution, because these acids do not oxidize fats. Mangelson did find that $HNO_3$ used in conjunction with $H_2O_2$ decomposes the fat at high pressure.

High-temperature dry ashing should be avoided as a method for sample preparation because of the selective loss of volatile elements. Nadkarni and Haldor (1972) reported that low-temperature plasma ashing is the best method to keep metal losses to a minimum. In terms of elemental recovery, the low-temperature technique is quite comparable with the high recovery rates of the wet digestion method. However, caution should be exercised when using low-temperature ashing, because the ashing power (in watts) is proportional to the surface temperature of the sample and should be kept low (100 W) to minimize losses due to volatility. Ashing parameters for a variety of foods and beverages are shown in Table 3.

Solvent evaporation involves placing the liquid in a Teflon evaporating

**TABLE 3   Ashing Parameters of Various Samples[a]**

| Sample material | Ashing time (hr) | Radio-frequency power (W) | Avg. temp. (°C) |
|---|---|---|---|
| Egg albumin | 125 | 190 | 84 |
| Beef liver[b] | 4 | 200 | 200 |
| Alfalfa | 26 | 300 | 223 |
| Rye | 48 | 200 | 185 |
| Casein | 141 | 190 | 120 |
| Instant tea | 24 | 250 | 108 |
| Orange juice | 30 | 300 | — |
| Cane syrup | 40 | 300 | — |
| Mazola corn oil | 24 | 200 | 160 |
| Seaweed | 28 | 175 | 228 |
| Raw sugar | 16 | 263 | 270 |
| Milk[c] | 10 | 166 | 122 |
| Ice cream paste | 25 | 207 | 164 |

[a] Data from LFE Corporation (1978), Waltham, Massachusetts.
[b] Lyophilized.
[c] Powdered.

dish and evaporating the solution to a small volume (0.5 ml). Subsequently, 5–40 μl of the concentrate can be deposited onto a sample backing, vacuum dried, and irradiated. An internal standard is often added prior to the evaporation to determine the preconcentration factor. Rapid evaporation should be avoided and the temperature kept below the boiling point.

In chelation, a complexing agent is added to the sample, after proper pH adjustment, and the solution is filtered through a membrane filter. The entire membrane filter is then irradiated. Leyden *et al.* (1982) compared several preconcentration methods for the determination of trace elements in water by WDXRF and EDXRF. They obtained the best results with sodium dibenzyldithiocarbamate (DBDTC) and WDXRF.

For solids, a typical sample preparation is as follows. Sample is obtained and freeze-dried if it contains moisture. The residue is ground to a fine powder in a mullite ($3Al_2O_3 \cdot 2SiO_2$) mortar so that it will pass through a 270-mesh nylon sieve. Two subsamples ($> 250$ mg) are transferred to acid-washed borosilicate glass vials, and the sample is ashed in a low-temperature asher. To the residue, 6-F Ultrex nitric acid containing 200 ppm yttrium as the internal standard is added. With a micropipette, 5–40 μl of the solution is transferred to sample backing, dried under vacuum, and irradiated. It is important to note that, when liquids are deposited on sample backings, the target should be dried under vacuum. This will produce a microcrystalline structure, thereby ensuring a homogeneous target.

For liquids of low organic content, a typical sample preparation would be as follows. For 100 ml of sample, adjust the pH to 4, add 1 ml of methanolic 1% (wt/vol) DBDTC, and shake for 30 min. Vacuum filter the solution through a 0.45 μm Nuclepore filter (specify low bromine content), attach to the target frame, and irradiate. An alternative to this procedure would be to freeze-dry 100 ml of the sample, dissolve the residue in 6-F Ultrex nitric acid, deposit the solution onto sample backing, vacuum dry, and irradiate. However, this method requires 24 hr for sample preparation. If the liquid has a high organic content, then filtering will be difficult. In this case, freeze-dry 100 ml of the sample, quantitatively transfer the residue to a mullite mortar, and grind it to a fine powder. The remaining procedure is identical to that described above for solids.

Wet digestion would be recommended only if the concentration of the elements to be determined were greater than 400 ppm. This is because wet digestion dilutes the sample. For example, if 250 mg of a sample were digested in a Teflon bomb and the contents were quantitatively transferred to a 5-ml volumetric flask and diluted to the mark with water, the sample would be diluted by a factor of 100. Because of this, the MDL of X-ray analysis would be approached. In addition, the concentration of toxic metals in foods is suspected to be less than 1 ppm, and this method of sample preparation would be inappropriate. Often, the preconcentration factors are high, and these lead to a higher sensitivity in the analysis.

Procedural blanks should be analyzed periodically throughout the analysis. Studies relating to elemental recoveries should be carried out so that the final calculated concentrations are adjusted for elemental losses. In addition, certified standards similar in composition to the sample (from the National Bureau of Standards' Standard Reference Materials) should be analyzed to document the accuracy of the analytical method.

## C. Sample Backing Materials

The sample backing material should be relatively free of trace elements (especially of those under analysis), ultrathin, resistant to chemical and radiation attack, strong, inexpensive, readily available, and easy to prepare targets from. Thinness of the backing material is stressed, because a thick backing can absorb protons and X rays and decrease the sensitivity of the analysis.

Mangelson *et al.* (1977) and Campbell *et al.* (1981) discuss the advantages and disadvantages of different backing materials. Campbell recommends that Kimfol be used because of low bremsstrahlung and trace impurities. In XRF analysis, the samples are irradiated in polypropylene cups with a thin ultrapure mylar window through which the X rays pass. In a few commercial XRF instruments, the experimenters can attach their own backing material to slide frames and irradiate the samples.

## D. Minimum Detection Limits

The MDL is important in quantitative analysis. An X-ray spectrum may contain several characteristic X rays that can be identified qualitatively, but analysis of the data may not be pushed to obtained quantitative information. The MDL is approached when the statistical uncertainty in the area of the background is of the same approximate magnitude as the area of the overlying X-ray peak. It is generally agreed that the MDL can be represented by the expression

$$N_z \text{ (MDL)} \geq 3\sqrt{N_{bkd}}$$

where $N_z$ is the area of the designated X-ray peak and $N_{bkd}$ is the area of the underlying background (Fig. 2). The MDL is also influenced by the analysis time, sample preparation, detector resolution, beam intensity, count rate, target thickness, and operational parameters of the system. For a typical analysis time of 5 min with an external beam system, the MDLs for the elements iron through uranium has been shown by Asch (1982) to be less than 60 ppb. For a typical XRF system, MDLs have been reported to be less than 100 ppb.

### E. Computerized Data Analysis

Once the appropriate X-ray spectra have been obtained, the next task is to convert the characteristic X rays into qualitative and quantitative information.

An X-ray analysis system should have a good computer program for spectrum analysis. The commercial systems for XRF contain software packages that can perform complex multielemental spectrum analysis on line.

The computer program should be capable of (1) locating and identifying all possible X rays, including multiplets, (2) accurately determining the concentration of the elements in the sample, (3) making corrections for elemental interferences, (4) making corrections for matrix effects, and (5) performing the preceding four operations under a variety of experimental conditions. In addition, the program should be free of operator attention and should perform the analysis in a short time.

In PIXE, the X-ray spectrum (Fig. 2) consists of a hump at the low-energy end that flattens out at the high-energy end. Several laboratories have used modified $\gamma$-ray spectrum analysis computer programs to analyze PIXE spectra. Several programs in current use include HEX (Kaufmann *et al.,* 1977), RACE (Cahill, 1975), and TOURIX (Monaro *et al.,* 1978).

The best method of analysis, whenever applicable, is to use an internal standard. The sample should not contain any traces of the internal standard. An internal standard that has been used extensively in X-ray analysis is yttrium. When an internal standard is being used, all calculations are made relative to the internal standard, and any fluctuations in the instrumentation will be taken into account. The concentration of an element in a sample can be determined by comparing the area of the $K_\alpha$ (or $L_\alpha$) peak of the element of interest with that of a reference element that contained yttrium as the internal standard. The concentration of an element in the sample, in parts per million in dry weight, can be finally calculated by

$$[Z_s] = V\,(\mathrm{ml})\,\frac{[Z_{st}]N_Z^s N_Y^{st}[Y]_s}{g N_Y^s N_Z^{st}[Y]_{st}}$$

where $V$ is the final sample volume, $g$ is the weight of the sample, $N_Z$ is the number of characteristic X rays for the element with atomic number $Z$, and $N_Y$ is the number of yttrium X rays. The symbols s and st refer to the sample and the standard, respectively. The standard reference spectra are obtained by bombarding high-purity atomic absorption standards under the same conditions (with respect to target thickness, type of backing and its thickness, integrated charge, and current) as the samples.

## V. APPLICATIONS

The majority of the published reports of PIXE and XRF applications have been applied to trace element analysis of biological, environmental, geological, and metallurgical samples.

### A. Foods

For PIXE, Szymczyk (1981) determined the concentrations of lead and bromine in lettuce, parsley, and onion leaves and in apples. The study was conducted to determine the influence of traffic and industrial pollution on vegetables and crops. Legge *et al.* (1980) used the proton microprobe for the distribution of trace elements in mature wheat seeds. Trace element analysis of panax ginseng was studied by Lecomte *et al.* (1981). Campbell *et al.* (1975) determined the concentrations of several trace elements in chicken meat and reported lead to be 54 ppm.

For XRF, Peto *et al.* (1982), using a $^3H/Ti$ source for excitation, determined low $Z$ elements ($Z < 17$) in cabbage and milk powder.

### B. Beverages

For PIXE, Campbell *et al.* (1977) determined trace elements in wines from different countries. The analysis was performed by depositing a small drop of wine on a thin carbon foil and irradiating the sample with 2.3-MeV protons. Chu *et al.* (1977) determined the concentrations of calcium, manganese, iron, copper, zinc, and bromine in human and cow's milk. Hall (1982) has used an external beam to determine trace metal concentrations in breast milk. The emphasis was on heavy metals. Using a proton beam of 3.5 MeV, Ishi *et al.* (1975) determined the concentrations of several metals in canned tomato juice. They reported cadmium at 13 ppm, tin at 57 ppm, and lead at 1.7 ppm.

The limited publications involving PIXE and XRF for trace element analysis of foods and beverages are probably attributable to a variety of causes, such as protection of proprietary information in the food and beverage industry, use of traditional methods of analysis (i.e., atomic absorption and colorimetry), and the limited availability of PIXE installations.

## VI. ADVANTAGES AND DISADVANTAGES OF X-RAY ANALYSIS

In the choice of an analytical technique, several criteria such as instrument cost and maintenance, need for support personnel, instrument availability, instrument sensitivity, instrument accuracy, number of elements determined, turnaround time, and versatility should be considered.

### Comparison with Other Analytical Techniques

A comparison of X-ray analysis with other analytical techniques is shown in Table 4. The main advantages of X-ray analysis over other analytical techniques, excluding instrumental neutron activation analysis (INAA), are turnaround time and the ability to achieve multielemental analysis nondestructively while analyzing small samples in either liquid, solid, or gaseous form. Another advantage is the ability of X-ray analysis to obtain a "fingerprint" analysis of the sample. This can be used to locate a source of contamination in the food or beverage or to monitor the uniformity of different batches of products. In comparison with INAA, turnaround time and high sensitivity for lead and iron are the advantages. Another factor in the advantage of X-ray analysis over the other techniques is the cost-to-sample ratio. The initial cost for an XRF system is approximately $100,000, but this cost can be justified when a large number of samples have to be analyzed for many elements. It should be noted that PIXE and XRF can determine suspected and unsuspected elements in a sample whereas atomic absorption (AA) is an *a priori* method. However, the detection limits in AA and atomic

TABLE 4   Comparison of X-Ray Analysis with Other Analytical Techniques

| Method | Sample preparation[a] | Multielemental | Nondestructive | Availability | Operators |
|--------|------------------------|----------------|----------------|--------------|-----------|
| PIXE | None | Yes | Yes | Limited | >2[b] |
| WDXRF | None | Sequential | Yes | Commercial | 1 |
| EDXRF | None | Yes | Yes | Commercial | 1 |
| AA[c] | Yes | No | No | Commercial | 1 |
| ICPAES[d] | Yes | Sequential | No | Commercial | 1 |
| INAA[e] | None | Yes | No | Limited | >2[b] |
| SIMS[f] | Yes | Yes | No | Commercial | 1 |

[a] For liquids and solids.
[b] Large support personnel needed to maintain operation of facility.
[c] Includes electrothermal and flame atomic absorption.
[d] Inductively coupled plasma atomic emission spectrometry.
[e] Instrumental neutron activation analysis.
[f] Secondary ion mass spectrometry.

emission spectrometry are one to two orders of magnitude better than in X-ray analysis.

In comparison with XRF, the disadvantage of PIXE is the availability of the instrumentation. The advantage of PIXE over XRF is its greater sensitivity for certain elements and its higher flux for the irradiating particles. Comparing the wavelength dispersive mode with the energy dispersive mode, the former is more advantageous because of the much better resolution (2 eV as compared with 150 eV). As a result of this large difference in resolution, a weak analyte can be analyzed in the presence of an overlapping line from a major element.

## VII. CONCLUSION

Proton-induced X-ray emission and X-ray fluorescence analysis are very powerful and versatile analytical techniques for quantitative–qualitative analysis of foods and beverages. Often, sample preparation can be eliminated and rapid analysis performed. The techniques are multielemental and nondestructive, and small samples can be analyzed. Percent relative standard deviations (%RSD) are often less than 5%, and accuracies are often less than 10%. PIXE and XRF complement each other.

Care should be taken when applying sample preparation techniques. Low-temperature ashing is the best means of preparing samples to produce homogeneous samples and to eliminate the light elements (carbon, hydrogen, oxygen, and nitrogen). Targets should be kept as thin as possible to minimize X-ray and particle absorption. Whenever possible an internal standard should be incorporated into the sample.

The linear dynamic range is greater in WDXRF as a result of the type of detector employed in the instrumentation. The superior resolution of the analyzing crystal in WDXRF systems makes easy the analysis of the material in the presence of a strong compound that has a major X-ray line close to it.

The limited availability of PIXE installations and the lack of knowledge of the technique has led to few applications in the food and beverage industry. However, as people become aware of the full potential and versatility of PIXE, its applications in the food and beverage industry will increase.

## REFERENCES

Asch, J. (1982). B. A. thesis, Dept. of Chemistry, Rutgers, State Univ. of New Jersey, New Brunswick.

Cahill, T. A. (1975). "New Uses of Ion Accelerators." Plenum, New York.

Campbell, J. L., Orr, B. H., Herman, A. W., McNelles, L. A., Thompson, J. A., and Cook, W. B. (1975). *Anal. Chem.* **47,** 1542–1553.

Campbell, J. L., Orr, B. H., and Noble, A. C. (1977). *Nucl. Instrum. Methods* **142,** 289–291.

Campbell, J. L., Russell, S. B., Shatha, F., and Schulte, C. W. (1981). *Anal. Chem.* **53,** 571–574.

Chaturverdi, R. P., Wheeler, R. M., Liebert, R. B., Miljanic, D. J., Zabel, T., and Phillips, G. C. (1975). *Phys. Rev. A.* **12A,** 52–60.

Chu, T. C., Navarret, V. R., Kajl, H., Izawa, G., Shiokawa, T., Ishii, K., and Tawara, H. (1977). *J. Radioanal. Chem.* **36,** 195–207.

Gorsuch, T. T. (1970). "The Destruction of Organic Matter." Pergamon, Oxford.

Hall, G. (1981). "182nd American Chemical Society National Meeting Book of Abstracts." Am. Chem. Soc., Washington, D. C.

Harris, T. H., and Cummings, J. G. (1964). *Residue Rev.* **6,** 104–110.

Ishii, K., Morita, S. Tawara, H., Chu, T. C., Kajl, H., and Shiokawa, T. (1975). *Nucl. Instrum. Methods* **126,** 75–80.

Iyenger, G. V., Kasperek, K., and Feinendegen, L. E. (1980). *Analyst (London)* **105,** 794–801.

Jenkins, R., and de Vries, J. L. (1975). "Practical X-Ray Spectrometry." Springer-Verlag, Berlin and New York.

Johansson, S. A. E., Johansson, T. B., and Akselsson, R. (1970). *Nucl. Instrum. Methods* **84,** 141–165.

Johansson, S. A. E., and Johansson, T. B. (1976). *Nucl. Instrum. Methods* **137,** 473–516.

Kaufmann, H. C., Akselsson, K. R., and Courtney, W. J. (1977). *Nucl. Instrum. Methods* **142,** 251–273.

Khan, M. R., and Crumpton, D. (1981a). *CRC Crit. Rev. Anal. Chem.* **2,** 103–160.

Khan, M. R., and Crumpton, D. (1981b). *CRC Crit. Rev. Anal. Chem.* **3,** 161–193.

Lecomte, R., Landsberger, S., and Paradis, P. (1981). *Radiochem. Radioanal. Lett.* **50,** 167–176.

Legge, G. J. F. (1980). *Nucl. Instrum. Methods* **168,** 563–569.

Leyden, D. E., Ellis, A. T., Wegscheider, W., Jablonski, B., and Bodnar, W. (1982). *Anal. Chim. Acta* **142,** 73–87.

LFE Corp. (1978). Technical Report, Waltham, Massachusetts.

Lindstrom, R. M., and Moody, J. R. (1977). *Anal. Chem.* **49,** 2264–2267.

Mangelson, N. F., Hill, M. W., Nielson, K. K., and Ryder, J. F. (1977). *Nucl. Instrum. Methods* **142,** 133–142.

Mangelson, N. F., Hill, M. W., Eatough, D. J., Christensen, J. J., Izatt, R. M., and Richards, D. O. (1979). *Anal. Chem.* **51,** 1187–1193.

Monaro, S., Lecomte, R., Paradis, P., Barrette, M., Lamoureux, G., and Menard, H. A. (1978). *Nucl. Instrum. Methods* **150,** 289–299.

Nadkarni, R. A., and Haldor, B. C. (1972). *Anal. Chem.* **44,** 1504–1512.

Natrella, M. G. (1963). "Experimental Statistics," NBS Handbook 91. U.S. Govt. Printing Office, Washington, D.C.

Olsen, E. D. (1975). "Modern Optical Methods of Analysis." McGraw-Hill, New York.

Ortec (1976). AN 34 "Experiments in Nuclear Science." Ortec, Inc., Oak Ridge, Tennessee.

Peto, G., Numba, N. K., and Csikai, J. (1982). *Radiochem. Radionanal. Lett.* **52,** 373–382.

Proc. Int. Conf. PIXE, 2nd (1981). *Nucl. Instrum. Methods* **181.**

Rhodes, J. R. (1977). "Spectroscopy." Int. Sci. Commun., Inc., Connecticut.

Röntgen, W. C. (1898). *Ann. Phys. (Leipzig)* **64,** 1–23.

Sparks, C. J. (1980). *Synchrotron Radiat. Res.* pp. 459–512.

Surkov, Y. A., Scheglov, O. P., Moskalyeva, L. P., Kirchenko, V. S., Dudin, A. D., Gimadov, V. L., Kurochkin, S. S., and Rasputny, V. N. (1982). *Anal. Chem.* **54,** 975A–960A.

Szymczyk, S., Kajfosz, J., Hrynkiewicz, A. Z., and Curzydlo, J. (1981). *Nucl. Instrum. Methods* **181,** 281–284.

Uemura, Y. J., Kuno, Y., Koyama, H., Yamazaki, T., and Kienle, P. (1978). *Nucl. Instrum. Methods* **153,** 573–579.

Walpole, R. E., and Myers, R. (1972). "Probability and Statistics." Macmillan, New York.

Woldseth, R. (1973). "X-Ray Energy Spectrometry." Kevex Corp., Burlingame, California.

# 13

# Scanning Electron Microscopy in the Analysis of Food Microstructure: A Review*

SAMUEL H. COHEN

*Science and Advanced Technology Laboratory*
*United States Army Natick Research and Development Laboratories*
*Natick, Massachusetts*

## I. INTRODUCTION

In the definitive review of scanning electron microscopy (SEM) in food science and technology, Pomeranz (1976) explored nearly every aspect of SEM for the study of foods. From tracing the history of SEM, to detailing preparative techniques, to discussing special topics, he showed how SEM has played, and will continue to play, an important role in food research.

---

* The views of the author do not purport to reflect the positions of the Department of the Army or the Department of Defense.

ANALYSIS OF FOODS AND BEVERAGES

Since that review was written, there have been a number of advances in the science of SEM, including the construction of instruments with better resolution, a better understanding of preparative techniques, and the increased use of microcomputers. Although some of these advances will be mentioned, the purpose of this review is to outline briefly how SEM has been used either alone or as a corroborative tool for the elucidation of food microstructure. It is impossible to cover every aspect of food-related SEM uses, as it is impossible to cite every reference, so only a few selected areas will be highlighted.

## A. Historical Perspective

When Pomeranz's review article was written, commercial scanning electron microscopes had been in use for approximately 10 years. However, 46 years ago, following Knoll's (1935) suggestion that if a scanning electron beam could be focused onto a specimen surface the emission current could be recorded, von Ardenne (1938a,b) built the first SEM instrument. It used two sets of magnetic coils to scan the beam across the specimen surface in a television rasterlike fashion. The method used to record the image photographically was inefficient, and focused micrographs could only be obtained by trial and error. Even though the instrument's performance was unfavorable, von Ardenne's suggestions regarding improvements were used by Zworykin et al., (1942) to construct an instrument whose resolution was approximately 500 Å, but with impaired contrast as a result of contamination.

After World War II Bernard and Davoine's (1957) improvements in resolution, and work at Cambridge, England, under Oatley and at JEOL in Japan, led to an instrument whose resolution was approximately 250 Å.

Some of the other early pioneers of the science of SEM were McMullan (1952), Smith (1956), Oatley and Everhart (1957), Wells (1957), Everhart (1958), Davoine et al. (1960), Thornley (1960), Pease (1963), and Kimura et al. (1966). By reading Oatley et al. (1965), Everhart (1968), as well as other similar books, one can gain a historical perspective of the establishment of SEM as an invaluable tool for the study of microstructure.

## B. Operation of the SEM

The method of generating an image in the SEM (Fig. 1) is by irradiation of the specimen with electrons in sequence instead of by simultaneous irradiation as in the light microscope (LM) or transmission electron microscope (TEM) (Coates and Brenner, 1973).

According to Thornton (1968), the source of electrons (a tungsten, lanthanum hexaboride, or field emission source) is centrally mounted in a shield

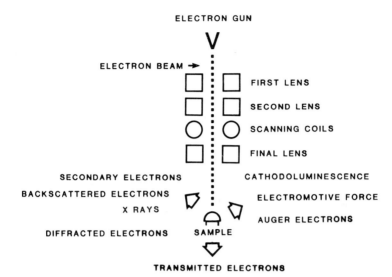

**Fig. 1.**   A schematic representation of an SEM with analytical capabilities.

that has a small aperture just below the filament point. The filament and shield form the cathode assembly and have a high negative potential. When heat is applied, electrons leave, are accelerated, and then are focused by the lenses. When the electrons pass through scanning coils, they are made to move across or "scan" a specimen. The scanning of the electron beam forms a pattern or a series of closely packed parallel lines called a raster.

The secondary electrons bounce off the surface of the specimen and are pulled toward a positively charged Faraday cup. Some electrons penetrate through the metal screen of the cup and are accelerated toward the scintillator. Because of the kinetic energy of the electrons used to excite the scintillator material, photons are produced, some of which travel via a light-pipe to a photomultiplier, where they are converted to a photocurrent that modulates the cathode-ray tube (CRT).

The SEM has magnetic lenses that demagnify the effective spot size of the electron beam so that, by the time the beam reaches the specimen, the spot size is reduced significantly.

Theoretical resolution in an SEM is limited by several factors, including chromatic aberration, spherical aberration, diffraction, and contamination from deposition of amorphous material around the point of impact of the focused beam, from the specimen itself, or from oils and polymeric materials (Love *et al.,* 1981).

When electrons impinge upon a sample they may interact with the atomic nuclei, either with the inner orbital electrons or with electrons in the outer bands (Chandler, 1980). The resulting reactions produce emissions or alterations in electron energy that can be measured to give information about the sample. Some of these reaction products are as follows.

1. X rays that are characteristic of the emitting atom can be used to identify specific elements within the specimen. There are two types of X ray collection detectors: wavelength dispersive crystal spectrometers and energy dispersive solid-state detectors.

2. Cathodoluminescence is produced by the electron beam, causing the emission of light in the visible, near-ultraviolet (near-UV), or far-infrared (far-IR) regions.

3. Auger electrons are produced by surface excitation. Although their application in food sciences is limited, they may, in the future, be used to study lighter elements on tissue surfaces.

4. Diffraction techniques can be used to obtain the crystal structure of a sample and can provide information regarding the quality of the diffraction patterns.

5. Backscattered electrons are produced when primary incident electrons are scattered back from the specimen surface. A better visualization of subsurface layers can be obtained.

6. Secondary electrons escape from the region in the immediate vicinity of the electron beam and can be used to detect topographical differences in a sample.

7. Electromotive force is used mainly by those in the semiconductor industry to obtain information about the flow of current due to a photovoltaic effect.

8. Transmitted electrons make it possible to see the internal structure of a thin specimen when an electron beam passes through. The TEM mode of an SEM gives higher contrast and brightness to an image obtained from a thicker specimen than normally prepared for TEM.

## 1. Electron Sources

In a TEM a primary electron beam of approximately 100 KV or more passes through a thin specimen. The transmitted and forward-scattered electrons form a diffraction pattern in the back focal plane and an image, which is magnified, in the image plane. With the aid of intermediate lenses, the image or diffraction pattern is projected onto either a fluorescent viewing screen or a photographic surface.

An SEM uses a fine electron probe to illuminate the surface of a bulk specimen. The probe beam is scanned across the specimen, and an image is formed by detecting the low-energy secondary or high-energy primary backscattered electrons.

Today there are three major electron sources for an SEM. The first is a tungsten (W) hairpin filament, a thermionic source that is the one most commonly used. The second, lanthanum hexaboride ($LaB_6$), can be either polycrystalline or a single crystal, which may perform better. It offers higher brightness than W and a longer life, but a better vacuum is required, and because $LaB_6$ is a very reactive material, several drawbacks still exist. The third type of electron source is the field emission (FE) gun, which uses a pointed needle of some metal, usually tungsten. When a strong electric field is applied to the tip, the resulting electron emission is very bright as compared to the W and $LaB_6$ sources. However, several problems need to be overcome before FE sources are more widely used. The first is tip fragility, which makes breakage all too easy. The second is the amount of time required to change tips. Should a W tip blow, the SEM can become operational within an hour, while the replacement of an $LaB_6$ tip and SEM turn-on is somewhat longer. But, in order to replace an FE tip, the new tip needs to be "baked" for about 2 hr, followed by cooling for several hours, before the SEM can be turned on again. Therefore, there is a trade-off— brighter, potentially longer-lasting tips with relatively long replacement time, or inexpensive tips with not quite so bright emission but fast replacement time.

## 2. Literature Related to SEM Physics

The list of references dealing with the physics of SEM is quite extensive. Examining some of the following reference material will assist the reader in gaining a larger overall perspective of this field. Broers (1965a,b), Danilatos (1981), Goldstein and Yakowitz (1975), Hold et al. (1974), Johari (1968 – present), Johari (1971, 1974), Kimoto and Russ (1969), Parsons (1978), Wells et al. (1974), Williams (1980), and Williams and Edington (1981, 1982). Also, the magazines published by the various electron micrograph companies provide a great deal of information. Some of these are *Hitachi Instrument News, JEOL News, Norelco Reporter* (Philips), and *Zeiss Information*.

## II. WHY SEM FOR FOOD MICROSTRUCTURE?

Among the most valuable tools used to study the microstructural characteristics of foods is the SEM. Although the TEM has the advantage of achieving higher resolution, the SEM, because of its ability to elucidate three-dimensional surface morphology, has recently enjoyed wider acceptance. Several other factors have enabled the SEM to become an even more valuable instrument. First, the preparative techniques are less tedious and complicated for SEM. Second, the resolution of the newer SEM instruments, while still not as good as in the TEMs, nevertheless has improved significantly. Third, the relatively large sample size permits the examination of a greater proportion of a sample, as compared with a TEM grid-mounted sample. Fourth, the SEM sample image (with a higher contrast) can be seen on a television monitor, whereas the TEM image can be seen only on a phosphorescent screen. Fifth, photomicrographs of an SEM sample can be made very rapidly with Polaroid film, while instant pictures cannot be made of TEM images. From sample preparation to photomicrography, SEM has been a most effective and efficient method for observing microstructural changes in foods, feeds, and ingredients.

## A. Quality of Photomicrographs

As mentioned above, one of the major advantages of SEM over other types of microscopy is the three-dimensional appearance of a particular structural feature. Even the nondescript surface of a compressed carrot (Fig. 2) has depth. In Fig. 3, which is a photomicrograph of sauerkraut, even

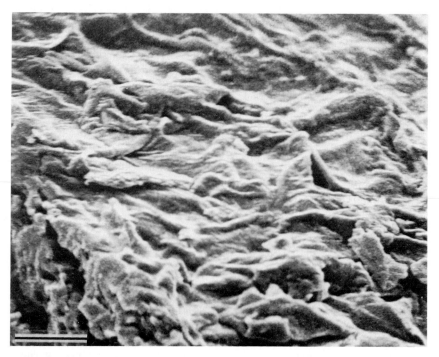

**Fig. 2.** Scanning electron micrograph of the surface of a compressed carrot. The bar equals 0.002 mm.

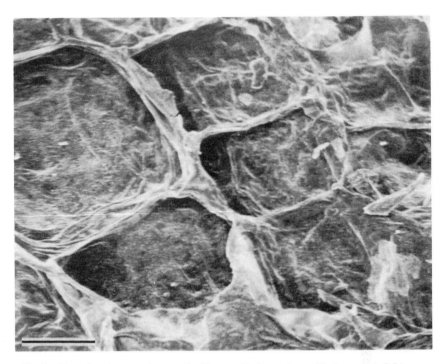

**Fig. 3.**   Scanning electron micrograph of freeze-dried sauerkraut. The bar equals 0.04 mm.

though the picture was taken at 0° tilt, the shadows caused by the bent and twisted cell walls give the picture quite a bit of depth.

The surface of a pea (Fig. 4) has some interesting structures that can be seen at a low magnification; however, when the magnification is increased (Fig. 5), not only are the structural details more clearly seen, but the three-dimensional quality is significantly improved. It is improved because of better contrast in the photomicrograph due partly to the bright areas produced by a buildup of electrons in an insulated area. Sometimes this buildup of electrons (charging) enhances a photomicrograph; however, in Fig. 6, a photomicrograph of chicken muscle fibers, the buildup of electrons was too great and the details of the surface microstructure were obscured, rendering the photomicrograph unsatisfactory.

In spite of some charging, when the contrast and focus are just right, a photomicrograph has depth to it, and even though it is not viewed in stereo, it has a three-dimensional quality. For example, in Fig. 7, some of the bovine muscle fibers are in a different plane than the rest. What usually happens in cases like this is that there is a slow buildup of electrons until the specimen charges quite severely. However, if a picture is taken of the

**Fig. 4.** Scanning electron micrograph of the surface of a pea. The arrows are pointing to the structures of interest. The bar equals 0.2 mm.

specimen before there is any charging, the quality of the photomicrograph can be quite good.

## B. Uses of SEM in Food Microstructure

From the time the SEM was first introduced, food scientists recognized the importance of the instrument; consequently, major food journals began including articles dealing with SEM. Whether it was in the *Journal of Food Science, Cereal Chemistry, Milchwissenschaft,* or *Food Technology,* readers began seeing an exponential increase in food–SEM articles. In 1982 a new journal, *Food Microstructure,* appeared. Whereas most of the other food-related journals published papers in which SEM was used, *Food Microstructure* was the first journal to emphasize the microstructure of food. The importance of SEM was now firmly established for food scientists.

Of the hundreds of food-related papers from dozens of journals that have used SEM, it is an impossibility to choose every one for inclusion within a

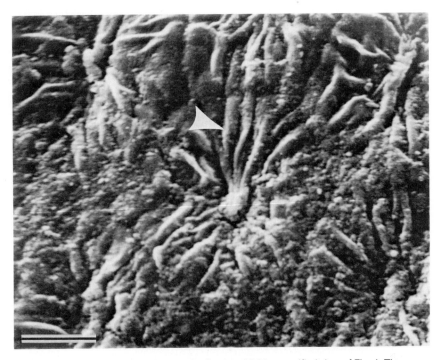

**Fig. 5.** Scanning electron micrograph of a more highly magnified view of Fig. 4. The arrow is pointing to a ridge where there is a buildup of electrons. The bar equals 0.04 mm.

review article; therefore, the following papers are presented here as examples of how some food scientists have used SEM to help study a variety of material.

## 1. Muscle Foods

Schaller and Powrie (1971) used a freeze-fracture technique to prepare skeletal muscle from rainbow trout, turkey, and beef for SEM. Although there were some differences between the three groups of animals, microstructural features were basically similar. Freeze-fracturing was also employed by Jones *et al.* (1976). Bovine semitendinosus and longissimus dorsi muscles were prepared by fixation in a glutaraldehyde–phosphate buffer solution followed by freeze-fracturing and then were critical-point dried or air-dried. Scanning electron microscopy was used to examine the muscle samples prepared by these methods. Careful preparation of the samples was essential. Fixation followed by freeze-fracturing from ethanol appeared to be the most dependable method. Micrographs of freeze-dried and air-dried

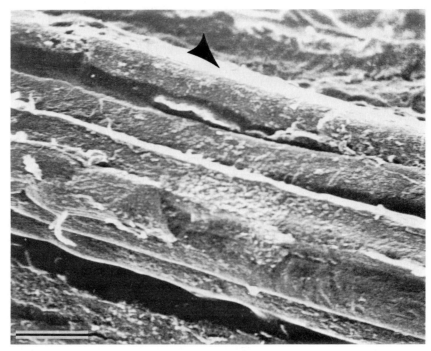

**Fig. 6.** Scanning electron micrograph of chicken muscle fibers. Note the area where charging is taking place (arrow). The bar equals 0.04 mm.

(from ethanol) samples showed that muscle fiber morphology was crisp. Even though each method had some faults, the authors concluded that the best results were obtained with smaller samples.

Although muscle that has been prepared for SEM is altered as a result of fixation, etc., the normal ultrastructural appearance of muscle should serve as a baseline for examination of meat ultrastructure (Geissinger and Stanley, 1981). The authors detailed several preparative techniques, including freeze fracture, dry fracture, staining, replication, etc. This paper serves as an excellent guideline for preparing samples not only for SEM but for correlative SEM–LM–TEM studies. In a paper by Geissinger *et al.* (1978) the correlative interpretation of LM, TEM, and SEM micrographs taken of the same sample were discussed. Some modifications in sample preparation are required; however, these preparation techniques are applicable for any skeletal muscle tissue.

Since the textural quality of meats and meat products is closely related to structure, SEM analysis is of major importance in examining the morphological characteristics associated with toughness, tenderness, granularity, or

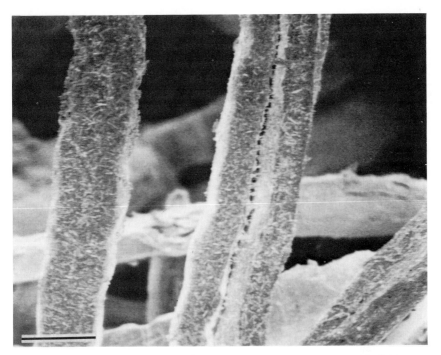

**Fig. 7.** Scanning electron micrograph of bovine muscle fibers. The bar equals 0.04 mm.

smoothness (Voyle, 1981a). In most instances, the SEM appearance of meat morphology is complementary to that with LM and TEM.

Cohen and Trusal (1980) examined the effects of a group of lysosomal enzymes called cathepsins on the microstructure of chilled bovine muscle. Cold shortening is a reversible process whose reversal is hastened by the addition of catheptic enzymes. Using SEM, TEM, and other analytical techniques, the authors showed how the use of corroborative evidence enhances the understanding of a process more than if evidence obtained from just one method is used.

The compactness of myofibrillar structure can be related to water loss in dry cooked bovine muscle (Godsalve *et al.*, 1977). Looser myofibrillar structure (as seen in the SEM) could account for high water loss rates in dry cooked muscle. A higher loss rate was found in muscles that were restrained while entering rigor than in those muscles that were unrestrained. Also, water loss was greater for muscles whose fibers were oriented perpendicularly to hot air flow.

Although Varriano-Marston *et al.* (1976) were unable to distinguish patterns of degradation between free and restrained aged bovine muscles, differences in states of contraction and degradation of Z lines and sarco-lemma were observed in SEM photomicrographs.

In a study of the relationship of the structure of contracted porcine psoas muscle to texture, the various microstructural components were observed in the SEM by Stanley and Geissinger (1972). The acceptability of the meat was significantly affected by the degree of postmortem shortening; however, sarcomere length of fresh muscle was not significantly related to subjective or objective measurements of tenderness.

Scanning electron microscopy was used to demonstrate the degradative changes that occur to normal and PSE (pale, soft, exudative) porcine muscle (Cloke et al., 1981). The changes were found to occur in proportion to the severity of the PSE condition brought about by stress.

Using a multitensile stage, Carroll et al. (1978) subjected raw and heated bovine semitendinosus muscle to tensile stress and then observed the changes in the SEM. The authors also used a video camera coupled to a stereo microscope to record the effects of stressing the muscle.

In a study of the effects of chopping temperature on the surface micro-structure of meat emulsions, SEM micrographs clearly showed differences between temperature treatments (Jones and Mandigo, 1982). There was strong evidence of the relationship of interfacial film thickness and emulsion stability. Also, a decrease in film elasticity coincided with a thick interfacial protein film surrounding fat droplets.

Extensive morphological degradation occurred to chicken, beef, and rainbow trout muscles when they were heated to 97°C (Schaller and Powrie, 1972). At 60°C, however, microstructural changes were relatively small in chicken and beef, while at that temperature trout muscles could be seen in SEM micrographs to be extensively damaged. Jones et al. (1977) also found that bovine semitendinosus heated for 45 min to 60 and 90°C exhibited significant structural changes, including increased coagulation and com-pactness of the A-band area and the disintegration of the I band. There were minor changes to the microstructure when the muscle was heated to 50°C. The alterations in microstructural topography during cooking were shown by Leander et al. (1980) to be progressive. Transmission electron micros-copy and scanning electron microscopy were used to compare the structural changes in bovine longissimus and semitendinosus muscles during cooking. A decrease in tenderness accompanied an increase in cooking temperature.

In another interesting study involving cooking, Otwell and Hannan (1979) used SEM to examine progressive structural alterations in cooked squid mantle tissue. A combination of microscopy with rheological and sensory evaluation provided a more complete textural characterization of the mantle as a food.

Papers by Ray et al. (1979) and Carroll and Lee (1981) used correlative microscopy in meat emulsion analysis. Ray's group found that, in a study of fat and protein components of meat emulsions, LM of stained samples provided more information. However, Carroll and Lee used TEM, LM, and SEM to determine an increase and a decrease in emulsion stability, which is a function of processing temperature.

Compared with conventional cookery, microwave cookery produced (as seen in TEM and SEM photomicrographs) less tearing and fragmentation in prerigor muscle, and in cold-shortened prerigor muscle it showed a more uniform alternating pattern of small stretched areas and dense contracted areas (Hsieh *et al.*, 1980).

When bovine muscle was subjected to high hydrostatic pressure, the interactions between the chemical and physical effects of pressure treatment were responsible for tenderization (Elgasim and Kennick, 1982). Prerigor pressurization has a profound influence on the disruption of myofibrillar structure that is clearly seen in SEM micrographs. The disruption of myofibrils causes an increase in lysosomal enzyme activity that could contribute to tenderization.

Electrical stimulation of a freshly slaughtered beef carcass produces a more tender carcass than nonstimulation and limits the adverse effects of cold shortening. Although dense zones of contraction were observed in muscle fibers from electrically stimulated carcasses 24 hr postmortem, they were not seen immediately following electrical stimulation (Voyle, 1981b). Although TEM micrographs showed zones of contraction, severe contraction was dramatically seen in SEM micrographs.

The effect of aging on microstructure has been the subject of countless studies, including several in which SEM was used to show aging-related changes. Johnson and Bowers (1976) correlated electrophoresis and SEM to study the influence of aging on turkey breast muscle. Typical postmortem changes in myofibrillar morphology were observed and correlated more with tenderness than with banding patterns on electrophoresis gels. Suderman and Cunningham (1980) looked at the effect of age, chilling, and scald temperature on the ultrastructure of poultry skin as related to factors affecting batter and breading adhesion. While neither chilling nor age contributed significantly to coating adhesion, increased scalding temperature did.

There were significant differences between cathepsin- and collagenase-treated bovine and rabbit muscles (Eino and Stanley, 1973a,b). Scanning electron micrographs showed considerable degradation of collagenase-treated muscle fibers; however, even though there was some degradation in the cathepsin-treated fibers, the catheptic enzymes produced surface structural changes similar to those found in naturally aged meat. This action by catheptic enzymes was also demonstrated by Robbins *et al.* (1979). In their study of the action of proteolytic enzymes on bovine myofibrils, they used SEM and LM to show the selective action of catheptic enzymes. When myofibril suspensions were treated with a catheptic enzyme solution at postmortem pH (5.1–5.3) conditions, it appeared from SEM micrographs that the Z bands were degraded as in aged meat. The selectivity of catheptic enzymes toward certain myofibrillar and sarcolemmal proteins points to their potential use as an exogenous meat tenderizer. This potential was clearly demonstrated when several sensory and instrumental methods, including SEM, were used by Cohen *et al.* (1982) to elucidate the effects of catheptic enzymes on flaked and formed (restructured) beef. Their most interesting finding was that catheptic enzymes reduce the amount of connective tissue while, at the same time, improving the textural quality of the meat.

In an earlier study, Theno *et al.* (1978) studied the effects of salt and phosphate on the ultrastructure of cured porcine muscle. They found that the massaging of the muscle during mixing increased the surface area, thereby increasing protein extraction with the addition of salts and phosphates. The resulting degradative action on the myofibrils is clearly shown in the SEM micrographs.

Although a larger surface area can result in increased protein extraction, it can also mean an increased bacterial population. To demonstrate this, Schwach and Zottola (1982) used SEM to observe microbial attachment to beef surfaces. They not only showed micrographs of bacteria with attachment fibrils directly on meat surfaces but also showed growing bacteria adhering to stainless steel surfaces that had been in contact with the meat.

## 2. Plant Foods

In their review paper Carroll and Jones (1979) described how the SEM could be used to solve specific problems related to agricultural products. Utilizing several different preparative techniques, including a freeze-fracture technique developed by Humphreys *et al.* (1974), they investigated three problems: (1) host–pathogen interaction where the fungus *Phytophthora infestans* caused late blight in potato plants, (2) the cracking of sweet cherries in the orchard before harvest, and (3) alterations related to heat and tensile stress in meat structure. They concluded that the use of SEM by food scientists should give information regarding the textural properties of foods and, when used in conjunction with other analytical methods such as X-ray and TEM or chemical analysis, should give researchers a clearer understanding of food systems.

In the study of plant foods, SEM can be used to examine the morphological structure of a food crop. For example, the SEM microstructure of the rice kernel and the implication of its characteristics to storage, marketing, and processing are discussed in a review by Bechtel and Pomeranz (1978). Components of plant tissue can be examined by different analytical methods, such as X-ray fluorescence and SEM, which can be used to identify specific cellular or tissue structures. Medina *et al.* (1978) used these methods to identify starch and protein components of green banana tissue.

The microstructural changes due to cooking can clearly be seen with the use of SEM. In a study of the effect of soaking time and cooking conditions on the textural and microstructural properties of cowpeas, Sefa-Dedeh *et al.* (1978) could correlate these properties with water absorption characteristics.

Chemical additives cause changes to tissue microstructure that can be readily observed in the SEM. Evans *et al.* (1977) showed that SEM was effective in resolving structural differences due to supplementation of bread dough with a yeast single-cell protein and an emulsifying agent, and Evans *et al.* (1981) used SEM to study the effects of oxidizing and reducing agents during various stages of dough development.

The advantage of SEM over TEM for the examination of cereal specimens is that flour and dough samples are especially sensitive to standard TEM fixation–dehydration methods (Aranyi and Hawrylewicz, 1969). Scanning electron microscopy was used by Stevens (1973a,b) in studies of the effects of aqueous extraction on the aleurone layer of wheat. Direct observation by SEM is an especially useful method for examining wheat endosperm and wheat flour dough and for studying the processing of milling wheat (Moss *et al.*, 1980). The efficiency of floury endosperm removal from the overlying bran and the degree of bran damage or powdering are influenced (during milling) by the manner in which bran fractures. This in turn is determined by the hardness and moisture content of the grain.

Hansen and Jones (1977) used LM and SEM to obtain information on the starch granule structure of wheat subjected to heat and enzyme treatment. They also discussed the possible application of heated starches in food systems.

Scanning electron photomicrographs of starch isolated from various baked products revealed that the proportion of folded and deformed granules varied from very few in sugar cookies to nearly all granules in angel food cake (Lineback and Wongsrikasem, 1980). Also, the proportion of granules was an indication of the extent of gelatinization and pasting.

Unlike LM with its low resolution and TEM with its tedious preparative techniques, SEM makes it possible to observe the three-dimensional structure of soybeans at a practical magnification range of from 20 to 20,000 diameters with a minimum of time devoted to sample preparation (Wolf and Baker, 1972). Wolf and Baker (1975) extended their earlier SEM studies on cellular structure of soybean cotyledons by examining a variety of commercially available soybean protein products, including flours, protein concentrates, and protein isolates. Their SEM micrographs revealed the internal structure of soybean cotyledon cells more clearly than previously observed. As a result of what they observed, the authors concluded that SEM is a valuable research tool for detecting processing-related changes in size and shape.

Using SEM, Saio (1981) studied the microstructure of tofu and kori-tofu, two traditional Japanese soybean foods. With the help of a cryounit, Saio was able to show a honeycomb-like structure with oil drops in tofu, and she was able to clarify the structural changes taking place during the preparation of kori-tofu.

The way in which the structure of soybeans changes during commercial processing of the seed into protein products was reviewed by Wolf and Baker (1980). Because of the low moisture content of the soybean products, the preparation of samples was quite easy and amounted merely to coating the samples with a thin layer of gold palladium. In addition to corroboration of the TEM morphology of soybean seeds, useful new information was obtained on the modification of native structure during processing into flours, concentrates, and isolates. In a related study of soybean seed coat, Wolf *et al.* (1981) correlated seed coat morphology with storage stability and oil yield. They concluded that seed cracking affects grading, handling, storage, and quality of finished soy food products. In addition, as the authors state, cracking lowers germination and seed vigor and is therefore of major economic importance.

In another, related study, Bair and Snyder (1980) examined ultrastructural changes in soybeans during processing. Flaking of conventionally processed soybeans was sufficient to disrupt nearly all the cotyledon cells but had very little effect on protein and lipid bodies. After hexane extraction the surface of the protein bodies was changed from smooth to granular as seen with SEM; however, when solventized soybeans were toasted, the cell walls were disrupted and the protein bodies were agglomerated.

With TEM, LM, and SEM, Taranto and Rhee (1978) studied the ultrastructural changes in defatted soy flour induced by nonextrusion texturization, and, with similar correlative microscopy, Taranto *et al.* (1978) analyzed textured cottonseed and soy flours in a study using defatted native and steam-heated glandless cottonseed and soy flours to make texturized products with a laboratory extruder. Also based on microstructural observation Soetrisno *et al.* (1982) found that the definition of soybean cell structure is decreased by soaking and cooking. Structural differences that could be correlated with rheological properties were investigated by Kuo *et al.* (1978) with papain used to induce morphological changes in textured soy protein.

Scanning electron microscope observations of different varieties of carrots and of diseased and healthy specimens were consistent with textural differences (Davis and Gordon, 1977). Xylem and phloem of raw and cooked carrots were prepared by mechanical excision and cryofracture in liquid $N_2$ (Davis et al., 1976a). Critical structures still maintained rigidity in cooked samples. Therefore, SEM, coupled with appropriate preparative methods, is a very useful tool that gives significant information on the physical structure of carrot tissue (Davis et al., 1976b). Microstructural changes during cooking by steaming, boiling, and pressure cooking were characterized. In another study of carrots, SEM was used to measure the effect of NaCl on the rehydration of freeze-dried carrots (Curry et al., 1976). Experiments confirmed that excess salt inhibits rehydration regardless of cell disruption.

Both LM and SEM were used by Walter and Schadel (1982) to evaluate the effects of lye peeling treatments on sweet potato tissue as a means of skin removal prior to canning. Discoloration, extent of starch gelatinization, cell wall detachment, and chloroplast disruption were all studied. The authors concluded that the best product is one that does not have rough xylem elements exposed (as seen with SEM) and that has had polyphenoloxidase (which reacts with o-dihydroxyphenols and causes discoloration) inactivated.

Employing histochemistry, LM, TEM, and SEM, Fretzdorff et al. (1982) investigated the microstructural modification in a kilned malt sample. They showed that primary changes occur in the near-epithelial starchy endosperm and to a limited degree in the starchy endosperm near the aleurone layer and that they were more pronounced in the dorsal, nonfurrowed part of the kernel than in the ventral portions.

Scanning electron microscopy was used by Allen and Arnott (1982) to investigate protein body structure in imbibed and dormant sunflower cotyledons. The purpose of the study was to visualize morphological changes in protein bodies from germination to seedling growth. Although TEM has been used extensively for comparison of SEM findings, the surface texture and fibrous nature of protein vacuoles are best seen with TEM. The same authors (Allen and Arnott, 1981) duplicated some stages of germination in experiments on the treatment of seed sections with proteolytic enzymes. By using SEM, they observed the morphology of yucca and sunflower seed protein bodies during the course of enzymatic digestion by either papain, pancreatin, or Adolf's Meat Tenderizer, which contains papain. Degradative changes to seed structure were observed.

Although the chemical composition of horny and floury corn kernel endosperm is similar, the morphological structure, as seen in the SEM, is different (Kikuchi et al., 1982). In baking tests gelatinization temperatures were different; however, X-ray diffraction patterns were similar. Also, sensory evaluation of cookies showed that those with added floury endosperms were better than those with added horny endosperms.

Christianson et al. (1982) were able to correlate the SEM microscopic structure of cornstarch granules with rheological properties of cooked pastes. Since processed starch is used to thicken foods in which starch occurs in low concentrations, the authors studied uncooked starch and starch dispersions cooked at low concentrations. They found the morphology of starch granules to be related to the rheological properties of cooked paste.

Using particle-to-particle heat transfer to roast navy beans, Aguilera et al. (1982) examined the effect of roasting on the morphological characteristics of the beans. Cross-sectional SEM photomicrographs showed that the percentage of beans with cracked hulls increased with higher roasting temperature. Black beans were the subject of a study by Varriano-Marston and De Omana (1979). Scanning electron microscopy was used to examine the effects of soaking in sodium salt solutions on the morphology of the beans. The morphological changes in micrographs were compared with X-ray microanalysis. Not only is SEM an excellent tool for studying surface features of beans, but the characteristics of bean starch are easily visualized in an SEM (Lai and Varriano-Marston, 1979).

One contribution to the complexity of water loss in the later stages of baking might be the nonuniform character of starch gelatinization (Gordon et al., 1979). Scanning electron micrographs of inhomogeneous cake crumb structure indicated degrees of starch gelatinization and gluten development in different locations within the cake. This was related to temperature, flow characteristics, and level of starch substitution.

In testing the effects of flour anticaking agents on the bulk characteristics of ground sugar, Hollenbach et al. (1982) demonstrated by SEM the adherence of anticaking agent particles to the sugar surface. This affinity, they concluded, was the cause of the noticeable effect on bulk characteristics.

The starch from black gram, an important legume crop found in India and several other countries, was investigated by chemical as well as LM and SEM methods by Sathe et al. (1982). Characterization of the external morphology of starch, although not necessarily correlating with the plant's chemistry, nevertheless provided much useful information. Starch was also examined by LM and SEM by Hahn et al. (1977) in a study of intracellular gelatinization of starch in large dry lima bean cotyledons. Configurational changes of starch granules during gelatinization were observed and related to differences in gelatinization temperatures. In another study of starch Rockland (1977) characterized the various stages of extracellular gelatinization of dry lima bean starch in water and in a mixed salt solution. The morphological changes, as seen by LM and SEM, were directly related to the stages of gelatinization. Scanning electron micrograph studies on starch granules of red kidney beans and bean sprouts revealed changes to the surface starch granules during seed germination (Silva and Luh, 1978). When the microstructure of cooked dry beans was compared with those germinated, there were significant differences in the appearance of the starch granules. Also, the cooked dry beans were resistant to fracture.

Textured vegetable protein from cottonseed flours was subjected to texture profile and ultrastructural analysis with SEM by Cabrera *et al.* (1979). Similarities were found between this product and soybean extrudates. Taranto and Rhee (1979) used TEM and SEM to study the morphological changes to glandless cottonseed flour during nonextrusion texturization. Photomicrographs showed how the deformation, rupturing, and fusion of protein bodies led to the formation of the products' protein-insoluble carbohydrate matrix. Noguchi *et al.* (1982) used SEM to examine the textural characteristics of extruded rice flour and rice flour fortified with soybean protein isolate. The rice exudate had a filmlike structure that was composed of gelatinized starch with insoluble starch skeletons, and fine strings that were contributed by the soy protein isolate in the fortified extrudate.

In a paper on the extrusion of spaghetti, Donnelly (1982) used SEM to show that spaghetti that was extruded through Teflon-lined dies had a better appearance and better cooking qualities than spaghetti that had been extruded through dies not lined with Teflon.

### 3. Dairy Foods

Kalab (1979) reviewed the use of SEM in dairy research and concluded that it is an invaluable tool for relating manufacturing processes to microstructure. The visualization of spray-dried skim milk and whole milk particles helps in the identification of various brands. Scanning electron microscopy can be used to correlate microstructure with water-holding capacity of dried whey protein. Also, lactic bacteria can be graphically visualized in yogurt by means of SEM. In cottage cheese, various structures are revealed with SEM, depending upon how the curd granules are fused and how the manufacturing process alters the protein matrix within the curd. Kalab (1981) presents an excellent review of the techniques used to prepare dairy samples for electron microscopic observation.

The importance of correlative studies is demonstrated by Kalab *et al.* (1981), who used SEM, TEM, sensory evaluation, and instrumental compressibility testing to perform analyses on several dairy products. Although there were correlations between some of the instrumental and sensory values, the microstructural differences were more difficult to perceive.

By using a freezing technique, Schmidt and van Hooydonk (1980) were able to study the microstructural details of both normal and homogenized whipping cream. The size differences of air bubbles in the homogenized product were significantly smaller than the normal whipping cream. The morphological changes conformed to the current theory of the whipping process.

Kalab (1980) used SEM to detect the presence of buttermilk that had been made from sweet cream in adulterated skim milk. Since buttermilk is lower in price than skim, the blending of small amounts of buttermilk into skim milk is illegal. Therefore, as the author claims, there is a need for the development of an SEM survey in this area.

In a study of processed cheese, Rayan et al. (1980) observed reduced dimensions of fat masses and increased emulsification. They found that the degree of emulsification, which corresponds to the fineness of fat particles, was related to firmness, meltability, and elasticity.

Buchheim (1981) examined the microstructure of various dried milk products and concluded that a combination of SEM and TEM methods is neccessary for a complete structural characterization of powdered food systems. In another dry food system, the effect of lactose crystallization on the properties of spray-dried whey was examined with SEM by Saltmarch and Labuza (1980). The authors evaluated the effects of water activity and temperature on the transition of amorphous lactose to its crystalline state. Their SEM photomicrographs of hydroscopic whey powder at various $H_2O$ activities and temperature and storage levels showed differences that indicated that lactose crystallization seemed to be related to the browning reaction.

Scanning electron microscopy was used by Lee and Merson (1978) to examine fouling substances on membranes used in the production of cottage cheese whey. They found that these substances could be reduced by prefiltration. Also, SEM and analysis of protein complexes were used to determine the best type of filtration system.

In two studies of mozzarella cheese, SEM was used to examine the structural morphology. In the first paper, Taranto et al. (1979) described the relationship of rheological properties of cheddar and mozzarella cheeses to their SEM and TEM microstructure. He found that there is a mimimum correlation between some rheological parameters of the mozzarella versus the cheddar cheese and that other properties, especially chemical, need to be assessed before making a determination of differences between cheeses. So even though SEM provides useful information, in this study it was not enough. In the second paper, Taranto and Tom Yang (1981) compared the morphological and textural characteristics of soybean mozzarella cheese analogs and found that a soy protein concentrate made into a soybean analog had structure as seen with SEM and texture more like natural mozzarella cheese than any of the other analogs examined.

The addition of acidulants to milk leads to the gelation of milk proteins. Harwalkar and Kalab (1981) studied how the microstructure of gels was affected by the type of acidulant used as well as by the temperature of the system. The importance of temperature was noted by Hokes *et al.* (1982), who, in a study of a model system for curd formation and melting properties for calcium caseinates, employed polarized LM and SEM to examine the structural transition from discrete particles to agglomerated curd.

Although there has been a tremendous amount of material written about human milk as a food, there has not been much written about its microstructure. Ruegg and Blank (1982) have written an excellent review on the structure and properties of the particulate constituents of human milk. They examined different types of colloidal or coarsely dispersed particles such as casein micelles, membrane fragments, fat globules, and cells. With the SEM microstructure, there is considerable variation both structurally and compositionally, depending upon the state of lactation.

The structure of milk during its passage through the digestive tract was the subject of a study by Berendsen (1982). Using suckling rats as test animals, he found that the appearance of rat milk curd was structurally similar to that of bovine milk curd and cottage cheese.

## 4. Miscellaneous

The proper preparation of food samples for SEM is critical for their evaluation. One must understand the methods to select the one most suited for a particular study (Chabot, 1979). Since an SEM cannot tolerate hydrated specimens (except in the case of a specially constructed stage) each sample must be dry. So, air-drying, freeze-drying, or critical-point drying are three choices. Air-drying is usually most unsatisfactory because of gross distortions; however, grains, flours, seeds, etc., might not be damanged by air-drying. Freeze-drying (specimens are placed under vacuum until $H_2O$ has been removed by sublimation) is commonly used, especially with a cryoprotectant to prevent ice crystal damage. The drying method of choice for most food samples is critical-point drying. For example, as seen by Moncur (1979) although there was uniform shrinkage in critical-point drying of wheat, freeze-drying caused some morphological details to be lost. With this method, samples are dehydrated in a graded series of alcohol or acetone that, in turn, is replaced with liquid Freon or $CO_2$ and then is heated. The point at which the liquid becomes a gas is the critical point. Once the gas has been removed, a dried sample remains that, after the deposition of a thin metal or carbon film, can be observed in the SEM.

Chabot discussed the vicissitudes of sample preparation, of limited $H_2O$ systems, and liquid food systems. She expounded on the preparative nuances in studying bread and starch as examples of a limited $H_2O$ system and gelatinized starch as an example of a liquid food system. In addition, she briefly described the preparation of seeds, textured protein, and muscle.

Using cake batters as examples of formulated food and bovine muscle as an example of a biologically intact nonformulated food, Davis and Gordon (1982) showed how total food systems can be evaluated in terms of overall properties at the component and molecular levels. Information at all levels within a food system must be integrated for proper evaluation. In other words, the complete interpretation of a food system requires analysis not just by SEM or TEM but by other corroborative evaluative methods. If possible LM, TEM, and SEM should be used to draw optimum conclusions regarding food microstructure (Buchheim, 1974). The SEM, although it can show surface characteristics, cannot show subsurface features unless samples have been prepared properly.

Davis and Gordon (1980) evaluated the effects of several different preparative methods on carrot tissue structure and showed that, when an optimum method is found for preparing samples, the visualized results can be applied to food systems at various stages in processing.

As mentioned earlier, critical-point drying is the method of choice for many food scientists. An extensive discussion of tissue preparation, especially critical-point drying from ethanol, can be found in Humphreys *et al.* (1974) and Humphreys (1975).

After a sample has been critical-point dried or freeze-dried, it must be coated with a thin conductive layer prior to insertion into the SEM. Sputter coating is probably the easiest method of metal deposition, but because of the higher resolution capabilities of the newer analytical SEM instruments, ion beam sputtering is recommended by Evans and Franks (1981). The advantage of this type of sputtering is that the size of metallic particles deposited over the surface of a food sample is small enough so as not to obliterate the fine structure of the sample.

Using several examples of research being carried out at the Leatherhead Food Research Association in England, Lewis (1981) explains how microscopy can be used to explain the behavior of foodstuffs. He discusses the various techniques that have been used and comments on how, for the role of microscopy (SEM included) in food science to be significant, its findings must have applicability.

Just such applicability was shown by Soo *et al.* (1978), who used SEM to examine texture definition in natural and fabricated shrimp and were able to relate textural to microstructural differences. They concluded that the effect of binding agent matrix composition contributed significantly to the texture and microstructure.

By using LM and SEM, one can begin to understand the spatial distribution and morphological characteristics of complex multiphase systems (Gejl-Hansen and Flink, 1976). This is quite useful for describing the structure of dehydrated food systems. For example, the types of molecular interactions stabilizing the gel structure of nonmeat proteins are more important than a three-dimensional network of protein fibers (Siegel et al., 1979a). These interactions must be the same as myosin-formed gels, so that interactions between nonmeat protein molecules and myosin molecules will be possible. Siegel and Schmidt (1979) studied the ionic pH and temperature effects on myosin-binding ability and found that, in the presence of salt and phosphate, the ultrastructure of gels formed by myosin showed a three-dimensional network of overlapping fibers. However, with no salt or phosphate present, the gel ultrastructure appeared spongelike.

A similar three-dimensional network structure, which has the ability to hold $H_2O$ in a less mobilized state, results when myosin gelation (which is of some significance in meat comminution) is induced by heat (Yasui et al., 1979). The formation of a finer protein matrix of heat-induced myosin gels, as seen in an SEM, is related to the higher shear modulus of the gelatin (Ishioroshi et al., 1979).

Gel characteristics and structure of blood plasma gels as seen in the SEM showed that, although some increased random aggregation decreased the water-binding ability and degree of elasticity of gels, it was not possible to generalize about the firmness and gel structure relationships (Hermansson, 1982).

Using SEM to examine the microstructure of isolated soy protein in combination ham, Siegel et al. (1979b) observed a denser gel structure formed by isolated soy protein compared with a gel formed by a crude myosin preparation. Also, a mixture of myosin and isolated soy protein formed a gel whose structure was similar to that of a crude myosin gel.

In another study of gels, Colombo and Späth (1981) used SEM to examine the morphology of three different kinds of gels — polyacrylamide, agarose, and alginate — that were subjected to various dehydration procedures prior to examination with the SEM.

Scanning electron microscopy was also used to characterize the morphology of omelets in a study of $H_2O$-holding capacity and textural acceptability of precooked frozen whole-egg omelets (O'Brien et al., 1982). The authors concluded that the more compact the ultrastructure, the greater the $H_2O$-holding capacity and textural acceptability.

Morphological modifications caused by heating and/or freezing of whole-egg magma were pointed out by Torten and Eisenberg (1982). Increase in apparent viscosity accompanied increased surface tension and occurred during heating and freezing. Scanning electron photomicrographs of non-frozen samples showed that pasteurization-related structural changes were minor and involve long-range network formation. After frozen storage, more dramatic morphological changes could be observed.

The structure of fat and carbohydrate in a freeze-dried matrix of oil-in-water model emulsion was studied using LM, SEM, and electron probe microanalysis by Gejl-Hansen and Flink (1977). The authors concluded that various microscopic techniques can be used to characterize the surface and encapsulation properties, including lipid distribution, and to relate it to certain morphological features.

The droplet sizes and overall morphology of mayonnaise and salad dressing were examined with LM, TEM, and SEM by Tung and Jones (1981). A lower concentration of lipid droplets was found in the mayon-naise, and amorphous material that was assumed to be cooked starch paste was observed between the lipid droplets in the salad dressing. SEM was an excellent method of elucidating the microstructure of salad dressing and mayonnaise.

Scanning electron microscopy was used to study the effects of heat and coagulating agents on the microstructure of soybean protein aggregates to help understand their relationship to the physical and textural properties of curd (Lee and Rha, 1978). The effects of heat and coagulating agents were analyzed from photomicrographs. Scanning electron microscopy was also used by Tsintsadze et al. (1978), who examined the microstructural and physical properties of yeast protein curds. Comparisons were made between the properties of yeast protein curds and of soybean protein curds. Scanning electron microscopy was a valuable tool in both of these studies and provided a corroborative method for other analytical techniques.

Lee and Rha (1979) employed SEM to show how, by introducing specific, well-defined interaction mechanisms to a model system, the degree to which interaction has caused changes to the system can provide the basis for controlling the microstructure and texture of fabricated foods. They used dehydration and extraction as interaction mechanisms. Luh et al. (1976) used sodium alginate gels to produce a fruitlike texture in fabricated nu-trient-controlled foods. Examination of the gel matrix with SEM can help in explaining why some samples are tough and difficult to rehydrate.

Processing of foods can be studied quite easily by SEM. Giddings and Hill (1976) were able to use SEM in a study of processing and preservation effects on crab muscle tissue. Lawrence and Jelen (1982) were able to demonstrate textural differences in protein extracts from residues of mechanically sepa-rated poultry. The authors were able to show freeze-fiber formation in acid-precipitated chicken protein that was alkali extracted from the bone and contained residues of deboned poultry. They studied the effects of freezing rates and pH of the protein extracts on the appearance of freeze-texturized heat-set protein. Although most of their observations were readily explainable from current knowledge, the formation of cross-linked, sponge-like structures seen primarily in the high pH samples needs further elucidation.

Scanning electron microscopy has also been used to aid in the identification of foreign matter in foods. Stasny *et al.* (1981) used SEM, LM, and diffraction to examine several different extraneous fragments found in foods, including glass, metal, asbestos, pest fragments, paint, and dust. Although other physical methods could identify the chemical composition, SEM was shown to be an excellent method for identification of many types of sample.

## III. CONCLUSION

In Everhart's (1968) keynote paper delivered at the first Scanning Electron Microscopy Symposium, he stated, "I rejoice that science and technology will gain from this instrument's application—I must confess a slight nostalgia now that the literature increases faster than one can follow. . . . "

In the years after Everhart's address and Pomeranz's (1976) review, significant improvements in the design and function of SEMs have resulted in instruments that are considerably better than those built during the 1960s and 1970s. With the typical SEM of the 1960s, 250-Å resolution was attainable, but only under ideal conditions, which, it seemed, occurred only when the service representative used the instrument. The typical SEM of the 1970s could resolve 100 Å, and by the late 1970s improvements in pumping systems, electron tip configurations, and design led to 50-Å-resolution SEMs.

At the beginning of the 1980s analytical electron microscopes with SEM (20-Å resolution) and TEM capabilities began to be introduced. They are able to perform a multitude of analytical tasks, a feature that Johari (1971), in his paper on total characterization of materials with SEM, could only speculate about.

It is hoped that the brief description of papers given within this review will give the reader an appreciation of some of the food-related SEM work that has taken place in the intervening years since Pomeranz's review.

In writing this review, I gained an appreciation of the fact, as mentioned by Everhart, that one cannot possibly keep up with the rapidly increasing body of literature related to SEM and food microstructure. I trust the literature on this topic will continue to grow.

## ACKNOWLEDGMENT

I wish to express my appreciation to my wife, Marilyn, for her help in preparing the review.

## REFERENCES

Aguilera, J. M., Lusas, E. W., Uebersax, M. A., and Zabik, M. E. (1982). *J. Food Sci.* **47**, 977–1000, 1005.
Allen, R. D., and Arnott, H. J. (1981). *Scanning Electron Microsc.* **III**, 561–570.
Allen, R. D., and Arnott, H. J. (1982). *Food Microstruct.* **1**, 63–73.
Aranyi, C., and Hawrylewicz, E. J. (1969). *Cereal Sci. Today* **14**, 230–233, 253.
Bair, C. W., and Snyder, H. E. (1980). *J. Food Sci.* **45**, 529–533.
Bechtel, D. B., and Pomeranz, Y. (1978). *J. Food Sci.* **43**, 1538–1542, 1552.
Berendsen, P. B. (1982). *Food Microstruct.* **1**, 83–90.
Bernard, R., and Davoine, F. (1957). *Ann. Univ. Lyon* **10**, 78–86.
Broers, A. N. (1965a). Ph.D. thesis, Cambridge Univ., Cambridge, England.
Broers, A. N. (1965b). *Microelectron. Reliab.* **4**, 103–104.
Buchheim, W. (1974). *Proc., Int. Congr. Food Sci. Technol. 2nd* pp. 5–12.
Buchheim, W. (1981). *Scanning Electron Microsc.* **III**, 493–502.
Cabrera, J., Zapata, L. E., de Buckle, T. S., Ben-Gera, I., de Sandoval, A. M., and Shomer, I. (1979). *J. Food Sci.* **44**, 826–830.
Carroll, R. J., and Jones, S. B. (1979). *Scanning Electron Microsc.* **III**, 253–259.
Carroll, R. J., and Lee, C. M. (1981). *Scanning Electron Microsc.* **III**, 447–452.
Carroll, R. J., Rorer, F. P., Jones, S. B., and Cavanaugh, J. R. (1978). *J. Food Sci.* **43**, 1181–1187.
Chabot, J. F. (1979). *Scanning Electron Microsc.* **III**, 279–286, 298.
Chandler, J. A. (1980). *Proc. R. Microsc. Soc.* **15**, 117–122.
Christianson, D. D., Baker, F. L., Loffredo, A. R., and Bagley, E.-B. (1982). *Food Microstruct.* **1**, 13–24.
Cloke, J. D., Davis, E. A., Gordon, J., Hsieh, S.-I., Grider, J., Addis, P. B., and McGrath, C. J. (1981). *Scanning Electron Microsc.* **III**, 435–446.
Coates, V. J., and Brenner, N. (1973). *Res./Dev.* **24**, 32–34.
Cohen, S. H., and Trusal, L. M. (1980). *Scanning Electron Microsc.* **III**, 595–600.
Cohen, S. H., Segars, R. A., Cardello, A., Smith, J., and Robbins, F. M. (1982). *Food Microstruct.* **1**, 99–105.
Colombo, V. E., and Späth, P. J. (1981). *Scanning Electron Microsc.* **III**, 515–522.
Curry, J. C., Burns, E. E., and Heidelbaugh, N. D. (1976). *J. Food Sci.* **41**, 176–179.
Danilatos, G. D. (1981). *Scanning* **4**, 9–20.
Davis, E. A., and Gordon, J. (1977). *Home Econ. Res. J.* **6**, 15–23.
Davis, E. A., and Gordon, J. (1980). *Scanning Electron Microsc.* **III**, 601–611.
Davis, E. A., and Gordon, J. (1982). *Food Microstruct.* **1**, 25–47.
Davis, E. A., Gordon, J., and Hutchinson, T. E. (1976a). *Home Econ. Res. J.* **4**, 163–166.
Davis, E. A., Gordon, J., and Hutchinson, T. E. (1976b). *Home Econ. Res. J.* **4**, 214–224.
Davoine, F., Pinard, P., and Martineau, M. (1960). *J. Phys. Radium* **21**, 121–124.
Donnelly, B. J. (1982). *J. Food Sci.* **47**, 1055–1058, 1069.
Eino, M. F., and Stanley, D. W. (1973a). *J. Food Sci.* **38**, 45–50.
Eino, M. F., and Stanley, D. W. (1973b). *J. Food Sci.* **38**, 51–55.

Elgasim, E. A., and Kennick, W. H. (1982). *Food Microstruct.* **1,** 75–82.

Evans, A. C., and Franks, J. (1981). *Scanning* **4,** 169–174.

Evans, L. G., Volpe, T., and Zabik, M. E. (1977). *J. Food Sci.* **42,** 70–74.

Evans, L. G., Pearson, A. M., and Hooper, G. R. (1981). *Scanning Electron Microsc.* **III,** 583–592.

Everhart, T. E. (1958). Ph.D. thesis, Cambridge Univ., Cambridge, England.

Everhart, T. E. (1968). *Scanning Electron Microsc.* **I,** 1–12.

Fretzdorff, B., Pomeranz, Y., and Bechtel, D. B. (1982). *J. Food Sci.* **47,** 786–791.

Geissinger, H. D., and Stanley, D. W. (1981). *Scanning Electron Microsc.* **III,** 415–426, 414.

Geissinger, H. D., Yamashiro, S., and Ackerly, C. A. (1978). *Scanning Electron Microsc.* **II,** 267–274.

Gejl-Hansen, F., and Flink, J. M. (1976). *J. Food Sci.* **41,** 483–489.

Gejl-Hansen, F., and Flink, J. M. (1977). *J. Food Sci.* **42,** 1049–1055.

Giddings, G. G., and Hill, L. H. (1976). *J. Food Sci.* **41,** 455–457.

Godsalve, E. W., Davis, E. A., and Gordon, J. (1977). *J. Food Sci.* **42,** 1325–1330.

Goldstein, J. I., and Yakowitz, H. (1975). "Practical Scanning Electron Microscopy." Plenum, New York.

Gordon, J., Davis, E. A., and Timms, E. M. (1979). *Cereal Chem.* **56,** 50–57.

Hahn, D. M., Jones, F. T., Akhavan, I., and Rockland, L. B. (1977). *J. Food Sci.* **42,** 1208–1212.

Hansen, L. P., and Jones, F. T. (1977). *J. Food Sci.* **42,** 1236–1242.

Harwalkar, V. R., and Kalab, M. (1981). *Scanning Electron Microsc.* **III,** 503–513.

Hayat, M. A. (1978). "Introduction to Biological Scanning Electron Microscopy." Univ. Park Press, Baltimore, Maryland.

Hermansson, A.-M. (1982). *J. Food Sci.* **47,** 1965–1972.

Hokes, J. C., Mangino, M. E., and Hansen, P. M. T. (1982). *J. Food Sci.* **47,** 1235–1240, 1249.

Hold, D. B., Muir, M. D., Grant, P. R., and Boswarua, I. M. (1974). "Quantitative Scanning Electron Microscopy." Academic Press, New York.

Hollenbach, A. M., Peleg, M., and Rufner, R. (1982). *J. Food Sci.* **47,** 538–544.

Hsieh, Y. P. C., Cornforth, D. P., Pearson, A. M., and Hooper, G. R. (1980). *Meat Sci.* **4,** 299–311.

Humphreys, W. J. (1975). *Scanning Electron Microsc.* **III,** 707–714, 762.

Humphreys, W. J., Spurlock, D. O., and Johnson, J. S. (1974). *Scanning Electron Microsc.* **I,** 275–282.

Ishioroshi, M., Samejima, K., and Yasui, T. (1979). *J. Food Sci.* **44,** 1280–1284.

Johari, O., ed. (1968–present). "Scanning Electron Microscopy Symposium." SEM, Inc., AMF O'Hare, Chicago, Illinois.

Johari, O. (1971). *Res./Dev.* **22,** 12–20.

Johari, O. (1974). *Res./Dev.* **25,** 16–18, 22, 23.

Johnson, P. G., and Bowers, J. (1976). *J. Food Sci.* **41,** 255–261.

Jones, K. W., and Mandigo, R. W. (1982). *J. Food Sci.* **47,** 1930–1935.

Jones, S. B., Carroll, R. J., and Cavanaugh, J. R. (1976). *J. Food Sci.* **41,** 867–873.

Jones, S. B., Carroll, R. J., and Cavanaugh, J. R. (1977). *J. Food Sci.* **42,** 125–131.

Kalab, M. (1979). *Scanning Electron Microsc.* **III,** 261–272.

Kalab, M. (1980). *Scanning Electron Microsc.* **III,** 645–652.

Kalab, M. (1981). *Scanning Electron Microsc.* **III,** 453–472.

Kalab, M., Sargant, A. G., and Froehlich, D. A. (1981). *Scanning Electron Microsc.* **III,** 473–482, 514.

Kikuchi, K., Takatsuji, I., Tokuda, M., and Miyake, K. (1982). *J. Food Sci.* **47,** 1687–1692.

Kimoto, S., and Russ, J. C. (1969). *Am. Sci.* **57**, 112–133.
Kimura, H., Higuchi, H., Maki, M., and Tamura, H. (1966). *J. Electron Microsc.* **15**, 21–25.
Knoll, M. (1935). *Z. Tech. Phys.* **11**, 467–475.
Kuo, C. M., Taranto, M. V., and Rhee, K. C. (1978). *J. Food Sci.* **43**, 1848–1852.
Lai, C. C., and Varriano-Marston, E. (1979). *J. Food Sci.* **44**, 528–530, 544.
Lawrence, R. A., and Jelen, P. (1982). *Food Microstruct.* **1**, 91–97.
Leander, R. C., Hedrick, H. B., Brown, M. F., and White, J. A. (1980). *J. Food Sci.* **45**, 1–6, 12.
Lee, C. H., and Rha, C. (1978). *J. Food Sci.* **43**, 79–84.
Lee, C. H., and Rha, C. K. (1979). *Scanning Electron Microsc.* **III**, 465–471.
Lee, D. N., and Merson, R. L. (1978). *J. Food Sci.* **41**, 403–410.
Lewis, D. F. (1981). *Scanning Electron Microsc.* **III**, 391–404.
Lineback, D. R., and Wongsrikasem, E. (1980). *J. Food Sci.* **45**, 71–74.
Love, G., Scott, V. D., Dennis, N. M. T., and Laurenson, L. (1981). *Scanning* **4**, 32–39.
Luh, N., Karel, M., and Flink, J. M. (1976). *J. Food Sci.* **41**, 89–93.
McMullan, D. (1952). Ph.D. thesis, Cambridge Univ., Cambridge, England.
Medina, M. B., Greenhut, V. A., and Lachance, P. A. (1978). *J. Food Sci.* **43**, 116–120.
Moncur, M. W. (1979). *Scanning* **2**, 175–177.
Moss, R., Stenvert, N. L., Kingswood, K., and Pointing, G. (1980). *Scanning Electron Microsc.*
    **III**, 613–620.
Noguchi, A., Kugimiya, W., Hague, Z., and Saio, K. (1982). *J. Food Sci.* **47**, 240–245.
Oatley, C. W., and Everhart, T. E. (1957). *J. Electron.* **2**, 568–570.
Oatley, C. W., Nixon, W. C., and Pease, R. F. W. (1965). *Adv. Electron. Electron Phys.* **21**,
    181–247.
O'Brien, S. W., Baker, R. C., Hood, L. F., and Liboff, M. (1982). *J. Food Sci.* **47**, 413–417.
Otwell, W. S., and Hannan, D. D. (1979). *J. Food Sci.* **44**, 1629–1635, 1643.
Parsons, D. F. (1978). "Short Wavelength Microscopy." NY Acad. Sci., New York.
Pease, R. F. W. (1963). Ph.D. thesis, Cambridge Univ., Cambridge, England.
Pomeranz, Y. (1976). *Adv. Food Res.* **22**, 205–307.
Ray, F. K., Miller, B. G., Van Sickle, D. C., Aberle, E. D., Forrest, J. C., and Judge, M. D.
    (1979). *Scanning Electron Microsc.* **III**, 473–478.
Rayan, A. A., Kalab, M., and Ernstrom, C. A. (1980). *Scanning Electron Microsc.* **III**,
    635–643.
Robbins, F. M., Walker, J. E., Cohen, S. H., and Chatterjee, S. (1979). *J. Food Sci.* **44**,
    1672–1677, 1680.
Rockland, L. B., Jones, F. T., and Hahn, D. M. (1977). *J. Food Sci.* **42**, 1204–1207.
Ruegg, M., and Blanc, B. (1982). *Food Microstruct.* **1**, 25–47.
Saio, K. (1981). *Scanning Electron Microsc.* **III**, 553–559.
Saltmarch, M., and Labuza, T. P. (1980). *Scanning Electron Microsc.* **III**, 659–665.
Sathe, S. K., Rangnekar, P. D., Deshpande, S. S., and Salunkhe, D. K. (1982). *J. Food Sci.* **47**,
    1524–1533, 1602.
Schaller, D. R., and Powrie, W. D. (1971). *J. Food Sci.* **36**, 552–559.
Schaller, D. R., and Powrie, W. D. (1972). *Can. Inst. Food Sci. Technol. J.* **5**, 184–190.
Schmidt, D. G., and van Hooydonk, A. C. M. (1980). *Scanning Electron Microsc.* **III**,
    653–658, 644.
Schwach, T. S., and Zottola, E. A. (1982). *J. Food Sci.* **47**, 1401–1405.
Sefa-Dedeh, S., Stanley, D. W., and Voisey, P. W. (1978). *J. Food Sci.* **43**, 1832–1838.
Siegel, D. G., and Schmidt, G. R. (1979). *J. Food Sci.* **44**, 1686–1689.
Siegel, D. G., Church, K. E., and Schmidt, G. R. (1979a). *J. Food Sci.* **44**, 1276–1279, 1284.
Siegel, D. G., Tuley, W. B., and Schmidt, G. R. (1979b). *J. Food Sci.* **44**, 1272–1275.
Silva, H. C., and Luh, B. S. (1978). *J. Food Sci.* **43**, 1405–1408.

Smith, K. C. A. (1956). Ph.D. thesis, Cambridge Univ., Cambridge, England.

Soetrisno, U., Holmes, Z. A., and Miller, L. T. (1982). *J. Food Sci.* **47**, 530–534, 537.

Soo, H. M., Davis, E. A., and Sander, E. H. (1978). *J. Food Sci.* **43**, 202–204.

Stanley, D. W., and Geissinger, H. D. (1972). *Can. Inst. Food Sci. Techn. J.* **5**, 214–216.

Stasny, J. T., Albright, F. R., and Graham, R. (1981). *Scanning Electron Microsc.* **III**, 599–610, 560.

Stevens, D. J. (1973a). *J. Sci. Food Agric.* **24**, 847–854.

Stevens, D. J. (1973b). *J. Sci. Food Agric.* **24**, 307–313.

Suderman, D. R., and Cunningham, F. E. (1980). *J. Food Sci.* **45**, 444–449.

Taranto, M. V., and Rhee, K. C. (1978). *J. Food Sci.* **43**, 1274–1278.

Taranto, M. V., and Rhee, K. C. (1979). *J. Food Sci.* **44**, 628–629, 631.

Taranto, M. V., and Tom Yang, C. S. (1981). *Scanning Electron Microsc.* **III**, 483–492.

Taranto, M. V., Cegla, G. F., Bell, K. R., and Rhee, K. C. (1978). *J. Food Sci.* **43**, 767–771.

Taranto, M. V., Wan, P. J., Chen, S. L., and Rhee, K. C. (1979). *Scanning Electron Microsc.* **III**, 273–278.

Theno, D. M., Siegel, D. G., and Schmidt, G. R. (1978). *J. Food Sci.* **43**, 488–492.

Thornley, R. F. M. (1960). Ph.D. thesis, Cambridge Univ., Cambridge, England.

Thornton, P. R. (1968). "Scanning Electron Microscopy—Applications to Materials and Device Science." Chapman & Hall, London.

Torten, J., and Eisenberg, H. (1982). *J. Food Sci.* **47**, 1423–1428.

Tsintsadze, T. D., Lee, C. H., and Rha, C. (1978). *J. Food Sci.* **43**, 625–630, 635.

Tung, M. A., and Jones, L. J. (1981). *Scanning Electron Microsc.* **III**, 523–530.

Varriano-Marston, E., and De Omana, E. (1979). *J. Food Sci.* **44**, 531–536.

Varriano-Marston, E., Davis, E., Hutchinson, T. E., and Gordon, J. (1976). *J. Food Sci.* **41**, 601–605.

von Ardenne, M. (1938a). *Z. Tech. Phys.* **19**, 407–416.

von Ardenne, M. (1938b). *Z. Phys.* **109**, 553–572.

Voyle, C. A. (1981a). *Scanning Electron Microsc.* **III**, 405–413.

Voyle, C. A. (1981b). *Scanning Electron Microsc.* **III**, 427–434.

Walter, W. M., and Schadel, W. E. (1982). *J. Food Sci.* **47**, 813–817.

Wells, O. C. (1957). Ph.D. thesis, Cambridge Univ., Cambridge, England.

Wells, O. C., Boyde, A., Lifshin, E., and Rezanowich, A. (1974). "Scanning Electron Microscopy." McGraw-Hill, New York.

Williams, D. B. (1980). *Norelco Rep.* **27**, 11–21.

Williams, D. B., and Edington, J. W. (1981). *Norelco Rep.* **28**, 1–25.

Williams, D. B., and Edington, J. W. (1982). *Norelco Rep.* **29**, 28–41.

Wolf, W. J., and Baker, F. L. (1972). *Cereal Sci. Today* **17**, 125–130, 147.

Wolf, W. J., and Baker, F. L. (1975). *Cereal Chem.* **52**, 387–396.

Wolf, W. J., and Baker, F. L. (1980). *Scanning Electron Microsc.* **III**, 621–634.

Wolf, W. J., Baker, F. L., and Bernard, R. L. (1981). *Scanning Electron Microsc.* **III**, 531–544.

Yasui, T. Ishioroshi M., Nakano, H., and Samejima, K. (1979). *J. Food Sci.* **44**, 1201–1204, 1211.

Zworykin, V. K., Hillier, J., and Snyder, R. L. (1942). *ASTM Bull.* No. 1177, pp. 15–23.

# 14

# Atomic Spectrometry for Inorganic Elements in Foods

JAMES M. HARNLY
WAYNE R. WOLF

*Nutrient Composition Laboratory*
*Beltsville Human Nutrition Research Center*
*United States Department of Agriculture*
*Beltsville, Maryland*

ANALYSIS OF FOODS AND BEVERAGES
ISBN 0-12-169160-8

## I. INTRODUCTION

By strictest definition, atomic spectrometry is the scientific discipline dealing with the measurement of spectra of atoms. Depending on which section of the energy spectrum is examined, atomic spectrometry can include optical (visible energies), X-ray (X-ray energies), and neutron activation (gamma energies) spectrometry among others. A more conventional definition of atomic spectrometry is the study of atomic spectra in the optical region (ultraviolet [UV], visible, and near-infrared [near-IR]) of the energy spectrum. This chapter is intended to overview the current state of the art of the three main types of atomic spectrometry: atomic absorption spectrometry (AAS), atomic emission spectrometry (AES), and atomic fluorescence spectrometry (AFS). It will present a simplistic explanation of the theoretical principles and a consideration of the application of the newest methods to the determination of inorganic elements in foods and beverages.

### A. Biological Significance

The determination of inorganic elements in foods and biological materials has received considerable recent interest because of the importance of many of these elements to human health. This interest arises from two areas of concern, nutritional and toxicological. Nutritionists are interested in the elements known, or suspected, to be necessary for maintenance of optimum health. Toxicologists are concerned with elements that are detrimental to optimum health. The former are interested in whether levels of intake are adequate, the latter with whether the levels are too high.

Elements of health interest can be divided into three major groups based upon concentration levels in biological material: major ($100\ \mu g/g$ or higher), intermediate (between $100\ \mu g/g$ and $1\ \mu g/g$), and trace (less than $1\ \mu g/g$). Because of the higher level of concentration of the major and intermediate-level elements, their role in human health is better defined. Nutritionally, recommended dietary intakes have been established and the major food sources identified. Toxicologically, the presence of many elements at these levels can be harmful. Toxic effects of the elements are generally well defined, and often only qualitative analytical identification of their presence is required. There are a number of analytical methods available that can provide adequate determinations with regard to accuracy and precision. The analytical problems are usually those of properly defining the required accuracy and precision and establishing proper quality assurance procedures to attain the required data.

Trace elements, on the other hand, represent much more poorly defined biological concerns and more difficult analytical challenges. In many cases, the nutritionists and toxicologists are concerned with the same trace elements, because many of the trace elements exhibit dual biochemical effects (Wolf, 1981). The difference between deficiency and toxicity may be only one or two orders of magnitude. Consequently, it is no longer sufficient to identify the presence of these elements qualitatively. Instead it is necessary to quantitatively establish the levels of deficiency and toxicity and to characterize the "normal" range of variation.

With increased awareness of the role of many trace elements in health, a demand has been created for more information on the levels of these elements in foods and beverages. For most people, the intake of foods and beverages constitutes the major source of exposure to most of these elements. Data on the elemental composition of foods is of interest to nutritionists and toxicologists alike. These data are necessary to establish adequate dietary intakes and to reduce exposure to harmful elements.

The ultimate goal is to develop an information base that establishes (1) human nutritional requirements and toxicity levels, (2) the flow of elements through the food supply, and (3) sources of "safe" foods that provide required nutrients and are free of harmful effects (Stewart, 1980). At the present time, the extent and quality of the inorganic composition data for foods and beverages are far from satisfactory. These insufficiencies exist because of inadequate analytical methodology, ignorance of the importance of many of these elements (until recently), and the large and constantly changing food supply. The demand for these data places pressure on modern atomic spectrometry for instrumentation capable of more sensitive and more rapid determinations. Many of the new spectrometer systems available today are capable of meeting one or both of these requirements.

## B. Modern Atomic Spectrometry

Tremendous advances have been made in the last 25 years in the field of atomic spectrometry. Prior to this period, elemental determinations were made by colorimetry, gravimetry, titrimetry, electrochemistry, arc or spark source emission spectrometry, and spark source mass spectrometry. While these methods are still used today, the bulk of elemental determinations are currently being made using a new generation of analytical instruments. These instruments are atomic absorption spectrometers using either furnace, flame, or chemical vapor generation atomization, atomic emission spectrometers using either an inductively coupled plasma (ICP) or a direct-current plasma (DCP), and atomic fluorescence spectrometers using an inductively coupled plasma as an atomizer.

The new atomic spectrometers have several aspects in common. First, all the spectrometers require the sample to be in a liquid form. Consequently, a large body of literature has accumulated on the subject of sample digestion and solubilization. These procedures are generally transferable between instruments, since the physical requirements (total salt and viscosity) are not considerably different. There is almost always a potential sample preparation method to be found in the literature for any type of sample.

Each type of spectrometer has current models that have been designed for full automation and computerization. The sample preparation step remains one of the few phases requiring manual operation. Beyond this step, samples, standards, and blanks are automatically introduced to the atomization source of the spectrometer. The analytical signal is detected by photoelectric sensors, the signal is amplified and processed by sophisticated electronics, and the analog electronic signal is converted to a digital signal, which is fed into a computer. Most spectrometers now come with a dedicated micro- or minicomputer. The computer can perform a wide variety of tasks, such as blank subtraction, calibration, computation of the sample concentration, correction for interferences, statistical analyses, and finally report generation. If these options are not available, the digitized data can usually be transferred to a larger computer that offers sophisticated statistical packages. Thus, modern spectrometers offer the ability to handle large numbers of determinations very rapidly with very sophisticated data-handling techniques and a great deal of versatility.

All of the atomic spectrometric methods require sample destruction and are only capable of determining total elemental concentrations. None of the methods can discriminate between the chemical states of any of the elements. Hyphenated methods, in this case, the use of atomic spectrometry with separation methods, have shown great promise for obtaining speciation information. Atomic spectrometers have been used for specific detectors for gas chromatography (GC), liquid chromatography (LC), high-performance liquid chromatography (HPLC), and flow-injection analysis (FIA) techniques (Horlick, 1982).

Finally, none of the atomic spectrometric methods is a panacea. No one method can provide the solution to every problem. The analyst must first clearly establish the analytical needs of the laboratory and then decide on the method that best meets those needs.

## II. LITERATURE REVIEW

There is an extensive body of literature available on the determination of elements in foods and beverages. In addition, there is a considerable volume of literature dealing with the determination of elements in biological and environmental samples that can be applied directly, or with slight modification, to foods. Similarly, approaches developed for nonorganic matrices can sometimes be quite useful for foods. For example, dissolution techniques developed for geological samples have proved quite useful for the analysis of foods with a high silicon content. Research on the determination of elements in foods and biological systems is progressing rapidly (Wolf, 1982). New developments are being reported continually. This section is aimed at providing the reader with direction to general reviews (when possible) and to those sources that will allow the reader to keep aware of the latest developments.

Several general reviews have appeared on the analysis of trace elements in foods (Crosby, 1977; Koirtyohann and Pickett, 1975) and in biological materials (National Bureau of Standards [NBS], 1976, 1977; *Advances in Chemistry Series,* 1979; Berman, 1980; International Atomic Energy Agency [IAEA], 1980; Bratter and Schramel, 1980; Veillon and Vallee, 1978). Three excellent reviews have been published dealing with the application of specific atomic spectrometric methods to the analysis of foods. Fricke *et al.* (1979) and Ihnat (1981) reviewed the use of flame AAS, while Jones and Boyer (1979) considered the application of inductively coupled plasma atomic emission spectrometry (ICP/AES) to the determination of elements in foods.

The spectroscopic literature is an excellent source for methods and periodical reviews. For current research dealing with the analysis of metals in foods the reader should consider *Analyst, Analytica Chimica Acta, Analytical Chemistry, Applied Spectroscopy, Atomic Spectroscopy, Fresenius Zeitschrift fuer Analytische Chemie,* the *ICP Information Newsletter, Progress in Analytical Atomic Spectroscopy, Science of the Total Environment, Talanta,* and the *Journal of Analytical Chemistry of the USSR. Atomic Spectroscopy* publishes a comprehensive bibliography every 6 months with sufficient indexing to permit easy location of papers related to the determination of metals in foods. *Annual Reports on Analytical Atomic Spectroscopy* (1971–1983) presents an annual comprehensive review with excellent indexing. *Analytical Chemistry* offers application reviews and fundamental (instrumentation) reviews in alternate years.

Journals oriented toward food and nutrition have proven to be less frequent sources of articles dealing with the determination of metals. Journals to be considered are the *Journal of Agriculture, Food Chemistry,* the *Journal of the Association of Official Agricultural Chemists,* and the *Journal of Food Science.*

A number of computerized literature search data bases are available.

Collections of journal article titles are available in *Current Contents. Chemical Abstracts* offer biweekly collections of abstracts on selected topics; for the determinations of metals in foods, "Atomic Spectroscopy" and "Trace Elements" are the two most useful categories.

## III. SAMPLE PREPARATION

Determination of elements in foods and beverages by atomic spectrometry requires a pretreatment of the sample to remove the bulk of the organic matter and to convert the sample to a liquid form. The pretreatment, or preparation, of organic samples usually consists of drying the sample followed by either dry ashing or wet oxidation.

Dry ashing can be accomplished at either high or low temperature. High-temperature ashing is usually carried out at temperatures between 400 and 1000°C, depending on the elements to be determined. Small quantities of acid are sometimes added to assist in destruction of the organic material. After the ashing process, the ash is dissolved in a small amount of concentrated acid and is then brought to volume. Dry ashing permits a large number of samples to be handled with a minimum of the analyst's attention.

Low-temperature or plasma ashing utilizes radiofrequency energy, rather than thermal energy, for destruction of the organic material (Hollahan, 1974). Plasma ashing is usually employed when the volatility of one of the elements of analytical interest prohibits high-temperature ashing. Thermal temperatures during plasma ashing do not exceed 150°C. As in high-temperature ashing, the ash is dissolved in a small volume of concentrated acid and is then brought to volume.

Wet oxidation is another method commonly used for removing bulk organic material from the sample. Wet oxidation involves heating or refluxing the sample (at less than 100°C) in the presence of a strong mineral acid, such as nitric, sulfuric, phosphoric, or hydrochloric acid, with strong oxidizing agents, such as hydrogen peroxide or perchloric acid. Again, a large number of samples can be handled with a minimum of the analyst's attention. The liquid sample is frequently suitable for analysis directly or can be diluted as necessary.

The major problem with dry or wet oxidation methods is that silicates are not solubilized. Usually hydrofluoric acid is used to ensure that the entire sample is dissolved. Many variations of the generalized dry and wet oxidation schemes are used. For methods for specific samples, the reader is referred to the literature.

## IV. ATOMIC ABSORPTION SPECTROMETRY

### A. Principles of Atomic Absorption Spectrometry

Modern atomic absorption spectrometry developed from the work of Walsh (1955). It achieved almost immediate commercial success and is today the most popular method for metal determinations. The popularity of AAS arises from its analytical specificity, low detection limits, excellent precision, and relatively low cost. The main drawbacks of AAS have historically been its limited calibration range and its inability to analyze more than one element at a time.

Although many varieties of atomic absorption spectrometers are available, the basic design is quite simple. Each spectrometer consists of a light source, an atomization source, and a dispersion–detection device. These three components are optically aligned in a straight line, usually on an optical rail.

The light source is a hollow cathode lamp (HCL). The cathode is constructed of the element to be determined. Electrical excitation produces a cloud of excited atoms in the hollow cathode that emits the distinctive elemental atomic spectrum. The internal low pressure of HCLs results in the emitted atomic spectra being extremely narrow (0.0004–0.002 nm full width at half the maximum height, between 200 and 600 nm). For this reason HCLs are often referred to as line sources, being considered monochromatic in nature. Monochromaticity is necessary to achieve the maximum sensitivity and calibration range for AAS; both of these are adversely affected by running the HCL at currents above the recommended maximums.

The atomization source produces a cloud of vapor-phase atoms from a liquid sample. The process usually occurs in distinct steps. Nebulization is the conversion of the liquid sample to an aerosol, desolvation is the evaporation of the solvent, and atomization is the thermal process whereby molecular components are reduced to individual atoms. The AAS atomizer is designed to generate free ground-state atoms. These ground-state atoms are then capable of absorbing light emitted by the HCL. The resultant emission and absorption spectra of the ground-state atoms are quite simple. In general, there are three types of atomizers currently being used in AAS: flame atomizers, furnace atomizers, and chemical vapor generation atomizers. Each of these will be discussed in detail in later sections.

Finally, the dispersion–detection device is used to isolate the wavelength of interest and measure the intensity. This is traditionally accomplished by a monochromator and a photomultiplier tube (PMT), respectively. Because the HCL is essentially a line source, high resolution of the monochromator is not critical. A pair of measurements must be made for each analytical signal, a reference intensity ($I_0$) and a transmitted intensity ($I$). The reference intensity is the intensity of the HCL source before the emitted beam is passed through the atomizer. The reference intensity can be measured before atomizing the solution of interest or "pseudosimultaneously," by periodically diverting the light around the atomizer. The transmitted intensity is the source intensity after it has passed through the atomizer and is equal to the source intensity minus the fraction absorbed by the atoms in the atomizer.

Absorbance $A$ is defined by the Beer–Lambert equation:

$$A = \log_{10}(I_0/I) = abc$$

where $I_0$ and $I$ were previously defined, $a$ is the absorption coefficient, $b$ is the path length of the atomizer, and $c$ is the concentration of the element. In general, the Beer–Lambert equation is valid ($A$ varies directly with $c$) up to absorbances of 0.4–0.5. At higher values, bending toward the concentration axis is observed.

## B. Atomizers

### 1. Flame Atomizers

Flame atomization has been employed since the inception of AAS. A chemical flame is the simplest, oldest, and best characterized of all the atomizers. The study of flame atomization is a rather static field, with no major breakthroughs having occurred in the last 10–15 years. During this time, flame atomization has been used for the determination of almost every element in almost every conceivable sample matrix.

Physically, the commonly used laminar-flow, premixed burner assemblies consist of a nebulizer, a mixing chamber, and a burner head. The nebulizer is a Venturi tube that aspirates the sample and produces an aerosol. The aspirator tube is simply dipped into the solution to be determined. The aerosol is mixed with the combustion gases in the mixing chamber and is carried upward to the burner head. The aspiration rate is typically 5–10 ml/min. This process is not very efficient in converting the solution into an aerosol, because only 5 to 10% of the aspirated solution typically reaches the flame.

A wide variety of flames have been used for AAS, but the air–acetylene and nitrous oxide–actylene mixtures are the most popular. The air–acetylene flame is the cooler of the two, about 2300°C. It is more suitable for elements that ionize easily but, as a result, suffers more interferences for elements that tend to form stable compounds. The nitrous oxide – acetylene flame, about 2700°C, supplies the thermal energy needed to bring about dissociation of the compound-forming elements but is more difficult to use. The high burning velocity of nitrous oxide – acetylene flame necessitates close control of the gas flows to prevent the flame from flashing back into the mixing chamber. Automated sensors and flow controls have largely eliminated this problem in the last 10 years.

In Table 1, flame atomization has been compared with other AAS atomization modes and with AES and AFS with respect to general usage. The use of HCLs for exact wavelength sources and the simplicity of the emission and absorption spectra result in flame AAS being a highly specific method. The detection limits range from 0.001 to 0.1 $\mu$g/ml and precisions of 0.1 to 0.3% can be attained. While the detection limits are very reasonable, they have received a poor mark in Table 1 because they are approximately an order of magnitude worse than those for ICP/AES or direct-current plasma atomic emission spectrometry (DCP/AES) and two or three orders of magnitude worse than those for furnace AAS.

Matrix interferences are common for flame AAS but have been well characterized over the last 25 years. An extensive body of literature now exists for many elements analyzed in a wide variety of materials. The analyst can almost always find a pertinent reference for every new sample type to be

**TABLE 1  A Comparative Evaluation of Spectrometers**

|  | AAS | | ICP/AES | ICP/AFS |
|---|---|---|---|---|
|  | Flame | Furnace | | |
| Specificity | +++ | +++ | +++[a] | +++ |
| Precision | +++ | ++ | +++ | +++ |
| Detection limits | ++ | +++ | ++ | ++ |
| Interferences | ++ | ++ | +++[a] | ++ |
| Calibration range | ++ | ++ | +++ | +++ |
| Multielement capability | + | + | +++ | +++ |
| Available literature | +++ | ++ | ++ | + |
| Ease of operation | +++ | + | ++ | ++ |
| Automation – computerization | +++ | ++ | +++ | +++ |
| Expense[b] | +++ | ++ | + | ++ |

[a] There are many documented line overlap interferences, but for food samples these interferences will be significantly less severe.

[b] In keeping with the symbolism, +++ is the best rating, indicating the lowest relative cost.

analyzed. Interferences will be discussed in more detail in the next section.

Flame AAS is easy to operate. After the initial instrument setup procedure the analyst need only aspirate the standards, samples, and blanks in a systematic manner. Readout of the sample signal is instantaneous, allowing the analyst to detect immediately any irregularities that might occur.

Flame AAS, in general, has proven readily adaptable to automation. Few modern spectrometers are available without a dedicated microprocessor. The analyst can now load the automatic sampler with standards, samples, and blanks, start the operation, and walk away. The spectrometer will measure each solution, calibrate, compute concentrations, and print the results for each sample. A sequential multielement instrument is commercially available that will repeat this automated determination for up to six elements without requiring the analyst to intervene. Presently, no simultaneous multielement AAS system is commercially available, although a variety of experimental systems have been reported. The authors of this chapter have demonstrated the use of continuum source atomic absorption with wavelength modulation for improved background correction for the analysis of a wide variety of food and beverage samples (Harnly et al., 1979; Harnly and Wolf, 1981). This system has successfully been used for the simultaneous determination of up to 16 elements and is in daily, routine use for the analysis of foods and biological materials. A commercially available simultaneous multielement AAS system will likely be seen in the future.

On the whole, flame AAS is the least expensive atomic spectrometric method. However, the cost of atomic absorption instruments covers a wide range, depending upon the optional features, computerization, and versatility desired. The simplest and least expensive instruments consist of the basic light source, flame atomizer, and dispersion–detector configuration with electronics for computing absorbance and a means of displaying the value. The most complex systems offer scanning monochromators; flame, furnace, or ICP atomization–excitation sources; HCL or electrodeless discharge lamp light sources; dedicated microprocessors for spectrometer control; and sophisticated data stations for controlling automatic sampling, spectrometer operation, data processing, and display. There are obviously many levels of complexity between these two extremes. The analyst must decide what capabilities and costs best meet the laboratory's needs.

### 2. Furnace Atomizers

Furnace atomization was first introduced by L'Vov in 1961. The growth of the method of furnace atomization has been quite rapid, with new developments occurring at frequent intervals. There are indications that the study of furnace atomization is just now approaching a period of stability, allowing for more complete characterization of its abilities and limitations.

Consequently, while furnace atomization is presently a dynamic and exciting research field, it is simultaneously being employed for routine applications. Koirtyohann and Kaiser (1982) and Slavin (1982) have recently reviewed the state of the art of furnace atomization.

A furnace atomizer is quite simple in design. A hollow carbon tube, open at both ends, is positioned horizontally, and an electrical current source is attached at both ends. A small sampling port is located in the middle of the tube, which allows the solution to be injected into the tube. An electrical current passed through the tube produces resistive heating. Atomization of the solution occurs in discrete steps. First, a current is passed through the tube, which raises the temperature high enough to dry the sample by evaporating the solvent. Next, in the ashing or charring step, a higher current is employed to destroy the organic matter and evaporate the more volatile, undesirable inorganic components. Finally, a high current is applied to atomize the element of interest. Temperatures as high as 3000°C can be reached in the atomization step with a carbon furnace.

There are two methods of injecting the solution into the furnace. The most common approach is to pipette a discrete volume into the furnace at room temperature. The second method is to make an aerosol of the solution, as in flame atomization, and then blow a jet of the aerosol into the tube, which has been heated above 100°C. Sample volumes required range from 5 – 100 $\mu$l for discrete deposition and are comparable to flame atomization for aerosol deposition.

Discrete deposition may place the solution either directly on the inside wall of the furnace (opposite the sampling port) or on a "platform." A "platform" was first suggested by L'Vov (1976) and consists of a flat or slightly curved piece of carbon that bridges across the bottom of the furnace directly below or opposite the sampling port. Material on the furnace wall heats at the same rate as the wall and is atomized into an atmosphere whose temperature lags behind the wall temperature and is changing rapidly with time. The platform is heated primarily by radiation from the furnace wall and therefore lags behind the wall temperature. When the delayed atomization from the platform occurs, the atmosphere in the furnace has had a chance to reach the wall temperature and is no longer changing rapidly. Platform atomization into an "isothermal" atmosphere reduces many interferences that have been observed for atomization from the wall.

Furnace atomization retains the specificity of flame atomization (Table 1) and offers the best detection limits of any of the atomic spectrometry methods. Detection limits range from 0.01 – 10.0 ng/ml. The precision, however, is not as good as that of flame AAS, ranging from 1 to 4% for an optimized automated system, primarily because of inherent lack of reproducible atomization from the carbon surface.

It is currently acknowledged that interferences are worse for furnace atomization than for flame atomization. Since the entire sample is atomized into a small volume, nonspecific background absorption and light refraction are much worse for furnace atomization. In addition, severe matrix effects have also been reported. However, the use of platform atomization, faster heating rates, and matrix modification (chemical additions to the solution in the furnace that alter the atomization characteristics of the analytical element and/or the interferent) has served to improve the accuracy of furnace determinations.

The literature on furnace atomization is presently far from complete or conclusive. This is not surprising considering the stage of development of furnace atomization. In most cases, the analyst must reevaluate existing methods or develop new methods for each new sample matrix. Published literature may or may not be applicable to the furnace atomizer being used or the material being analyzed.

Furnace atomization is not easy to use for two major reasons: the inherent nature of the atomizer and the concentration levels being determined. The furnace atomizer requires several minutes between individual determination and, because of its less precise nature, requires two or three determinations for each solution. The surface of the carbon tube steadily degrades as a function of the number of determinations. Coating the tube with pyrolytic carbon helps in slowing the degradation but does not cure the problem. Consequently furnace atomization produces results at a slow rate and is subject to low-frequency noises of both a random and a systematic nature. Furnace atomization is not well suited for the routine determination of a large number of samples.

The concentration levels being determined by furnace atomization are extremely low. The useful calibration range for most elements falls between 1 and 100 ng/ml. Contamination of the standards or the samples can lead to highly erroneous results. Realistically, the detection limits for most elements are limited by the contamination of the blanks. Detection limits based on repeated atomization of a low standard can be greatly misleading, since they do not reflect the contamination variation for each solution. Successful and consistent trace element determinations by furnace atomization require extreme care in handling of the samples and standards; high-purity water, acids, and reagents; metal-free plasticware; and clean air hoods in which to work. Without these precautions, trace metal analyses are not possible.

Many aspects of furnace atomization, like flame atomizations, have been automated and computerized. In fact, automated solution delivery to the furnace is mandatory in order to achieve the best precisions of 1 to 4%. However, the authors' experience has been that it is impossible to start a furnace analysis and then walk away, leaving the spectrometer unattended. The mechanical furnace operations are reliable, but unexpected fluctuations, lack of agreement of repeated determinations, and/or systematic changes in sensitivity inevitably result in additional determinations or redeterminations of blanks, samples, or standards. Close attention by the analyst is required to produce reliable results.

The price of the furnace atomizer and power supply is approximately the same as that of the rest of the spectrometer. In most cases, however, as discussed for flame spectrometers, the major determining factor in the cost is the optional features.

Again, there is no commercially available multielement furnace AAS. Preliminary data from the authors' laboratory suggest that multielement furnace AAS can be accomplished and that the compromises for efficient simultaneous multielement atomization are less for the furnace than for the flame. However, significant modifications of presently available furnaces to prevent cross-contamination problems (Harnly, 1982) and improvements in data acquisition methods to allow extended calibration ranges (Harnly and O'Haver, 1981) are necessary before multielement furnace AAS is likely to be developed commercially.

## C. Interferences

Interferences are common for both flame and furnace atomization. In general there are two basic types of interference: additive interference arising from spectral sources and multiplicative interference arising from the chemical and physical characteristics of the sample solution (O'Haver, 1976).

### 1. Additive Interferences

Spectral interferences add to the apparent analytical signal; that is, they produce no change in the slope of the calibration curve but contribute a positive and equal shift of the calibration curve at all points. Thus, the absolute amount of interference is not affected by the size of the analytical signal, and the severity of spectral interferences is inversely proportional to the analytical signal.

There are two types of spectral interference: direct overlap of narrow atomic spectra, or line overlap, and nonspecific broad-band background spectra. Line overlap interferences are extremely rare for AAS because of the simplicity of the spectra of the ground-state atoms, whereas background absorption is common in complex matrices, primarily below 400 nm. Background absorption is particularly severe for furnace atomization, and background correction is essential for accurate determinations. For flame atomization, background absorption is considerably less severe, and, consequently, background correction is frequently not used. However, it must be remembered that the level of interference is relative. Even though background absorption may be very low, if the analytical signal is small, then the interference may still be significant. The newer background correction methods are easy to employ and should be used for all flame determinations for the best possible accuracy.

Correction for background absorption is conceptually quite simple. The transmitted intensity ($I$) is measured in the normal manner, at the analytical line, and reflects the source intensity minus the sum of the atomic and background absorption. The reference intensity ($I_0$) is redefined slightly and is measured so that it reflects the source intensity minus the background absorption. Since the background absorption reduces $I_0$ and $I$ by the same percentage, the computed absorbance, using the new definition of $I_0$, will be the true, or corrected, absorbance.

The classic method for correcting for background absorption was to employ a hydrogen or deuterium continuum source, in addition to the HCL, and a wide spectral bandpass (Koirtyohann and Pickett, 1966). The beams from the HCL and the continuum source were superimposed through the atomizer. The transmitted intensity of the HCL reflected the atomic and background absorption. The transmitted intensity of the continuum source was decreased only by the background absorption since the spectral band pass is sufficiently wide to render the atomic absorption component insignificant. The computed absorbance for the continuum source was then subtracted from the HCL absorbance to obtain the true absorbance.

The two-source method of background correction was not completely satisfactory. Superimposing the two beams was critical but very difficult to do accurately. Correction was not adequate for high background absorption (0.5 or greater) or for low analytical absorption. In addition, a tungsten halide source was necessary for wavelengths greater than 300 nm.

In the past 5 years significantly improved background correction methods have become available. Two new methods that have been commercially developed are the Zeeman splitting (Hadeishi *et al.,* 1975; Koizumi and Yasuda, 1975) and the Smith–Hieftje (Smith *et al.,* 1982) methods. In addition, a third, experimental method, the continuum source–wavelength modulation (CS/WM) method, has been reported (Harnly and O'Haver, 1977). All three methods use a single source, measure $I$ normally, and then shift 0.002–0.1 nm from the analytical line to measure $I_0$. Each method shifts rapidly between the two positions, measuring $I_0$ and $I$ many times each second. The limit of the background correction, for each method, is determined by the electronics. In addition, the calibration curves for all three methods can be expected to bend toward the concentration axis and then continue downward, totally reversing at high absorbances. This reversal of the calibration curves is a result of the reference intensity being measured only a short distance from the analytical line. The further from the analytical line that $I_0$ is measured, the less the tendency toward reversal.

The Zeeman splitting method uses a magnetic field around the flame or furnace to shift the absorption profile away from the HCL emission line. Thus $I$ is measured with no magnetic field and reflects atomic and background absorption. $I_0$ is measured with full magnetic strength and, with the absorption profile shifted away from the HCL line, reflects only background absorption.

The Smith–Hieftje method overdrives (uses too high a current for) the HCL at regular intervals. This broadens the HCL emission line and causes a reversal in the HCL intensity at the line center. Consequently, in the overdriven phase, the HCL has almost no intensity where it previously had a maximum and has two maxima at wavelengths to either side of the original maximum. Consequently, $I$ is measured during the normal current phase and reflects atomic and background absorption while $I_0$ is measured during the overdriven phase and reflects only background absorption.

The CS–WM method uses wavelength modulation to scan the absorption profile. This method was first proposed by Snelleman (1968). The continuum source provides a constant intensity over the narrow region around the wavelength of interest. Absorption appears as an inverted Gaussian trough in the continuum. $I$ is determined at the center of the absorption profile, measuring both atomic and background absorption. $I_0$ is determined first to one side of the absorption profile and then to the other, measuring only background absorption. The computed absorbance is corrected for background absorption. If the measurements and intensity ratios are made at a high enough frequency (greater than 40 Hz), the continuum source flicker noise is almost eliminated, making continuum source atomic absorption practical. Since the light-source currents are not changed for the CS–WM and Zeeman splitting methods, the computed absorbances result in both having double-beam operational characteristics even though only a single light path is being employed.

## 2. Multiplicative Interferences

Multiplicative interferences arise from chemical and physical differences between the samples and the standards. These interferences are proportional to the analytical signal. Thus, the calibration curve will still intercept the absorbance axis at 0, but the slope will change.

There are two types of chemical interferences in the flame. Stable compounds that fail to be thermally dissociated and atoms that are ionized reduce the free atom population in the flame. Compound formation is prevalent for aluminum, silicon, and the alkaline earth elements in the lower-temperature, air–acetylene flame. The hotter, nitrous oxide–acetylene flame eliminates this interference. Many transition metals display interelement effects that do not appear to be related to compound formation. These interferences are also eliminated by a nitrous oxide–acetylene flame, but there is also an accompanying loss in sensitivity. The alkaline metals and alkaline earth elements are easily ionized in an air–acetylene flame. The problem is even more serious in a nitrous oxide–acetylene flame.

Physical interferences arise from physical differences, such as salt concentration and viscosity, between the samples and standards. These physical effects lead to different atomization rates.

There are several solutions to these problems, aside from altering the flame characteristics. The most obvious correction is to make up standards that exactly match the samples. Matrix matching eliminates compound formation, ionization, and physical interferences but presupposes a thorough knowledge of the sample matrix and is not suitable for large numbers of samples.

A more general correction method is to dilute the samples and standards in a solution that suppresses the interference. For example, a 0.5% lanthanum matrix (usually as a nitrate or chloride) is used to eliminate the phosphate suppression of magnesium and calcium, while a 0.1 – 1.0% cesium solution (usually as a chloride) eliminates ionization interferences. This approach is readily adopted as a routine method for analyzing large numbers of samples.

Finally, there is the method of standard additions. The method of standard additions is based on the incorporation of a series of added concentrations (the spikes) of the element of interest into equilibrium with the endogenous species. The spikes, therefore, are suppressed to the same extent as the endogenous species. Ratioing of the analytical signals of the spiked and unspiked samples produces a corrected concentration for the sample. In order for this technique to be valid several conditions must be met: (1) the spikes must be of sufficiently small volume that there is no significant change in the sample volume (less than 5%), (2) the spiked species must equilibrate with the endogenous species, (3) if no calibration standards are used, the analytical signals must lie in the linear region of the instrument's response. The requirement for linearity may be avoided if calibration standards are also determined. The recovery of the spikes, based on concentrations of the spiked and unspiked samples computed from the calibration function, can be used to correct the unspiked sample concentration. Frequently, only a single spike is employed. Spike recoveries are commonly used to evaluate sample preparation and handling methods. Multielement spikes are not theoretically valid, since the matrix must be constant except for the element being tested.

Furnace multiplicative interferences are difficult to generalize because of the many conflicting reports. This is not too surprising considering the variety of furnace designs available from different companies and the rapid rate of furnace design changes within each company. However, contradictory results have been reported by analysts using the same equipment. This suggests that interferences may be related to specific operating parameters (Slavin and Manning, 1982). The latest improvements in furnace AAS have served to clarify some of the problems. Longer-lived carbon, pyrolytic surfaces, faster heating rates, faster electronic circuits, and platform atomization have eliminated many of the problems that previously led to conflicting results. As furnace atomization reaches maturity, the classes and causes of interferences will become clearer. Slavin and Manning (1982) have compiled a literature guide to graphite furnace interferences on an individual element basis.

Ionization interferences are relatively minor for furnace AAS. Sturgeon and Berman (1981) have shown that ionization does not occur to a great extent in the furnace. Errors due to sample delivery to the furnace are usually insignificant, since only a single pipetting (usually automatic) is involved. But transfer of a highly viscous solution can produce highly erratic results. Proper pretreatment (ashing and dissolution, enzyme digestion, or addition of a surfactant) is required for viscous solutions in order to obtain reproducible results.

Free atoms are formed in the furnace through chemical reactions on the carbon surface and in the the gas phase, as the material is dried, ashed, and then atomized. Chemical differences between the samples and the standards can lead to differences in the atom formation mechanisms that affect the free atom population and the signal shape. Integration of the furnace absorbance pulse can eliminate some interferences that arise for peak height measurements.

A major vapor-phase interference mechanism is the diffusion loss of nonabsorbing compounds from the furnace before a temperature high enough to bring about thermal decomposition into individual atoms is reached. L'Vov (1978) has shown that this interference occurs only for metal monocyanides and monohalides. Use of argon as a sweep gas significantly reduces the probability of monocyanide formation, but halides, especially when present as a major component of the sample (such as high NaCl in the sample), can cause significant suppression of the analytical signal. As a result, nitric acid is the preferred acid matrix for sample dissolution.

Loss of nonabsorbing metal monohalides can be reduced and/or eliminated by atomization into a hotter atmosphere. This can be done in several ways. Faster heating rates result in the furnace's reaching higher temperatures before the halides are lost through diffusion. Platform atomization retards the sample atomization until higher furnace temperatures are reached. Insertion of the sample on a carbon probe into a preheated constant-temperature furnace can eliminate halide interferences in the presence of large excesses of a halide (Giri *et al.,* 1982).

Another means of eliminating interferences is by chemical, or "matrix," modification of the sample (Ediger, 1975). The principle involves shifting the atomization temperatures of the element of interest and the interferent away from one another. For example, if NaCl is a major constituent of the sample (which gives rise to halide suppression and high background absorption), the addition of $NH_4NO_3$ results in the formation of $NaNO_3$ and $NH_4Cl$, which volatilize during the ashing step.

## D. Applications

Both flame and furnace AAS are used for the determination of inorganic elements in foods and beverages. The ease of operation and familiarity of flame atomization make it the method of choice where large numbers of samples are to be analyzed and when furnace detection limits are not required.

Flame atomization has been used for the analysis of almost every metal and nonmetal in the periodic table by direct and indirect methods. For methods for specific elements in foods, the reader is referred to several excellent sources. Koirtyohann and Pickett (1975) and Crosby (1977) provide general overviews for metal determinations in foods. Both devote more attention to the analysis of harmful metals than those of nutritional interest. Crosby presents a listing of analytical methods for the determination of As, Cd, Hg, Pb, Se, and Sn and Koirtyohann covers Cd, Hg, Pb, and Sn.

A more comprehensive survey of methodologies is presented by Fricke *et al.* (1979). This survey is a fairly comprehensive bibliography, organized by element, for the determination of trace metals in foods over the past 15 years. Each reference is listed according to the food material analyzed and includes a brief summation of the ashing, separation – preconcentration, and atomization methods. A range of results is reported and recovery data are included when available.

Ihnat (1981) has written an extremely detailed chapter covering the analysis of foodstuffs by AAS. This work is essentially a how-to manual and covers all aspects of the physical sample preparation, sample treatment, and analytical process. Physical sample preparation and homogenization methods are recommended for each type of foodstuff (cereals, dairy products, meat, fish, etc.). Sample digestion procedures, reagents, apparatus, and standards preparation are discussed. Finally, general analytical protocol and specific techniques are recommended for 21 elements on an individual basis.

The preceding reviews are excellent references for methods for the determination of elements in materials new to the analyst. Because of the maturity of flame AAS (it has been used for 25 years), methods for many metals in many materials have been tested, and many of these methods are still current and applicable. A survey of the AAS literature for the last 2 years, as it applies to foods and beverages, reveals mostly refinements of long-accepted sample preparation methods. Heanes (1981), Feinberg and Ducauze (1980), and Rowan *et al.* (1982) used dry ashing to determine a variety of elements in food and plant materials. Agemian *et al.* (1980), Borriello and Sciandone (1980), Evans *et al.* (1980), and Jackson *et al.* (1980) explored wet digestion methods for the analysis of fish, beer, and assorted foodstuffs. Each of these methods represents a fine tuning of existing methods to obtain more accurate results in the materials of interest.

Methods development for furnace AAS is a much more dynamic field. This is a result of the improving furnace designs and the application of furnace atomization to many areas for the first time. As mentioned previously, there are several sources of AAS reviews that permit papers on the determination of metals in foods to be located. These reviews are the *Analytical Chemistry* application reviews, which are published every other year, the *Atomic Spectroscopy* bibliographies, which appear every 6 months, and the *Annual Reports on Analytical Atomic Spectroscopy.*

In the last 2 years, furnace AAS has been used to determine trace metals in fish (May and Brumbaugh, 1982; Langmyhr and Orre, 1980), yeast and nutritional supplements (Thompson and Allen, 1981), edible oils (Slikkerveer *et al.,* 1980), orange juice (Nagy and Rousseff, 1981; Harbach *et al.,* 1980), oysters (Chakrabarti *et al.,* 1980b), legumes (Rockland *et al.,* 1979), vegetarian products (Freeland-Graves *et al.,* 1980), plants (Duprie and Hoenig, 1980), bovine liver (Chakrabarti *et al.,* 1980a), and canned foods (Dabeka, 1979). In approximately half of these cases, the studies originated primarily from instrumentation development goals.

## V. ATOMIC EMISSION SPECTROMETRY

Atomic emission spectrometry has experienced a rejuvenation in the last 10 years with the development of the ICP and the stabilized DCP as atomization–excitation sources. With these sources, AES offers freedom from chemical interferences, precision comparable to that of flame AAS, detection limits superior to those of flame AAS (but not as good as those of furnace AAS), calibration ranges covering six orders of magnitude, and the ability to determine from 20 to 60 elements simultaneously. Emission spectra are more complex than absorption spectra, leading to the greater possibility of spectral interferences.

## A. Principles of Atomic Emission Spectrometry

Conceptually, AES is a two-component system, consisting of an atomization–excitation source and a dispersion–detection device. The atomization–excitation source generates a cloud of excited atoms that emit energies distinctive to each element. The atomization process occurs in the ICP and DCP sources in much the same manner as in flame atomization, except that the source temperature is much higher. A liquid sample is first nebulized into the source, where it is desolvated, and compounds are vaporized and then broken down into individual atoms. At this point, the high energy of the source also excites the atoms, which then emit their characteristic spectra. Excitation of the atoms is frequently sufficient to produce ionization. In many cases, the spectra of the excited ions are measured. Both the ICP and the DCP are designed so that clouds of cooler sample gases (which can cause self-absorption and reversal of the emission line) do not form in the optical path. As a result, the intensity of the emission spectrum is directly proportional to the concentration of the element for up to six orders of magnitude.

The dispersion–detection device is used to isolate the spectral line for each element and to measure the intensity of the emitted light. This is generally done with either a scanning monochromator, which employs a single PMT and scans rapidly through the wavelengths of interest, or a polychromator, which employs a separate exit slit and PMT for each wavelength. The polychromator is preferred for routine determinations of large numbers of samples, whereas the scanning monochromator is more versatile and is more useful for fewer samples when the elements of interest might vary. The resolution of the dispersion device is important, since the high-energy excitation sources produce complex spectra. Resolving overlapping emission lines, or correcting for overlapping lines, is necessary for accurate determinations.

## B. Atomization – Excitation

### 1. Inductively Coupled Plasma

The ICP was developed simultaneously in the mid 1960s by Wendt and Fassel (1964) in the United States and Greenfield et al. (1964) in Great

Britain. After a period of rapid growth in the early 1970s the ICP is now reaching a period of maturity. ICP/AES has been applied to almost every general type of sample matrix, and its advantages and limitations have been well characterized.

The ICP torch consists of a series of concentric argon flows. The plasma gas flow is passed through an induction coil. The induction coil magnetically induces a radiofrequency plasma in the argon. A second flow, the carrier gas flow, lies inside of and parallel to the plasma gas flow and injects the sample into the middle of the plasma. The carrier gas flow punches a hole through the middle of the plasma, giving the plasma a doughnut (toroid) shape.

A third argon flow, the coolant gas flow, parallel to and outside the plasma gas, prevents the plasma from melting the surrounding quartz tube. The temperature of the argon ICP is approximately $10,000°C$. The atomized solution and carrier gas flow reach temperatures of $6000-7000°C$. Emission measurements are usually made $1-2$ cm above the induction coil to obtain the maximum signal-to-noise ratio. Other gases and minitorches have been used with varying success.

Two different designs of pneumatic nebulizers, the concentric nebulizer and the cross-flow nebulizer, are currently popular. These nebulizers are limited to a maximum of 1% salt and are more consistent at lower salt concentrations. A third nebulization source, an ultrasonic nebulizer, based on a pizoelectric crystal vibrating at high frequencies (several million cycles per second), has been used but has not found widespread acceptance. The ICP, using pneumatic nebulization, consumes slightly less volume than flame AAS with a nebulization efficiency of approximately 1%. Consequently, the ICP is even less efficient than flame AAS with respect to atomization. The problem is exacerbated when a wavelength scanning detector, which may require several minutes of sample nebulization, is used.

Several approaches to introducing microsolutions have been proposed (for polychromatic detection systems), including funnels, microcups, flow injection, and furnace atomization (Boyko et al., 1982). The detection limits vary as a function of the method of introducing the sample and the sample volume used.

In Table 1, the performance characteristics of ICP/AES have been compared with those of AAS and ICP/AFS. In general, the specificity of ICP/AES is not as good as that of either flame or furnace AAS because of the frequency of line overlap interferences. However, the organic matrix of foods and beverages generally produces relatively simple spectra. Consequently, the specificity of ICP/AES is comparable to that of AAS for the determination of metals in food. As mentioned previously, the detection limits for ICP/AES are, on the average, better than for flame AAS but worse than for furnace AAS. Detection limits for refractory elements are much better than those for flame AAS and are comparable to those for furnace AAS. The precision of ICP/AES is comparable to that of flame AAS.

Chemical interferences are essentially nonexistent for ICP/AES, but spectral interferences are quite common. Direct line overlap interferences have been extensively investigated in recent years, and several spectral line atlases are now available (Boumans, 1980; Parsons and Forster, 1980).

Mechanically, the operation of ICP/AES is quite similar to that of flame AAS. The nebulizer tube needs only to be removed from the blank to the standards and samples. However, ICP/AES lacks the single-element simplicity of flame AAS. The operator must interact with the computer to identify all data collected. It is not possible to follow each element simultaneously. In addition, calibration over six orders of magnitude of concentration creates more potential for cross contamination. Time is required between atomizations to allow the nebulizer to clear. Thus, in many respects, ICP/AES more closely resembles furnace AAS as far as ease of operation is concerned.

Computerization is a necessity if the full analytical capabilities of ICP/AES are to be realized. Scanning monochromators and direct reading systems require computer operation in order to collect large amounts of data in short periods of time. The automation and computerization features of ICP/AES should not be regarded as optional features but, instead, as requirements for effective utilization of the instrument.

ICP/AES is more expensive than flame or furnace AAS. The least expensive scanning ICP systems are roughly twice the cost of the most elaborate AAS instruments. An ICP source with a dedicated minicomputer and simultaneous detection capabilities for more than 60 elements costs four or five times as much as the scanning ICP systems. However, the cost aspect is not all negative. With this expenditure for equipment, and because of its more complex computerization and electronics, organizations assign only experienced analysts to the instrument operation. As a result, ICP results reported in the literature have been of a consistently high quality.

Finally, the real strength of ICP/AES is its multielement capability. The large calibration range and the high-energy source, free of chemical interferences, allow the same instrumental parameters to be used for all elements. There is little compromise in the detection limits for multielement determinations.

## 2. Direct-Current Plasma

The DCP is a controlled direct-current arc with temperatures comparable to those of the ICP. Although a number of experimental systems have been reported, only one design is currently commercially available. This design employs three electrodes for stability. The plasma looks like an inverted Y. The sample carrier gas originates from below the inverted Y and directs the nebulized material toward the junction of the two legs. Nebulization is accomplished with a conventional flame AAS nebulizer. Emission measurements are made at a point just below the junction.

The DCP is sold with an Echelle polychromator. The Echelle polychromator offers resolution an order of magnitude better than that of most conventional monochromators and allows separation of many emission lines that cause overlapping interferences on other spectrometers. The Echelle polychromator design allows the positioning of up to 20 end-on-style PMTs behind the focal plane.

The operational characteristics of DCP/AES are similar to those of ICP/AES (see Table 1). The difference lies in the introduction of the nebulized material into the plasma. For ICP/AES the carrier gas flows through the hole of the doughnut-shaped plasma, but for DCP/AES the carrier gas flows into the junction of the plasma legs and then around the plasma to either side. As a result, the nebulized material is exposed to lower temperatures in the DCP than in the ICP. Chemical interferences, which are relatively nonexistent in the ICP, are more troublesome in the DCP, although still of less concern than they are for flame AAS. Since DCP/AES is capable of determining only 20 elements simultaneously, its costs lie toward the lower end of the range for ICP/AES instruments.

## C. Interferences

### 1. Additive Interferences

Additive, or spectral, interferences are common and can be severe for high-energy plasma AES. The recombination of electrons and cations gives rise to a broad-band background continuum under normal operating circumstances. Specific matrices can produce significant increases in the background continuum intensity. Transition and rare earth elements produce extremely complex emission spectra with dozens of major line overlap interferences for the commonly analyzed elements.

Broad-band interferences are corrected for AES as they are for AAS, by means of an off-line method. Emission measurements, made just off the analytical line, reflect the background emission, while measurements on the analytical line reflect background and atomic emission. Subtraction of the off-line intensity from the on-line intensity produces the corrected atomic emission. Moving off the analytical line can be more difficult for AES than it is for AAS, since considerable emission line structure can be found superimposed upon the background continuum. The position of the off-line measurements can be critical in making an accurate correction. In many cases, the optimum position may vary with each element. The spectrum to either side of each analytical line must be scanned in order to determine the correct location to make the off-line measurements.

Line overlap interferences are corrected using an on-line method. The concentration of a potential interferent is determined from the intensity of the interferent at its primary analytical line. A mathematical factor is then used to correct the concentration of the element of interest. This method involves anticipating the presence of the interferent or detecting its presence by scanning across the analytical line. Several recently published atlases of elemental emission spectra allow the analyst to identify potential interferences. Scanning systems can be altered to make measurements at the wavelength of potential interferents; however, this is not possible for direct readers. If the interfering element is not included in a group of elements being measured, the possibility exists that the interference will go undetected.

### 2. Multiplicative Interferences

Multiplicative chemical interferences are relatively nonexistent for the high-energy plasma as compared to flame or furnace AAS. Ionization enhancement can be a problem for the DCP. An appropriate buffer is necessary to obtain accurate results for the alkaline earth elements and alkaline metals in complex matrices. The buffer also enhances the linearity of the calibration curves at higher concentrations. Ionization does not appear to be a problem for the ICP under normal operating conditions. Data suggest that the ICP serves as an electron reservoir that is not significantly altered by standard or sample material.

Physical interferences are more a problem for the ICP than for the DCP. Inductively coupled plasma nebulizers have been optimized to provide maximum sample efficiency with a minimum of argon flow. As a result, both the concentric and cross-flow nebulizers are sensitive to the salt content and viscosity of the solution being nebulized. High-salt or acid solutions can produce changes in the rate of nebulization and can cause slow, systematic changes by depositing solid material around the nebulizer orifices. These effects can be avoided though a routine cleanout of the nebulizer and/or through the use of internal standards. The DCP is much less susceptible to changes in the nebulization rate, since a peristaltic pump is used to feed the solution, at a set rate, to a concentric AAS-type nebulizer.

### D. Applications

The ICP has not been applied to the determinations of metals in foods to the extent of flame AAS. However, the applications that have been made are impressive for the number of elements simultaneously determined and the number of samples analyzed.

The first extensive report on the determination of metals in biological materials was published by Dahliquist and Knoll (1978). This study looked at the determination of 19 elements in a variety of reference materials and botanical samples. More recently, Jones *et al.* (1980), Jones *et al.* (1982), and Kuennen *et al.* (1982) have used the ICP to determine from 14 to 25 elements in biological samples, foods, liquid protein, and raw crop materials.

Two reviews on the application of ICP/AES to food samples have appeared in the last 2 years. Jones and Boyer (1979) reviewed the determination of metals in food by ICP/AES, and Mermet and Hubert (1982) reviewed the application of ICP/AES to the analysis of biological materials. The latter review was more concerned with the usefulness of ICP/AES in determining trace elements in biological materials than with methodology or results. The use of ICP/AES for the determination of metals in biological materials has also been reported by Barnes (1982) and Barnes *et al.* (1982). Black *et al.* (1981) used the ICP as a detector for gas chromatography after the metal species were selectively removed from biological samples through direct chelation. In general, ICP/AES does not have the detection limits to determine many trace-level elements of biological interest. However, trace metals in foods and beverages are more concentrated, and ICP/AES promises to be a powerful analytical tool.

An emerging area of application for the ICP, which may have considerable impact in the near future, is the determination of nonmetals. Heine *et al.* (1980) and Windsor and Denton (1979) used the far-ultraviolet region to determine C, H, I, and Cl, while Fry and his group (Northway *et al.*, 1980; Fry *et al.*, 1980; Northway and Fry, 1980; Hughes and Fry, 1981; Brown and Fry, 1981) used the visible–near-infrared region to determine Br, Cl, F, N, and O.

DCP/AES is not quite as popular as ICP/AES but has been used effectively for the determination of metals in foods and biological materials. Melton *et al.* (1978) used DCP/AES for the determination of boron in plant material. DeBolt (1980) measured 11 elements simultaneously in plant tissue. Woodis *et al.* (1980) and Hunter *et al.* (1981) analyzed fertilizers. McHard *et al.* (1979) used DCP/AES for the determination of 10 elements in orange juices.

## VI. ATOMIC FLUORESCENCE SPECTROMETRY

Atomic fluorescence spectrometry, with an ICP atomizer, is the newest atomic spectrometric method on the market. Currently, only a single commercial manufacturer is producing an ICP/AFS instrument (Demers and Allemand, 1981). ICP/AFS offers the precision, detection limits (for the nonrefractory elements), and calibration range of ICP/AES and the spectral simplicity of AAS. In addition, up to 12 elements can be determined simultaneously. Since ICP/AFS has been commercially available for only slightly more than a year, there is little information available other than the manufacturer's publications and data obtained for similar, experimental instruments.

## A. Principles and Instrumentation

Like AAS, AFS is composed of three components, a light source, an atomization source, and a dispersion – detection device. These components cannot be in a straight line, because transmission, reflection, or refraction of the source into the detector produces erroneous results.

The atomization source is responsible for converting a solution into a cloud of free atoms through the same nebulization, desolvation, and atomization steps described previously for AAS and ICP/AES. Like AAS, the atomization source is designed to generate free atoms in the ground state. These free atoms are excited spectroscopically by absorbing light emitted by the light source. The atoms then re-emit the light, or fluoresce, at the same wavelength or at a series of longer wavelengths. The intensity of the fluorescence is proportional to the concentration of the atoms in solution and the intensity of the source. The resolution required by the dispersion—detection device is dictated by the nature of the light source. The more specific the light source, the simpler the fluorescence spectra, and the cruder the dispersion device can be.

ICP/AFS employs an ICP as an atomization source. Unlike AES, which uses the region 1–2 cm above the induction coil, AFS uses the region 6–8 cm above the coil. At this height most ions have recombined, producing elements in the atomic state.

Hollow cathode lamps are used as highly specific excitation sources for ICP/AFS. The emitted light of the HCL, with the characteristic spectrum of a single element, can be absorbed only by that element in the ICP. Fluorescence then occurs at the same wavelengths. As a result, the fluorescence spectrum is as simple as the HCL spectrum. The optical properties of the ICP that produce large calibration ranges for ICP/AES provide calibration ranges covering four to five orders of magnitude for ICP/AFS.

The high specificity of the HCL means that a very simple dispersion—detection system can be used. Photomultiplier tubes are mounted behind optical notch filters, which transmit light only over regions covering several nanometers around the wavelength of interest. The HCL and the filter—

PMT for each element are mounted together. The HCL is angled upward at approximately a 45° angle. The filter–PMT is mounted directly above the HCL, even with the height at which the HCL beam hits the plasma. As many as 12 of these single-element HCL–filter–PMT mounts can be positioned around the ICP. The HCLs are pulsed in sequence to further eliminate spectral interferences.

In Table 1, ICP/AFS has been compared in terms of performance characteristics with AAS and AES. ICP/AFS has the specificity of AAS and the precision and detection limits (for nonrefractory elements) of ICP/AES. The detection limits for the refractory elements range from equal to those for ICP/AES to two orders of magnitude worse than those for ICP/AES.

Spectral interferences for ICP/AFS, like those for AAS, are relatively inconsequential. Chemical interferences are worse than those for ICP/AES, since a much cooler region of the plasma is being used. However, the chemical interferences for ICP/AFS are fewer than those for flame and furnace AAS. Again, documentation of methods and interferences is minimal because of the newness of the instrumentation.

The ease of operation of ICP/AFS is approximately the same as that of ICP/AES. The smaller number of elements that can be simultaneously determined by ICP/AFS (12 at most) further simplifies operation. Data acquisition, storage, processing, and reporting are accomplished by a microprocessor. As with AAS and AES, automatic sampling and computerization permit the analyst to set up the standards, samples, and blanks; start the operation; and return later to a finished experiment and a printed report.

The cost and the multielement capabilities of ICP/AFS are comparable to those of DCP/AES. Both lie in the lower range of ICP/AES multielement capabilities and are priced in the region of the least expensive scanning ICP instruments and the most sophisticated AAS systems.

## B. Interferences

Spectral interferences for ICP/AFS are not significant. Direct line overlaps of the fluorescence spectra are rare because of the simplicity of the spectra. Broad-band interferences are eliminated by pulsing the HCL and using synchronous detection. The fluorescence signal will pulse in phase with the HCL, while background emission from the ICP will remain constant and therefore will not be detected. The most serious possibility for a spectral interference is the reflection of the HCL beam back into the PMT by particulate matter in the ICP. However, the high temperatures of the ICP and a complete sample digestion make this type of interference unlikely. Tests have shown that percentage concentrations of most elements produce no reflection interferences.

Physical interferences will be equivalent, but chemical interferences will be worse for ICP/AFS than for ICP/AES. A cooler portion of the plasma is used for AFS to allow the metallic ions and electrons to recombine to form free atoms. However, the cooler environment also allows compounds to re-form. Several suppression and ionization effects have been reported by the manufacturer (Demers and Allemand, 1981). At present, it is too soon to judge the merits of ICP/AFS performance.

## C. Applications

There has always been a strong interest in AFS as an analytical method. In the past 2 years, a pair of excellent reviews has appeared on this topic (Ullman, 1980; Van Loon, 1981). Most AFS research is aimed at the development of instrumentation and answering questions of a more fundamental nature. There has been little work of an applied nature done with respect to food or with a "standard" AFS system. One of the few applications of AFS to the determination of metals in food was the determination of 11 metals (in the single-element mode) in orange juice by McHard *et al.* (1979) using an experimental fluorescence spectrometer. With a commercial instrument now available, more applied research should be appearing.

## VII. CHEMICAL VAPORIZATION

The chemical vaporization technique has been widely used for atomic spectroscopic determinations of a number of metals that have generally proved difficult to atomize by more conventional means. In chemical vaporization the element of interest is chemically reduced to either a volatile covalent hydride (As, Bi, Ge, Pb, Sb, Se, Sn, and Te) or the volatile metal Hg. These volatile species are then introduced into the atomization source of the spectrometer in the gas phase. This process generally leads to a greater sensitivity for these elements, since all of the element in the sample is introduced to the atomization source, whereas conventional nebulization normally exhibits only a 1–10% efficiency in aerosol formation. Hydride generation requires atomization of the vapor by a thermal source to complete the generation of the atomic species, whereas the "cold" vapor of Hg is suitable for direct determination. Hydride generation has been combined with atomization by various chemical flames, electrically heated silica tubes, graphite furnaces, ICPs, microwave induced plasmas, and DCPs.

In general, the chemical reduction step is specifically designed for the element of interest and the sample type. Interferences for the chemical treatment step are common and highly specific to the element, the method of reduction, and the sample matrix. There is little standardization of equipment and methods. Only in recent years has equipment specifically designed for vapor generation appeared on the market. Automation of the vapor generation process has led to improved accuracies and precisions.

A general review of chemical vaporization has been published by Godden and Thomerson (1980). Critical appraisals of methods have been presented by Snook (1981) and Ihnat and Thompson (1980). The latter article is of particular interest. After extensive statistical evaluation, the authors concluded that chemical vaporization was unsuitable as a reference method because of the lack of standardization of the basic method.

Chemical vaporization has been used for the determination of As, Bi, Ge, Hg, Pb, Sb, Se, Sn, and Te in a wide variety of matrices (Horlick, 1982). The specificity of chemical vaporization has generally made it a single-element method. However, Wolnick *et al.* (1981) demonstrated that multielement determinations could be obtained using an ICP. The same group obtained multielement detection limits an order of magnitude better using a cryogenic trap (Hahn *et al.,* 1982).

## REFERENCES

*Adv. Chem. Ser.* (1979). *In* "Ultratrace Metal Analysis in Biological Sciences and Environment" (T. H. Risby, ed.), *Adv. Chem. Ser.* No.172, Am. Chem. Soc., Washington, D.C.

Agemian, H., Sturtevant, D. P., and Ansten, K. D., (1980). *Analyst (London)* **105**, 125.

*Ann. Rep. Anal. At. Spectrosc.* (1971–1981). The Chemical Society, Burlington House, London, England, Vols. 1–10.

Barnes, R. M. (1982). *9th Annu. Meet. Fed. Anal. Chem. Spectrosc. Soc.* Pap. No. 62.

Barnes, R. M., Mian-zhi, Z., and Fodor, P. (1982). *9th Annu. Meet. Fed. Anal. Chem. Spectrosc. Soc.* Pap. No. 61.

Berman, E. (1980). "Toxic Metals and Their Analysis." Heyden Press, London.

Black, M. S., Thomas, M. B., and Browner, R. F. (1981). *Anal. Chem.* **53**, 224.

Borriello, R., and Sciandone, G. (1980). *At. Spectrosc.* **1**, 131.

Boumans, P. W. J. M. (1980). *"Line Coincidence Tables for Inductively Coupled Plasma Atomic Emission Spectrometry."* Pergamon, New York.

Boyko, W. J., Kebber, N. K., and Patterson, J. M., III (1982). *Anal. Chem.* **54**, 188R.

Bratter, P., and Schramel, P., eds. (1980). *Proc. Workshop, Trace Elem. Anal. Chem. Med. Biol., Neuhenberg, 1980.*

Brown, R. M., Jr., and Fry, R. C. (1981). *Anal. Chem.* **53**, 532.

Chakrabarti, C. L. Wan, C. C., and Li, W. C. (1980a). *Spectrochim. Acta, Part B* **35B**, 93.

Chakrabarti, C. L., Wan, C. C., and Li, W. C. (1980b). *Spectrochim. Acta, Part B* **35B**, 547.

Crosby, N. T. (1977). *Analyst (London)* **102**, 225–268.

Dahlquist, R. L., and Knoll, J. W. (1978). *Appl. Spectrosc.* **32**, 1–29.

Dakeba, R. W. (1979). *Anal. Chem.* **51**, 902.

DeBolt, D. C. (1980). *J. Assoc. Off. Anal. Chem.* **63**, 802.

Demers, D. R., and Allemand, C. D. (1981). *Anal. Chem.* **53**, 1915.

Duprie, S., and Hoenig, M. (1980). *Analysis* **8**, 153.

Ediger, R. D. (1975). *At. Absorpt. Newsl.* **14,** 127.

Evans, W. H., Dellan, D., Lucas, B. E., Jackson, F. J., and Read, J. (1980). *Analyst (London)* **105,** 529.

Feinberg, M., and Ducauze, C. (1980). *Anal. Chem.* **52,** 207.

Freeland-Graves, J. H., Ebangit, M. L., and Bodze, P. W. (1980). *J. Am. Diet. Assoc.* **77,** 648.

Fricke, F. L., Robbins, W. B., and Caruso, J. A. (1979). *Prog. Anal. At. Spectrosc.* **2,** 185–286.

Fry, R. C., Northway, S. J., Brown, R. M., and Hughes, S. K. (1980). *Anal. Chem.* **52,** 1716.

Giri, S. K., Littlejohn, D., and Ottaway, J. M. (1982). *Analyst (London)* **107,** 1095.

Godden, R. G., and Thomerson, D. R. (1980). *Analyst (London)* **105,** 1137.

Greenfield, S., Jones, T. L., and Berry, C. T. (1964). *Analyst (London)* **89,** 713.

Hadeishi, T., Church, D. A., McLaughlin, R. D., Zak, B. D., Nakamura, M., and Chang, B. (1975). *Science* **187,** 348.

Hahn, M. H., Wolnick, K. A., Fricke, F. L., and Caruso, J. A. (1982). *Anal. Chem.* **54,** 1048.

Harbach, D., Diehl, H., Timis, J., and Huntemann, D. (1980). *Fresenius Z. Anal. Chem.* **301,** 215.

Harnly, J. M. (1982). *9th Annu. Meet. Fed. Anal. Chem. Spectrosc. Soc.* Pap. No. 297.

Harnly, J. M., and O'Haver, T. C. (1977). *Anal. Chem.* **49,** 2187.

Harnly, J. M., and O'Haver, T. C. (1981). *Anal. Chem.* **53,** 1291.

Harnly, J. M., and Wolf, W. R. (1981). *15th Mid. Atl. Reg. Meet. ACS.* Pap. No. 29, Am. Chem. Soc., Washington, D.C.

Harnly, J. M., O'Haver, T. C., Golden, B. M., and Wolf, W. R. (1979). *Anal. Chem.* **51,** 2007.

Heanes, D. C. (1981). *Analyst (London)* **106,** 182.

Heine, D. R., Babis, J. S., and Denton, B. M. (1980). *Appl. Spectrosc.* **34,** 595.

Hollahan, J. R. (1974). *In* "Techniques and Applications of Plasma Chemistry" (J. R. Hollahan and A. T. Bell, eds.), pp. 229–253. Wiley, New York.

Horlick, G. (1982). *Anal. Chem.* **54,** 276R.

Hughes, S. K., and Fry, R. C. (1981). *Anal. Chem.* **53,** 1111.

Hunter, G. B., Woodis, T. C., Jr., and Johnson, F. J. (1981). *J. Assoc. Off. Anal. Chem.* **64,** 25.

IAEA (1980). *Tech. Rep. Ser. IAEA* No. 197.

Ihnat, M. (1981). *In* "Atomic Absorption Spectrometry. V. Techniques and Instrumentation in Analytical Chemistry" (J. E. Cantle, ed.), pp. 139–210. Elsevier, Amsterdam.

Ihnat, M., and Thompson, B. K. (1980). *J. Assoc. Off. Anal. Chem.* **63,** 814.

Jackson, F. J., Read, J. T., and Lucas, B. E. (1980). *Analyst (London)* **105,** 359.

Jones, H. O. L., Jacobs, R. M., Fry, B. E., Jr., Joes, J. W., and Gould, J. H. (1980). *Am. J. Clin. Nutr.* **32,** 2545.

Jones, J. W., and Boyer, K. W. (1979). "Applications of Inductively Coupled Plasmas to Emission Spectroscopy" (R. M. Barnes, ed.), p. 83. Franklin Inst., Philadelphia, Pennsylvania.

Jones, J. W., Capar, S., and O'Haver, T. C. (1982). *Analyst (London)* **107,** 353.

Koirtyohann, S. R., and Kaiser, M. (1982). *Anal. Chem.* **54,** 1515A–1524A.

Koirtyohann, S. R., and Pickett, E. E. (1966). *Anal. Chem.* **28,** 585.

Koirtyohann, S. R., and Pickett, E. E. (1975). *In* "Flame Emission and Atomic Absorption Spectrometry. III. Elements and Matrices" (J. A. Dean and T. C. Rains, eds.), Chap. 17. Dekker, New York.

Koizumi, H., and Yasuda, K. (1975). *Anal. Chem.* **47,** 1679.

Kuennen, R. W., Wolnick, K. A., Fricke, F. L., and Caruso, J. A. (1982). *Anal. Chem.* **54,** 2146.

Langmyhr, F. J., and Orre, S. (1980). *Anal. Chimi. Acta* **118,** 307.

L'Vov, B. V. (1961). *Spectrochim. Acta* **17,** 761.

L'Vov, B. V. (1976). *3rd Annu. Meet. Fed. Anal. Chem. Spectrosc. Soc.* Rep. No. 214.

L'Vov, B. V. (1978). *Spectrochim. Acta* **33B,** 153.

McHard, J. A., Foulk, S. J., Nikdel, S., Ullman, A. H., Pollard, B. D., and Winefordner, J. D. (1979). *Anal. Chem.* **51**, 1613.

May, T. W., and Brumbaugh, W. G. (1982). *Anal. Chem.* **54**, 1032.

Melton, J. R., Hoover, W. L., Morris, P. A., and Gerald, J. A. (1978). *J. Assoc. Off. Anal. Chem.* **61**, 504.

Mermet, J. M., and Hubert, J. (1982). *Prog. Anal. At. Spectrosc.* **5**, 1.

Nagy, S., and Rousseff, R. L. (1981). *J. Agric. Food. Chem.* **29**, 889.

NBS (1976). "Accuracy in Trace Analysis: Sampling, Sample Handling, Analysis" (P. D. Lafleur, ed.), Vols. 1 and 2. *NBS Spec. Publ. (U.S.)* No. 422. U.S. Govt. Printing Office, Washington, D.C.

NBS (1977). "Procedures Used at the National Bureau of Standards to Determine Selected Trace Elements in Biological and Botanical Materials" (R. Mavrodineua, ed.), *NBS Spec. Publ. (U.S.) No. 492.* U.S. Govt. Printing Office, Washington, D.C.

Northway, S. J., and Fry, R. C. (1980). *Appl. Spectrosc.* **34**, 332.

Northway, S. J., Brown, R. M., and Fry, R. C. (1980). *Appl. Spectrosc.* **34**, 338.

O'Haver, T. C. (1976). *In* "Trace Analysis: Spectroscopic Methods for Elements" (J. D. Winefordner, ed.), pp. 15–62. Wiley, New York.

Parsons, M. L., and Forster, A. (1980). "An Atlas of Spectral Interferences in ICP Spectroscopy." Plenum, New York.

Rockland, L. B., Wolf, W. R., Hahn, D. M., and Young, R. (1979). *J. Food Sci.* **44**, 1711.

Rowen, C. A., Zajicek, O. T., and Calabrese, E. J. (1982). *Anal. Chem.* **54**, 149.

Slavin, W. (1982). *Anal. Chem.* **54**, 685A.

Slavin, W., and Manning, D. (1982). *Prog. Anal. At. Spectrosc.* **5**, 243.

Slikkerveer, F. J., Broad, A. A., and Hendrikse, P. W. (1980). *At. Spectrosc.* **1**, 30.

Smith, S. B., Schleicher, R. G., Pfeil, D. L., and Hieftje, G. M. (1982). *Pittsburgh Conf. Expo. Anal. Chem. Appl. Spectrosc.* Pap. No. 442.

Snelleman, W. (1968). *Spectrochim. Acta, Part B* **23B**, 403.

Snook, R. D. (1981). *Anal. Proc.* **18**, 342.

Stewart, K. K. (1980). *In* "Nutrient Analysis of Foods: The State of the Art for Routine Analysis," (K. K. Stewart, ed.), pp. 1–19. Assoc. Off. Anal. Chem., Washington, D.C.

Sturgeon, R. E., and Berman, S. S. (1981). *Anal. Chem.* **53**, 632.

Thompson, D. D., and Allen, R. J. (1981). *At. Spectrosc.* **2**, 53.

Ullman, A. (1980). *Prog. Anal. At. Spectrosc.* **3**, 87.

Van Loon, J. (1981). *Anal. Chem.* **53**, 322A.

Veillon, C., and Vallee, B. L. (1978). *In* "Methods in Enzymology, Vol. 54: Biomembranes, Part E" (S. Fleischer and L. Packer, eds.), pp. 446–484. Academic Press, New York.

Walsh, A. (1955). *Spectrochim. Acta.* **7**, 108–117.

Wendt, R. H., and Fassel, V. A. (1964). *Anal. Chem.* **37**, 920.

Windsor, D. L., and Denton, B. M. (1979). *Anal. Chem.* **51**, 1116.

Wolf, W. R. (1981). *NBS Spec. Publ. (U.S.)* No. 618, pp. 235–242.

Wolf, W. R. (1982). *In* "Clinical Biochemical and Nutritional Aspects of Trace Elements" (A. Prasad, ed.), pp. 427–466. Liss, New York.

Wolnick, K. A., Fricke, F. L., Hahn, M. H., and Caruso, J. A. (1981). *Anal. Chem.* **53**, 1030.

Woodis, T. C., Jr., Holmes, J. H., Jr., Ardis, J. P., and Johnson, F. J. (1980). *J. Assoc. Off. Anal. Chem.* **63**, 1245.

# 15

# Quality Assurance for Atomic Spectrometry

JAMES M. HARNLY
WAYNE R. WOLF

*Nutrient Composition Laboratory*
*Beltsville Human Nutrition Research Center*
*United States Department of Agriculture*
*Beltsville, Maryland*

## I. INTRODUCTION

The most important aspects of any analytical method are the steps taken to ensure the quality of the results. Quality assurance is the system of activities whose purpose is to ascertain that the overall quality control limits are being met (Taylor, 1981). The quality control limits (established by the analyst, the laboratory, or an outside organization) define the expected

ANALYSIS OF FOODS AND BEVERAGES
ISBN 0-12-169160-8

accuracy and precision of the result. Quality assurance must cover all aspects of the analytical process, from sampling, sample treatment, standards preparation, signal measurement, and methods validation to data handling and evaluation. Without a defined quality assurance program, all analytical results must be suspect.

Although many aspects of quality assurance are method independent, this chapter will deal specifically with the application of quality assurance to atomic spectrometry, primarily to atomic absorption spectrometry (AAS) and to inductively coupled plasma atomic emission spectrometry (ICP/AES).

With the recent advances in atomic spectrometry, the need for quality assurance is even more critical. The current trend toward multielement determinations, automation, and computerization (topics discussed in detail in other chapters) has resulted in the analyst's being less directly involved in the actual analytical measurement process. Deviations from normal operation for automated, computerized procedures, which previously were noted and corrected by the analyst as a part of operator interaction throughout the procedure, must now be corrected by an appropriate computer algorithm. All too frequently the analyst accepts computerized results without question. This point is well made by Frank and Kowalski (1982) in their review on chemometrics, which should be required reading for everyone involved with automated and computerized methods. Without the proper quality assurance, the latest advances in atomic spectrometry will only have the effect of turning out erroneous results at a faster rate.

## II. SAMPLING

Sampling is the selection, from a population, of a finite number of individuals to be analyzed. The object is to choose the individuals such that the results obtained for the analyses are an accurate characterization of the entire population. A bias in sampling can produce a bias in the characterization of the population. Proper sampling techniques must be used to ensure that the individuals are representative of the whole. The problem, for foods, has been stated succinctly by Horwitz and Howard (1979):

> . . . Throughout this discussion, keep in mind the significance of the sample — it is only a means to an end. You are usually only incidentally interested in the sample per se; most of the time you are really interested in the composition of the lot that the sample represents. To paraphrase a profound advertisement . . . depicting a drug dosage form with the caption "The one you take is never tested," *"the sample you analyze is never eaten."*

The problems in establishing a valid sampling of foods are considerable because of the many variables. For processed foods the analyst must consider the wide variety of brand names available, geographical location, seasonal variation, and the effect of shelf life. There are also longer-term variations associated with the development of new processing methods and new sources of raw materials. The number of variables associated with fresh foods and commercially prepared foods are even greater. Unless valid statistical methods are used, the results of a study may not be representative beyond the limited population analyzed.

The first step in designing any sampling procedure is to ask the question "What is the purpose of the study?" This question is best discussed with everyone involved in the project, since there often are conflicting views and too many questions to be answered in a single study.

Once a concise statement of the purpose of the study has been made, the design of the sample collection can begin. Proper use of statistical methods allows the minimum number of samples to be collected and the minimum number of measurements to be made to reach a conclusion with the desired confidence level. Thus statistical methods allow the greatest efficiency.

Statistically valid sampling is a complex science. This review cannot begin to cover the topic in any depth. The reader is referred to the many excellent publications on the topic for detailed discussions in areas of interest (Youden and Steiner, 1967; National Bureau of Standards [NBS], 1976; Kratochvil and Taylor, 1981). Perhaps the best advice to the reader is to make use of the statistical resources that are available. Most organizations have statisticians on their staffs. If not, a statistician should be hired as a consultant if considerable sampling is to be done. However, this does not relieve the analyst of all responsibility. Some background in statistics (the more, the better) is necessary for effective communication with statisticians.

## III. VALIDATION OF ANALYTICAL DATA

There are three general aspects to validating the analytical data. These are (1) developing a method of analysis, (2) validating the method, and (3) ensuring that the method remains valid, or in control, throughout its use. To develop a meaningful routine method of analysis, the whole analytical process must be critically examined and understood. The strengths, weaknesses, limitations, and ruggedness of the analytical aspects of a procedure must be documented by appropriate research and development. Potential matrix effects must be identified for each type of sample. Each type of food is basically a different chemical matrix. For a particular nutrient, a method that is valid for one type of food may not be valid for another. Certain aspects of the instrumentation and/or sample preparation might cause significant deviations for different matrices and different elements. Such effects must be identified and corrected in the development of an accurate procedure.

Once the analytical aspects of the procedure are well known, and the method is under control (i.e., once reproducible results with acceptable precision can be obtained), the method must be shown to give results in agreement with the "true" value of the analyte concentration in the samples.

After a method has been developed, validated, and established as routine, it must be monitored by appropriate procedures that are established to ensure that the data obtained in the use of the method remain valid. The key to this quality control is the availability and use of appropriate control samples.

## A. Analytical Aspects

The analytical method will be considered under four broad headings: (1) sample treatment, (2) standards preparation, (3) signal measurement, and (4) optimization of analytical parameters.

### 1. Sample Treatment

Sample treatment includes collection, storage, homogenization, division (into subsamples), further storage, and digestion (or solubilization). To obtain accurate results, all aspects of the sample treatment must be designed to avoid either contaminating or losing any of the elemental components. For intermediate and major constituents (greater than 1 $\mu$g/g), this task is important, but it can be accomplished without extreme measures. For trace elements (less than 1 $\mu$g/g), however, every phase of the sample treatment is critical. An attitude of "useful paranoia" toward contamination is a prerequisite for accurate trace metal determinations. Laboratory techniques to avoid trace element contamination have been well reviewed by Zeif and Mitchell (1976). In the analysis of trace elements, as little sample treatment as possible as desirable.

The bulk sample must be collected in a manner that ensures a representative sample (as previously discussed) and avoids contamination from the sampling devices or the storage containers. In the case of foods, a fairly large sample (several hundred grams) is usually required to guarantee a representative sample. Fortunately, there are seldom limits on sample availability for foods. In the laboratory, the sample must be homogenized and divided into subsamples for storage and analysis. The homogenization step is particularly susceptible to contamination, since the sample must be physically degraded. Commercial grinders and blenders can be significant sources of contamination for their constituent metals and should be chosen to avoid metals of interest. For example, stainless steel, which contains percentage levels of Cr, Ni, and Mn alloyed with the iron, should be avoided when these elements are to be determined.

Preanalysis storage of food samples can often be for extended periods. Freezing of the bulk or subsample is usually a reasonable approach to preserving samples if care is taken to monitor the moisture content so that results can be referred back to the correct fresh weight. Freeze-drying of samples gives good, stable storage, but potential contamination must be monitored.

Atomic spectrometric methods predominately require a pretreatment of the sample to remove the bulk of the organic matter and to convert the sample to a liquid form. The pretreatment usually consists of drying the sample, followed by either dry ashing or wet oxidation. Dry ashing can be accomplished at either high or low temperatures. High-temperature dry ashing is carried out at temperatures up to 500°C in a muffle furnace. Usually, either platinum or quartz crucibles are used to avoid contamination. Small quantities of acid are sometimes added to assist in destruction of the organic material. Upon completion, the ash is dissolved in a small amount of concentrated acid and is then brought to volume. The advantages of high-temperature ashing are its ability to handle large numbers of samples at one time with a minimum of the analyst's attention, and minimal reagent contamination of the sample. The disadvantages are its potential for airborne contamination, since the sample must be open to the atmosphere during ashing, and its potential for volatilization of an element of interest at the high ashing temperature. The ashing temperature must be carefully selected to suit the metal species of interest. Volatilization losses may be particularly severe for Hg and Se. Ashing at temperatures slightly in excess of 500°C can lead to losses of As, Cd, Pb, and Zn. The anions present can also influence volatilization. Halides are especially notorious for aiding losses through the formation of volatile compounds (Thiers, 1957). Muffle furnaces with sealed ceramic interiors are now available to prevent contamination from the heating coils. Air circulated through the furnace, to aid in complete ashing of the sample, should be filtered to avoid airborne contamination. For trace metal determinations, adsorption onto the crucible wall is potentially a problem, especially with porcelain.

Low-temperature ashing is achieved through the use of a radiofrequency-generated oxygen plasma, rather than heat, to destroy the organic matter (Hollahan, 1974). The procedures, advantages, and disadvantages are quite similar to those of high-temperature ashing, except that volatile elements are generally not lost . The ashing efficiency of these systems limits their use for larger samples (samples must be less than 100 mg) and for samples containing significant levels of fat.

Wet oxidation – digestion is another method commonly used for removing bulk organic material from the samples. Wet oxidation involves heating or refluxing the sample in the presence of strong acids, such as $HNO_3$, $H_2SO_4$, and/or $H_3PO_4$, with strong oxidizing agents, such as $H_2O_2$ or $HClO_4$. The advantages of this method are that a large number of samples can be digested simultaneously, with a minimum of the analyst's attention; that the finished sample is in a liquid form, often suitable for analysis with no further treatment other than dilution; and that there is little loss due to volatilization. The major disadvantages are that the method is extremely susceptible to contamination from the reagents and that oxidative methods are not successful in solubilizing the siliceous fraction. The latter problem can be overcome by a postdigestion treatment with hydrofluoric acid.

Reagent purity is critical to all procedures for trace element determinations. All reagents should be analyzed for the levels of impurity for all the elements of interest. In many cases, the variation of the reagent blank signal is the limiting uncertainty that determines the detection limit of a method. High-purity acids and chemicals are available from a number of commercial sources. In addition, some high-purity acids can be prepared by subboiling distillation (Keuhner et al., 1972), and ultrapure HCl, $NH_4OH$, and $CH_3OH$ can be obtained quite easily in the laboratory by the process of isothermal distillation (Veillon and Reamer, 1981). High-purity water is available by means of any of several commercially available filtration–ion-exchange systems that generate 18-m$\Omega$-cm water (Veillon and Vallee, 1978).

A major source of contamination is the glassware and plasticware. For trace element determinations, plasticware offers less contamination than glassware. Essentially metal-free plasticware is commercially available. Plastic pipette tips, beakers, sample cups, and cups for automatic samplers come in sealed bags and can often be used (after testing the batch) without even rinsing. The level of cleanliness varies with the manufacturer and with each shipment. The analyst should use caution when changing brands or opening new shipments. Cleaning and recycling of the plasticware can be easily accomplished using pH-neutral chelating detergents. The chelating agent is usually present as an alkali metal salt (Na or K). Multiple rinsings with metal-free water are necessary to remove excess chelating agent from the surface, especially if Na or K are to be analyzed. A common mistake is to clean plasticware in the same manner as glassware, using a harsh acid treatment. For plasticware, the acid treatment converts the inert surface into an efficient ion exchanger (Veillon and Vallee, 1978), which can lead to highly irregular analytical results.

Another major source of contamination is airborne particulate matter. Generally, sample handling in a laboratory is carried out in an area open to the laboratory atmosphere, except when fumes must be drawn off and a traditional fume hood is used. These conditions are not adequate for trace element analysis. Sample treatment for trace element determinations should be carried out in laminar-downflow clean air hoods. In these hoods, the air is passed through high-efficiency particulate air (HEPA) filters before being blown down over the work area. These filters remove 99.97% of the particulate matter greater than 0.3 $\mu$m in diameter. This results in Class 100 air, which contains fewer than 100 particles per cubic foot. Conditions that exist on an open bench top for most *clean* laboratories are Class 10,000 air. Conditions of over 1,000,000 airborne particles per cubic fact have been measured in a "normal" laboratory. Moody (1982) has presented a very extensive discussion on clean laboratories.

## 2. Calibration Standards

All atomic spectrometric methods determine the concentration of samples by comparing their analytical signals to those of a series of calibration standards. As a result, the determinations are only as accurate as the standards.

The difficulty of preparing calibration standards increases as one moves from the single-element mode of analysis, atomic absorption spectrometry (AAS), to the multielement mode, atomic emission spectrometry (AES). The increased difficulty is a function of the number of elements determined and the method of analysis. For AAS, the main requirement of the calibration standards is accuracy and stability. The total contamination of commercially available single-element stock solutions (generally 1000 $\mu$g/ml) is usually below the 1% level. Because of the simplicity of atomic absorption spectra (there are only rare occurrences of spectral overlap interferences) and because calibration standards are made for only a single element, low-level contamination of the standards does not produce erroneous results. The calibration standards are usually made up in weak (0.1–1.0%) HCl or $HNO_3$ to provide long-term stability.

For calibration of AAS systems, routine practice usually involves preparation of the samples and standards in a common matrix. A detailed matrix matching of the standards to the samples is usually avoided. Matrix matching is undoubtedly the most accurate means of calibration, especially when a sample is closely bracketed by a pair of standards. However, this approach has two serious drawbacks. First, it presupposes a detailed knowledge of the sample composition, and, second, it requires a pair of standards to be made for each sample. Consequently, matrix matching is not suitable for unknown samples or for the determination of a large number of samples where the matrices will vary. Matrix matching is used only when the importance of the results warrants the extensive custom preparation of the standards.

Preparation of samples and standards in a common matrix is used to match the matrices crudely and/or to eliminate specific interferences. For

example: (1) samples prepared using the lithium metaborate fusion procedure, followed by dissolution into 5% HCl, are analyzed using standards
with the same concentration of lithium metaborate and HCl with no
attempt made to duplicate any other component of the sample matrix; (2)
both samples and standards may be made up in 0.5% La (usually as a Cl or
an $NO_3$) to eliminate suppression of Ca and Mg in an air–acetylene flame;
(3) samples and standards may be prepared in 0.1–1% Cs to prevent
ionization interferences for Na, K, Rb, Sr, and Ba. The common matrix
approach is most suitable for large numbers of samples.

Standards for multielement methods such as high-temperature plasma
emission must be accurate and also very pure, if a range of concentration
levels of different metals is used in the same solutions. For example, a
standard for food analysis might contain 100 $\mu$g/ml of Fe, Ca, and Mg,
1 $\mu$g/ml Cu, and 0.1 $\mu$g/ml of Cr. A Cr impurity in the Fe stock solution,
although less than 0.01%, would still be significant with respect to the known
Cr concentration. Additionally, spectral overlap interferences are common
with high-temperature plasmas because of the complex spectra of many
elements. The unsuspected presence of a metal with an interfering emission
line in either the samples or standards can lead to erroneous results for other
elements.

Since the high temperatures of plasmas eliminate most matrix effects, the
ratios of the metals in solution are less important. It has been suggested that
multielement standards be prepared, with each standard solution the same
concentration for all metals (McQuaker *et al.,* 1979). The mono-concentra-
tion approach eliminates cross contamination due to impurities but requires
a larger number of standards.

The long-term stability of multielement standards requires additional
precautions compared to single-element standards. This is a result of the
additional mass in solution and of the high concentrations for AES methods,
which, in general, greatly exceed those for AAS. Many times the individual
requirement of different metals for stability in solution are not compatible.
In addition, an acid concentration that is sufficient to produce stability for a
single element may not be concentrated enough for multielement standards.
The presence of additional components alters the equilibrium of each metal.
Higher acid concentrations, for example, are necessary to prevent the
hydrolysis of Fe when there are additional metals in solution.

## 3. Signal Measurement

An accurate and precise measurement of the analytical signal is crucial to
every determination. There are two major sources of measurement error
that are of concern to the analyst: those errors that arise from the detection
system and those that arise from the chemical and physical nature of the

sample (O'Haver, 1976). Detection system errors arise from uncertainties inherent in the atomization and detection processes and are usually random in nature. Errors due to the chemical and physical nature of the sample are systematic in nature and are usually called interferences. There are two types of interference: additive interference and multiplicative interference. Spectral interferences (broad-band and line overlap interferences) are examples of the additive type, while chemical and physical interferences associated with the sample matrix (matrix effects) are examples of the multiplicative type. In general, interferences must be corrected in order to obtain accurate determinations, whereas random errors are evaluated using statistics.

Detection errors are inherent in the instrument operation and are present in the measurement of every analytical signal, regardless of whether it is a blank, standard, or sample. Considerable research has been devoted to identifying the many sources of detection error in atomic spectrometry. The identification of these sources of error is not pertinent to this discussion, although familiarity with the various sources is an aid in evaluating and improving instrumentation. The uncertainty of a measurement (noise) can be evaluated by making multiple measurements of the same signal and computing the standard deviation. The signal-to-noise ratio or, inversely, the percentage coefficient of variation (the noise-to-signal ratio expressed as a percentage) can be used to establish the reliability of a signal.

The most widely recognized noise is the uncertainty in measuring the baseline or blank (these may or may not be the same). The random noise in the baseline or the blank determines the lowest analytical signal that can be detected, the detection limit, with a specified level of confidence. It is generally accepted that the *baseline* noise establishes an instrumental detection limit while the *blank* noise establishes a method detection limit. For analytical purposes, the method limit is the most useful to the analyst. However, the instrumental limit is useful for comparing spectrometers and for evaluating instrument performance for each experiment.

The detection limit is defined as the smallest analytical signal, $X_{DL}$, that can be detected (Currie, 1968):

$$X_{DL} = 2ts_m = 2ts \sqrt{\frac{1}{n_B} + \frac{1}{n_S}}$$

where $t$ is the Student's $t$ value for the specified degrees of freedom ($n_B + n_S - 2$) and confidence level, $s$ is the computed baseline (or blank) standard deviation, $s_m$ is the standard deviation from the mean of the baseline (or blank), and $n_B$ and $n_S$ are the number of measurements of the baseline (or blank) and signal, respectively. This expression assumes that the standard deviation for the signal and the baseline are the same. The factor 2 ensures that there is an equal improbability of the anlytical signal being mistaken for baseline noise as there is of baseline noise being mistaken for an analytical signal. The IUPAC (Commission of Spectrochemical and Other Optical Procedures, 1976) has adopted a standard of $2t = 3$. Thus, for well-characterized baseline noise, there is only a 7% chance that baseline noise will be mistaken for an analytical signal and only a 7% chance that the reverse will happen. The minimum detectable signal can be translated to a minimum detectable concentration by the calibration function (see Section IV,A).

The quantitation level is defined as the minimum analytical signal for which quantitation can take place. The quantitation level has been defined as 10 times the standard deviation (American Chemical Society [ACS], 1980) from the mean $(s_m)$. As a result, an analytical signal may be classified as not detected ($<3s_m$), detected ($>3s_m$ but $<10s_m$), or quantitative ($>10s_m$). The detection and quantitation limits do not mark discontinuities but instead denote signal levels where the coefficients of determination are 33% and 10%, respectively. These values allow the analyst to flag samples or standards that have poor signal-to-noise ratios. Samples falling below the quantitation limit should be rerun at a higher concentration. For auto-mated–computerized systems, these detection and quantitation limits can and should be reported with each calibration.

A full discussion of additive (spectral) and multiplicative (chemical and physical) interferences and the methods of correction is contained in Chapter 14 (this volume). In general, correction for interferences is not automatic. The analyst must either anticipate the interference and take corrective action or prove that interferences are absent. The types and severity of interference differ according to the mode of atomization and/or excitation and the nature of the sample.

For AAS, chemical and broad-band spectral interferences are most com-mon. Chemical interferences are almost always anticipated at a nonspecific level by crudely matching the samples and the calibration standards, dilut-ing both in the same acid or the same general solution. Specific interferences (such as phosphate suppression of Ca) are treated through the use of specific diluents, matrix matching, or the method of standard additions. Spike recoveries (a single standard addition) and/or analyses of reference materials closely matching the samples are generally employed to validate the accu-racy of the sample preparation and analytical methods. Broad-band spectral background interferences are difficult to generalize for flame atomization and can be expected always to be present for furnace atomization. The new methods of background correction are easily employed and should be used for all atomic absorption determinations.

For AES, broad-band and line overlap spectral interferences are of pri-mary concern. Broad-band background interferences are very common and are corrected routinely using the off-line method (measuring the background signal just off the analytical wavelength). Line overlap interferences are highly dependent on the chemical makeup of the sample. With a knowledge of the elemental components and rough concentrations, the analyst can anticipate line overlap interferences using any of the atlases of spectral lines that are now available. Correction for these interferences is accomplished using the on-line method (determining the concentration of the interfering element at its analytical wavelength and then making a mathematical correction for the element of interest). Spike recoveries are not affected by line overlap interferences and will not indicate the presence of spectral interferences. Reference materials that *closely match* the samples can be used to determine the presence of spectral interferences; however, the variety of reference material presently available for food analysis is very limited. Consequently, anticipation and correction for line overlap interferences are primarily the responsibility of the analyst.

## 4. Optimization of Analytical Parameters

One of the major advantages of the single-element mode of analysis is that all aspects of the analytical method can be optimized for the element of interest. The sample treatment can be customized to prevent contamination or loss of the element. The sample size, the dilution factor, and all the instrument parameters can be selected to maximize the signal-to-noise ratio. Interferences can be closely investigated by matrix matching, by scanning the vicinity of the analytical spectra, and/or by using the method of standard additions. The computed results can be critically evaluated, the method refined, and the determination repeated. This reiteration process can continue until the analyst is satisfied with the result or until funding is exhausted. The single-element mode is well suited for samples that require high precision or whose concentrations fall close to the detection limit.

Multielement determinations require compromises in the analytical parameters, including the sample treatment, the optimization of instrument parameters, and the handling and critical evaluation of the data. The sample treatment must be carried out in a manner that is either the most suitable for all elements or the best suited for the most sensitive element; that is, the most volatile element determines the maximum temperature for a dry ashing method. The sample size, the dilution factor, and the instrumental parameters cannot be optimized for all elements simultaneously. Parameters can be selected that compromise all elements equally, favor the element with the worst signal-to-noise ratio, or are weighted in any manner the analyst desires. Finally, handling and critical evaluation of the data are compromised with respect to time. As the data are produced at faster rates, it becomes impossible for the analyst to compute individually the concentration for every element for every sample and to examine critically the accuracy of each result. Instead, these tasks are turned over to a computer.

The usefulness of a multielement method is determined by the severity of the compromises in the analytical parameters. The extent of the compromise is dependent on the accuracy and precision requirements, the elements to be analyzed, and the method of determination. First, few determinations in foods require accuracy and precision better than ±5%. Optimization of all possible analytical parameters is usually not absolutely critical in these cases. Samples requiring high precision or those with concentrations falling close to the detection limit are usually run in the single-element mode. The chemical nature of the elements can cause significant compromises in the analytical parameters, depending on the method of atomization and excitation. Elements forming stable refractory compounds and highly volatile elements are not best analyzed in the same atomization source. In general, most elements can be categorized such that multielement determinations are feasible and reasonable compromise parameters can be achieved. The broader the categories, or the fewer required, the more applicable the multielement method. Finally, the nature of the analytical method also determines the extent of compromise. The success of the inductively coupled plasma (ICP) can be attributed to the minimum amount of compromise required for the analysis of almost all of the metals and many of the nonmetals. The high temperature of the ICP makes it insensitive to matrix effects, thus providing excellent detection limits and eliminating the compromise atomization parameters that might be encountered in flame AAS. The ICP calibration range of six orders of magnitude for each element minimizes compromise with respect to the sample size and dilution factors.

## B. Method Validation

Having established an analytical method, the analyst must validate the accuracy of the method. This is most effectively done by verifying the results using an independent method or by obtaining accurate results for a certified reference material of the same composition. Other, less conclusive methods are the comparison of results with values reported in the literature and the comparison of results with those of other laboratories for the analysis of an exchange sample.

### 1. Two Independent Methods

The establishment of two independent methods for determining the same elements is the ideal way of validating analytical results. Independent methods, each based on a different physical principle, seldom suffer from the same systematic biases or interferences. If different sample preparation methods are employed, then analytical agreement of the two methods places a great deal of confidence in the result. This approach is not often employed, since it is beyond the capability of most laboratories and/or is not economically feasible. This is especially true for the analysis of trace elements. The development of a single state-of-the-art method can be quite draining of time, money, and resources. Development of a second method is not usually considered unless the purpose of the laboratory is to produce highly reliable values for a reference material.

## 2. Reference Materials

The most common means of validating results is to obtain accurate determinations for a reference material of the same chemical composition. Reference materials are substances for which "true" values have been determined for one or more components by extensive analytical efforts. Agreement with the reference values is necessary, but not sufficient for proof of the validity of the method.

There are two types of reference materials, certified and noncertified. Certified reference materials are reference materials accompanied by a certificate issued by a recognized official standards agency. These certificates give the reference value of the component plus confidence limits. The National Bureau of Standards (NBS) is the United States source of certified reference materials. The NBS Standard Reference Materials (SRMs) are carefully prepared for homogeneity and stability and are characterized by at least two independent analytical methods (NBS, 1977). The NBS currently lists 10 biological SRMs that are appropriate for foods, although a number of these (such as the botanical SRMs) have significantly different elemental matrices than most typical foods consumed in the United States (Wolf, 1982).

The International Atomic Energy Agency (IAEA) also has several biological matrices certified for a number of inorganic elements including trace elements (Parr, 1980). Several other laboratories have issued biological reference materials, and a directory of these materials is available (International Standardization Organization [ISO], 1982). There is promise for several new materials in the near future (Ihnat et al., 1982; Muntau, 1980).

The major problem associated with reference materials is finding one similar to the samples to be analyzed. The limited number of currently available food reference materials means that, in many cases, a reference material must be used that only superficially resembles the sample. Cluster analysis procedures have been used to show that, for metals, there is relatively little composition overlap betwen the available food reference materials (from the NBS and the IAEA) and 160 of the foods most commonly consumed in the United States (Wolf, 1982). In addition, the range of concentrations of the elements in the reference materials is not compatible with the range of concentrations found in foods. There is currently a great need for a wider range of food reference materials, with respect to both food types and elemental concentrations.

### 3. Other Validation Methods

In the absence of a second, independent method or a standard reference material, results can be compared to values previously reported in the literature. Agreement with the published values lends strong support to the accuracy of the method, but disagreement can be very unsatisfying. In most cases, there is no way to resolve differences between the data without the development of an independent method. It is also possible that the analytical differences arise from a difference in the samples. This would be best resolved by an exchange of samples.

One of the few documented cases of the stability of the metal content of a food material is the work of McHard *et al.* (1980) on orange juice. They have shown that the trace metal contents of oranges from Florida, California, Mexico, and Brazil are stable and different and that trace metal content of the orange juice can be used to identify its source. Thus, the determination of trace metals in orange juice would be an excellent means of evaluating an analytical method for fruit juices.

Analysis of a common exchange sample by a group of laboratories is another means of evaluating analytical results. As in the previous method, agreement with other laboratories lends strong support to the validity of the method, but differences may be hard to resolve. If the group exchanging samples is not sufficiently large or if everyone employs the same analytical method, the consensus result may not be the most accurate value. The agreement between laboratories is more significant if more than one independent method is employed.

### C. Quality Control Samples

After a method has been developed, validated, and put into use, it is necessary to ensure that results continue to be valid by periodically determining a sample of known value. Primary or certified reference materials can be expensive and are frequently not available in large amounts. These materials are usually used only for the initial validation of the method. For routine usage, it is necessary for laboratories to develop secondary reference materials or quality control samples (QC samples). The QC samples usually consist of a particularly large homogeneous amount of an individual sample or several samples pooled together. By their nature, the QC samples are usually much more representative of the samples to be analyzed than the reference materials. The QC samples are carefully characterized using certified materials (if possible) and the best possible quality assurance methods to determine the "true" values of their components and the normal ranges of variation.

The QC samples are used in a number of ways. They may be analyzed in conjunction with the samples in a random or systematic pattern. For atomic spectrometric methods, where the instrument must be recalibrated at frequent intervals, it is desirable to analyze at least one QC sample for each calibration. The QC samples may be labeled or analyzed "blind" (unidentified to the analyst). The results of the QC samples are compared with the predetermined "true" values and ranges of variation. Poor accuracy or precision of a QC sample throws doubt on the validity of the results of all the samples analyzed during that calibration and usually leads to rejection of the results for that batch of samples. Rejection criteria must be established by the analyst ahead of the time of actual analysis.

## IV. DATA HANDLING AND EVALUATION

After an analysis, the raw data must be used to convert the sample readings into concentrations. A calibration function that best fits the calibration standards must be determined, the sample concentrations must be computed, and the results must be evaluated with respect to their reliability and with respect to the experimental design. The recent boom in computer technology has had a dramatic impact on all aspects of data handling and evaluation.

### A. Calibration Functions

The historical approach to calibration was to restrict analysis to the linear range, plot the calibration standards on graph paper, throw out those points that did not look right, draw the best straight line to fit the remaining points, and then convert the sample signals to concentrations using this linear calibration. With a computer, much more sophisticated mathematical operations are possible. Complex equations can now be used for calibration functions, and statistical methods of analysis can be applied to the raw data and the computed concentrations.

The first step in any calibration procedure is to inspect the calibration data. The standards must show good agreement between repeated determinations. A systematic change usually indicates drift of one of the analytical parameters. The standards must also have an acceptable signal-to-noise ratio. Standards whose signals lie below the quantitation limit can deviate considerably from the true value unless determined a sufficient number of times to obtain an accurate mean value. Finally, the standard signals must progress logically in the order and proportion of their concentrations. Although a standard may appear in error to visual inspection, elimination on the basis of a statistical test may prove difficult. Many times the best approach is to redetermine the standard or make up a new standard.

In general, there are two types of calibration curves encountered with atomic spectrometry: curves characteristic for the absorption mode and curves characteristic for the emission (and fluorescence) mode of operation. Absorption curves are governed by the Beer–Lambert equation ($A = abc$, where $A$ is absorbance, $a$ is the absorptivity coefficient, $b$ is the absorption cell path length, and $c$ is the concentration of the absorber). Absorbance curves are usually linear at lower concentrations and begin to deviate from the Beer–Lambert equation by bending toward the concentration axis above 0.4 or 0.5 absorbance units. The entire curve, from the detection limit to the point where absorbance no longer increases with concentration covers 2.5–3 orders of magnitude of concentration. The linear portion of the curve, which covers a much smaller range, has been used primarily for analytical work. Derivations show that, when the limiting error is reading the signal from a strip chart recorder, the maximum signal-to-noise ratio occurs at an absorbance of 0.43. With computerization, the reading noise is no longer limiting, and more detailed calculations show that the maximum signal-to-noise ratio lies at much higher absorbances. Thus, restricting calibration to the linear range uses only a fraction of the best signal-to-noise region. The use of more complex calibration functions to fit the nonlinear regions of the curves allows the entire curve to be used.

Use of the nonlinear range is dependent on an accurate fit of the calibration function to the calibration standards. Any calibration function must have sufficient degrees of freedom to fit the shape of the curve but must not have so many degrees of freedom as to be influenced by random deviations. Numerous calibration functions have been used. These functions include the linear regression of polynomial equations, the rational equation (Rowe and Routh, 1977), the Stineman (1980) function, and the cubic spline function (Hornbeck, 1975). The former two employ least-squares fitting, while the latter two are iterative techniques. For the least-squares methods, weighting of the data points is usually desirable; otherwise, the highest concentrations are weighted most heavily. In addition, for least-squares methods, conversion of the linear absorbance and concentration coordinates to the logarithmic domain smooths the curvature and provides better fits. Recent data suggest that four standards (Miller-Ihli et al., 1981), evenly dispersed over each order of magnitude (in logarithmic coordinates), can be fit as well by straight lines between each pair of points as by much more sophisticated functions.

Emission curves (for the ICP and direct-current plasma [DCP] sources and for fluorescence) are essentially linear for five to six orders of magnitude of concentration. This linearity requires fewer standards for calibration and needs only a first-degree linear regression to compute a calibration function. It has been suggested that two standards can be used to calibrate the entire six orders of magnitude, but the usefulness of this approach depends on the accuracy required. Most analysts use one or more standards per decade and a second-order linear regression to account for the slight deviations from linearity that do occur.

The most accurate means of calibration, in either the absorption or the emission mode, is to use a pair of standards that bracket the sample as tightly as possible. This approach eliminates any concern about linearity and reduces the time interval between the determination of the standards and the determination of the sample, minimizing fluctuation, or drift, errors. Of course this approach is not practical when a large number of samples is being used, when the samples include a wide range of concentrations, or when operation is in the multielement mode.

In cases of nonlinear calibration, care must be taken in the blank subtraction. For linear calibration, it makes no difference whether the blank is subtracted as the analytical signal or as a concentration. In general, blank subtractions are made as analytical signals prior to computing a calibration function. However, if the calibration is nonlinear, subtraction of the blanks as analytical signals can lead to biased results. The blank value must be converted to a concentration value before subtraction.

## B. Data Evaluation

### 1. Uncertainty of the Data

Computation of sample concentrations is never the final step in the analytical process. Even though the analyst may have no further involvement in the project, the results will always be used (or misused) by someone. All too often, the users of the data lack an analytical background and tend to interpret the results as absolute, forgetting the inherent uncertainty in each value. It is, therefore, the duty of the analyst to assign a value of uncertainty to data in order to ensure their proper usage.

In Section III,A,3, random detection system errors and systematic interference errors were discussed. Systematic errors can be detected and corrected, but the random errors remain and can only be statistically evaluated. Ignoring the sampling process, the major sources of random errors are signal measurement, calibration, and sample preparation. The random errors, from each source, are summed quadratically ($S_T^2 = S_1^2 + S_2^2 + \cdots$) to yield the total analytical uncertainty.

Repeated determinations of the same sample can be used to compute a standard deviation that reflects the measurement error. If the analytical signals fall in the linear range of a calibration curve, either the signals or the resultant concentrations can be used to compute the standard deviations. However, if the signals are in the nonlinear range, only the concentrations will yield an accurate standard deviation.

The significance of the standard deviation can be expanded by including more variables. If the repeat determinations include different preparations of the sample, then the standard deviation will reflect measurement and sample preparation uncertainty. If the repeat measurements are performed on different days, the standard deviation will reflect the uncertainties mentioned above and those uncertainties arising from daily variations in instrument performance and calibration. The best method of characterizing analytical errors is to include multiple determinations made on different days by different analysts of different sample preparations. For this reason, the standard deviation of the QC sample is usually the most accurate measurement of analytical uncertainty.

## 2. Filtering the Data

When data are to be reported outside the laboratory, there is a tendency to report only "reliable" results. It is not unusual, in a table of published results, to see the notations "not detected," "less than the detection limit," or "less than the quantitation limit." In each case, data were omitted whose coefficient of variation (CV) exceeded (or whose signal-to-noise ratio was less than) a predetermined limit. This limiting, or filtering, of the data, if not applied properly, can lead to biased interpretation of results.

The use of the signal-to-noise ratio to define "reliable" data has found wide acceptance, since specific ratios, the detection limit, and the quantitation limit are now defined by the IUPAC (1976) and other organizations (ACS, 1980). These limits divide all analytical data into three categories: undetectable (CV > 33%), detected or semiquantitative (33% > CV > 10%), and quantitative (CV < 10%). Such a classification of the data can create problems when repeat determinations of a sample fall into different categories. If the true value of a sample falls close to either the detection limit or the quantitation limit, then the normal distribution expected for repeated determinations will yield results falling on both sides of the limit. If this limit is applied as a filter, then a fraction of the determinations will be discarded. Averaging the remaining fraction will produce a biased result. This bias will increase as the discarded fraction increases. Additional problems arise when repeat determinations are made during different experiments. Calibration and noise variations can lead to different detection and quantitation limits for each experiment. As a result, different filters may be applied to various fractions of the data.

A more intuitive limit is zero. Most analysts will throw out zeros or negative values. Horwitz *et al.* (1980) have shown, however, that when the CV is 100%, 17% of the data will fall below zero for a normal distribution. It is intuitive that negative concentrations have no meaning. Yet, failure to include negative and zero values would produce an average that was biased high.

The signal-to-noise ratio for individual measurements is extremely poor below the quantitation limit. However, for a sufficiently large number of determinations a reliable estimate of the mean value can be obtained. The standard deviation of the average value improves with the square root of the number of determinations. It is seldom worthwhile to make the number of determinations necessary to obtain an acceptable standard deviation of the mean, but, in all cases, all positive, negative, and zero values must be used in order to obtain an unbiased average of the sample concentration. The average and the standard deviation of the average can then be used to calculate a coefficient of variation. The same filters (CV = 33% or 10%) can now be applied without biasing the results, although they are no longer strictly "detection" or "quantitation" limits. These limits do allow a categorization of the reliability of the data. However, if the data are to be subjected to further statistical analysis, it may be best to avoid any filtering.

## 3. Statistical Analysis

Depending on the purpose of the study and the sampling scheme employed, the data may be tested using a wide variety of statistical methods. Some of the simpler tests are the *t* test to determine whether data or groups of data are significantly different, the *F* test to determine whether the variances of the data are different, one-way and two-way analysis of variance to test for one- and two-dimensional patterns in the data, and multiple regression analysis to determine whether the data are dependent on preselected variables. These tests are described in most statistics texts. In addition, statistics packages are available for almost every micro- and minicomputer and for every large computer.

With the current emphasis on multielement determinations, automation, and computerization, it is now possible to generate data faster than ever before. The weakest link in the analytical process is increasingly becoming the interpretation of the data. Development of the field of chemometrics has offered great potential for strengthing this data evaluation link. This discipline uses mathematical and statistical methods to design an optimal measurement procedure and to provide maximum chemical information by analyzing chemical data (Frank and Kowalski, 1982). A number of chemometric methods are available for evaluation of multicomponent data, such as that generated in multielement atomic spectroscopic studies. The techniques of cluster analysis, pattern recognition, and factor analysis are very useful in determining and sorting out multifactorial relationships within groups of samples. These chemometric concepts have begun to be applied to development of automated computerized analytical atomic spectroscopic systems to generate trace element data in foods. Only a few studies have so far used these techniques to evaluate trace elements in foods.

## C. Computerization

The computer offers the analyst tremendous advantages with respect to time and computing power. Data that previously kept the analyst busy for weeks can now be processed in a matter of minutes using mathematical approaches that until now were too complex and/or time-consuming to be used. However, computerization does create additional quality assurance problems for the analyst. Because a computer is not flexible, deviations from the expected data patterns can lead to erroneous results. This is not because of inaccuracies in the number handling of the computer but because of the rigid nature of the computer logic. Once established, a program will treat each piece of data in an identical fashion. Whereas the analyst might note that one of the standards in a calibration was reading slightly high or low, the computer, unless specifically programmed to check the standards (which would be difficult), will accept the data as absolute. This problem is compounded in multielement spectrometers, where most of the collected data have never been seen by the analyst. Unless the computer is programmed to detect a wide variety of errors and to evaluate the validity of the data, significant analytical errors can result.

To automate completely the data analysis process, each step that was previously performed by the analyst must be translated into a computer algorithm or a series of algorithms. This is an extremely complex undertaking. As programs become more versatile, they become more complex and cumbersome. The logic of more complex programs becomes difficult to follow and more susceptible to errors. On the other hand, conceptually simple programs are often too narrow in their application. As a result, most programs are a compromise, logically as simple as possible but able to handle the most common problems. Programs are also compromised by the amount of memory available in the computer and the development costs.

It is imperative that the analyst be familiar with how the data are handled by a computer program. This knowledge allows the analyst to inspect the data and results, detect errors, and anticipate conditions for which the program algorithms are not appropriate. Obtaining listings of commercial programs is not always possible. Most manufacturers do not make original program, or "source," listings available to the users. Thus, the analyst is able to obtain only a general idea of how the data are being handled. Certainly, the analyst must try to evaluate the spectrometer's performance under as many different circumstances as possible. In most cases, analyses of reference materials and QC samples are the best methods for evaluating the accuracy of the computer programs.

## V. SUMMARY

The value of proper quality assurance in analysis of food samples by atomic spectroscopy cannot be overstated. The imperative for the analysis of foods originates from a wide variety of purposes that reflect a common, basic, day-to-day necessity of life: a safe and nutritious food supply. Horwitz and Howard (1979), in their paper on sampling, make some very pertinent comments on the validity of conclusions drawn from analysis of representative food samples:

> The capital and operating costs of analysis for trace constituents in food are extremely high. Very often the decisions made as a result of these analyses have an important significance on human health as well as social, economic, and political implications. Analysis of samples from which a valid conclusion cannot be made is not only a waste of resources but a potential cause of considerable social and political mischief.

This statement is equally true when directed toward the analysis itself. The ability of modern techniques of atomic spectroscopy to generate very large amounts of analytical data very rapidly demands good quality assurance practices to avoid waste and potential mischief.

## REFERENCES

Am. Chem. Soc. (1980). *Anal. Chem.* **52**, 2242.

Currie, L. A. (1968). *Anal. Chem.* **40**, 586.

Frank, I. E., and Kowalski, B. R. (1982). *Anal. Chem.* **54**, 232R–243R.

Hollahan, J. R. (1974). *In* "Techniques and Applications of Plasma Chemistry" (J. R. Hollahan and A. T. Bell, eds.), pp. 229–253. Wiley, New York.

Horwitz, W., and Howard, J. W. (1979). *NBS Spec. Publ. (U.S.)* No. 519, pp. 231–242.

Horwitz, W., Kamps, L. R., and Boyer, K. W. (1980). *J. Assoc. Off. Anal. Chem.* **63**, 1344.

Ihnat, M., Cloutier, R. A., and Wolf, W. R. (1982). *9th Annu. Meet., Fed. Anal. Chem. Spectrosc. Soc.* Pap. No. 459 (Abstr.).

ISO (1982). Directory of Certified Reference Materials. Int. Standard. Org./Remco, Geneva, Switzerland.

IUPAC Commission of Spectrochemical and Other Optical Procedures (1976). *Pure Appl. Chem.* **45,** 99.

Keuhner, E. C., Alvarez, R., Paulsen, P. J., and Murphy, T. J. (1972). *Anal. Chem.* **44,** 2050–2056.

Kratochvil, B., and Taylor, J. K. (1981). *Anal. Chem.* **53,** 924A.

McHard, J. A., Foulk, S. J., Jorgensen, J. L., Bayer, S., Winefordner, J. D. (1980). *In* "Citrus Nutrition and Quality" (S. Nasy and J. Attaway, eds.), *ACS Symp. Ser.* No. 143, pp. 363–392 Am. Chem. Soc., Washington, D.C.

McQuaker, N. R., Brown, D. F., and Kluckner, P. D. (1979). *Anal. Chem.* **51,** 1082–1084.

Miller-Ihli, N. J., O'Haver, T. C., and Harnly, J. M. *8th Annu. Meet. Fed. Anal. Chem. Spectrosc. Soc.* Pap. No. 286.

Moody, J. R. (1982). *Anal. Chem.* **54,** 1358A–1376A.

Muntau, H. (1980). *In* "Trace Element Analytical Chemistry in Medicine and Biology" (P. Bratter and P. Schramel, eds.), pp. 707–726. de Gruyter, Berlin.

NBS (1976). "Accuracy in Trace Analysis: Sampling, Sample Handling, Analysis" (P. D. LaFleur, ed.), *NBS Spec. Publ. (U.S.)* No. 422. U.S. Govt. Printing Office, Washington, D.C.

NBS (1977). "Procedures Used at the National Bureau of Standards to Determine Selected Trace Elements in Biological and Botanical Materials" (R. Mavrodineau, ed.), *NBS Spec. Publ. (U.S.)* No. 492. U.S. Govt. Printing Office, Washington, D.C.

O'Haver, T. C. (1976). *Trace Anal.: Spectrosc. Methods Elem., 1976* pp. 15–62.

Parr, R. M. (1980). *In* "Trace Element Analytical Chemistry in Medicine and Biology" (P. Bratter and P. Schramel, eds.), pp. 631–655. deGruyter, Berlin.

Rowe, C. T., and Routh, M. W. (1977). *Res./Dev.* November, pp. 24–30.

Stineman, R. W. (1980). *Creative Comput.* **6,** 54–57.

Taylor, J. K. (1981). *Anal. Chem.* **53,** 1588A–1596A.

Thiers, R. E (1957). *In* "Trace Analysis" (J. H. Yoe and H. J. Koch, Jr. eds.), p. 636. Wiley, New York.

Veillon, C., and Reamer, D. C. (1981). *Anal. Chem.* **53,** 549–550.

Veillon, C., and Vallee, B. L. (1978). *In* "Methods in Enzymology, Vol. 54: Biomembranes, Part E" (S. Fleischer and L. Packer, eds.), pp. 446–484. Academic Press, New York.

Wolf, W. R. (1982). *East. Anal. Symp., November 1982, N.Y.* (Abstr.)

Youden, W. J., and Steiner, E. H. (1967). *"Statistical Manual of the Association of Official Analytical Chemists."* Assoc. Off. Anal. Chem., Washington, D.C.

Zeif, M., and Mitchell, J. W. (1976). "Contamination Control in Trace Element Analysis." Wiley (Interscience), Wiley New York.

# 16

# Near-Infrared Reflectance Analysis of Foodstuffs

## T. HIRSCHFELD

*Lawrence Livermore National Laboratory*
*Livermore, California*

## EDWARD W. STARK

*Technicon Industrial Systems*
*Tarrytown, New York*

ANALYSIS OF FOODS AND BEVERAGES

## I. INTRODUCTION

During the past 10 years, the use of near-infrared reflectance analysis (NIRA) has become popular as a technique for the analysis of agricultural products in their raw and processed forms. The great rapidity of the technique, capable of measuring a sample and producing final results in seconds after only minimal sample preparation, has been the driving force of this expansion.

Today, the technique has been exploited at more than 5000 installations worldwide; it is used for the routine handling of an enormous variety of samples, from grains to dairy products, from meat to oils, and for a great number of nonfood applications.

Historically, NIRA was developed as an empirical technique, with a sketchy base in fundamental spectroscopic theory, which is only now being supplied. In its heavy dependence on computers, it serves as a proving ground for quite sophisticated mathematical tools of spectroscopic analysis. Today, as rapid progress in its technology continues, it is introducing into the chemical laboratory some of the concepts and techniques of artificial intelligence research.

Extensive experience with NIRA techniques has demonstrated them to be highly accurate and reliable if properly used, and many publications describing its methodology are now appearing in the literature.

In this description of NIRA as applied to food analysis, we discuss first the special features of the near-infrared (near-IR) spectral region and then those of correlation transform spectroscopy, the technique underlying NIRA. The development and nature of an analytical method are then described, followed by a review of the technique's application in the food industry.

## II. THE NEAR-IR SPECTRAL REGION

### A. Spectral Features

The near-IR spectral region lies between the approximately 0.8-$\mu$m edge of the visible spectral region and the beginning of the mid-infrared (mid-IR) region at 2.7 $\mu$m. Spectroscopically it may be defined as the region lying between the electronic and the fundamental vibration bands of molecules. On one hand, only exceptional organic molecules and a few ions have significant electronic absorption beyond 0.8 $\mu$m. On the other, the shortest fundamental vibration absorption band is H–F at 2.65 $\mu$m.

What, then, is the origin of molecular absorption in the near-IR spectral region? Here we are looking mainly at the overtone and combination bands arising from vibrational bands in the mid-IR region. For each infrared (IR) absorption band, there will be a series of overtones at successively higher multiples of the fundamental frequency whose intensity diminishes very rapidly as the multiplier increases. Different bands can also combine to generate one band at the new frequency, or even the sum of multiples of the frequencies. Here again the intensity drops sharply as the frequency increases.

The near-IR spectral region thus is characterized by weak band intensities from 10 to 100 times lower (at best) than those common in the fundamental IR. These bands repeat several times throughout the region with gradually decreasing intensity. For bands whose IR fundamental frequency lies beyond 5.4 $\mu$m, only the second overtone will fall in the near-IR region, and for bands whose fundamental is beyond 8.1 or 10.8 $\mu$m, only the third, fourth, etc. Since the higher overtones are very weak, near-IR absorption will be moderately strong only for bands having fundamental absorption in the 2.7- to 5.4-$\mu$m region.

Since absorption in this region arises mainly from X–H bonds (where X = C, N, O, etc.) and C–O bonds, strong absorption will come only from compounds containing these bonds (Wheeler, 1959; Goddu and Delker, 1960). Fortunately, this includes most organic compounds. Those bands whose near-IR overtones or combinations are too weak to be detected in the near-IR can often influence the measurement by shifting, changing the intensity of, or otherwise perturbing the detectable bands produced by X–H bonds.

As in the fundamental IR, spectral interpretation in the near-IR is possible by comparison and matchup with reference spectral collections, or by comparison of individual spectral bands with tabulations or charts that allow correlation with specific molecular structural features.

## B. Usefulness of the Near-IR Region

Historically, exploitation of the near-IR region developed faster than that of the mid-IR, the former having been discovered earlier and being more compatible with early instrumentation technology (Kaye, 1954, 1955; Wheeler, 1960). Thus many hundreds of papers, most of which applied the technique to quantitative work (both analytical and physicochemical), were published on near-IR spectroscopy during the 1950s and early 1960s.

However, it was realized quite early that the near-IR region is not optimally suited for qualitative analysis. Because many bands appear in the near-IR for every fundamental band, near-IR spectra tend to be crowded, with extensive overlap of the various bands, which complicates interpretation. This is further complicated by the many combinations and overtones that can be involved in explaining a particular band. Also, since many structural features of the molecule do not contribute to the near-IR spectrum, it becomes nearly impossible to reconstruct the structure from spectral measurement. The wide range of intensities of the various overtones and combinations, in the presence of spectral overlap, makes it easy to miss significant bands altogether.

These drawbacks of the near-IR region for qualitative analysis strongly favored the use of the mid-IR region for this purpose. Economics then motivated analysts to use the mid-IR for quantitative purposes as well, even though the instrumental state of the art at that time favored near-IR for the purpose. And even when a second instrument optimized for quantitative use was affordable, the analytical community nearly universally adopted gas chromatography (GC) for that purpose, because it could not only quantitate a pure sample in a solvent but also resolve a mixed sample.

A negative feedback loop now developed. Since the analytical chemist preferred mid-IR and GC for qualitative analysis and for quantitative analysis and mixture resolving, respectively, this is where the budget went. And since, then as now, analytical chemists economically dominate the instrumentation marketplace, that is where the instrument sales and therefore development dollars went. The state of the art of mid-IR instrumentation thus advanced relative to that of the near-IR, whose quantitative advantage shrank rapidly. A similar feedback loop, operating among the manufacturers and compilers of data bases, quickly led to the growth of mid-IR data bases and to the disappearance of the near-IR ones.

During all this time, there existed some intrinsic, fundamental advantages for quantitative analysis in the near-IR spectral region. The first of these was the relative weakness of near-IR spectral features. The apparent disadvantage of at least a tenfold (and possibly a hundredfold) reduction in band intensity was readily compensated by a proportional increase in the thickness of the sample cell. In the vast majority of cases in which enough sample was available, all of the lost sensitivity was recovered and sample thickness changed from microns to millimeters, actually a more convenient region. Cells in the millimeter range are easier to build and maintain at a precise thickness, are more resistant to clogging and less sensitive to surface contamination layers, and can be cleaned and assembled far more easily than micron-sized cells. In these thicker cells, uniformity tolerances to avoid wedging errors in spectra are far less stringent. In fact, some deliberate wedging can be built into cells to avoid the presence of interference fringes, one of the serious problems in quantitating mid-IR spectra.

The low intrinsic absorptivity of near-IR bands has a number of other benefits. Because the anomalous dispersion of the refractive index is proportional to the peak band absorptivity, the effect will be far less in the near-IR region. This in turn removes narrow features from the scattering spectra of solids, such as the Christiansen bands of powders. At the same time, this absence of refractive index dispersion makes absorption a more linear phenomenon and enhances the additivity of superimposed spectral features.

The intrinsic weakness of the bands has the further advantage of weakening a number of error mechanisms, such as the resolution error, spectral distortion in diffuse reflectance spectra, and multipassing errors.

A number of instrumental factors also favor the near-IR spectral region, given equally developed instrumentation. Broad-band blackbody spectral sources as commonly used for spectroscopy are much brighter in this spectral region. Photoconductive detectors in this spectral region, owing to low background emission, have far higher sensitivity than mid-IR ones. The availability of superior-performance optical components, such as glass sample cells, multielement optical glass lenses, and holographic gratings, further enhances the intrinsic potential of near-IR instruments.

The shorter wavelengths of the near-IR region also enhance the NIRA technique's suitability for sampling by diffuse reflectance analysis. At these shorter wavelengths, light-scattering coefficients are much larger, which in combination with the weaker absorption makes the scatter–absorption ratio very large. This in turn makes diffuse reflectance a highly linear phenomenon and reduces many of its error mechanisms, thus rendering this particularly simple technique suitable for quantitative analysis.

These potential capabilities were actualized by the development in the late 1960s of NIRA techniques by Karl Norris, of the United States Department of Agriculture (USDA), and his co-workers. Working on exceptionally difficult analyses such as the determination of protein in wheat, they combined near-IR spectroscopy, reflectance analysis, and correlation spectroscopy (Norris and Hart, 1965).

This combination technique benefited from reflectance analysis, to allow rapid measurements with sample preparation limited to mere grinding, and from correlation spectroscopy, to allow accurate quantitative resolution of complex mixtures of ill-defined compounds. The success of the procedure then encouraged rapid development of advanced instrumentation, which in turn accelerated the present boom in the use of NIRA (Watson, 1977; Day and Fearn, 1982a,b; Wetzel, 1983).

## C. Instrumentation

Traditionally, near-IR instrumentation has originated in minor modifications, sometimes available as accessories, to existing commercial ultraviolet (UV)–visible instruments. By exchanging gratings and using PbS detectors, conventional three-digit absorbance reproducibility measurements are readily obtainable.

Such performance, however, is wholly incompatible with NIRA techniques, as we shall see later. Extensively reconstructed instrumentation, still based on commercial grating spectrometers and advanced electronics, was used to develop this technique and is capable of performances in the four-digit range.

The success of NIRA has spawned several generations of highly advanced instrumentation specially designed to match its requirements (Stark, 1983; Rotolo, 1979; Landa, 1979; Hunt *et al.,* 1977). These include two types of spectrophotometers. The first type uses scanning monochromators with very high aperture holographic gratings and very careful electronic design to approach five digits of accuracy.

A crucial requirement in such a scanning spectrometer is of course a very high signal-to-noise level (and there the difference between four and five digits of accuracy is significant). A reasonably low stray light, adequate for measurements in fairly low reflectivity samples, is also needed. Surprisingly, so is a wavelength reproducibility of at least 0.05 nm, to enable the use of the high photometric accuracy. As will be seen later, the use of computers in data processing is extensive, thereby making the computing power of the system quite important. So is the nature of the software, which may represent a 10:1 difference in capability (mainly in rapid methods development) between instruments. Figure 1 shows a monochromator-based computerized NIRA instrument, the InfraAlyzer 500, with these features.

While such general-purpose instruments are quite versatile, they are also expensive and leave more decisions to the operator's discretion. In routine applications, considerably simpler systems, in which the monochromator is replaced by a set of filters, may be used. This allows measurements to be performed at a discrete set of wavelengths that are determined from the literature or preferably via methods development using either the filter instrument or the more flexible monochromator machines. One of these simpler machines is shown in Fig. 2.

In order to observe samples without preconditions on transparency,

**Fig. 1.** The InfraAlyzer 500.

**Fig. 2.** The InfraAlyzer 300.

flatness, or uniformity, all these machines use diffuse reflectance sampling. Here a beam of light is incident on a powdered scattering sample that diffuses the light, part of which emerges from the sample in the opposite direction. This diffuse reflectance samples the powder particles by transmission and will give a well-behaved spectrum that can be used quantitatively if three conditions are met. First, the sample grain size must be so small that no single grain has high opacity. This, by the way, would require often impractically fine powders if the measurement was done in the mid-IR. Second, the specular reflection off the surface of the sample, which carries no information (and would carry erroneous information for the intense bands of the mid-IR), must be rejected as efficiently as possible by sampling optics, and the residual must be corrected for. Third, the diffuse reflectance must be well averaged over the surface of a possibly inhomogenous sample and over the various reflection angles at which light leaves the sample.

To gather a good average of the reflected light it is possible to surround the sample with several large area detectors or to use the National Bureau of Standards technique of putting the sample in an integrating sphere that averages all sample locations and reflection angles.

Diffuse reflectance sampling is an efficient way of dealing not only with scattering powders but also with various opaque liquids or solids, such as dairy products (Hecht, 1968). Even transparent samples can be handled by putting them in a diffusively reflecting cup (the transflectance method). For some liquid samples, those whose absorption has a strong temperature dependence, it is important to thermostat the sample quite accurately. Open and closed sampling cups, for powders and pastes, and a thermostated sample holder, for liquids, can be seen in Figs. 3 and 4.

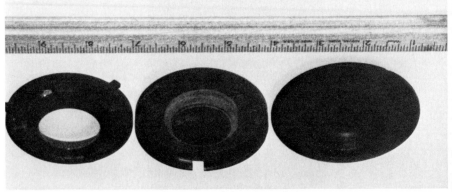

**Fig. 3.**   Open and closed sample cups.

**Fig. 4.**   A liquid sample drawer.

Another important consideration is the need for often-repeated double-beam referencing, which can be met by periodically inserting blank samples or by switching the incidence beam back and forth between the sample and the sphere wall for more accurate reference measurement.

While the large gap in performance between instruments for NIRA and classical ones is not obvious, it is essential to the success of the technique. Only one instrument not designed with NIRA applications in mind has a good enough performance for these tasks. There is a current boom in Fourier transform – infrared (FT/IR) instruments in the mid-IR region, where they have combined exceedingly high performance with a powerful built-in computer. What is not generally known is that some of these instruments have an even higher performance in the near-IR, which they can reach with a commercial accessory. With slight modifications, the more expensive FT/IR instruments can handle NIRA methods. However, data analysis software for NIRA is not commercially available with FT/IR instruments. In recent times, instrument usage in NIRA has evolved toward a two-tier system. Expensive, highly flexible monochromator or FT/IR instruments are used for methods development, which is greatly accelerated by their unlimited choice of wavelengths. The methods then developed define a set of wavelengths for measurement that is then embodied in the filter choice for low-cost instruments dedicated to specific areas.

## III. CORRELATION TRANSFORM SPECTROSCOPY

### A. Requirements for NIRA of Complex Samples

The spectroscopic problems posed by the quantitative analysis of complex mixtures by NIRA are quite forbidding. This can be seen by considering the case of the first-developed NIRA technique, the analysis of protein in wheat (Law and Tkachuk, 1977; Massie and Norris, 1965).

The major constituents of wheat are starch, water, oil, and protein; their spectra can be seen in Fig. 5. The extreme overlap of all these spectra is all too apparent, and there seem to be no interference-free bands available. In fact, the spectrum of whole wheat shows no outstanding bands whatsoever. Furthermore, since protein is a relatively minor constituent of weak relative absorption, its contribution to the total spectrum will be minor. As an example, Fig. 6 shows the spectra of two wheat samples with extremely low

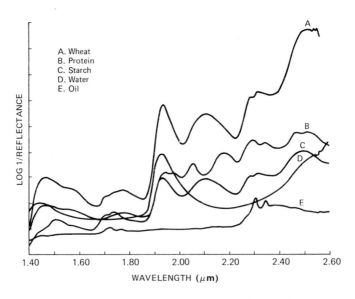

**Fig. 5.**   Spectra of starch, water, oil, protein, and wheat.

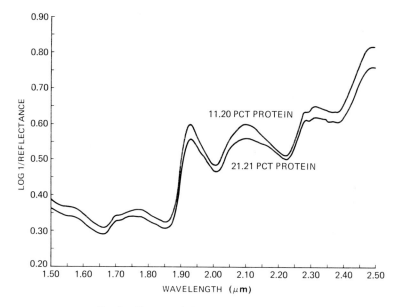

**Fig. 6.** Spectra of high- and low-protein wheat.

and high protein contents (some 11 and 21%, respectively). Even for this extreme compositional difference, the spectral difference can barely be seen. In fact, the entire analytical excursion in the data compares unfavorably with background fluctuations. This can be seen in Fig. 7, showing the spectra of three varieties of wheat with the same protein content. The spectral changes we see here are clearly larger than those of Fig. 6.

A further problem is that there is no possibility of calibrating this technique with pure standards, since pure preparations of the extracted constituents of wheat are quite unlike chemically and spectroscopically to their native form in the wheat itself. As if this were not enough, the sample is an irregular coarse powder, as seen in Fig. 8, which is no help for accurate spectroscopy either. In fact, it is quite astonishing that anyone dared to try this approach. Be that as it may, the technique was found to be quite successful, as can be seen in Fig. 9, showing the excellent correlation between NIRA and the far more laborious wet chemical determination. The success of the procedure is of course based on the ability of diffuse reflectance spectroscopy to sample irregular powders and on the high performance of the instrumentation used. However, a decisive additional contribution is made by the chosen data reduction procedure, correlation transform spectroscopy.

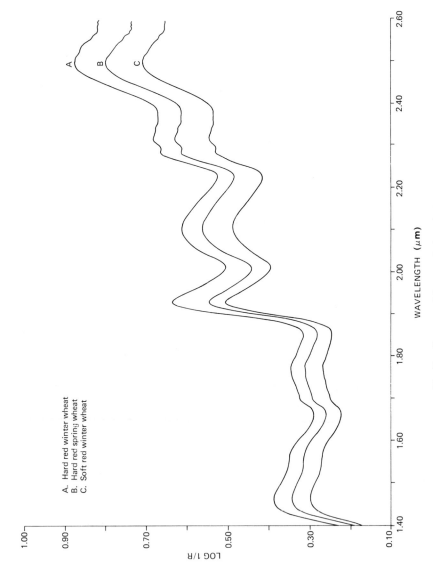

**Fig. 7.** Spectra of three varieties of wheat.

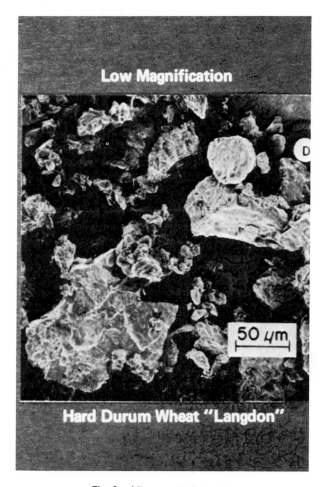

**Fig. 8.** Micrograph of wheat.

## B. Spectroscopic Analysis of Mixtures

Under the proper conditions, which prevail in the near-IR spectral region as discussed above, the spectrum of a mixture will be the exact sum of the spectra of its constituents, provided no chemical interactions occur between them. Even in the latter case, strict spectral additivity can be satisfactorily approximated by operating across limited concentration ranges.

Given these conditions, it will always be possible to reconstruct the amounts of all absorbing constitutents of a mixture by measuring a number of points in the spectrum and solving a system of equations for the concen-

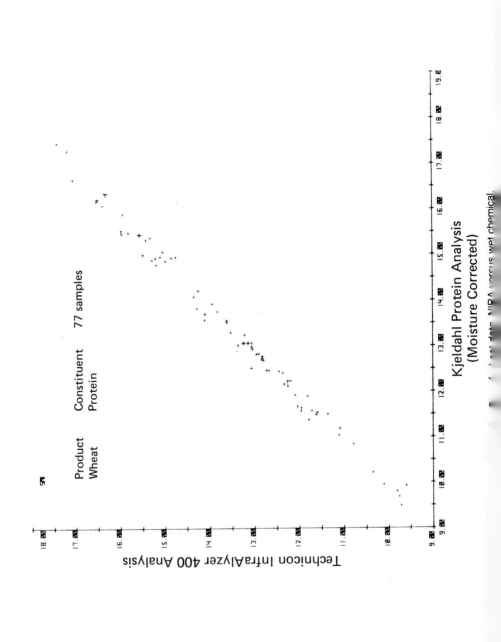

Kjeldahl Protein Analysis
(Moisture Corrected)

trations. To determine all the constituents, the number of wavelengths used in these measurements must at least equal the number of independently variable constituents in the mixture (Hirschfeld *et al.,* 1983). For any given constituent, however, the number of wavelengths required may be reduced to as little as one analytical wavelength and sufficient other wavelengths to compensate for the spectral interferences at this analytical wavelength. If the number of wavelengths is larger than the minimum required, then this system of equations can be solved in several ways and is said to be overde-termined. When the measurement contains errors or noise, these various solutions will be different. By a least-squares fit it is then possible to obtain a result with minimal sensitivity to errors and noise.

While it would seem that it is best to use all near-IR wavelengths for this calculation, a number of limiting factors must be considered. In the first place, the use of a very large number of wavelengths would preclude the use for NIRA of low-cost filter instruments. Since some wavelengths are better than others, and since the additional usefulness of further wavelengths drops very rapidly if we use the best ones first, the benefits of using so many wavelengths are not obtained in practice.

The reduction in the utility of additional wavelengths in a rank-ordered set also means that the expenditure of additional measurement time on these extra wavelengths is a progressively less rewarding operation. In a similar way, the additional chance of running into an unsuspected interfer-ent (or component not allowed for in the original calculation), which increases with the number of wavelengths, is no longer justified by the reduced benefits of further wavelengths.

All of this combines to keep the number of wavelengths used low. In special instances, where one of the constituents happens not to absorb at any of the wavelengths chosen, one can even use fewer wavelengths than constit-uents. However, there is an additional factor that provides even stronger encouragement to stick to few wavelengths, based on the method we use to generate the concentration calculation equation, as will be described later.

This concentration calculation equation can always be written as the summation of a constant plus a set of products, each of which is a coefficient multiplied by the measured reading at a given wavelength. We now have to determine the number of wavelengths to use, which wavelengths to use, the type of measured reading to use, and the multiplying calibration coeffi-cients.

Some approximate criteria exist that can be used to guess the correct wavelengths to use. Specific absorption features of the compounds and their background may be used, or wavelengths where the compounds and their background show maximal differences may be used, or occasionally sample bands in low-interference regions may be used, etc. The spectroscopic

**TABLE 1   Typical Data Transformations**

| | Exact form | $f[R(\lambda)]$ approximate form |
|---|---|---|
| "Absorbance" | $-\log[R(\lambda)]$ | — |
| "Absorbance" ratio | $\log[R(\lambda_i)]/\log[R(\lambda_j)]$ | — |
| First derivative | $d\,\log[R(\lambda)]/d\lambda$ | $\log[R(\lambda + \Delta\lambda/2)] - \log[R(\lambda - \Delta\lambda/2)]$ |
| First-derivative ratio | $\dfrac{d\,\log[R(\lambda_i)]/d\lambda}{d\,\log[R(\lambda_j)]/d\lambda}$ | $\dfrac{\log[R(\lambda_i + \Delta\lambda_i/2)] - \log[R(\lambda_i - \Delta\lambda_i/2)]}{\log[R(\lambda_j + \Delta\lambda_j/2)] - \log[R(\lambda_j - \Delta\lambda_j/2)]}$ |
| Second derivative | $d^2\,\log[R(\lambda)]/d\lambda^2$ | $-2\log[R(\lambda)] + \log[R(\lambda + \Delta\lambda)] + \log[R(\lambda - \Delta\lambda)]$ |
| Second-derivative ratio | $\dfrac{d^2\,\log[R(\lambda_i)]/d\lambda^2}{d^2\,\log[R(\lambda_j)]/d\lambda^2}$ | $\dfrac{-2\log[R(\lambda_i)] + \log[R(\lambda_i + \Delta\lambda_i)] + \log[R(\lambda_i - \Delta\lambda_i)]}{-2\log[R(\lambda_j)] + \log[R(\lambda_j + \Delta\lambda_j)] + \log[R(\lambda_j - \Delta\lambda_j)]}$ |
| Kubelka–Munk | $[1 - R(\lambda)]^2/2R(\lambda)$ | — |
| Log Kubelka–Munk | $\log[(1 - R(\lambda))^2/2R(\lambda)]$ | — |

experience and judgment required for this are quite demanding and may be hard to fulfill for these complex samples in a new field with a small data base where fast methods development is at a premium. For the time being, consider as a workable alternative the use of a computer to try out all possible wavelengths and choose the few best using a set of known samples.

The next item to be settled is what to measure. Since spectral additivity is a necessity for these calculation procedures, and since this exists only for the absorbance function in transmission measurements, this appears to be a good first choice. A number of possible functions to be measured can also be considered, as seen in Table 1. For diffuse reflectance spectroscopy, the Kubelka–Munk function seems to have the best theoretical basis and should give the best additivity and therefore the best calculation results. In practice, and over the narrow ranges of absorption usually encountered in the near-IR region of most of these complex samples, absorbance does just as well as any of the functions tried.

A more complex matter is the choice of the coefficients. In past practice in other spectral regions, when the method was applied to simple mixtures, the necessary coefficients were quickly obtained by calculations starting from the spectra of the pure components. But for the analysis of these complex organic samples, the pure components either are not separable or are unrepresentative after the separation has been accomplished. A new approach is clearly required here.

## C. Numerical Regression Techniques

The unavailability of pure standards can be counterbalanced by the use of a collection of samples of known composition, and by application of a set of algorithms known as numerical regressions (Watson, 1977). Conceptually it is easy to see how to construct a calibration curve by measuring a learning calibration set of preanalyzed samples. As shown in Fig. 10, one plots readings at one wavelength against concentration, does a mathematical least-squares fit of a straight line for all the points, and has a calibration line that can be expressed as a constant plus a coefficient multiplied by the reading at the chosen wavelength. The value of this calibration line can be estimated by measuring the residuals of the least-squares fit of it to the experimental points. This is the internal consistency of the calibration, and the standard deviation of these residuals has been called the standard error of estimate (SEE). One can confirm this estimate and verify the calibration by performing additional measurements of a second set of preanalyzed samples (the verification set) and comparing them with the calibration line. The second set of least-squares residuals gives the error of measurement (which should be no better than the internal consistency of the calibration). The standard deviation of these residuals is called the standard error of prediction (SEP).

If there are several wavelengths that could be used here, this procedure can be repeated and the optimum wavelength chosen on the basis of reduced SEE, which should be verified by checking SEP as well.

It is easy to see how this can work in the presence of an interfering compound, which we try to compensate for by measuring a second wavelength, as a "guard channel." The reading at this second wavelength indicates how much interference absorption there is at the first wavelength and can be used to correct for it. The mathematical expression would then be different from the first one, in that another term would be present, containing the reading at the second wavelength with an appropriate coefficient.

How do we carry this out in practice? Here we again use a learning or calibration set, but this time we measure its reading at the two wavelengths. We now plot for each sample the concentration, the absorption at one wavelength, and the absorption at the other, in three dimensions as shown in

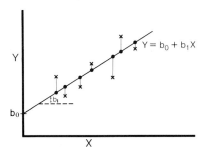

**Fig. 10.** Linear regression.

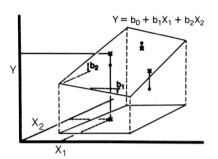

$$Y = b_0 + b_1 X_1 + b_2 X_2$$

**Fig. 11.**   Bilinear regression.

Fig. 11. We now define a calibration plane on the plane that has the least-squares fit to the plotted points. The residuals of this fit define an SEE, and a second least-squares fit of this calibration plane to a verification set defines an SEP. The geometrical calibration plane can be expressed mathematically as the summation of a coefficient and two terms containing the readings at each of the two wavelengths multiplied by a coefficient for each. All of these calculations are efficiently done by the numerical regression algorithm. If there is more than one pair of wavelengths to choose from, again we can compare candidates by test calculations and SEE and SEP comparisons. This procedure is known as a bilinear regression.

If there is more than one interferent, many wavelengths may be necessary. Then, if, for example, six wavelengths are required, all we need is to make a plot of our learning set data in seven-dimensional hyperspace and fit a six-dimensional calibration hyperplane to it. This sounds impressive, but computers can represent data in seven-dimensional hyperspace just as readily as in everyday space. What is more, they will grind out optimal wavelength choices, calibration coefficients, formulas for calculating concentration, and SEE's and SEP's in just the same way. This procedure is known as multilinear regression.

Here we must steer clear of a pitfall. If we should choose a set of wavelengths as large as that of samples in the learning set, the mathematics will always give a perfect fit, and SEE will always equal 0. Unfortunately, it would still equal 0 if the measurements were no good, if we picked the wrong wavelengths, if the samples were bad, or if the noise was overwhelming. Then, when one is testing for SEP, very bad results can be obtained. One must therefore always have more samples in the learning set than the number of wavelengths one intends to use for calibration. That way, only good data on good samples can make the SEE's look good, and this can be verified by the SEP's also being good (Hirschfeld *et al.*, 1983; Honig *et al.*, 1983). However, there is a further restriction. If any two samples are identical, if a given sample can be obtained by interpolating between two others, or if that sample has a spectrum that is an exact multiple of another's, then the above restriction operates as if the learning set had one fewer sample. In fact, only "independent" samples (those not related to each other as described above) in the learning set can be counted to define the maximum number of wavelengths that can be used.

Since it is hard to determine whether samples are independent of each other, it is necessary to use many more samples in the learning set than the number of wavelengths one wishes to use. If one were to use too many wavelengths, the required number of samples in the learning set might be impractical, which limits the size of the wavelength set.

This simple procedure is readily applied to systems that have only a limited number of possible wavelengths for measurement. For modern systems using monochromators or FT/IR systems, the number of possible wavelength combinations is too much even for the dedicated computer. The alternative is not to try all possible wavelength combinations but to try only all possible pairs or triplets of wavelengths. Then additional wavelengths are added one at a time to this set by testing one by one all the available further wavelengths. After the best (as defined by SEE and SEP) additional wavelength has been added, and if SEE and SEP have improved over those with one wavelength less, the procedure can be repeated and yet another wavelength added. This approach is usually called a stepwise regression. New algorithms now in testing promise further improvement (Honig *et al.,* 1983; Hamid *et al.,* 1981; Martens, 1979).

While all of this sounds complicated, it is so mainly for the computer and the manufacturers who program it, and many of the above points need not be dealt with by the operator. We shall see how the method is used in practice.


## IV. SETTING UP AN NIRA PROCEDURE


### A. Requirements of Reference Sample Sets

Near-infrared reflectance analysis in its present state of development is a secondary analytical method. As currently used it does not analyze an individual parameter directly but requires a mathematical expression derived from a known set of samples in which the values for a particular parameter are correlated to the near-IR optical data as provided by these learning samples. Analysis is performed by applying this mathematical expression to the optical data from the unknown samples. It is intuitively obvious that the performance of the method depends on the similarity between the learning set and the unknown samples and on the accuracy of the values of the parameters for the learning set. Verification of performance requires a second reference set of samples. These reference sets must therefore conform to the following criteria, which are presented as guidelines in the ideal situation. Practical considerations may require modification and adaptation of these principles in specific cases. This can often be done without significantly degrading the analytical results so long as care is taken and cross-checking of effects is carried out.

The first important factor to consider in developing a reference set is to choose a comprehensive set of samples such that the significant variations possible for future samples are present in this sample set (Watson et al., 1977; Osborne et al., 1982a). These variations include the range of the parameters, their intercorrelation, and sample morphology. Sufficient samples must be used to provide the requisite variations.

One must select for the reference set a group of samples that essentially evenly spans the probable concentration range of the constitutent being analyzed. The simplest explanation for the criticality of this criterion is to imagine the large effect on the calibration line of a limited number of nonrepresentative, erroneous samples at the extremes of the range. With no continuum of samples across the concentration range, such samples could drastically change the slope of the calibration line, pivoting it about a point within the concentration range of the bulk of the samples. To avoid this potentially serious error it is necessary to have the requisite number of independent samples represented at the extremes.

In process control, this range should extend to the limits of the production specification, and then the range should be increased on both sides to include the probable range of sample concentrations caused by out-of-control conditions. The addition to the range on either side of the production specification will not necessarily be equal in magnitude, since it depends on the nature of the process and the factors that go into setting the target concentration. It is important to have as many samples at the extreme regions of the range as within the production-specified range.

The ideal set minimizes the intercorrelation of all the chemical variables. At the gross level one has a variety of chemical constituents: protein, water, starch, oil, long-chain carboxylic acids, and so on. As one constituent varies in concentration so should the others, but in a nonpredictable pattern. If in the learning set constituent A always increases in the same proportion to the increase or decrease in the concentration of B, it is possible to predict the concentration of A based on the optical absorbance provided by B. This is obviously to be avoided unless the relationship between A and B must by its very nature remain fixed.

Intercorrelation is a particular problem in artificial sample sets. For example, as enticing as it may be to create a doctored sample set for a moisture calibration (i.e., by drying a sample down and adding known amounts of water), it is an exercise in futility, since all constituents will follow the moisture variation. Unless this is the rare situation where every chemical constituent in the sample population "never" undergoes a change in its concentration relative to any other constituent, the calibration will be worthless. This absolute consistency is not a reality in natural products; thus it becomes mandatory that real samples be used, and these by their very nature have at least a minimum level of relative constituent concentration variations. The random variation of these constituents and consequently their absorbances is necessary in order to force the calibration expression to compensate for these nonuniform overlapping interferences. In a natural product it is sometimes hard to break the natural intercorrelation of constituents; however, it is worth trying where possible in order to provide a more robust calibration. Conversely, if it is close to impossible to break this intercorrelation, then by implication the overwhelming majority of samples to be analyzed by the NIRA method will also have this intercorrelation and the method will not suffer from this constraint.

Another of the variables to be included in the reference sample sets is the morphological changes that are often due to particle size variations (Williams and Thompson, 1978). These may be caused by variations in processing from one plant or process line to another. It is strongly recommended that, just as different starting batches are used for the samples in the reference set, effort be made to include the variations caused by different plants, different sources of supply, and slightly different processing parameters. In some cases a valuable by-product to this inclusion of diverse sample sources is the method's mathematical highlighting and compensation of heretofore unknown differences between process lines (e.g., heating or pressure differences). If the difference between plants is too great, the derived calibration will make the best statistical compromise mathematically possible; however, this may be done at the expense of the accuracy of the fit of the calibration expression. The most efficient way to evaluate and benefit from the above situation is to include these samples, while recording the source differences. The regression mathematics is then capable of using this information to evaluate whether any effect is present and then to provide a mathematical compensation for this variation or to call for separate calibrations, if required.

While the concentration of the constituent of interest is represented in this controlled and specific fashion, it is also necessary to have the other constituents varying through their normal concentration range. Because of practical logistics, stringent conditions are imposed on one constituent at a time, but the ultimate set would vary all other constituents through their ranges at all points in the range of the constituent of interest. For example, at each protein level samples should have some low, some medium, and some high moisture levels so that the protein and moisture are not intercorrelated.

The age of the sample and storage conditions play an important part at the time of assembling a set. Many natural products degrade or otherwise change with time, heat, humidity, light, or other factors. If they are packed in plastic bags, they may come to moisture equilibrium with their surroundings. At this point in time their chemical composition and physical state may no longer be representative of the sample variables present in unknown samples, nor may the actual samples correspond to the reference assays done at an earlier time.

The number of samples needed for a learning set is the next obvious question. As a rule of thumb, the more the better, because this increases the probability of including samples that contain the significant sample variations. From statistical as well as practical considerations, a set of 30 samples is usually a good beginning for the calibration of a filter instrument that contains a limited number of wavelength choices. After the calibration has been worked out it should be evaluated using an additional set of at least 10 verification samples, meeting the same criteria as the learning set. In the case of a scanning spectrophotometer the wavelength search encompasses 250–350 possible wavelengths. The fact that only a limited number of wavelengths is used in the final calibration and that a stepwise down regression is not used on 350 wavelengths helps to diminish the expected increase in the number of samples required. However, with 50 wavelengths per sample, there is still a reasonable probability that a choice of wavelengths will be found that fits only the specific data set and its errors rather than the general case of samples. Unless the multiple correlation statistic is very high ($>0.93$) additional samples should be used to improve the reliability of the wavelength selection. The final sample set size indicated may be as much as 50–100 samples and represents a compromise between development effort and method reliability.

The criteria for a reference set, an even distribution of samples throughout the concentration range and low intercorrelation of the parameters, are probably the hardest requirements in the NIRA technology scheme. This difficulty in attaining a large range of independent samples is understandable when the goal for production is a composition within a narrow range of a target value. Often, an initial usable calibration is developed and additional samples are later added to the learning set as they become available through production deviations, thereby improving the method with time.

The instrument manufacturers maintain applications laboratories that should be consulted for guidance and techniques useful in overcoming these difficulties. They have developed methods for creating artificial learning sets where required.

This brings the discussion to the second of the important factors to be considered in assembling a reference set: the accuracy and precision of the manual analysis. The introduction of NIRA into the laboratory analysis repertoire poses an interesting situation and opportunity. It is at present a secondary analytical method that correlates to and creates its calibration equation from the existing laboratory analysis. The effect of the experimental errors of the manual analysis on the calibration have been repeatedly described in most discussions of NIRA calibration techniques. A working laboratory faced with multiple analysis on a variety of products, changing priorities, heavy work loads, and rapid turnaround of analytical results is not easily in a position to cross-check the accuracy of its manual assays, its instrumentation (from ovens and balances to GC and nuclear magnetic resonance [NMR] instruments), or the proficiency of its personnel, with independent methods. A quality control-conscious laboratory manager might, on a regular basis, monitor the overall reproducibility of the analyses performed in the lab through the use of blind check samples assayed by all involved under different laboratory conditions. This would give lab precision but no indication of the accuracy of the analysis method.

In preparing a calibration reference set, it is important and necessary that the methods development scientist critically examine the reference method and the environment in which it occurs. There is no substitute for on-the-spot observation. It has been found that an oven analysis with only three samples gives results at variance with a fully loaded oven as a result of airflow capacity, hot spots, cycling, or other factors. The bias between analysts has shown up dramatically within sample sets. Analysis of the same sample on different days and using different reagents has shown reproducibility differences. Sample inhomogeneities are the traditional bane of any analytical chemist's existence. A recent problem was traced to sampling method. One analyst used a spatula to sample from a bottle, another poured from the same bottle. The distribution of fine particles varied between these two methods, as did the concentration of the constituents under investigation. Factors affecting analysis reproducibility are the ones that deserve the most thought and effort. There are literally hundreds of variations on this lab error theme, from carry-over in Soxhlet thimbles to nonlinearities in a balance. It is important to eliminate those that can be eliminated and to keep the others as constant as possible.

A method that begins to address, recognize, and quantitate these realities of laboratory analysis is for a single analyst to assay the learning and verification set samples in blind duplicate, identified by coded random numbers, on two different days. Repeating this procedure with a second analyst will help identify possible biases in the procedure. It is generally valuable to discuss the NIRA calibration project with the analyst(s) to gain insights into problem spots in the manual method, as well as to enhance the alertness and care applied to the analysis of the learning sample set. Calculation of the pooled standard deviation of the blind duplicates based on 30 duplicates will give a good indication of the reference method reproducibility under the given conditions. If the duplicate analyses of a particular sample differ by more than is expected or desired, it is worthwhile to investigate the history of that sample as well as to repeat the analysis. There are statistical criteria that define how large the difference must be before it is scientifically justifiable to discard a particular analysis in a set of three, four, or five replicate analyses. Of course these same criteria of analysis apply to validation and unknown samples. However, there is a bright side to this difficulty. Near-infrared reflectance analysis calibrations are based on a statistical manipulation of all the data. This means that it is possible for the calibration equation to be better than the individual data points that went into creating it, because of the beneficial effect of mathematically creating the best compromise equation derived from many independent samples. Near-infrared reflectance analysis, therefore, becomes a tool to evaluate the quality of the reference laboratory analyses.

Once a preliminary NIRA calibration has been obtained, the differences between the reference laboratory and NIRA results should be examined. Samples showing significant errors should be reanalyzed by the reference method. Biases or other patterns in the errors should be investigated. After corrections to the data have been made, a new calibration is performed and verification repeated. In many cases, introduction of NIRA has dramatically improved the quality of the reference method laboratory results.

## B. Sample Preparation and Presentation

There are two general classes of sample matrices, solids and liquids. A solid will be defined as any substance that appears to retain its shape at room temperature. Within this group are grains, formulated feeds, cereal products, cheeses, meats, butters, and margarines. A liquid could be one of the three types: a true solution with no scattering particulate matter present, an emulsion, or a suspension with particles of varying sizes.

The first essential requirement for sample preparation and presentation is obtaining a representative sample. Sampling is a critical technique requiring protocols carefully tailored to the sample matrix under study. Sampling from a continuous process (under or out of control) may be different from sampling from a batch process, which is different from sampling from a boxcar filled by individual suppliers. Sampling devices may be manual or automatic; they may include sample probes, stream splitters, or sampling cups. No matter how sampling is done, one must obtain a representative small composite sample from varying depths and positions in the larger main body of sample.

Once this small composite sample is obtained, it must be mixed and blended prior to analysis. For solid samples, this should be done prior to grinding and again after grinding so that subsequent subdivisions for multiple NIRA and reference analysis are truly representative and equivalent to each other in chemical composition, particle size, and general morphology. There are a number of laboratory-size mixing units that tumble the sample, stir it, split it, and then repeatedly recombine it to give homogeneous subsamples (Hunt *et al.,* 1978). Proper sampling is essential to both the learning sample set and the analytical samples. One can use the fact that NIRA is a nondestructive analytical method and initially measure the sample aliquots by NIRA and then submit these same samples to the reference assay to ensure equivalence of samples for calibration and verification. After an accurate calibration has been obtained, NIRA can be used to monitor sampling inhomogeneities.

The success of the chosen grinding and blending method to give a homogeneous sample can be qualitatively assessed by recording the optical data at all available wavelengths. Compare these values for repacks of the same aliquot of sample material to repacks of different aliquots from the same sample batch. If the optical data remain consistent from repack to repack, then one can have a degree of confidence in the experimental techniques used to prepare the samples. The absolute magnitude of numbers that represent optical reading consistency varies depending on the instrument and the sample matrix. However, one can get a feel for this magnitude by experimenting with known product applications and methodologies.

Many solid samples require being ground to a small and relatively uniform particle size. In general, particles on the order of $150-250$ $\mu$m are acceptable; even wider ranges have proved serviceable. Routinely those particles that upon grinding pass through a 20-mesh or a 1-mm screen are suitable for NIRA analysis. It is important to maintain consistency in particle size between the learning set samples and the analysis samples. This can be done by using a particular type of grinder in good operating condition.

There are at least six particle size diminution principles used in grinders, either singly or in combination. Specifically they are:

1. A knife mill, in which sharpened blades in various configurations effect the grinding either alone or by impacting against another knife blade or screen.

2. A cyclone mill, in which rapid rotation of the sample causes centrifugal force to impel the sample outward against an abrasive surface that grinds the sample down in size.

3. A hammer mill, which relies on the mechanical collision of rotating hammer blades with the sample to cause particle breakage.

4. A disk mill, which uses the shearing action on a sample caught between two rotating disks each with teeth. The resulting particle size can be varied by adjusting the gap between the disks.

5. A ball mill, which reduces particle size by crushing the sample between the individual balls and between the ball and the container wall as it is being shaken or rotated.

6. An extrusion grinder, which forces a sample via a worm feed or centrifugal force through a screen that serves to break up the particles.

The choice of grinder is usually determined by the physical and chemical properties of the samples, the quantity available per sample, and the number of samples to be ground, particularly as it affects the intersample cleanout time. Those products with high oil contents are often done most satisfactorily on a knife mill. Conversely, grinders like the knife mill are in certain instances avoided because of their rapid heat buildup during grinding. It is prudent to check the chosen grinder for heat buildup, as well as to monitor other aspects of its design that might be problematic. Heating of the sample will cause moisture losses as well as possible sample degradation through chemical reaction. Some cyclone grinders require strong air currents, which may result in evaporative moisture losses. Some grinders collect particles and dust from previous samples on filters that periodically dislodge their loads, resulting in cross contamination of samples. When choosing a grinder, look for the following:

1. A uniform and finely ground product.

2. A minimum intersample carry-over or an acceptable cleanup procedure.

3. No adverse effect due to heat buildup or airflow.

Temperature is a useful variable when recognized and controlled. Cooling in the refrigerator, in the freezer, with dry ice, or with liquid nitrogen can make many products more amenable to grinding. However, it generally produces a sample not suitable for original moisture determination, since most samples undergo large moisture changes with extreme changes in temperature. Similarly, many calibrations are sensitive to the temperature of samples submitted for analysis. The NIRA calibration by its nature is based on correlation to and compensation for overlapping peaks. The NIRA water spectrum is quite temperature sensitive, undergoing varying amounts of absorbance band shift depending on the temperature and the wavelength being measured. Therefore, one should not take a calibration developed at ambient temperature and expect to analyze with the same accuracy a sample at 32°F. Some natural or artificial emulsions like butter or margarine, comminuted meats, and cheese provide better results when assayed either colder or hotter than ambient. This can be accomplished with a thermo-stated cup.

Uniformity of sample preparation (i.e., type of grinder, grinding time, quantity of sample, how many samples are run consecutively in time, cleanout procedure, size and shape of screen) is mandatory once a grinding procedure is established. It must be maintained on all calibration samples as well as on *all* analysis samples.

Most manufacturers currently have closed and open sample cells or cups for solid products. Closed cups are the method of choice for those products that can be easily removed without significant contamination of the sample cup. Among its advantages is a closed system that prevents evaporation and other changes at the sample's surface. The window provides a flat sample surface for measurement. Various techniques from springs to sponges are used to provide a uniform sample packing density. An open sample cup finds utility for analyzing pastes and other messy samples or those samples that separate or decompose upon compression. The open cup need not be cleaned between samples but only emptied, since NIRA looks at the surface layers of the sample and does not penetrate through the sample to the walls or bottom of the cup. Products like butter, cheese, and meat are usually analyzed in an open cup. New solid-sample cups are currently being developed in NIRA applications labs to solve specific application needs and will soon reach the users in the field.

Within the realm of liquid samples there are three recognizable subdivisions:

1. True solutions of totally miscible components containing no particulate matter.

2. Suspensions, containing particulate matter that collects and separates from the liquid over a relatively short period of time.

3. Emulsions, the in-between class containing particles of colloidal or slightly larger size such that when separation does occur it requires a long period of time to become observable.

Liquid matrices with large particles have at least three inherent restrictions that must be accounted for and if possible removed. Are the particles too large to pass through the gap of the sample cell without clogging it? Do they rise, settle, or otherwise change in a manner that prevents obtaining a stable representative sample? How does the sample change during the measurement time? The same concepts as those concerning solid-sample homogeneity apply to liquid samples. In general, the recommended approach is to minimize the particle size through homogenization or other means.

There are a number of types of homogenizers on the market: ultrasonic wave-action types, high-pressure shearing-action types, and a combination of ultrasonic and rotary shear homogenizers. The specific product matrix will determine the particle size reduction technique. Currently, milk, ice cream mixes, and cream products are being homogenized by a high-pressure multistage shear-action homogenizer that feeds directly into the transflectance cell of the NIRA instrument, as shown in Fig. 12. Even particles of submicron size, as are present in milk after homogenization, undergo noticeable separation during the time required for measurement. This effect can be minimized by utilizing standard instrument features: an operator-specified uniform reading time from the time the sample enters the cell to the initializing of the measurement and a forward- and reverse-order average of the reading of the measurement filters, which is used to average out time drifts.

The weaker band strength on the near-IR overtone region is quite useful in liquid analysis because it avoids saturation of the water or other strong absorbance bands. This weaker band strength also allows liquid samples to be assayed undiluted in larger-path-length cells than is possible in the

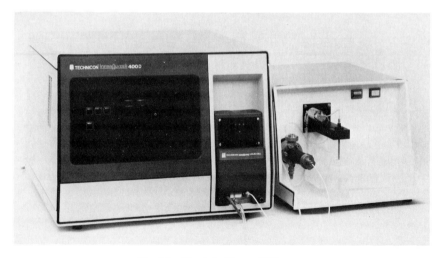

**Fig. 12.** The InfraAlyzer 400D.

mid-IR. A larger-path-length cell is easier to fill and easier to clean and maintains better path-length consistency under continuous-flow condition (McGann, 1978).

Near-IR liquid cells can be designed in either a transmission or transmission-reflectance (transflectance) geometry. The latter geometry has certain advantages over classical transmission. The time and work involved in changeover from solid reflectance measurements to liquid measurements is significantly diminished since realignment of the optics is not necessary. The ability to use a true dual-beam measurement without special accessories is another advantage of the transflectance measurement mode. This also eliminates the tight tolerances and special care necessary when working with matched liquid cells. A transflectance signal collection geometry (identical to the solid-sample reflectance situation) eliminates pseudoabsorbance errors due to particle scattering of the light out of the beam in the analysis of highly scattering liquids.

Aqueous samples are very temperature sensitive and require very accurate temperature control. The exact position of the water band is highly temperature dependent, allowing uncontrolled tempratures to produce serious errors. A well-engineered heated liquid cell provides a constancy and reproducibility of sample temperature to better than 0.1°C. Aqueous product matrices, like those of interest to the dairy industry, are assayed at 45°C, which keeps the butterfat in its liquid form above the upper limit of its melting point at 38°C.

Temperature control also allows analysis of oils and fats, which run the gamut from liquid to solid at room temperature. Edible oils are routinely being analyzed using a liquid cell at 65°C, which effectively liquefies most oils of current interest. Increasing the temperature may also be used to decrease sample viscosity to the point where it can be pumped through a homogenizer or injected by syringe.

Intersample carry-over and washing behavior with various cleaning solutions must be addressed for successful liquid applications. The most straightforward approach uses each succeeding sample to wash out the previous one. This is a successful technique in a system that does not build up contaminants, especially on the cell window. If buildup occurs, the system will require periodic washing with a suitable cleaning solution. The instrument manufacturers maintain applications labs to assist the NIRA user in the nuances of liquid methods development.

## C. Measurement

The prepared samples of the learning set are then measured to determine their reflectance at multiple wavelengths. In the absence of any *a priori* information, it is desirable to use a monochromator-based scanning instrument and to determine the reflectance over a major portion of the near-IR region so that computerized wavelength search can find the appropriate analytical wavelengths. Scans from 1400 to 2400 nm are usually sufficient, although data are often collected over the range of 1100–2500 nm or more. A 10-nm spectral bandwidth has been found adequate, with data points spaced 4 nm apart. Therefore, 250–350 independent measurements are made of each sample with the scanning instrument. As discussed previously, use of this many measurements places significant requirements on the learning set of samples. It is preferable to use *a priori* information to reduce the number of measurements made and/or analyzed.

Most users of NIRA have filter-based spectrophotometers rather than the more sophisticated monochromator-based scanning instruments. In this case, the available wavelengths have been determined by the instrument manufacturer based on experience with a broad range of applications and spectroscopic knowledge. The majority of instrument manufacturers use a number of fixed-wavelength filters on a rotating filter wheel to establish the measurement wavelengths. As demonstrated by the applications discussed later, selection from up to 19 well-chosen filters has proved adequate for analysis of foodstuffs and for many other analyses. One manufacturer favors the use of absorption differences, which requires measurement of two or three closely spaced wavelengths for each term of the analysis equation and variation of the analytical wavelengths as part of the calibration process. These additional requirements are met by tilting the filters to vary the wavelength over a finite range as multiple measurements are made. More data points must be measured on the learning set of samples than for fixed filters to allow for calibration by varying the selected wavelengths.

It is important when measuring the learning set of samples to provide data so that the various errors can be quantified. As a minimum, all samples should be prepared and measured in duplicate. The standard deviation and bias calculated on the differences between the first and second readings of each sample then give measures of the total precision of the method. This total precision includes sampling error in separating the two aliquots, sample preparation variations, and instrument measurement errors.

If the total precision error is significant (i.e., if the standard deviation is equal to or greater than one-half the SEE), further measurements to define the error source are justified. The minimal precision measurement is repeat reading of the same sample, without reloading into the sample cup. It is particularly important to look for drift in the measurement, which can be indicative of sample change due to temperature, separation, reaction, or other causes. Control of these variables is important in many cases. These errors can also be minimized by maintaining consistent timing of the sample preparation and measurement and, in certain instruments, use of a measurement cycle that automatically cancels linear drift of the sample absorbance during the measurement cycle. The standard deviation of the optical readings on repeat reading of the same sample should approach that due to instrument noise alone. The instrument noise levels should be obtained from the manufacturer as a basis for comparison.

The total precision discussed initially also includes sampling errors, sample preparation variation, and differences in sample presentation to the instrument. Differences in sample presentation can often be quantified by repacking the same sample aliquot in the sample cup and repeating the measurement. This minimizes sampling error and sample preparation variations. This process of quantifying error sources can be continued right back to the original sampling error, provided the errors introduced as a result of the later steps in the process are not overwhelming. Mathematically the procedure known as analysis of variance is applied to quantify the various error sources.

Determining the cause of these errors may require additional investigation. A useful approach is deliberate introduction of variables, such as particle size, temperature, and packing pressure, to determine their effect on the data. Given this information, it is often possible to identify these effects in the data from the learning set. Principal-components analysis is a statistical technique that, when applied to the error data, separates, quantifies, and characterizes the errors by their effect on the spectral data.

## D. Wavelength Selection and Calibration

The general principles of calibration by multilinear regression are discussed in Section III,C. In the case of simple "absorbance," $\log 1/R$, or $-\log R$ optical data, the calibration equation is

$$Y = F_0 + F_1 \log 1/R(\lambda_1) + F_2 \log 1/R(\lambda_2) + \cdots + F_n \log 1/R(\lambda_n)$$

where there are $n$ wavelengths used. The coefficients $F_n$ are determined directly by the multilinear regression mathematics. The predication equation for difference data treatment is of the form

$$Y = F_0 + F_1 f[(\log 1/R(\lambda_1)] + F_2 f[\log 1/R(\lambda_2)] + \cdots + F_n f[\log 1/R(\lambda_n)]$$

where $f(\log 1/R) = \log 1/R(\lambda + \Delta\lambda/2) - \log 1/R(\lambda - \Delta\lambda/2)$ for the first difference and $f(\log 1/R) = \log 1/R(\lambda + \Delta\lambda) - 2 \log 1/R(\lambda) + \log 1/R(\lambda - \Delta\lambda)$ for the second difference.

In these cases, $\Delta\lambda$ becomes a calibration variable in addition to $F_n$. Each term is the sum of two or three $\log 1/R$ terms.

Also, second-difference ratios have been used as the variable. Each term then involves two wavelengths ($\lambda_N$ and $\lambda_D$), two wavelength differences ($\Delta\lambda_N$ and $\Delta\lambda_D$), and the coefficient $F_n$. The numerator (N) and denominator (D) are determined sequentially.

The wavelength search for difference equations must include determination of $\Delta\lambda$ as well as $\lambda$. The $\Delta\lambda$ parameter may be preselected by the operator and the data transformed to first- or second-difference format. Search and calibration then proceeds as for log $1/R$ data. Optimization of $\Delta\lambda$ may be performed manually by repeated computer runs. Automated methods have been developed, but the algorithms have not been published.

Returning to the simpler log $1/R$ case, the question is selection of a minimum set of analytical wavelengths from a larger possible set, which may also have been selected from all possible wavelengths, by either the instrument manufacturer or the user.

In some cases, appropriate analytical wavelengths have already been determined by previous applications development on similar products. A single multilinear regression using these known wavelengths will produce a calibration. If the results are acceptable when tested for prediction, no additional analysis is required, even though somewhat better results might be possible.

Wavelength selection typically involves generating calibrations for many possible wavelength combinations. From the calibration statistics, such as the multiple correlation coefficient, standard error of estimate, and the $F$ test, the best wavelength selection from all the trials is determined.

In the extreme case, it is possible to take 350 or more data points for each sample and to let the computer select wavelengths without human intervention. It would apear desirable to try all possible combinations of wavelengths so as to find the "best" ones. However, this is impractical. To select all possible combinations of six wavelengths from 350 data points using 100 samples requires $7.7 \times 10^{15}$ computer multiplications and divisions. Even with a computer capable of $10^5$ such floating point operations per second and 256 kilobytes (KB) of data memory, so as to hold all required information in fast-access memory, it would require over 300 years(!) to accomplish this task. Fortunately, there are a number of approaches to reduce the task of wavelength selection to manageable proportions.

In many applications, especially in the case of foodstuffs, the major constituents of interest have known absorption bands. Using this spectroscopic knowledge, an initial limited set of wavelengths can be chosen. The analysis wavelengths can then be selected from this restricted set. For example, Karl Norris, of the USDA, originally selected six wavelengths for the analysis of protein, moisture, and oil in grains and oilseeds on the basis of the spectra of gluten, vegetable oil, starch, and water. He also selected two reference wavelengths to account for scattering differences.

It has been demonstrated that these wavelengths are suitable for analysis of a wide variety of products despite small variations in the band positions in various materials. Based on further spectral analysis, Stark added 13 wavelengths to produce a general-purpose filter instrument for NIRA research. These wavelengths were chosen to include the repeated bands at shorter wavelengths and absorption bands of other constituents, such as fiber, sugars, and alcohols. The power of the NIRA correlation transform methodology is demonstrated by the successful use of these wavelengths in the broad range of applications discussed later.

When a limited set of wavelengths are chosen as candidates, it is practical to perform an all-possible-combinations search for 6 or even more wavelengths in a reasonable time period. The best strategy is to search for the optimum combinations of increasing numbers of wavelengths (e.g., 3, 4, 5, and 6), stopping when statistical tests indicate that no further improvement is obtained. The analysis time and memory requirements both grow rapidly with the number of data points used and the number of wavelengths selected. Determination of 6 wavelengths from 50 or 3 or 4 from 100 is a practical limit for all-combinations searches.

Stepwise down regression has also been applied in cases where there are a limited number of data points. In this method, an initial regression including all the wavelengths is performed, and statistics defining the significance of each wavelength are calculated. The least significant wavelength is then dropped and the process repeated until all remaining wavelengths are significant and removing one more decreases the accuracy substantially. In most cases, this method is quite satisfactory; however, it is possible that a wavelength will be dropped from consideration at an early stage that would improve the result at a later stage in the process. The method has the advantage of requiring relatively few regressions, so that results can be obtained in reasonable time using a desk-top calculator. A practical limit is the size of the cross-product matrix for the initial, largest regression and the cubic increase in the calculations required with the number of data points. In addition, it is essential to have many more samples than wavelengths in the regression, which also limits the number of wavelengths that can be used as a starting point.

While the above methods are useful for analysis of data from filter instruments, scanning monochromators can produce 700 or more data points per spectrum. All possible combinations of over three wavelengths or stepwise down wavelength searches are impossible with this much data. Fortunately, much of the data is redundant and can be eliminated.

The first approach to reduction of the number of data points is to consider the information content of the spectrum. The half-power spectral bandwidth of the instrument puts a limit on the data resolution, regardless of the original spectral content of the sample. Furthermore, most natural products, such as foodstuffs, have absorption bands much broader than the 10-nm spectral resolution typical of NIRA scanning monochromators. This limits the information content still further. If the spectral features are on the order of 20 nm half width and greater, as is quite common, the information content is equivalent to that which would be obtained with a 20-nm resolution instrument.

Nyquist's criteria from information theory indicate that only two data points are required per resolution element. Therefore, 5-nm spacing of the data points is sufficient for a 10-nm resolution instrument, and 10-nm spacing for typical spectra with 20-nm width spectral features. The obvious choice is to reduce the data to 10-nm increments. It should be noted that this result in a data point within ±5 nm of any wavelength. This is the center wavelength tolerance of filters that have successfully performed the applications discussed later and have demonstrated the transferability of calibrations from instrument to instrument. Data may be collected at better resolution and narrow wavelength interval to allow test of intermediate wavelengths; however, it should be smoothed and sampled at a wider interval, such as 10 nm, for initial wavelength search.

The second approach to reduction of the number of data points is to restrict the wavelength range. The 1400- to 2400-nm region covers two sets of absorption bands for the molecular bonds of interest in natural products. The next set below 1400 nm is so much weaker than the bands from 1900–2400 nm that combining data from the 1000- to 1400-nm region with data from the 1900- to 2400-nm region is usually futile. Therefore, most work can be done with a 1000-nm range, usually 1400–2400 nm with an alternative of 900–1900 nm. With 10-nm data intervals, this results in the use of 100 data points. Depending on the computer algorithms used, a four-wavelength all-possible-combination search can be completed within hours. A three-wavelength search takes only minutes.

The question arises as to how to perform a search to test whether additional wavelengths or intermediate wavelengths would improve the accuracy. These possibilities can be investigated using stepwise up regression methods. In this approach, an initial selection of two or more wavelengths is made. Additional wavelengths are added by searching for which wavelength of the data set provides the most improvement in accuracy when added to the previous wavelengths. In order to minimize the chance of missing a better combination, it is desirable to iterate the selection process. When a new wavelength has been selected, one of the previously selected wavelengths is removed and another search performed to determine whether a better wavelength can be found. This iteration is continued until the new trials always result in the same wavelength selection. Then, an additional wavelength is added by trial of all remaining data points, and the iteration is performed again.

It is vital to start with the combination of at least two wavelengths and preferably three or more using "absorbance" or log $1/R$ data and stepwise up regression. Near-infrared reflectance analysis spectra show very high inter-correlation between the data at various wavelengths as a result of scattering variations common to all wavelengths, which can be as large as or larger than the analytical signal. This completely defeats stepwise up regression starting with one wavelength, since this wavelength will usually be selected to reduce background variations independent of the constituent data. Even two variables will sometimes be selected solely for background rejection. With three or more variables, it is possible simultaneously to handle the background variation and to predict constituent data. Successful data analysis strategies to develop a wavelength selection and calibration include some or all of the following steps.

1. Perform an all-possible-combinations search starting with two or three wavelengths and increasing the number until the accuracy is sufficient or shows negligible improvement. The search can include: (a) the wavelengths available in an existing filter instrument to which the calibration will be transferred, (b) the more extensive set of filters available as standard wavelengths from the instrument manufacturer, or (c) evenly spaced wavelengths defined by a range and an interval, such as 1400–2400 nm with 10-nm spacing.

2. Use the wavelengths determined above as the starting point for a stepwise up search with iteration. This will determine whether one of the selected wavelengths should be replaced and/or whether additional new wavelengths should be added to improve results. For this search the range and/or the interval may be adjusted to include more data points in the analysis.

3. Investigate the possibility of alternative selections in any of several ways: (a) The all-possible-combinations search can be used to determine several combinations of wavelengths to be used as starting points for stepwise up selection or as alternative selections for evaluation. (b) It is sometimes useful to examine the multiple correlation coefficient as a function of wavelength for each term, holding the remaining wavelengths constant (Fig. 13). A narrow correlation peak at the selected wavelength is indicative of potential wavelength sensitivity problems. If an alternative broader peak with nearly as high correlation is found, it may be tested by forcing the wavelength of maximum correlation into the regression.

The most important aspect of the wavelength selection and calibration process is the elimination of erroneous data. It is usual, unfortunately, to

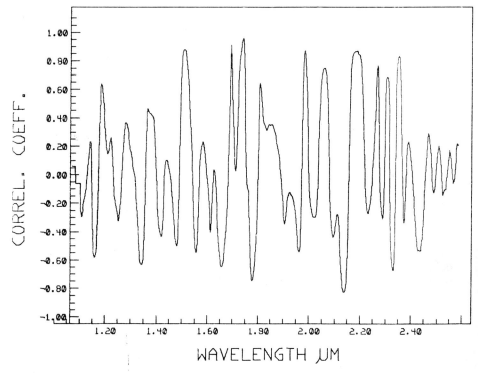

**Fig. 13.**  Correlation function versus wavelength.

find mix-ups in the sample numbers, incorrect laboratory values, transcription mistakes, and other errors in the data. Less frequently, the instrument measurements are wrong because of improper sample handling or equipment malfunction. A single bad data point, if the error is large, can completely change the regression and destroy the utility of the results. Therefore, wherever possible, data should be transferred electronically rather than manually. All manually entered data must be proofread carefully by a second person.

It is also desirable to examine carefully the results of the first calibrations on a data set and to search for erroneous data points, or "outliers." Examination of plots of the least-squares fit residuals as a function of each of the variables in the prediction equation, as well as the reference values, can aid in detecting errors. Patterns may appear in the data that provide clues to the solution of problems. It is important, however, to confirm that the data are actually erroneous, by reanalysis of the samples if necessary. Deletion of "outliers" that are actually correct data will lead to false expectations of the performance of the calibration is analyzing unknown sample values.

Finally, it is useful to examine the wavelengths selected based on knowledge of the molecular absorption regions of the constituents of interest or of the expected molecular bonds. Usually the selected calibration wavelengths can be understood on this basis. Occasionally, surprising information is revealed by the wavelength selection. For example, in analysis of the extraction of oil from rice bran, a calibration was performed for residual oil that used the expected oil absorption bands. When a calibration was performed for the fraction of oil recovery, however, the wavelengths of water were selected, and the relative error was reduced! The computer learned, on its own, that the moisture content of the rice bran controlled the oil recoverability.

## E. Calibration Verification

The calibrations obtained by the computerized wavelength selection and calibration process usually perform as expected from the calibration statistics. However, verification of the analytical performance is an essential step in developing an NIRA method. The choice between alternative calibrations should also be made based on these verification procedures.

Each calibration is tested by determining the values of a separate verification set of samples and comparing the results to the reference values. These samples should conform to the requirements for the learning set of samples discussed previously. Again, obtaining reference data of the highest accuracy and eliminating data errors are vital parts of the process. The verification statistics describe the differences between the NIRA results and the reference values. The accuracy of the results, therefore, can appear no better than the accuracy of the reference values. In practice, NIRA may actually be more accurate than the reference data, since many reference values are combined to determine the calibration.

## F. Fine Tuning the Procedure

Once the wavelength selection and calibration have been completed and the results satisfactorily verified, the NIRA method can be placed in routine use as an analytical tool. Fine tuning the procedure, however, can often improve the analytical results.

All instruments require occasional recalibration due to aging of components, repair or replacement of parts, and similar alterations. In addition, it may be necessary or desirable to transfer calibrations done on one instrument to a different instrument, since (1) the instrument manufacturer may provide the calibration, (2) the calibration may have been developed using a central methods development instrument, or (3) an instrument may be replaced. In general, this recalibration involves adjustment of the bias, and sometimes the slope, of the calibration equation. Certain types of instruments require additional, more complex corrections to transfer calibrations to a different unit.

Bias and slope adjustment does not require a completely new wavelength selection and calibration with an extensive learning set of samples. The multilinear calibration equation is converted to a simple linear equation by combining all the log $1/R$-dependent terms so that

$$Y = K + mX = F_0 + m[F_1 \log 1/R(\lambda_1) + F_2 \log 1/R(\lambda_2) + \cdots]$$

where $F_0$ is the bias or offset and $m$ is the slope. In principle, the bias and slope can be determined by analyzing two known samples. In practice, to minimize errors, it is preferable to use 10 representative samples covering the calibration range. A simple linear regression, performed in the instrument or with a calculator, then determines the bias and slope adjustment. Often a bias is all that is required. In this case, $F_0$ is adjusted by the mean difference between the NIRA results and the reference laboratory values, and a linear regression is not required.

As with any analytical method, laboratory quality control procedures should be applied. Blind check samples should be included in the routine work load and appropriate control charts and statistics generated. Bias and slope adjustments, when required, can be based on these check samples. The frequency of adjustment depends on the stability of the instrument, the error sensitivity of the calibration, the required accuracy of the analysis, and potential changes in the sample due to growing season, changing production parameters, and similar factors. Slope adjustment is rarely required after initial setup of the method unless significant changes are made in the instrument or sample processing. Bias adjustments are more common; however, it is important to have a valid basis before applying a bias change.

The initial calibration, while satisfactory, may have been based on a less than ideal learning set of samples. Adding more samples to improve the distribution over the range or to include additional variability may improve the accuracy or robustness of the calibration. For example, in process control applications the most valuable samples are those obtained when the process is operating near the permissible control limits or is actually out of control. Since these are undesirable conditions for plant production, such samples are presumably rare. They can be identified, when they occur, on the basis of NIRA data and subsequent analytical quality control results and then saved for future inclusion in the calibration data set.

Another example for which enhancement of the calibration may be desirable occurs with crops, which may vary from year to year with growing conditions. In this case, samples from the new crop may be added to the learning set to make the calibration less sensitive to year-to-year crop variations, although historical data seem to be indicating that calibrations, are insensitive to crop-year changes. Fortunately, it is not necessary to keep the previous samples in order to add new samples and enhance the calibration. The data from the previous learning and verification sets are saved and data from the new samples added. In order to account for possible bias differences between the old and new data, dummy indicator variables are included in the regression identifying each separate group of samples. This technique can even account for different instruments if the calibration is transferable from instrument to instrument with only a bias adjustment.

When the learning and verification data sets have been expanded, the calibration is repeated. Usually, wavelength selection is not required and a single multilinear regression is sufficient to obtain new calibration constants. If desired, however, a new wavelength search may be performed to determine whether significantly better results can be obtained. The coefficients of the indicator variables provide information on the bias between the new and old data sets. If this is significant, the calibration may be used with different biases to reflect the different conditions. This approach is particularly useful in the case of changes in the process or the addition of new processing lines. A single robust wavelength selection and calibration may be used with bias adjustment to reflect the particular process differences.

While the setting up of an NIRA method may appear complex, most of the effort is involved in obtaining the learning and verification samples with good reference values. The computer software included with the methods development instruments provides the wavelength selection, calibration, and verification processing and information to the user in a form suitable for decision making. The value of the effort is demonstrated by the applications, such as those discussed in the next section of this chapter.

## V. APPLICATIONS OF THE NIRA TECHNIQUE IN THE FOOD INDUSTRY

From the beginnings of commercial use in grains and oilseeds in the mid-1970s to the present, NIRA technology has expanded to include nearly every facet of food production and processing (Wetzel, 1983). The applications of NIRA in the food industry are increasing so rapidly that any listing is immediately out of date. In addition, many applications that have been developed by NIRA users are not published. Therefore, this discussion is intended to give only an overview of the breadth of NIRA applications in the food industry.

## A. Grains and Grain Products

The first widespread use of NIRA was in the determination of protein and moisture in hard red wheat. Research and development by Karl Norris, of the USDA, P. C. Williams, of the Canadian Grain Commission, and others have made this application by far the most developed and reported used of NIRA. Near-infrared reflectance analysis is now the official method used to determine protein in wheat for trading purposes in both the United States and Canada. Instruments are used at the country elevator to segregate incoming grain by protein content and to control blending to maximize profitability based on protein premiums. Terminal elevators use NIRA to confirm the protein content of incoming wheat and to control load-out to ensure meeting purchase contract requirements. The overwhelming preference for NIRA is based on its speed, accuracy, and low cost compared to other methods. The accuracy and cost are summarized in Tables 2, 3, and 4 based on information published by Williams (1977).

Well-established and experimental applications of NIRA for grains and grain products are listed in Table 5. The list reflects the growth of NIRA from the agricultural commodity market to grain processing, particularly flour milling, malting, and wet and dry corn milling (Watson *et al.*, 1976; Rubenthaler and Bruinsma, 1978).

## B. Oilseeds and Oilseed Products

Analysis of oil seeds and their products was another early application of NIRA, although originally restricted to solids, such as the meals and expeller

**TABLE 2   Protein Determination in Wheat: Cost Breakdown**

| | NIRA | | Kjeldahl | | Neutron activation | Kjel Foss | Kjel Tec | Thermal decomposition |
|---|---|---|---|---|---|---|---|---|
| | Large | Small | Large | Small | | | | |
| Daily capacity | 500 | 200 | 480 | 144 | 165 | 100 | 100 | 54 |
| Annual capacity | 130 K | 52 K | 100 K | 35 K | 43 K | 26 K | 26 K | 14 K |
| Cost = $ × 1000 | | | | | | | | |
| Staff | 45 | 15 | 60 | 30 | 25 | 20 | 30 | 15 |
| Space | 5 | 2.5 | 12.5 | 6.3 | 15 | 5 | 5 | 2.5 |
| Maintenance | 1 | 1 | 3 | 1 | 12 | 11 | 0.5 | 1 |
| Chemicals | 0 | 0 | 12.5 | 4 | 0 | 5 | 2.5 | 1.4 |
| Utilities | 0.5 | 0.5 | 5 | 1.5 | 2 | 0.5 | 0.5 | 0.5 |
| Depreciation | 2 | 2 | 5 | 1.5 | 9 | 3 | 1.5 | 5 |
| Cost/test | $0.39 | $0.40 | $0.98 | $1.27 | $1.23 | $1.36 | $1.54 | $1.81 |

TABLE 3   Protein Determination in Wheat: Methods, Performance, and Costs

| | Precision standard deviation | Standard deviation of difference | Accuracy correlation | Bias | Cost/test ($) |
|---|---|---|---|---|---|
| InfraAlyzer (NIRA) | 0.099 | 0.183 | 0.996 | −0.022 | 0.39–0.40 |
| Kjeldahl, two averaged | 0.118 | 0.153 | 0.997 | −0.012 | 0.98–1.26 |
| Neutron activation | 0.074 | 0.127 | 0.999 | −0.032 | 1.23 |
| Kjel Foss | 0.077 | 0.172 | 0.997 | −0.27 | 1.36 |
| Kjel Tec | 0.191 | 0.232 | 0.996 | −0.116 | 1.54 |
| Thermal decomposition | 0.185 | 0.216 | 0.996 | −0.162 | 1.81 |

cakes. The recent advent of liquid capability for NIRA instruments has accelerated development of analysis of the vegetable oils. Iodine value, as a measure of the degree of unsaturation, is an established application. Determinations of solid fat index (SFI) and free fatty acids (FFA) are in the experimental stage. NIRA applications in oilseeds and oilseed products are listed in Table 6 (Tkachuk, 1981; Williams *et al.*, 1978).

## C. Processed Food Products

The list of processed food products in Table 7 represents analysis parameters chosen by current users of NIRA to justify the purchase and use of their instrumentation. These analyses are being done not only in the plant quality control lab but also on the plant floor to monitor strategic points in the product processing. Methods development is occurring in both the corporate research centers and the plant quality control laboratories. Both sites are uniquely suited to this work. The corporate research center can work out research projects such as degree of cooking or gelatinization or moisture content in product packaging materials while the plant lab has an ample supply of current production samples undergoing time-consuming and expensive routine tests. In the area of food analysis constant advancement is

TABLE 4   Moisture Determination in Wheat: Methods, Performance, and Costs

| | Precision standard deviation | Standard deviation of difference | Accuracy correlation | Bias | Cost/test ($) |
|---|---|---|---|---|---|
| Oven | 0.171 | 0.271 | 0.92 | −0.26 | 0.25–0.45 |
| InfraAlyzer (NIRA) | 0.113 | 0.217 | 0.966 | 0.226 | 0[a] |

[a] When measured with protein.

**TABLE 5   Grain and Grain Product Applications**

| Food | Application |
|------|-------------|
| Wheat | |
|    Hard wheat | Protein, moisture |
|    Soft wheat | Protein, moisture |
|    Midds | Protein, moisture, fiber, oil |
|    Wheat bran | Protein, moisture, fiber, sugar, starch |
|    Hard wheat flour | Protein, moisture ash (bran cellulose), color index (Agtron measurement) |
|    Soft wheat flour | Protein, moisture, ash (bran cellulose) |
|    Semolina | Protein, moisture, ash (bran cellulose) |
|    Wheat germ | Protein, oil |
|    Defatted wheat germ | Fat |
| Barley | |
|    Barley | Protein, moisture, $\beta$-glucan |
|    Malt | Protein, moisture, extractables, soluble protein, diastatic power, free nitrogen extract |
| Rice | |
|    Rice (ground and whole grain) | Moisture, oil |
|    Rice bran–hulls mix | Protein, moisture, fat, fiber |
|    Rice mill feed | Protein, moisture, fat, fiber |
| Oats | |
|    Oats | Protein, moisture oil |
| Corn | |
|    Corn | Protein, lysine, oil, fiber, starch |
|    Corn grits | Moisture, fat |
|    Corn gluten meal | Protein, moisture, oil |
|    Corn flour | Oil |
|    Corn bran | Moisture, fiber, oil |
|    Furafil (corncobs) | Moisture |
|    Corn plant tissue | Nitrogen uptake in plant (to control fertilization) |
|    Corn gluten slurry | Protein (as is and dry basis), solids |
|    Corn starch | Ethylene oxide, propylene oxide, succinic anhydride |
|    Corn starch slurry | Protein (as is and dry basis), solids |
|    Corn germ | Moisture oil |
|    Corn germ expeller cake | Moisture oil |
|    Corn syrups | Dextrose equivalent, total solids |
| Milo | |
|    Sorghum (milo) | Protein, moisture, oil, fiber |

TABLE 6   Oilseeds and Oilseed Product Applications

| Food | Application |
|------|-------------|
| Cottonseed meal | Protein, moisture, oil, fiber |
| Cottonseed expeller cake | Protein, moisture |
| Peanut meal | Protein, moisture, oil, fiber |
| Rapeseed | Protein, moisture, oil |
| Rape meal | Protein, moisture, oil |
| Sunflower seed meal | Protein |
| Soybeans (whole) | Protein, moisture, oil |
| Spent soybean flakes | Oil |
| Soy isolates | Protein, moisture, soluble $N_2$ |
| Soy oil | Iodine value (degree of unsaturation), FFA |
| Soybean meal | Protein, moisture, oil, fiber |
| Margarines | Oil, moisture |

occurring within individual companies and plants often on a proprietary or research basis. Thus the list in Table 7 is meant to be illustrative rather than comprehensive. New applications are and will be determined by economic justification and user creativity (Osborne *et al.*, 1982b; Rotolo, 1978; Allison *et al.*, 1978; Park *et al.*, 1982).

## D.  Dairy Products

The dairy industry, starting with one natural raw material, cow's milk, produces a wide variety of solid and liquid products. Determination of fat, total solids, and protein in milk to AOAC specifications [Association of Official Analytical Chemists (AOAC), 1980] has been demonstrated (Biggs, 1983). These processed dehydrated powders, concentrated liquids, and ready-to-use liquids often require rapid analysis for multiple parameters, of incoming ingredients, process control, and final products. Solid or paste products are analyzed in an open or closed sample cup (Fig. 3), while liquids need only homogenization by passing through the high-pressure homogenizer, which feeds directly into the liquid drawer (Fig. 4) of the spectrophotometer (Fig. 12). In various companies throughout the world additional product applications (Table 8) are being researched and developed for yogurts and hard and soft cheeses, as well as for quantitation of casein in total protein content (McGann, 1978; Frank and Birth, 1981).

## E.  Meats and Meat Products

Applications of NIRA in the area of meats and meat products (Table 9) are actively being pursued at many sites in Europe. Because of the diversity

**TABLE 7   Processed Food Product Applications**

| Food | Application |
|------|-------------|
| Bakery products | |
|     Cake mixes | Fat |
|     Pie doughs | Fat, moisture, protein |
|     Cookie doughs | Fat, moisture |
|     Self-rising flour | Percentage of flour |
|     Pastas | Moisture, number of eggs |
|     Dried eggs (yolk and white) | Moisture, protein, fat |
| Snack foods | |
|     Crackers | Fat, moisture |
|     Corn chips | Fat, moisture |
|     Potato chips | Moisture |
| Ready-to-eat cereals | |
|     All types | Moisture, sugar, protein |
| Chocolate products | |
|     Cocoa beans | Shells, nibs |
|     Powdered mixes | Fat, moisture |
|     Molded chocolate | Fat, moisture |
| Miscellaneous | |
|     Desert mixes | Starch, emulsifiers |
|     Dehydrated fruits and vegetables | Moisture |

**TABLE 8   Dairy Product Applications**

| Food | Application |
|------|-------------|
| Solids and pastes | |
|     Dry milk powder | Moisture, fat |
|     Whey powder | Moisture, protein |
|     Butter | Moisture, fat |
|     Cheddar cheese | Moisture |
| Liquids | |
|     Milk (whole, low fat, skim) | Fat, protein, total solids, lactose |
|     Ice cream mixes | Fat, total solids |
|     Whey liquids | Fat |
|     Cottage cheese | Fat, total solids |
|     Cream cheese | Fat, total solids |

**TABLE 9  Meat and Meat Product Applications**

| Food | Application |
|------|-------------|
| Raw beek | Protein, moisture, fat |
| Raw pork | Protein, moisture, fat |
| Raw chicken | Protein, moisture, fat |
| Combined raw meats | Protein, moisture, fat |
| Sausage meats (e.g., franks and bolognas) | Protein, moisture, fat |
| Hams (cooked) | Protein, moisture |

of products and the perishability of the product it is best developed on site by trained users familiar with meat analysis, because representative sampling is a real problem. There are some challenging research products that suggest themselves in this area, for example, exploring the feasibility for quantitating bound versus free water, as well as discriminating between types of protein tissues (Massie, 1976).

## F.  Feeds and Forages

While feeds and forages are not strictly foodstuffs, their importance in the production of meat and dairy products warrants their inclusion in this discussion. Analysis of feed ingredients (Table 10), to provide the basis for

**TABLE 10  Feed and Forage Applications**

| Feed | Application |
|------|-------------|
| Mixed feeds | |
|   Cow, hog, and poultry | Moisture, protein, oil, fiber |
|   Dry and semimoist pet foods | Protein, moisture, fat, propylene glycol |
| Mixed feed ingredients | |
|   Fish meal | Protein, moisture, oil |
|   Midds | Protein, moisture, oil, fiber |
|   Meat and bone meal | Protein, meat, oil |
|   Grains and oilseed meal | See Tables 5, 6 |
| Forages | |
|   Hay | Protein, fiber |
|   Haylage | Protein, fiber |
|   Silage | Protein, fiber |
|   Grasses and legumes | ADF, moisture, protein |

least-cost formulation or mixed feeds, provides major economic benefits to the user. Analysis of finished mixed feeds is usually undertaken as a quality control or quality assurance measure, ensuring that label claims are met.

Forage analysis will become increasingly important if government marketing standards are established. It is also useful in designing a balanced feeding program for livestock. Research at a number of USDA sites, under the leadership of John Shenk of Penn State University, has indicated the potential of NIRA for forage measurements. On-site usage at the farm and the market has been demonstrated by development of a mobile NIRA laboratory in a small van (Norris *et al.,* 1976; Shenk *et al.,* 1979).

### ACKNOWLEDGMENT

The authors wish to express their appreciation to Karen Luchter of the Technicon Instrument Corporation InfraAlyzer Applications Laboratory for contributions to this chapter. Dr. Luchter prepared much of the material on reference sample sets, sample preparation and presentation, and applications of the NIRA techniques in the food industry, as well as reviewing the entire chapter.

### REFERENCES

Allison, M. J., Cowe, J. A., and McHale, R. (1978). *J. Inst. Brew.* **84,** 153.
Assoc. Off. Anal. Chem. (1980). "Official Methods of Analysis," 13th ed. Assoc. Off. Anal. Chem., Arlington, Virginia.
Biggs, D. A. (1983). *J. Assoc. Off. Anal. Chem.* (in press).
Day, M. S., and Fearn, F. R. B. (1982a). *Lab Pract.* **31,** 328.
Day, M. S., and Fearn, F. R. B. (1982b). *Lab. Pract.* **31,** 429.
Frank, J. F., and Birth, G. S. (1981). *J. Diary Sci.* **65,** 1110.
Goddu, R. F., and Delker, D. A. (1960). *Anal. Chem.* **32,** 140.
Hamid, A., McClure, W. F., and Whitaker, T. B. (1981). *Am. Lab. (Fairfield, Conn.)* **13,** 108.
Hecht, H. G. (1968). "The Present Status of Diffuse Reflectance Theory." Plenum, New York.
Hirschfeld, T., Honig, D., and Hieftje, G. (1983). *Appl. Spectrosc.* (in press).
Honig, D., Hieftje, G., and Hirschfeld, T. (1983). *Appl. Spectrosc.* (in press).
Hunt, W. H., Fulk, D. W., Elder, B., and Norris, K. (1977). *Cereal Foods World* **22,** 534.
Hunt, W. H., Fulk, D. W., Thomas, T., and Nolan, T. (1978). *Cereal Foods World* **23,** 143.
Kaye, W. (1954). *Spectrochim. Acta* **6,** 257.
Kaye, W. (1955). *Spectrochim. Acta* **7,** 181.
Landa, I. (1979). *Rev. Sci. Instrum.,* **50,** 34.
Law, D. P., and Tkachuk, R. (1977). *Cereal Chem.* **54,** 256.
McGann, T. C. A. (1978). *Ir. J. Food Sci. Technol.* **2,** 141.
Martens, H. (1979). *Anal. Chim. Acta.* **112,** 423.
Massie, D. R. (1976). *ASAE Publ.* **1–76.**
Massie, D. R., and Norris, K. H. (1965). *Trans. ASAE* **8,** 598.
Norris, K., and Hart, J. R. (1965). *In* "Humidity and Moisture" (P. N. Win, ed.), p. 19. Van Nostrand-Reinhold, Princeton, New Jersey.

Norris, K. H., Barnes, R. F., Moore, J. E., and Shenk, J. S. (1976). *J. Anim. Sci.* **43**, 889.
Osborne, B. G., Douglas, S., Fearn, T., and Willis, K. H. (1982a). *J. Sci. Food Agric.* **33**, 736.
Osborne, B. G., Douglas, S., and Fearn, T. (1982b). *J. Food Technol.* **17**, 355.
Park, J. W., Anderson, M. J., and Mahoney, A. W. (1982). *J. Food Sci.* **47**, 1558.
Rotolo, P. (1978). *Baker's Dig.* **52**, 24.
Rotolo, P. (1979). *Cereal Foods World* **24**, 94.
Rubenthaler, G. L., and Bruinsma, B. L. (1978). *Crop Sci.* **18**, 1039.
Shenk, J. S., Westerhaus, M. O., and Hoover, M. R. (1979). *J. Diary Sci.* **62**, 807.
Stark, E. W. (1983). "NIRA Instrumentation: State-of-the-Art," Proceedings of the 2nd Annual NIRA Symposium. Technicon, Tarrytown, New York (in press).
Tkachuk, R. (1981). *J. Am. Oil Chem. Soc.* **57**, 819.
Watson, C. A. (1977). *Anal. Chem.* **49**, 835A.
Watson, C. A., Etchevers, G., and Shuey, W. C. (1976). *Cereal Chem.* **53**, 803.
Watson, C. A., Shuey, W. C., Banasik, O. J., and Dick, J. W. (1977). *Cereal Chem.* **54**, 1264.
Wetzel, D. L. (1983). *In* "Instrumental Analysis of Foods," Vol. 1 (G. Charalambous and G. Inglett, eds.), pp. 183–202. Academic Press, New York.
Wheeler, O. H. (1959). *Chem. Rev.* **59**, 629.
Wheeler, O. H. (1960). *J. Chem. Educ.* **37**, 234.
Williams, P. C. (1977). "Proceedings of the 7th Technicon International Congress," Technicon, Tarrytown, New York.
Williams, P. C., and Thompson, B. N. (1978), *Cereal Chem.* **55**, 1014.
Williams, P. C., Stevenson, S. G., Starkey, P. M., and Houtin, G. C. (1978). *J. Sci. Food Agric.* **29**, 285.

# 17

# Applications of Fourier Transform Infrared Spectroscopy in the Field of Foods and Beverages

R. A. SANDERS

*Food Product Development Division*
*Winton Hill Technical Center*
*The Procter and Gamble Company*
*Cincinnati, Ohio*

ANALYSIS OF FOODS AND BEVERAGES

## I. INTRODUCTION

Over the past 10 years, Fourier transform infrared (FTIR) spectroscopy has grown from infancy to maturity as the standard to which other infrared (IR) techniques are compared. This maturing has resulted in the availability of new analytical approaches, such as gas chromatography/infrared (GC/IR) and liquid chromatography/infrared (LC/IR), which are all but impossible without Fourier transform methods. In addition, many other IR sampling techniques, such as attenuated total reflectance (ATR) and diffuse reflectance, can be made more sensitive, flexible, and convenient to use because of Fourier transform methods.

The recent literature reveals only a hint of the potential applications of FTIR in food and beverage analyses. The literature abounds with applications of near-infrared reflectance analysis (NIRA) of food products, because of that technique's advantages for rapid, quantitative analysis on relatively inexpensive instruments with minimal sample preparation. However, there has been a parallel, though less impressive, revival of interest in mid-infrared (mid-IR) quantitative analysis (Wilks, 1979). This is primarily due to IR instruments with dedicated computers that have capabilities (e.g., subtraction, regression analysis) not previously available. Such computerized infrared instruments (both dispersive and interferometric) cost more than $30,000 and, therefore, have not found widespread use in routine, quantitative applications. Because of many of the potential benefits discussed in this chapter, it is likely that FTIR will be judged to be worth the initial capital expense for many applications and that FTIR applications in general, and applications to food and beverage analyses in particular, will steadily increase. Among FTIR instruments of even greater sophistication (and expense), GC/IR shows promise of making an impact in trace analysis of complex flavors that could match the important contributions of gas chromatography/mass spectrometry (GC/MS). Though it is less far along, similar comments apply to LC/IR. As a molecular identification technique for pure compounds, the value of IR is well known, and the on-line coupling of IR to high-resolution chromatography is, perhaps, the most valuable capability that the Fourier transform approach makes possible. Food and beverage matrices are unsurpassed in their complexity. Meaningful measurements in such matrices often require the very best chromatography, and the very best chromatography usually requires limited sample loading. Thus, the dramatic recent progress in capillary GC/IR, microbore LC/IR, and thin-layer chromatography/infrared (TLC/IR) required significant IR sensitivity increases. These have come gradually as a result of better detectors, better optics, improved light-pipe designs, better flow cells, effective diffuse reflectance devices, and many other hardware improvements designed to increase the signal-to-noise ratio (SNR) obtainable on submicrogram quantities of chromatographically isolated components. Because there is little in the current literature concerning use of these techniques in food and beverage analysis, a few relevant nonfoods references are cited here. With only minor variations, such approaches can be used to solve food and beverage problems. Similarly, some dispersive IR applications to foods or beverages are cited when it is believed that FTIR could be used to enlarge or improve the application.

## II. PRINCIPLES OF FTIR

There are several excellent detailed descriptions of interferometry instrumentation and Fourier transform principles (Griffiths and de Haseth, 1983; Griffiths, 1975; Martin, 1980; de Haseth, 1982; Griffiths, 1978a). A very brief description is provided here to help in visualizing the origin of FTIR advantages.

Most FTIR instruments are based on the Michelson interferometer (see Fig. 1). A beam splitter allows approximately half of the source radiation to pass to a movable mirror while the other half is reflected to a fixed mirror. After reflection from the respective mirrors, the two beams are reflected and transmitted at the beam splitter again. The beam reaching the detector consists of approximately equal portions of radiation that traveled the two different paths. The manner in which these two beam intensities add depends upon the difference in path length for the two trajectories and the wavelength of the radiation. At zero path difference, all wavelengths interfere constructively, producing the center burst. As the movable mirror is translated, an interferogram (intensity $I$ versus path difference $\delta$) is produced. Fourier transformation of $I(\delta)$ converts the intensity measurement from the path difference domain to the wavelength $\lambda$ or wave number

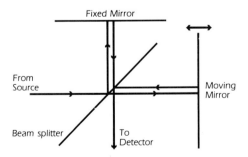

**Fig. 1.** Schematic of a Michelson interferometer.

($\bar{v} = 1/\lambda$) domain of the traditional IR spectrum:

$$I(\bar{v}) = I(\delta)e^{-2\pi i \bar{v} \delta} \, d\delta$$

Consideration of this equation allows one to assess the factors, so different from those in dispersive IR, that limit performance of FTIR instruments. Since it is impossible to digitize the spectrum at infinitely small retardation increments, it is not possible to obtain the spectrum from $0 - \infty$ cm$^{-1}$. It has been shown [Griffiths (1975), Chapter 2 and references therein] that a sampling frequency of twice the bandwidth of the system is sufficient to digitize sinusoidal FTIR waveforms. For a bandwidth of 4000 cm$^{-1}$ and a mirror velocity of 1.2 cm/sec, a digitization frequency of about 20 kHz is required. Such mirror velocities and digitization rates are appropriate for rapid-scan experiments such as GC/IR and are available on commercial FTIR instruments.

Strict use of the equation presented here would also require infinite path difference measurements or infinite mirror travel. The effect of the requirement of a finite mirror travel is to multiply $I(\delta)$ by a rectangular truncation function, which causes ringing or "feet" around the IR band. By replacing the rectangular function with "apodization" functions the ringing can be minimized, but at the cost of resolution. Resolution improves as the length of mirror travel increases, and high-resolution FTIR instruments are those that have solved the alignment and stability problems associated with long mirror travel.

There are many other factors arising from interferometry and Fourier transformation that must be addressed (e.g., phase errors). However, these subjects have been covered in detail in the references cited and are beyond the scope of this work. Suffice it to say that commercial FTIR manufacturers have dealt with such problems to a degree that alows the realization of the significant advantages discussed in Section III.

## III. ADVANTAGES OF FTIR

There are three FTIR advantages that arise from the use of interferometry: the Fellgett (or multiplex) advantage, the Jacquinot (or throughput) advantage, and the Connes (or wavelength accuracy) advantage.

1. Fellgett advantage. A dispersive spectrometer must observe a relatively narrow region of the spectrum at any particular time during a measurement. Thus, the SNR depends on the number $N$ of spectral elements ($N$ equals the bandwidth divided by the resolution). However, an interferometer collects data from all spectral regions simultaneously, throughout the measurement. It can be shown that the Fellgett advantage results in an SNR that is $\sqrt{N}$

times greater for interferometers compared with dispersive instruments for a fixed-time measurement in the mid-IR. Alternatively, if a particular SNR is desired, an interferometer will make the measurement faster by a factor of $N$.

2. Jacquinot advantage. Since an interferometer has no slits, there is no corresponding attenuation of radiation. The actual magnitude of the Jacquinot advantage is not easily determined but is not believed to be as significant as the Fellgett advantage (Griffiths, 1975, p. 205). Together, the Jacquinot and Fellgett advantages make FTIR instruments superior in those situations in which the amount of available sample is minute or when low light transmission through the measurement system limits sample detectability.

3. Connes advantage. Use of an He–Ne laser (which has a convenient emission at 15,800 cm$^{-1}$) to measure $\delta$ allows very accurate measurement of IR absorption frequencies. Measurements to 0.01 cm$^{-1}$ are typical. Thus, the rather tedious frequency calibration procedures of dispersive instruments are not necessary in an FTIR instrument. Additionally, the ability to make accurate frequency measurements makes spectral subtraction a much more useful feature.

Rapid-scan speed is not listed as a separate advantage, because its usefulness is a direct result of the multiplex and throughput advantages.

## IV. INFRARED TECHNIQUES THAT BENEFIT FROM FOURIER TRANSFORM METHODS

### A. Advantages Due to Data System

The need for a dedicated computer on practical FTIR systems to accomplish Fourier transforms, coadd spectra for SNR enhancement, correct for phase errors, apply apodization functions, etc., was obvious from the beginning. The presence of the computer for such necessities has led to its use for luxuries not associated with the basic Fourier transform technique [Koenig (1980,1981) and references therein]. Such capabilities are now available on both FTIR and dispersive instruments and include rapid conversion among absorbance, transmission, and other display formats; background subtraction; factor analysis for qualitative analysis of mixtures (Gillette *et al.*, 1982); regression analysis for quantitative analysis of mixtures (Haaland and Easterling, 1982); digitized IR libraries; library search algorithms; software for examining chromatographic data; and other valuable capabilities too numerous to discuss here.

Perhaps the best-known, and most abused, technique listed above is background subtraction. Absorbance subtractions are often able to reveal minor differences between samples when the samples are analyzed under carefully controlled conditions (Krishnan and Ferraro, 1982). However, one cannot expect to perform a successful subtraction without scaling on samples of widely differing concentrations, or on samples whose spectra were obtained at different thicknesses (e.g., typical film spectra). In addition, subtraction magnifies sampling and photometric artifacts (Koenig, 1981; Hirschfeld, 1978; Anderson and Griffiths, 1978). We have found that subtractions are most likely to be successful when equal-concentration, dilute solutions (maximum analyte absorbances should be less than 0.8 absorbance units) are examined in the same fixed-path-length cell. In this regard, single-beam FTIR instruments are superior to double-beam dispersive instruments in that the necessity for having perfectly matched cells is eliminated. Under the conditions described, an absorbance subtraction will often reveal a "pure" spectrum of a relatively minor contaminant, or it may enhance the sensitivity of a method for measuring a minor ingredient. An example of the latter is a comparison of two methods for measuring the level of fatty acid methyl esters with carbon–carbon double bonds in the trans configuration. Figure 2a is an absorbance spectrum in the trans region for a mixture of 1% methyl elaidate and 99% methyl stearate in carbon disulfide. This spectrum is ratioed to a pure carbon disulfide spectrum. Figure 2b is the result of subtracting a methyl stearate spectrum of equal concentration. In this way, a significant improvement in the detection limit for trans fatty acids can be achieved. For most vegetable oils, methyl stearate is not the ideal choice for the subtraction of interferences: methyl linoleate works

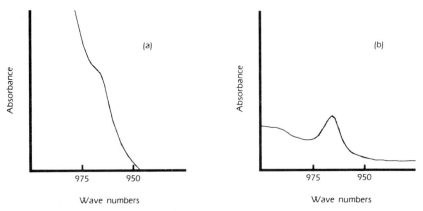

**Fig. 2.**   Absorbance spectra for a mixture of 1% methyl elaidate and 99% methyl stearate in carbon disulfide, before (a) and after (b) subtraction of a methyl stearate spectrum.

better for soybean oil. The perfect blank would be the unachievable: soybean oil with no trans components.

Regression techniques have been thoroughly exploited with NIRA, where the method of locating the "best" bands for quantitating specific components is sometimes called correlation transformation. Though NIRA has many quantitation advantages, the mid-IR is preferable for low-level components and for any analyses for which a complete reference set of samples is not available. Software for multicomponent analyses in the mid-IR is available from several vendors and often includes not only the capability of performing the extensive matrix algebra associated with multicomponent analyses but also the capability of selecting the optimum absorbance bands from the analysis of known mixtures (Antoon et al., 1977). Kisner et al. (1982) used extensive data analysis capability to develop an IR method for triglycerides, phospholipids, and cholesteryl esters in serum. Examples of quantitative mid-IR analyses of foods include determination of fructose and dextrose in corn syrups (Sleeter, 1978); determination of fat, protein, and lactose in milk (Biggs, 1979; Grappin et al., 1980; Robertson et al., 1981; Thomasow and Paschke, 1981); determination of trans fatty acid in vegetable oil (Madison et al., 1982); measurement of extent of fat oxidation (Maksimets, 1979; Lisitskii et al., (1981); determination of fatty oils in the presence of hydrocarbon oils (Kahn et al., 1977; Fukushi and Kondo, 1979); measurement of total unsaturation (Bernard and Sims, 1980; Ahmed and Helal, 1977); determination of individual flavor components of essential oils (Mizrahi and Juarez, 1980); measurement of protein, fat, and carbohydrate levels in meat products (Bjarno, 1981, 1982); and determination of lipids on food contact surfaces (Eugster et al., 1980). Wilks (1979) cites several examples of on-line IR monitoring of product streams in the food-processing industry.

## B. Advantages for Low-Energy Techniques

Fourier transform infrared excels in those analyses that result in low-energy throughput to the detector. Techniques that fall into this category include ATR or multiple internal reflectance (MIR), diffuse reflectance, and GC/IR. Since GC/IR is discussed at length in later sections, only ATR and diffuse reflectance are discussed here.

Attenuated total reflectance is a technique that has long enjoyed great success on dispersive and FTIR instruments alike (Fahrenfort, 1961; Harrick, 1960, 1979). Its use is normally restricted to major components (more than 1%), because the effective path length of the radiation is very short. Fourier transform infrared, in conjunction with ATR, offers the advantages of greater sensitivity for minor components and lower limits of detection for

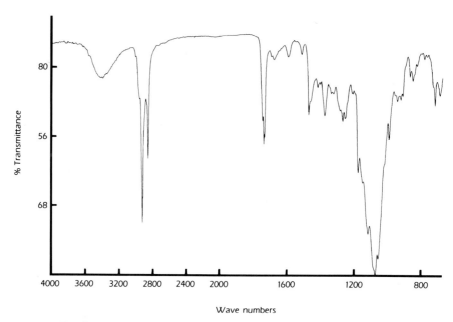

**Fig. 3.** The FTIR/ATR spectrum of a chocolate chip (germanium crystal).

trace components. These advantages are a direct result of the multiplex and throughput benefits of FTIR. Besides providing high-quality spectra in sample-limiting situations, FTIR also allows the user much greater flexibility in analyses in which it is desirable to vary the depth of radiation penetration of the sample. By varying either the refractive index of the substrate crystal or the angle of incidence of the radiation, the penetration depth can be changed. Low depths of penetration tax the sensitivity of dispersive instruments and, thus, can limit such applications. It should be noted that FTIR advantages in ATR applications require full utilization of the circular cross-sectional beam. Attenuated total reflectance crystals that effectively accomplish this have only recently become available (e.g., from Spectra Tech, Inc.*).

Figure 3 is an FTIR/ATR spectrum of a chocolate chip. Use of a germanium crystal (high refractive index, low penetration depth of approximately 0.5 $\mu$m) permitted acquisition of a spectrum of the coating with minimal interference from fat absorptions characteristic of the chocolate. This permitted identification of the hydroxypropyl cellulose coating, even though a spectrum of the bulk sample revealed nothing but fat and sugar.

* 652 Glenbrook Road, Stamford, Connecticut 06906.

An obvious application of ATR is the characterization of food packaging materials. Figure 4a is an FTIR/ATR spectrum of the outside surface of a plastic wrapper. This spectrum was obtained with germanium as the substrate crystal. Figure 4b is the same sample with KRS-5 (penetration depth of approximately 3 $\mu$m). The clear appearance of polypropylene bands in Fig. 4b suggests that we are observing a coating on the surface of polypropylene that is roughly 0.5 – 3 $\mu$m thick. That the major polymeric component was polypropylene was established from a transmission spectrum of a cast film of the sample.

Bruckner *et al.* (1972) and Charalambous *et al.* (1974) demonstrated the value of ATR for characterizing coatings on beer cans and for correlating beer flavor and aroma attributes with coating characteristics such as type of resin, presence or absence of additives, coating thickness, and degree of cure.

Wilks (1979) has enumerated the advantages of ATR for quantitative analyses. These advantages include fixed path length (as long as sample thickness exceeds penetration depth), short effective path length (assuring that absorbances are in the range where Beer's law holds), near immunity from effects of undissolved particles, and elimination of need to dilute viscous samples. We have successfully used ATR to measure the percentage of ethyl vinyl acetate (EVA) in EVA–polypropylene copolymers (from

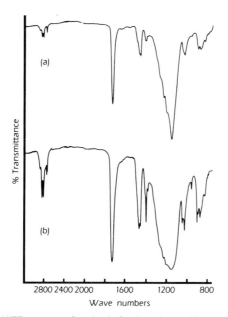

**Fig. 4.** The FTIR/ATR spectra of a plastic food package: (a) germanium crystal and (b) KRS-5 crystal.

0–17% EVA) by measuring the carbonyl (1740 cm$^{-1}$) and CH$_2$ scissors (1460 cm$^{-1}$) band intensities. Quantitation need not be limited to packaging materials, nor is it restricted to cases where band ratios can be conveniently measured. The major requirements are that sample mounting take place without causing misalignment of the ATR crystal and that reproducible, thorough contact with the ATR crystal be attained. Permanently installed, flow-through ATR crystals meet both these requirements for liquid samples and allow the use of ATR for on-stream quality control (Pochon, 1980).

Finn and Varriano-Marston (1981) found that ATR was useful for functional group characterization of food surfaces and suggested that this approach should prove useful for studying surface interactions of starches, fats, and emulsifiers.

For molecular identification purposes it is useful to correct ATR spectra for the wavelength dependence of the penetration depth. Once this is done, ATR spectra generally compare well with transmission spectra and can be used effectively in manual and digitized library searches. For the purpose of identifying unknowns, we often find ATR to be the most convenient technique to apply in a first look. Since the FTIR generally employs only a single beam, alignment (which is greatly aided by a digital display of the interferogram) is very straightforward, and samples can be analyzed nearly as rapidly in this way as by smearing them on a salt plate for transmission, with no concern about excessive absorption.

The diffuse reflectance technique for obtaining IR spectra is a much newer approach than ATR and takes full advantage of the throughput benefits of interferometers (Fuller and Griffiths, 1978, 1980). The sample of interest is thoroughly mixed with a nonabsorbing substrate (usually KCl or KBr) and is exposed to the source radiation, and diffusely reflected radiation is collected by ellipsoidal or paraboloidal mirrors. Because of the relatively low intensity of light reflected from powders, poor SNR has prevented the development of diffuse reflectance in the mid-IR. However, Fuller and Griffiths (1978) showed that the use of efficient collection optics, coupled with a high-sensitivity spectrometer, allows high-SNR diffuse reflectance spectra to be acquired from low-nanogram levels of organic compounds. This makes diffuse reflectance ideal for identifying trace components isolated by thin-layer or liquid chromatography (Kuehl and Griffiths, 1979). But, even more useful, diffuse reflectance is capable of routinely providing high-quality spectra on difficult solid samples with much less preparation effort than the KBr pellet or split mull techniques. In fact, preference for diffuse reflectance has nearly eliminated those techniques from our laboratory.

Photoacoustic spectroscopy (PAS) is another technique that benefits from the multiplex and throughput advantages of FTIR (Vidrine, 1982). The intrinsic insensitivity of the gas microphone detector is partially overcome by the inherent high sensitivity of FTIR. Photoacoustic spectroscopy is applicable to nearly all types of solid and liquid samples, including types that present sample preparation problems for more traditional techniques. The technique's relative insensitivity to surface morphology differences gives it an advantage over KBr pellet and diffuse reflectance approaches for certain samples. Optically opaque samples are also amenable to PAS. The possibility of depth profiling samples by varying the modulation frequency is an attractive one that is still being explored; even now, the fact that PAS samples at a depth intermediate between ATR and diffuse reflectance makes the combination of these techniques very useful for studying compositional changes as a function of depth from the surface.

## C. Advantages Due to Rapid-Scan Capability

Modern spectrometers have the capability of scanning the IR spectrum four times at 8-cm$^{-1}$ resolution, coadding the four interferograms, calculating a number related to the total absorbance (Gram–Schmidt reconstruction), and Fourier transforming the coadded interferogram — all in less than 1 sec. This includes the time it takes to make phase-error corrections and apply apodization functions. Such capability allows one to study a variety of time-dependent phenomena, the most general of which is the detection of components eluting from chromatographic columns. Other time-dependent phenomena that can be studied by FTIR include evolution of gaseous products during pyrolysis, deposition of proteinaceous material from serum onto artificial organ material (Gendreau et al., 1981), and characterization of deformation phenomena in polymers. The latter application requires the measurement of events that happen much faster (e.g., 100 $\mu$sec) than the scan time of even FTIR instruments. This technique is known as time-resolved spectrometry (TRS), and though there are still many problems with practical application of TRS, it shows much promise for following very rapid events by IR (Garrison et al., 1980).

The application of FTIR to the analysis of GC eluents (Erickson, 1979) is a subject that is treated in some depth in the next section. Packed-column GC/IR has been a possibility since the early 1970s; however, practical application of the technique had to await the development of better detectors, better gas cells for GC/IR, and better data systems for processing the vast amount of data generated. Important steps were taken by Azzaraga and McCall (1974) and Azzaraga (1976) in order to alleviate all of these problems. They used a mercury–cadmium–telluride (MCT) detector, developed light-pipe technology, and commissioned the development of GC/IR software, including Gram–Schmidt orthogonalization (de Haseth and Isenhour, 1977) and spectral search capability. Even so, those of us in the food and beverage industry who were looking for means of complementing GC/MS capability with GC/IR as late as 1980 were faced with commercially available GC/IR systems that either were too insensitive for use with capillary columns or caused too much chromatographic resolution degradation. The practical application of such systems to trace components in complex food and beverage matrices was further hampered by data systems that could not do Fourier transforms in real time (thus keeping the user in the dark until after the analysis, when several hours of waiting for transformation of a GC/IR run might be necessary) and could store only a few minutes of GC/IR data on the relatively small disks available. This situation has changed dramatically in the past 3 years, as Section V demonstrates.

The situation with respect to IR detection of components eluting from high-performance liquid chromatographic (HPLC) columns has been summarized by Vidrine (1979), by Kuehl and Griffiths (1979), and by Jinno *et al.* (1982a,b). Approaches to LC/IR can be divided into two groups: those that remove the mobile phase and those that utilize a flow-through cell. If the mobile phase is not removed, IR information is limited to those regions where the solvent does not absorb (or absorbs very weakly), which is extremely restrictive, often prohibitive, when reverse-phase liquid chromatography (LC) with its polar mobile phases is employed. On the other hand, at the time of this writing, there is no commercially available LC/IR accessory that removes the solvent prior to detection, so it is important to understand the potential applications of flow-cell LC/IR. For size exclusion or gel permeation chromatography (GPC), the solvent serves only the purpose of dissolving the sample, and there is, therefore, a high probability of finding a suitable solvent with windows in appropriate regions of the spectrum. Normal-phase HPLC is also amenable to IR detection because the nonpolar solvents most commonly used have broad IR windows. Reverse-phase HPLC presents as many problems for the IR spectrometer as it does for the mass spectrometer. Without solvent removal, reverse-phase LC/MS is limited to chemical ionization using the mobile phase as a reagent, which limits the molecular information obtained to a simple molecular weight determination. In reverse-phase LC/IR, the mobile phase blocks out most, if not all, of the spectrum and cell path lengths are so short (less than 0.1 mm) that sensitivity is poor. This is a result not only of the short path length but also of the fact that the resulting cell volume ensures that less than 1% of each component is present in the cell at any one time. Detection based on the appearance of one or two bands is the best we can hope for: identification of unknowns is a rare result of reverse-phase HPLC/IR with flow-through cells.

For the reasons just discussed, most applications of flow-through cell

LC/IR employ GPC or normal-phase separations. DiCesare and Ettre (1982) applied on-the-flow LC/IR to the detection of diolein, phosphatidylethanolamine, and cholesterol. Shafer *et al.* (1979) demonstrated the usefulness of LC/IR in separating and identifying chloronitrobenzene isomers. If our own laboratory is typical, however, reverse-phase HPLC is the most useful mode for general, organic LC separations. It is not reasonable to request the LC laboratory to develop normal-phase methods only. It is much more reasonable to collect the peaks of interest from reverse-phase analyses manually, remove the mobile phase, and analyze by IR. The advent of diffuse reflectance has made such identifications much more routine owing to that technique's sensitivity to trace components and its ease of implementation. The automation of this procedure (Kuehl and Griffiths, 1979), despite the mechanical complexity, shows promise as an LC/IR technique that will produce full IR spectra suitable for identification purposes. This approach eliminates solvent-related problems, such as background subtraction in solvent-programmed analyses. However, aqueous solvents still constitute a major problem for the solvent elimination system. Normal-phase applications of this technique demonstrate submicrogram sensitivity.

Brown *et al.* (1982) report that microbore (1-mm inside diameter [i.d]) columns have advantages for flow-cell LC/IR. The high resolution of these columns allows a higher percentage of sample in the flow cell than is possible on lower-resolution columns. Since these columns have substantially lower capacities than standard 5-$\mu$m-particle-size columns, it appears that this advantage is realized only in sample-limited analyses. The low flow rates of microbore LC allow the possibility of using deuterated solvents and other expensive solvents that may be more appropriate spectroscopically or chromatographically. These advantages, however, must be balanced against the following disadvantages: low dead volume equipment is required, column quality is poor, and particle sizes smaller than 10 $\mu$m are not available at this time.

## V. CAPILLARY GC/IR

### A. Introduction

The importance of trace amounts of certain volatile components to the flavor of foods and beverages has led to wide application of capillary gas chromatography (GC) in the food and beverage industry. Besides offering the highest resolution of the chromatographic techniques, GC benefits from the flame ionization detector (FID), which provides a sensitive and universal detector for organic molecules. In addition, the technique lends itself to organoleptic aroma profiling via the use of a sniffing port in parallel with the GC detector. After aroma assessment of the relative importance of components in a complex chromatogram, identifications are traditionally made by capillary GC/MS (Horman, 1981; Kolor, 1972, 1980). Interpretable mass spectra of low-nanogram quantities of most volatile organic compounds can be obtained in this way. By interpretable mass spectra we mean that the SNR is good enough that a mass spectral fingerprint, worthy of comparison to reference spectra, is obtained or that reliable mass fragment intensities allow confident manual interpretation procedures to be applied. We have in our files numerous examples of interpretable mass spectra of components that have not been identified. Identification of complete unknowns from their mass spectra alone is often impossible and always tedious if appropriate reference spectra are not available. The need for complementary spectroscopic data has been recognized for many years, and extensive off-line trapping procedures have been used in order to obtain IR and nuclear magnetic resonance (NMR) spectra of chromatographic effluents. The inappropriateness of such time-consuming procedures in a productivity-conscious environment is clear except for the most important components. Capillary GC of food and beverage headspace, extracts, and distillates present hundreds of unknown peaks to the detector. Many of these are judged as important by aroma profiling; many others are found to be significant by statistical analyses of chromatographic data, such as multivariate analysis and factor analysis. The effective and efficient identification of such components requires integration of spectroscopic data from on-line techniques such as GC/MS, GC/IR, and (in the future?) GC/NMR.

Identification of significant flavor contributors can lead to several viable options: synthetic reproduction of an expensive natural flavor, optimization of processing conditions to enhance positive flavors or remove negative ones, improvement of flavor or aroma impact by adding more of key ingredients, and documentation of improved flavor character as a result of patentable process changes. These are important contributions to flavor development efforts. However, it is often the defensive efforts that attract the most attention: identification of components responsible for a product's not meeting specifications and, from the identification, assessment of the steps required to eliminate the problem. Capillary GC/IR is becoming an important part of such contributions in the food and beverage industry.

## B.  Column Requirements for Capillary GC/IR

Capillary GC utilizes a variety of columns; some of these are suitable for GC/IR, while others are not. The most popular capillary columns at present are small-bore (0.20- to 0.25-mm), fused-silica, wall-coated open tubular (WCOT) columns, 10–60 m in length. With the standard film thickness (approximately 0.25 $\mu$m), these columns are almost useless for GC/IR, because they overload at sample loadings (25–50 ng) that are just becoming acceptable for GC/IR. The strongest IR absorbers, such as isobutyl methacrylate, produce acceptable spectra of 5- to 10-ng quantities if eluted rapidly. However, for most flavors, 50- to 100-ng quantities are desirable. If one is to observe both major and minor components in the same GC/IR analysis, it is clear that the dynamic range allowed by these small-bore, fused-silica columns with standard film thickness is prohibitive. The wider-bore (0.32-mm) fused-silica columns and the chemically bonded phases (J & W Scientific, Inc.*) with thicknesses as great as 1.0 $\mu$m go a long way toward alleviating this problem. Such phases have capacities (500 ng/component) that are compatible with GC/IR sensitivity and provide a reasonable dynamic range for GC/IR (and GC/MS, for that matter) analyses. These thick phases are presently available on 0.25-mm and 0.32-mm-i.d. columns and include the following phases: DB-1 (similar to SE-30), DB-5 (SE-54, 52), DB-17 (OV-17), DB-1701 (OV-1701), and DX-1. A bonded phase nearly comparable to Carbowax is offered (DX-4), but only with the standard (0.25-$\mu$m) film thickness. Since steady progress is being made in column coating technology, it is likely that 1.0-$\mu$m-thick films of Carbowax-type phases will soon be available on fused-silica columns.

Glass capillary columns of wider bore size and greater capacity are also useful for GC/IR work. In fact, as discussed below, most capillary GC/IR light-pipe designs are more compatible with the faster flow rates (3–7 ml/min) of 0.5-mm-i.d. columns than the 1 ml/min of 0.2-mm-i.d. columns. Such columns include WCOT and support-coated open tubular (SCOT) columns, which offer capacities in the microgram per component range. The sacrifice is efficiency, since these columns offer fewer plates per meter. However, by using longer columns, the same overall resolution can be obtained as from shorter, narrow-bore columns at the expense of analysis time.

Aroma profiling experts have used long (170-m), wide-bore (0.75-mm), glass capillary WCOT columns for many years in order to achieve wide dynamic range in organoleptic evaluations. These columns offer medium resolution (100,000 plates), great capacity, and more inertness than SCOT columns. The high flow rates used (8–15 ml/min) minimize the effect of light-pipe volume on resolution. Thus, these columns seem to be ideal for

---

* 3871 Security Park Drive, Rancho Cordova, California 95670.

GC/IR. According to the requirements discussed below, such columns require light-pipe volumes that are more in line with packed-column requirements. Thus, laboratories that do a lot of packed and capillary column GC would have the additional benefit of never needing to change light-pipes. Our own GC laboratory uses, almost exclusively, fused-silica capillary columns. Faced with reproducing chromatograms from such columns, we use a low-volume light-pipe that functions well with the fused-silica columns but causes significant sensitivity losses when the very wide bore (0.75-mm) columns are used. However, when high resolution and wide dynamic range are required and analyte levels are relatively high, these columns are useful in spite of the sensitivity losses. One major drawback is the very limited commercial availability of the long, wide-bore columns.

## C. Light-Pipe Volume Considerations

Griffiths (1978b) has dealt extensively with sampling and light-pipe design considerations for GC/IR. One major conclusion from his work is that, for optimum sensitivity, the light-pipe volume should be approximately equal to the half-width volume ($V_{1/2}$) of the GC peak. If the carrier gas flow rate is 2 ml/min and the width at half height is 3 sec, $V_{1/2}$ is 100 $\mu l$ and the optimum light-pipe volume is 100 $\mu l$. These flow and peak-width parameters are fairly typical for the 0.32-mm × 30-m fused-silica columns we commonly use. However, even a 100-$\mu l$ volume causes significant broadening of peaks, so that 2–4 ml/min of makeup gas flow is required to maintain optimal resolution. Since addition of makeup gas causes a significant loss in sensitivity ($V_{1/2}$ no longer equals the light-pipe volume), one must decide whether resolution or sensitivity is the critical parameter and use makeup gas accordingly. The 0.5-mm-i.d. glass capillary columns use flow rates that alleviate the need for makeup gas, and these columns offer the best combination of GC/IR sensitivity and resolution in systems with light-pipe volumes in the 100 $\mu l$ range. However, as indicated earlier, the wide-bore fused-silica columns are so widely used that minor degradation in performance may be tolerated in order to take advantage of their inertness and their convenience and the fact that GC/IR chromatograms can be correlated with GC chromatograms obtained on similar or identical columns.

So, with relatively minor trade-offs, a single light-pipe can accommodate a variety of capillary columns. The alternatives are to use a single type of column for all analyses (with a light-pipe designed for compatibility with this one type of column) or to change light-pipes when different columns are used. The former is probably too restrictive for most laboratories and the latter is probably too inefficient.

## D. Data System Considerations

Years of GC/MS experience have set a challenging standard for GC/IR data systems to match. Gas chromatography/mass spectrometry systems have long had the capabilities of acquiring data at high (200-kHz) digitization rates, displaying a total ionization chromatogram and/or mass spectra in real time, storing several long GC/MS runs on a single disk, massaging data (postrun) with a variety of flexible GC-oriented programs including automated background subtraction and library search routines, and operating in a foreground–background mode that allows simultaneous data acquisition and data massage.

Gas chromatography/infrared faces several data system problems that are absent in GC/MS, not the least of which is the need to carry out coaddition of several interferograms, a Gram–Schmidt computation, phase correction, apodization, Fourier transformation, and display of the current IR spectrum and Gram–Schmidt intensity in less than 1 sec. In this way, real-time data on capillary column peaks may be displayed, and when a run is completed, there is no need to wait for these operations to finish prior to data analysis. One vendor, Digilab, *utilizes a high-speed array processor for these calculations, but other vendors also perform these operations in real time.

Another consideration is available disk storage. A 5-megabyte (MB) disk cartridge can store five or six 1-hr GC/MS analyses, but less than 45 min worth of eight-wave-number GC/IR data even on systems that transform the data in real time and throw away the interferograms. For years, the solution has been to use a peak detection threshold that causes the data system to retain only the scans corresponding to chromatographic peaks. Having used a limited-storage, bench-top GC/MS system in a similar mode, we know that such an approach wastes very valuable information. There are several reasons for retaining all GC/IR scans in addition to the obvious reason that the thresholding method will often miss minor peaks of great interest. The ability to reconstruct frequency window chromatograms and, more importantly, the ability to reconstruct Gram–Schmidt chromatograms with more appropriate basis set vectors require retention of all scans. The benefit of the latter capability is illustrated in Fig. 5. The top chromatogram is the Gram–Schmidt trace that resulted from a basis set chosen immediately before the run began; the bottom trace is the result of a reconstruction using scans denoted by the arrow as the basis set. It is clear that Gram–Schmidt reconstruction can reveal the presence of components that would otherwise be missed and greatly facilitates the correlation of GC and GC/IR data. The availability of large storage modules alleviates the

* Division of Bio-Rad, 237 Putnam Avenue, Cambridge, Massachusetts 02139.

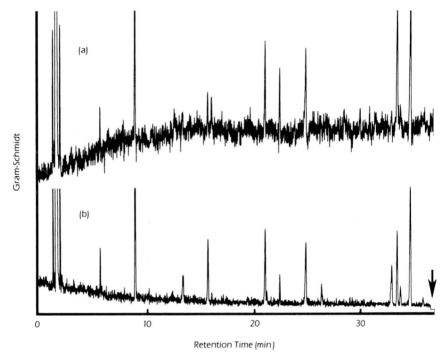

**Fig. 5.** Gram–Schmidt reconstructions using basis set vectors acquired (a) before injection and (b) at the retention time indicated by the arrow.

need to discard GC/IR data and is almost essential for making GC/IR a routinely applicable technique.

The large disks are valuable for other reasons as well. Digitized IR libraries are now available from Sadtler* and Aldrich. † These libraries, in conjunction with effective search algorithms from FTIR vendors, are an extremely valuable molecular identification tool, as will be demonstrated. The size of these libraries will eventually exceed 100,000 compounds, and efficient use of them will require large storage modules and means for accomplishing rapid searches (e.g., array processors). Sprouse (1982) revealed that a 5-MB disk will accommodate only 5770 four-wave-number spectra or 10,730 eight-wave-number spectra. It is clear that effective use of these growing libraries will require substantial disk storage or effective means of compressing IR or combined IR and MS libraries (Williams *et al.,* 1982; de Haseth and Leclerc, 1982).

A laboratory expecting to do extensive GC/IR analyses should also

---

* Sadtler Research Laboratories, Inc., 3316 Spring Garden Street, Philadelphia, Pennsylvania 19104.

† Aldrich Chemical Company, Inc., 940 W. St. Paul Avenue, Milwaukee, Wisconsin 53233.

consider the benefits of multiple grounds. An additional terminal on a mapped system with appropriate additional memory allows previously acquired data to be examined while data acquisition is occurring in the other ground. The time savings involved are substantial if several long GC/IR analyses are performed each day. Depending on how the data acquisition priorities are designed, extensive computations in one ground may cause scan delays during data collection in the other ground. Such delays may or may not be important depending on the time resolution required for the analysis.

Commercially available software permits a host of manual and automated, postrun data analysis techniques. These include graphics-oriented routines that permit thorough examination of the Gram–Schmidt chromatogram for selection of transformed scan sets to coadd, selection of appropriate background scans for subtraction, selection of appropriate basis vector scans for Gram–Schmidt reconstructions, and selection of peaks for which IR spectra are desired and for which library searches are desired.

### E. Comparison with GC/MS

Experience in the use of powerful GC/MS and GC/IR instruments to solve practical food and beverage problems has led to several conclusions concerning their relative merits. The most important conclusions are that the two techniques are powerful, reliable, and usually complementary to each other.

The systems on which this comparison is based are shown schematically in Fig. 6. The Kratos MS-30 is a medium-resolution ($M/\Delta M = 20,000$ maximum) mass spectrometer that has the unique capability of routinely making accurate mass measurements (typically within 15 ppm) on components eluting from capillary columns. The MS-30 is interfaced to a Data General Nova 3 computer with 64K words of mapped memory, 10-MB disk drive, two visual display units (VDUs), and one hard copy unit. At the time of this writing, the Nova 3 is being interfaced to a CDC Model 9766 300-MB storage module. The 300-MB drive is also interfaced to the Data General Nova 4X computer, which controls the operation of a Digilab FTS-15E capillary GC/FTIR system. This data system has 128K words of mapped memory, a fast array processor, two visual display units, an additional 10-MB disk drive, a magnetic tape unit for data archiving, two printers, and a plotter. The storage module alleviates data storage concerns for both systems, contains a 38,000-compound National Bureau of Standards mass spectral library and the 7100-compound Vap-IR library (from Sadtler), allows transfer of mass spectral data to magnetic tape for archiving, and permits the use of spectral search algorithms that use both IR and MS data in a combined search.

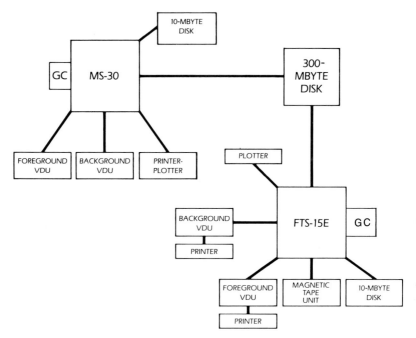

**Fig. 6.** Schematic of GC/MS and GC/IR systems.

This instrumentation allows one to compare attributes of GC/IR and GC/MS on instruments that are close to state of the art in the analyses for which they were designed. The MS-30 is less sensitive, by a factor of 10, than some currently available mass spectrometers, but in its mode of greatest usefulness (providing accurate mass data during capillary GC/MS) it is unsurpassed in sensitivity because of its unique ability to provide such data at relatively low mass resolution. Figure 7 is a GC/MS total ionization chromatogram of a mixture of eight common flavors obtained at a mass resolution of about 1200 and time resolution of 2.7 sec/scan. Each peak represents 20 ng on column; this GC/MS system produces interpretable spectra of such compounds to levels of about 2 ng eluting in less than 25 min from 0.31-mm × 30-m fused-silica columns. Figure 8a displays comparable data from a 38-ng/component injection on the GC/IR system. Figure 8b is the response of the GC FID located in series with and after the light-pipe.

All eight components are visible in the Gram–Schmidt chromatogram (Fig. 8a) although the p-cymene peak is barely distinguishable from noise. Figure 9 shows library search results for the p-cymene peak demonstrating that 38 ng of a poor absorber gives a spectrum that is "interpretable." This particular library search ignores frequencies less than 700 cm⁻¹, between 2100 cm⁻¹ and 2700 cm⁻¹, and greater and 3250 cm⁻¹. Elimination of these

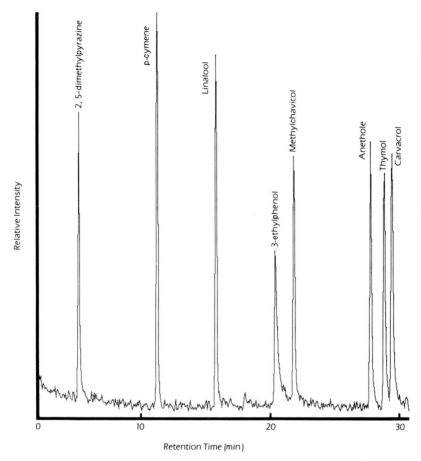

**Fig. 7.** A capillary GC/MS chromatogram of a flavor mixture (approximately 20 ng/component).

regions allows use of the fast array processor, which greatly speeds library searches. The regions below 700 cm$^{-1}$ and above 3250 cm$^{-1}$ are regions of poor response for the MCT detector used, and the 2100- to 2700-cm$^{-1}$ region is rarely informative. Figure 10 displays library search results for the 3-ethylphenol peak, demonstrating that 40 ng of a moderately strong absorber produces a high-quality IR spectrum.

Several conclusions are obvious from mere inspection of the chromatograms in Figs. 7 and 8. Compared to FID and MS, IR responses vary widely from molecule to molecule. This fact makes it difficult to compare IR and MS sensitivities although these data suggest that GC/MS sensitivity exceeds GC/IR sensitivity by about a factor of 20 for poor IR absorbers or a factor of

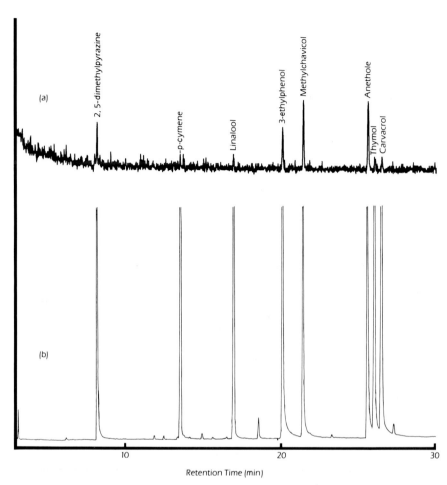

**Fig. 8.** (a) A capillary GC/IR chromatogram of a flavor mixture (38 ng/component). (b) Simultaneous response of the FID to the same injection.

4 for good absorbers such as 3-ethylphenol. Again, these analyses used conditions that we consider to be a practical compromise between chromatographic resolution and IR sensitivity. Elimination of makeup gas could significantly improve the GC/IR sensitivity.

Mass spectrometry sensitivity must be discussed with reference to two modes of operation. The scanning mode is used for identification purposes and is the mode assumed in the foregoing. If the need is not identification but detection or measurement of known compounds, selected ion monitoring (SIM) may be used to provide limits of detection 100–1000 times lower

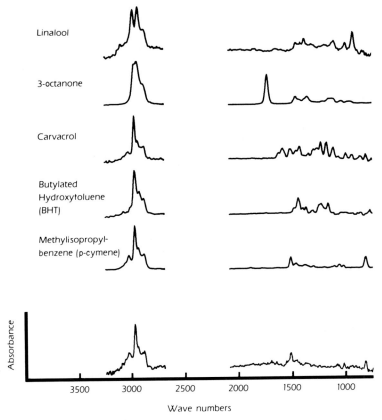

**Fig. 9.** Library search results for the 38-ng *p*-cymene peak. The bottom spectrum is that obtained from the flavor mixture; the five spectra above are the best library matches. The library used in this example and in Fig. 10 is a small in-house one.

than in the scanning mode. There is nothing comparable to SIM in FTIR since all frequencies are measured simultaneously. And even the procedure of reconstructing narrow window absorption ranges does not improve the SNR over what is available from the Gram–Schmidt chromatogram. Thus, in the target molecule detection mode, GC/MS has an advantage of 3–4 orders of magnitude.

Much has been said about the classical situations in which either IR or MS solves a problem easily that the other could not. For example, distinguishing species that differ only in aromatic substitution pattern is easy by IR and not by MS; distinguishing aliphatic homologs is easy by MS and not by IR. Another such example is the use of GC/IR to distinguish the cis and trans

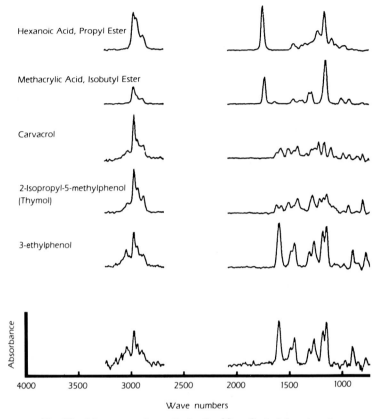

Hexanoic Acid, Propyl Ester

Methacrylic Acid, Isobutyl Ester

Carvacrol

2-Isopropyl-5-methylphenol
(Thymol)

3-ethylphenol

Absorbance

4000     3500     3000     2500     2000     1500     1000

Wave numbers

**Fig. 10.** Library search results for the 38-ng 3-ethylphenol peak.

isomers of cyclododecene. Separation was accomplished as shown in Fig. 11 using a DB-5 wall-coated, fused-silica column. The IR absorbance spectra of the two isomers are shown in Fig. 12. The out-of-plane hydrogen deformation at $975-cm^{-1}$ clearly pinpoints the first peak as the trans isomer. The mass spectra of these compounds are virtually identical, and even if they were reproducibly different, the differences would not offer any unambiguous clues to the structural feature causing the differences. This brings up an advantage of IR over MS: IR absorption bands are more directly relatable to structural features of the molecule than are mass spectral fragments. Certainly, functional-group information is much more readily available from an IR spectrum. In spite of recent progress in metastable ion scanning techniques and in the use of functional-group-specific chemical ionization reagents, GC/MS approaches are not nearly as effective as GC/IR when it

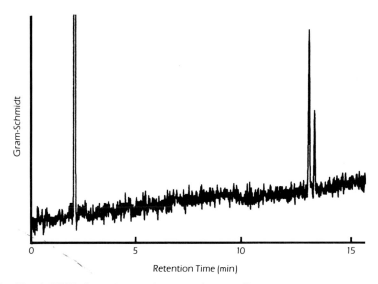

**Fig. 11.** A GC/IR chromatogram demonstrating a capillary column separation of cis and trans isomers of cyclododecene.

comes to delineation of the functional-group makeup of a molecule. On the other hand, MS has the capability of providing perhaps the most valuable piece of information about the molecule — its molecular formula. A molecular formula plus an IR spectrum is a very powerful combination for identification purposes. Throw in a few characteristic mass fragments and effective library searches, and most (but not all) peaks eluting from a GC can be identified.

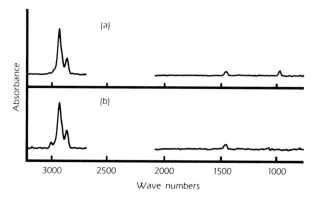

**Fig. 12.** The IR absorbance spectra of the first peak (a) and the second peak (b) in Fig. 11. The band at 975 $cm^{-1}$ indicates that (a) is *trans*-cyclododecene.

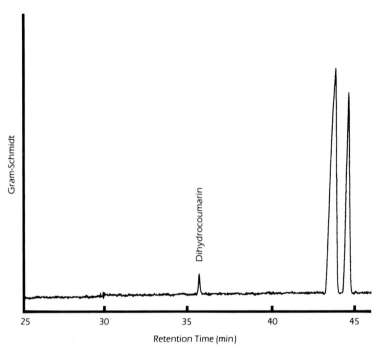

**Fig. 13.** A GC/IR chromatogram of a commercial vanilla flavor. Frontal tailing of the major (vanillin and ethyl vanillin) peaks is symptomatic of column overload.

Figure 13, a Gram–Schmidt chromatogram of a simple vanilla flavor, is shown to illustrate several features just discussed. The column used was a narrow-bore, fused-silica one, and it clearly does not have the dynamic range required to observe the minor (first) peak and the major peaks without resolution degradation. The frontal tails followed by sharp drops to the baseline are characteristic symptoms of exceeding the capacity of the column. For such a simple mixture, this presents no problems, but for more typical food and beverage matrices such overloading often obscures important components. This sample also provides an example of MS and IR complementarity. The small peak, dihydrocoumarin, was retrieved as the seventh-best match by the MS library search. There were several very similar mass spectra, but only one of them was a six-membered ring lactone, which the carbonyl absorption frequency said it must be. In fact, the IR library search retrieved only lactones, while the MS search retrieved a variety of functional groups. In our experience, it has been generally true that mass spectral library searches have been more effective in terms of retrieving compounds present in the library (the MS libraries are bigger than IR vapor-phase libraries and the search algorithms are excellent), but IR searches are much more likely to retrieve compounds of a functional-group nature similar to the unknown. For compounds not present in either library, IR searches are more useful. In the past 5 years, the value of automated library searches (MS and IR) has progressed from near uselessness to great value as larger and higher-quality libraries have been collected and better search algorithms have been developed. Approximately half the identifications made in our laboratory require a staff person only to review the results of a successful search.

Since MS and IR are so useful together, it is reasonable to consider connecting them. There are two approaches to GC/IR/MS: the parallel approach (Wilkins *et al.*, 1981; Crawford *et al.*, 1982), which utilizes a splitter to send 90% (typically) of the sample to the FTIR and 10% to the MS, and the series approach (Shafer *et al.*, 1981b; Wilkins *et al.*, 1982), which takes advantage of the nondestructiveness of IR by sending the effluent from the IR light-pipe to the mass spectrometer. The parallel approach suffers the obvious disadvantage of decreasing GC/MS sensitivity by a factor of 10, a luxury one can ill afford in trace analyses by GC/MS. The latter approach seems more reasonable; however, as mentioned earlier, even the low-volume light-pipes require about 4–6 ml/min of gas flow to avoid chromatographic band broadening. This flow is greater than most mass spectrometers can handle directly. One is then faced with splitting out 50–85% of the sample or letting the chromatographic resolution degrade by lowering the flow rate or using a jet separator with additional makeup gas as the IR/MS interface. The last option, which has not been reported for capillary GC/IR/MS, is one that has been used successfully in GC/MS with wide-bore capillary columns and may be the most practical solution. Wilkins *et al.* (1982) and Crawford *et al.* (1982) report greater success with the parallel approach, because it preserves chromatographic resolution and presents fewer plumbing problems.

The advantages of combining GC, IR, and MS are the following: sample and time savings by accomplishing both GC/IR and GC/MS in a single analysis and greater reliability in correlating peaks in GC/IR and GC/MS. The former advantage must be weighed against the cost of losing the freedom to use the GC/IR for the solution of one problem while using the GC/MS for another. The second advantage is not crucial if the GC/IR data is accompanied by the corresponding GC/FID chromatogram as described earlier. GC/MS data normally correlate very well with GC/FID data, which correlates one to one with the GC/IR data. For these reasons, in spite of the nearness of our GC/MS and GC/IR systems, we have decided that, for our particular analyses, the coupling of IR and MS is not worth the performance degradation incurred.

Given the immense value of knowing the molecular formula, it would seem wise for the mass spectrometers in GC/IR/MS systems to have the capability of measuring masses accurately (with less than 15-ppm error) for components eluting from capillary columns.

Some recent GC/IR references are cited (Kuehl *et al.,* 1980; Shafer *et al.,* 1981a; Azzaraga and Potter, 1981; Rossiter, 1982; Imre *et al.,* 1982). Some older references discuss flavor analysis using packed-column GC/IR (Low, 1971; Low *et al.,* 1974) or effluent trapping followed by IR (Shankaranarayana *et al.,* 1975). Capillary GC/IR flavor analyses are a largely untapped resource, but this situation should change in the near future.

## VI. FUTURE DEVELOPMENTS IN FTIR

Despite the major accomplishments of the last few years in FTIR, there is still much room to grow. Some expected developments are summarized briefly below.

1. In GC/IR, Hirschfeld (1982) expects a move to smaller detectors that are faster, easier to cool (no liquid nitrogen), and cheaper and have higher response and less noise than the standard MCT detectors currently being used for GC/IR. Such a move requires high-quality mirrors and elimination of aberration and alignment problems. He also has speculated that pressure-tunable lead selenide diode lasers might greatly improve GC/IR sensitivity (to a theoretical detection limit of 1 pg). There is probably more progress forthcoming in the design of low-volume, high-transmission light-pipes that will provide optimal chromatographic resolution without loss of sensitivity. Regardless of the development of these improvements, there is little doubt that GC/IR will become a widely used tool in food and beverage analysis.

2. In LC/IR, microbore columns (1-mm i.d.) are currently of unpredictable quality and are hindered by their incompatibility with particle sizes smaller than 10 $\mu$m. The technique also requires special pumps and very low volume plumbing. Some progress in correcting the column problems may be expected. In addition, 2-mm-i.d. columns have recently become available and show promise of providing the advantages of high analyte concentration and reduced solvent usage without the need for special LC equipment. Also, capillaries (100-$\mu$m i.d.) show similar promise (their extremely low flows are attractive for LC/MS), but they require special LC instrumentation not currently in widespread use. The hope is that one or more of these techniques will provide high enough analyte concentration to permit flow-through cells of very short path length, thus minimizing regions of solvent opacity. At least one vendor is pursuing the liquid chromatography/diffuse

reflectance/infrared approach of Griffiths, so that solution of the substantial mechanical problems of such an approach is a possibility on a commercially available system.

3. In general, the use of diffuse reflectance FTIR for the identification of chromatographically isolated trace food components is bound to increase dramatically. The study of food surface interactions using ATR/FTIR is a potentially valuable technique and one planned for implementation in our laboratory. Phenomena such as flavor binding to food substrates will be studied by FTIR. The use of small, relatively inexpensive FTIR instruments for quantitative multicomponent analyses for quality control is an area that will probably have widespread applications in the food and beverage industry.

Krishnan and Ferraro (1982) describe a host of FTIR techniques that are readily available on commercial FTIR spectrometers. These techniques make IR spectroscopy applicable to a much greater variety of problems than previously. The diamond anvil cell, for example, is a very useful microsampling device that makes tiny, opaque specks amenable to IR analysis. With the exception of ATR, the techniques described are quite new developments and are not widely used in the food and beverage industry. In the next several years, we should witness a great increase in diffuse reflectance, PAS, relection–absorption, and diamond cell applications in conjunction with FTIR.

## ACKNOWLEDGMENT

I wish to thank Dr. Curtis Marcott and Mr. Al Fehl for their helpful review of this manuscript.

## REFERENCES

Ahmed, N. S., and Helal, F. R. (1977). *Milchwissenschaft* **32**, 272–273.
Anderson, R. J., and Griffiths, P. R. (1978). *Anal. Chem.* **50**, 1804.
Antoon, M. K., Koenig, J. H., and Koenig, J. L. (1977). *Appl. Spectrosc.* **31**, 518.
Azzaraga, L. V., and McCall, A. C. (1974). EPA 660/1-73-034. U.S. Environ. Prot. Agency, Environ. Res. Lab., Athens, Georgia.
Azzaraga, L. V., and Potter, C. A. (1981). *HRC CC J. High Res. Chromatogr. Chromatogr. Commun.* **4**, 60–69.
Azzaraga, L. V. (1976). *Pittsburgh Conf. Anal. Chem. Appl. Spectrosc.* Pap. No. 334.
Bernard, J. L., and Sims, L. G. (1980). *Ind./Res. Dev.* **22**, 81–83.
Biggs, D. A. (1979). *J. Assoc. Off. Anal. Chem.* **62**, 1211–1214.
Bjarno, O. (1981). *J. Assoc. Off. Anal. Chem.* **64**, 1392–1396.
Bjarno, O. (1982). *J. Assoc. Off. Anal. Chem.* **65**, 696–700.

Brown, R. S., Johnson, C. C., and Taylor, L. T. (1982). *Am. Chem. Soc. Natl. Meet.* Pap. No. 58.

Bruckner, K. J., Charalambous, G., and Hardwick, W. A. (1972). *Tech. Q. Master Brew. Assoc. Am.* **9,** 47–53.

Charalambous, G., Bruckner, K. J., and Hardwick, W. A. (1974). *Tech. Q. Master Brew. Assoc. Am.* **2,** 26–30.

Crawford, R. W., Hirschfeld, T., Sanborn, R. H., and Wong, C. M. (1982). *Anal. Chem.* **54,** 817–820.

de Haseth, J. A. (1982). *In* "Fourier, Hadamard, and Hilbert Transforms in Chemistry" (A. G. Marshall, ed.), pp. 387–420. Plenum, New York.

de Haseth, J. A., and Isenhour, T. L. (1977). *J. Chromatogr. Sci.* **49,** 1977–1981.

de Haseth, J. A., and Leclerc, D. F. (1982). *9th Annu. Meet. Fed. Anal. Chem. Spectrosc. Soc.* Pap. No. 496.

DiCesare, J. L., and Ettre, L. S. (1982). *J. Chromatogr.* **251,** 1–16.

Erickson, M. D. (1979). *Appl. Spectrosc. Rev.* **15,** 261–325.

Eugster, K. E., Skura, B. J., and Powrie, W. D. (1980). *J. Food Prot.* **43,** 447–449, 464.

Fahrenfort, J. (1961). *Spectrochim. Acta* **17,** 698.

Finn, J. W., and Varriano-Marston, E. (1981). *J. Agric. Food Chem.* **29,** 344–348.

Fukushi, K., and Kondo, G. (1979). *Kobe Shosen Daigaku Kiyo, Dai-2-Rui* **27,** 207–212.

Fuller, M. P., and Griffiths, P. R. (1978). *Anal. Chem.* **50,** 1906–1910.

Fuller, M. P., and Griffiths, P. R. (1980). *Appl. Spectrosc.* **34,** 533–539.

Garrison, A. A., Crocombe, R. A., Mamantov, G., and de Haseth, J. A. (1980). *Appl. Spectrosc.* **34,** 399–404.

Gendreau, R. M., Winters, S., Leininger, R. I., Fink, D., Hassler, C. R., and Jakobsen, R. J. (1981). *Appl. Spectrosc.* **35,** 353–357.

Gillette, P. C., Lando, J. B., and Koenig, J. L. (1982). *Appl. Spectrosc.* **36,** 661–665.

Grappin, R., Packard, V. S., and Ginn, R. E. (1980). *J. Food Prot.* **43,** 374–380.

Griffiths, P. R. (1975). "Chemical Infrared Fourier Transform Spectroscopy." Wiley, New York.

Griffiths, P. R., ed. (1978a). "Transform Techniques in Chemistry." Plenum, New York.

Griffiths, P. R. (1978b). *In* "Fourier Transform Infrared Spectroscopy" (J. R. Ferraro and L. J. Basile, eds.), Vol. 1, pp. 143–168. Academic Press, New York.

Griffiths, P. R., and de Haseth, J. A. (1983). "Chemical Infrared Fourier Transform Spectroscopy," 2nd ed., Wiley, New York.

Haaland, D. M., and Easterling, R. G. (1982). *Appl. Spectrosc.* **36,** 665–673.

Harrick, N. J. (1960). *J. Phys. Chem.* **64,** 110.

Harrick, N. J. (1979). "Internal Reflection Spectroscopy." Harrick Sci. Corp., Ossining, New York.

Hirschfeld, T. (1978). *Appl. Opt.* **17,** 1400.

Hirschfeld, T. (1982). *9th Annu. Meet. Fed. Anal. Chem. Spectrosc. Soc.* Pap. Nos. 32 and 36.

Horman, I. (1981). *Biomed. Mass Spectrom.* **8,** 384–389.

Imre, L., Danoczy, E., Jalsolvszky, G., and Holly, S. (1982). *J. Mol. Struct.* **79,** 35–38.

Jinno, K., Fujimoto, C., and Ishii, D. (1982a). *J. Chromatogr.* **239,** 625–632.

Jinno, K., Fujimoto, C., and Hirata, Y. (1982b). *Appl. Spectrosc.* **36,** 67–69.

Kahn, L., Dudenbostel, B. F., Speis, D. N., and Karras, G. (1977). *Am. Lab. (Fairfield, Conn.)* **9,** 61–66.

Kisner, H. J., Brown, C. W., and Kavarnos, G. J. (1982). *Anal. Chem.* **54,** 1479–1485.

Koenig, J. L. (1980). *NATO Adv. Study Inst. Ser., Ser. C* **57,** 79–88.

Koenig, J. L. (1981). *Acc. Chem. Res.* **14,** 171–178.

Kolor, M. G. (1972). *In* "Biochemical Applications of Mass Spectrometry" (G. R. Waller, ed.), pp. 701–722. Wiley (Interscience), New York.

Kolor, M. G. (1980). *In* "Biochemical Applications of Mass Spectrometry, First Supplementary Volume" (G. R. Waller, ed.), pp. 821–857. Wiley (Interscience), New York.

Krishnan, K., and Ferraro, J. R. (1982). *In* "Fourier Transform Infrared Spectroscopy" (J. R. Ferraro and L. J. Basile, eds.), Vol. 3, pp. 149–209. Academic Press, New York.

Kuehl, D., and Griffiths, P. R. (1979). *J. Chromatogr. Sci.* **17**, 471–476.

Kuehl, D., Kemeny, G. J., and Griffiths, P. R. (1980). *Appl. Spectrosc.* **34**, 222–224.

Lisitskii, V. V., Taran, A. A., and Burtsev, V. (1981). *Rybn. Khoz.* **2**, 77–79.

Low, M. J. D. (1971). *J. Agric. Food Chem.* **19**, 1124–1127.

Low, M. J. D., Mark, H., and Chan, S. H. (1974). *J. Agric. Food Chem.* **22**, 488–490.

Madison, B. L., DePalma, R. A., D'Alonzo, R. P. (1982). *J. Am. Oil Chem. Soc.* **59**, 178–181.

Maksimets, V. P. (1979). *Izv. Vyssh. Uchebn. Zaved., Pishch. Technol.* **4**, 35–41.

Martin, A. E. (1980). *In* "Vibrational Spectra and Structure. VIII. Infrared Interferometric Spectrometers" (J. R. Durig, ed.), Elsevier, Amsterdam.

Mizrahi, I., and Juarez, M. A. (1980). *Riv. Ital. E.P.P.O.S.* **62**, 245–247.

Pochon, M. (1980). *Chimia* **34**, 385–396.

Robertson, N. H., Dixon, A., Nowers, J. H., and Brink, D. P. S. (1981). *S. Afr. J. Dairy Technol.* **13**, 3–6.

Rossiter, V. (1982). *Am. Lab. (Fairfield, Conn.)* **14**, 71–79.

Shafer, K. H., Lucas, S. V., and Jakobsen, R. J. (1979). *J. Chromatogr. Sci.* **17**, 464–470.

Shafer, K. H., Cooke, M., DeRoos, F., Jakobsen, R. J., Rosario, O., and Mulik, J. D. (1981a). *Appl. Spectrosc.* **35**, 469–472.

Shafer, K. H., Hayes, T. L., and Tabor, J. E. (1981b). *Proc. Soc. Photo-Opt. Instrum. Eng.* No. 289, pp. 160–164.

Shankaranarayana, M. L., Abraham, K. O., Raghavan, B., and Natarajan, C. P. (1975). *CRC Crit. Rev. Food Sci. Nutr.* **6**, 271–315.

Sleeter, R. T. (1978). U.S. Patent 4102646..

Sprouse, J. (1982). *9th Annu. Meet. Fed. Anal. Chem. Spectrosc. Soc. Pap.* No. 499.

Thomasow, J., and Paschke, M. (1981). *Milchwissenschaft* **36**, 65–68.

Vidrine, D. W. (1979). *In* "Fourier Transform Infrared Spectroscopy" (J. R. Ferraro and L. J. Basile, eds.), Vol. 2, pp. 129–164. Academic Press, New York.

Vidrine, D. W. (1982). *In* "Fourier Transform Infrared Spectroscopy" (J. R. Ferraro and L. J. Basile, eds.), Vol. 3, pp. 125–148. Academic Press, New York.

Wilkins, C. L., Giss, G. N., Brissey, G. M., and Steiner, S. (1981). *Anal. Chem.* **53**, 113–117.

Wilkins, C. L., Giss, G. N., White, R. L., Brissey, G. M., and Onyiriuka, E. C. (1982). *Anal. Chem.* **54**, 2260–2264.

Wilks, P. A. (1979). *Am. Lab. (Fairfeld, Conn.)* **11**, 69–77.

Williams, S. S., Lam, R. B., Sparks, D. T., Isenhour, T. L., and Hass, J. R. (1982). *Anal. Chim. Acta* **138**, 1–10.

# 18

# Quantitative Quality Control of Foods and Beverages by Laser Light-Scattering Techniques

PHILIP J. WYATT

*Wyatt Technology Company*
*Santa Barbara, California*

## I. INTRODUCTION AND LIGHT-SCATTERING FUNDAMENTALS

All visually sensed qualities of foods and beverages are in reality determined by the observation of light scattered by the substance. When we say that a particular beer, for example, looks cloudy, we really mean that the light passing through it has been scattered by some of the entrained particulates to a degree sufficient for them to be observed. A wine exhibiting a brilliant red color as it is poured has a high concentration of particulates that absorb green, orange, and blue strongly while scattering red. Naturally, there are various types of optical instruments available to make measurements of such phenomena more sensitive — far more than they are to the naked eye. With such instrumentation, the effects of molecular scattering and, under suitable conditions, various polarization phenomena may also be observed and quantified. The soluble fraction of many types of processed foods is amenable to similar optical studies and measurements. Indeed, optical techniques have long been a major tool in the armamentarium of the food and beverage industry, both for process control and for analytical research. Unfortunately, with few exceptions, the use and interpretation of optical instrumentation results are often faulty and erroneous.

In this chapter, we shall emphasize the use of laser light-scattering techniques for the quality control of beverages and foods. The qualities so measured include features not generally associated with visual effects, such as taste and the detection of adulterants and other impurities, toxicants, and general product formulation consistency. First, however, we must review some basic concepts and why they are misunderstood and, therefore, often misinterpreted.

The discussions and derivations are very elementary but may not be familiar to the reader. If one is to become involved seriously with optical instrumentation, especially as it is used to ensure food quality and purity, an understanding of the underlying physical principles of the instruments so used is essential. Such an understanding ensures that the instruments will not often be used beyond their intended range and that, if they are, the user will be aware of possible ambiguities and be able to recognize problems before they have resulted in tragic and/or financially catastrophic consequences.

## A. Turbidity

Consider a transparent liquid, such as water, whose constituent molecules scatter a negligibly small amount of the light incident on them. All particles in such a liquid, *including* the molecules of the liquid itself, will scatter some of the incident light, but we shall assume that the molecular contribution to the total scattering is small. Now assume that suspended in this liquid are $n$ identical particles per cubic meter, each being spherical, homogeneous, and nonabsorbing. If the radius of each sphere is $R$, then the projected area $A$ of the sphere is just $\pi R^2$. Let us also assume that this suspension of particles is

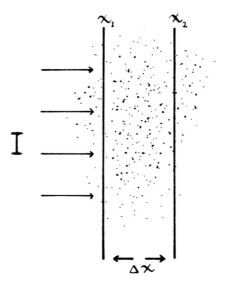

**Fig. 1.** The intensity of light $I$ changes by an amount $\Delta I$ as it passes through a thickness $x_2 - x_1 = \Delta x$ of the sample containing $n$ uniform spheres per unit volume.

uniformly illuminated by a beam of monochromatic light of wavelength $\lambda$. In passing through a thickness of suspended particles $\Delta x = x_2 - x_1$, the intensity of the transmitted light is changed by an amount $\Delta I$ (Fig. 1). This *change* of light intensity is certainly proportional to the number of particles in the beam between $x_1$ and $x_2$, namely, $n(x_2 - x_1) = n\Delta x$, since the amount of light scattered must increase as $n$ increases and as the path $\Delta x$ increases. Also, the total amount of light scattered by the particles in the region $\Delta x$ must be proportional to the intensity of light incident at $x_1$. Hence we write in general

$$\Delta I = -KIn\,\Delta x \tag{1}$$

where $K$ is a constant. Note the minus sign: it indicates that the particles *remove* light from the incident beam by scattering; that is, the change of $I$ is always decreasing. The constant $K$ has yet to be determined. We now rewrite Eq. (1) in the limit as $\Delta x \to 0$, that is, as a differential equation:

$$\lim_{\Delta x \to 0} \frac{\Delta I}{\Delta x} = \frac{dI}{dx} = -KIn \tag{2}$$

or

$$dI/I = -Kn\,dx \tag{3}$$

Integrating both sides yields

$$\int_{I_0}^{I} dI/I = -Kn \int_0^x dx \tag{4}$$

where the integration limits are obtained by noting that, at $x = 0$, the incident intensity is $I_0$ W/m². 

The final result of Eq. (4) is

$$\log(I/I_0) = -Knx \tag{5}$$

or

$$I = I_0 \exp(-Knx) \tag{6}$$

Equation (6) is the mathematical statement of Beer's law and forms the basis for most turbidimetric measurements when the following associations are made:

$$\text{Turbidity } (\tau) = Kn \tag{7}$$

$$\text{Transmittance } (T) = \exp(-\tau x) \tag{8}$$

$$\text{Absorptance } (A) = 1 - \exp(-\tau x) = 1 - T \tag{9}$$

$$\text{Optical density (O.D.)} = \tau x \tag{10}$$

$$\text{Scattering cross section } (\sigma) = K \tag{11}$$

$$\text{Percent transmission} = 100 \exp(-Knx) \tag{12}$$

We should now note some of the problems associated with the aforementioned definitions, and especially Eq. (6), which form the basis for commercial turbidimeters and spectrophotometers (a turbidimeter in which the wavelength $\lambda$ may be varied). Note first that $K$ has the dimensions of an area and is often (erroneously) associated with $\pi R^2$. (In the limit as $R \to \infty$, $K \to 2\pi R^2$, not counting certain pathological resonances.) Interestingly, when $\rho = kR = 2\pi R/\lambda$ (where $k$ is a propagation constant) is between, say, 1 and 10, the variation of $K$ with $R$ is not even monotonic. Thus, as particle size increases, the particle does not necessarily scatter more light, and vice versa. Figure 2 shows the variation of the normalized scattering cross section $Q = K/\pi R^2$, with $\rho$ for glass spheres of relative refractive index $m = 1.5$. Thus $K$ is in general a function of $R$ and the relative refractive index $m$ (the refractive index of the particle divided by the refractive index of the surrounding medium).

Since turbidimeters are used so often to monitor the optical density of a solution or its particle concentration $n$, one can see many other difficulties associated with the definitions of Eqs. (6) and (10). How do we interpret $n$ if there is a *distribution* of particle sizes, $n(R)$? If our spectrophotometer

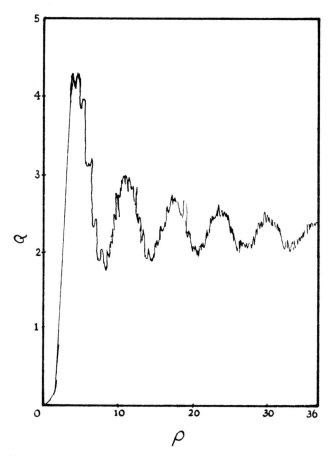

**Fig. 2.** Variation of normalized scattering cross section $Q$ for glass spheres with $\rho = kR$ and $m = 1.5$.

measures the optical density [Eq. (10)], has this value been averaged over $R$ per Eq. (8), namely,

$$\frac{I}{I_0} = \frac{\int_0^\infty n(R) \exp[-k\sigma(m, R)x] \, dR}{\int_0^\infty n(R) \, dR} \tag{13}$$

to yield O.D. $= -\log(I/I_0)$, where $I/I_0$ is given above? The problem of extracting $n(R)$ or even an average value of $R$ from an instrumental measurement of "transmittance" is by no means trivial, nor (in view of the above) does a turbidity reading yield any meaningful result other than that something is there [cf. Sigrist (1979)]. Now add a few complications: the particles are *not* all spherical, the particles have a range of refractive indices (different materials are present), the particles are inhomogeneous, and the incident light is not monochromatic. (Note also that the refractive index of a particle *varies* with wavelength; i.e., most materials are *dispersive.*)

Since turbidimetric measurements are so ambiguous, why then are such instruments used so often, especially in the food and beverage industries? There are two basic reasons: habit and hope. As far as habit is concerned, these instruments have been used for so long that a common point of departure for the industry exists. Why change? Measurements based on hope are even more precarious, because they are of little scientific validity. If one is working with the same type of solution day after day, then its turbidity as measured under the same conditions at different times *might* be used as a fingerprint related to the types of particles contained in a solution. If the fingerprint changes a little, it may be readily monitored as a change in turbidity, through trying to state unequivocally that it was due to a change in particle size, or shape, or number density is always risky. Consider haze in beer [cf. Tebeau (1976)]. As long as the brewery maintains tight control on its production, a turbidimetric measurement might well be used to characterize or fingerprint a certain measure of haze or haze potential (Erber, 1978; Moll and That, 1976). But if something has gone wrong with the process so that the distribution of haze particles changes, tremendous problems might arise from relying on one's spectrophotometric measurements. Indeed, under certain conditions, huge changes can occur in the physical properties of the suspended particulates in a solution with little or no change in turbidity.

Despite the foregoing discussion, there *still* remains a major problem with Beer's law [Eq. (6)] itself: the relation is only valid when the sample optical density is small enough to ensure *single* scattering. In other words, if a certain particle scatters some light, that light should not scatter from a subsequent particle before it is detected: no multiple scattering may occur. This is implicit in our derivation of Eq. (1): once light is removed from the incident beam, it can never reappear in a detector placed in the beam to measure $I(x)$. Consider that the sample is placed in a square glass cylinder of thickness $x$ as shown in Fig. 3. Neglecting internal reflections at the cell walls and assuming that parallel light is incident from the left with intensity $I_0$, the intensity of light detected at $D$ given by Beer's law is just

$$I = I_0 \exp(-n\sigma x) \tag{14}$$

where we have replaced $K$ of Eq. (6) by $\sigma$. The mean free path for light scattered by the $n$ particles per unit volume is readily seen to be given by

$$\lambda_m = 1/(n\sigma) \tag{15}$$

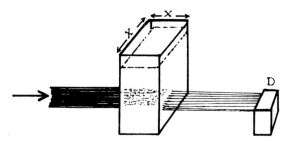

**Fig. 3.** Attenuation of light by a sample contained in a square cell of side $x$. The intensity of the transmitted light is detected at $D$.

This is the mean (average) distance that light must travel before it scatters from a particle. Note that Eq. (15) is also intuitively reasonable since larger particle scattering cross sections would be associated with a smaller distance traversed through a sample before a collision. Similarly, the greater the particle density $n$, the smaller the mean free path. In order to preclude multiple scattering, the mean free path must be somewhat greater than the dimensions of the cell, or

$$\lambda_m > x \tag{16}$$

Substituting Eq. (15) into Eq. (16) yields immediately the result

$$n\sigma x < 1 \tag{17}$$

For 1-$\mu$m-average-diameter particles, $\sigma \sim 10^{-8}$ cm$^2$, and for a standard 1-cm sample cell ($x = 1$ cm), Eq. (17) requires that $n < 10^8$.

When $n\sigma x \ll 1$, single scattering is always ensured throughout the scattering volume and Beer's law simplifies to

$$I/I_0 = \exp(-n\sigma x) \sim (1 - n\sigma x) \tag{18}$$

For the example of $\sigma \sim 10^{-8}$ cm$^2$ and $x = 1$, $n \ll 10^8$. This would be satisfied if $n < 10^6$/cm$^3$.

Multiple-scattering events often go unnoticed. This is easily demonstrated in the case of milk. Prepare a sample of milk diluted 1:1000 with distilled water. At $\lambda = 633$ nm, this should yield a 1-cm tube-held sample with a transmittance in excess of 0.97. Now continually add milk to the sample and the transmittance will decrease rapidly to about 0.5; then *increase* to about 0.6, and stay there irrespective of the milk particulate density. Here we see clearly the effects of multiple scattering. As the particulate concentration increases, the suspension finally becomes a *diffuser* of light; that is, it behaves as a translucent medium. Since the particles are nonabsorbing, all the light that comes in must eventually get out, even after many multiple collisions with the entrained particles. Diffused light per se has no specific direction associated with it: it diffuses equally in all directions and the apparent brightness of the emitting volume is independent of the angle of observation. (This is Lambert's law.) For the milk example, one should be able to observe roughly equal brightness at the cell walls through which light diffuses perpendicularly to the direction of the incident beam.

Sometimes samples may be relatively thin, that is, $x < 1/(n\sigma)$, yet deep. A typical sample tube such as shown schematically in Fig. 3 has a far greater length (depth) than width. Although single scattering will generally occur in the direction of the beam, some light will scatter upward (or downward) and pass through a far greater amount of sample, sufficient to result in multiple scattering, since such a path, say $y$, does *not* satisfy $y < 1/(n\sigma)$. Thus some of the light detected in the $x$ direction may contain a contribution of light multiply scattered in the $y$ direction and then back into the $x$ direction. Such effects are easily recognized if the recorded transmittance changes as the depth of the sample increases.

A final feature of turbidity measurements should be considered. Referring to Fig. 2, we see that the scattering cross section of a particle approaches *twice* its geometrical cross section as its size becomes large compared to the wavelength of the incident light [cf. Kerker (1969)]. This "additional" contribution has its origin in diffraction, that is, light scattered by the edges of the particle into the *forward* direction. By forward direction, we mean the same direction ($\theta \sim 0$) as the incident beam. Similar diffraction effects are readily apparent when the particle size becomes larger than about a tenth of the incident wavelength. Sometimes the diffracted light cancels out via interference some of the unscattered forward beam, and sometimes it adds significantly to it—far more so than the limiting 100% value. Figure 2 illustrates these two effects quite vividly via the oscillatory nature of the normalized scattering cross section. Although the diffraction effects may be separated from the scattering effects for very large particles (absent anomalous scattering), for smaller particles one cannot practically separate the effects; they are combined by general reference to "scattering," and this includes scattering, diffraction, and absorption. Because forward-scattering/diffraction effects clearly distort turbidity measurements, most turbidimetric instruments have provisions to block out much of the forward-scattered/diffracted light by means of masks subtending a few degrees from the $\theta = 0$ direction. By now the reader should recognize that the derivation of Beer's law was based on geometrical optics concepts *without* diffraction. To correct somewhat for this simplification (and the problems associated with diffraction contributions), turbidimeter manufacturers usually design instruments that prevent forward-scattered light from entering into the detection system, as mentioned above. However, then the resultant measurement of near turbidity is not quite a measurement of light removed from the forward incident beam, but it is almost that. Hence a conventional turbidimeter is a type of nephelometer, as we shall see by a more detailed consideration of the latter.

## B. Nephelometric Measurements

Because of the complex interaction of light with a particle, the scattered light intensity varies with angle. Ideally, one might measure the scattered intensity with an array of detectors such as shown in Fig. 4. The variation of scattered intensity with detector position is usually referred to as the differential light-scattered (DLS) intensity. Figure 5 presents a typical measurement from a suspension of bacterial cells. Note that the variation of scattered intensity with angle is *not* a monotonic function but rather exhibits maxima and minima suggestive of a diffraction pattern [cf. Ditchburn (1955)]. As the size distribution of the particles broadens, we shall see (Section I,C) that these features wash out and a monotonic variation appears, generally decreasing as the scattering angle increases.

It was recognized long ago that, since turbidity measurements conventionally tend to measure a *decrease* of the incident beam intensity and therefrom a decrease in the sample transmittance, one is often trying to detect very small changes in a very large signal. Because most measured samples are optically thin (with no multiple scattering), a collimated detector placed at some reasonably large angle to the forward direction would detect only light scattered by the illuminated particles, as illustrated in Fig. 6. The intensity of light $\Delta I$ detected at $\theta$ will be given by

$$\Delta I = g\sigma(\theta)I_0 n\,\Delta V\,\Delta\Omega \qquad (19)$$

where $I_0$ is the incident intensity, $n$ is the particle density, $\Delta V$ is the volume contributing to the signal at the detector, $\Delta\Omega$ is the solid angle subtended by the detector at $\Delta V$ about the direction $\theta$ shown, $\sigma(\theta)$ is the differential

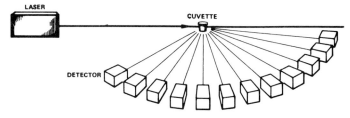

**Fig. 4.** Schematic measurement of scattered intensity from a suspension of bacteria as a function of detector position, $\theta$. (Reprinted by permission of Wyatt Technology Company, Santa Barbara, California.)

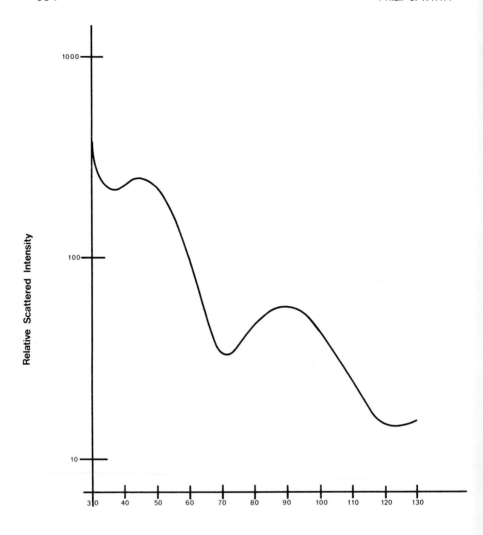

**SCATTERING ANGLE (degrees)**

**Fig. 5.** Differential light-scattering pattern from a suspension of young *Bacillus subtilis* (TKJ6321) cells in water ($\lambda = 632.8$ nm).

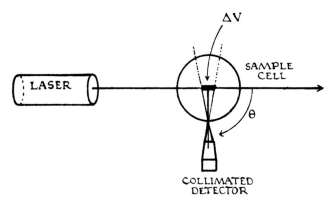

**Fig. 6.** Light scattered by the particles in sample volume $\Delta V$ into the collimated detector at an angle $\theta$ of $90°$ to the incident direction.

scattering cross section of each particle, and $g$ is a factor relating to geometrical constraints of the detector collimation. The quantity $\sigma(\theta)$ is related to the total scattering cross section of Eq. (11) by

$$\sigma = \int_0^{2\pi} \int_0^{\pi} \sigma(\theta) \sin d\theta \, d\varphi \tag{20}$$

Generally, $\sigma(\theta)$ is also a function of the azimuthal angle $\varphi$ measured *around* the direction of the incident beam. In the preceding discussion, we have assumed that all of the particles were identical and spherically symmetric, the latter so as to preclude any $\varphi$ dependence. Alternatively, if the number of particles contributing to the signal is large, then even if they are of irregular shape, their tumbling motion will average out any $\varphi$ dependence of $\sigma(\theta)$. If the particles are all of the same homogeneous material and spherical in shape but with a size distribution per unit radius increment per unit volume, $p(R)$, then the differential scattering cross section of Eq. (19) must be calculated as follows:

$$n\sigma(\theta) = \int_0^{\infty} p(R)\sigma(\theta, R) \, dR \tag{21}$$

where $\sigma(\theta, R)$ is the differential scattering cross section for a particle of radius $R$, and $\sigma(\theta)$, without an explicit $R$ dependence shown, means that this quantity has been averaged over $R$. Furthermore,

$$\int_0^{\infty} p(R) \, dR = n \tag{22}$$

where, as before, $n$ is the particle number density. If the particles are not spherical, then an average of $\sigma(\theta)$ over their orientations will be required.

Additionally, if the particles are not homogeneous and/or are of different refractive indices, then additional averagings will be required in order to describe theoretically what we are measuring experimentally. The purpose of going into the detailed formalism presented above and describing the various types of averaging procedures necessary to yield the correct $\sigma(\theta)$ for Eq. (19) is to make the reader aware of the complex origins of the nephelometric measurements to be decribed shortly. Details of the formalism above are not essential for a general understanding of the material of this section, but they should not be overlooked by those who wish to have a firm grasp of the strengths and weaknesses of the various types of optical measurements that are discussed in this chapter.

Equation (20) forms the basis of nephelometric determinations. Historically, a nephelometer ("cloudiness" meter) consisted of a tightly collimated detector fixed at 90° to the forward direction of a light beam that passed through a sample cell [cf. Moll *et al.* (1981)]. Often the light came from an incandescent tungsten source with crude collimating optics. The white light tended to average over refractive index variations of the particulates. If a white light source was used, then the scattering cross sections had to be averaged over wavelength as well. The term *nephelometer,* however, is now more generally applied to a device whose detector is set at any fixed angle, and quite often this is around 30°. Highly monochromatic light sources such as a laser [see Moll (Chapter 9, this volume)] permit size and refractive index changes to be more readily observed, or at least detected.

Since the basic objective of a nephelometer is to extract information or some kind of fingerprint indicative of the particle density $n$ and/or the average particle size (erroneously identified with $\sigma$) or changes therein, Eq. (20) is often applied to an integrating nephelometer. Such a device detects a large fraction of the energy scattered by the particles over broad ranges of $\theta$ and $\varphi$ to yield an estimate of or an approximation to Eq. (20). In other words, the intensity of scattered light collected by large-aperture optics might correspond to

$$I/I_0 = g \, \Delta V \int_0^\infty dR \, \rho(R) \int_{\varphi_1(\theta)}^{\varphi_2(\theta)} \sigma(\theta, R) \, d\varphi \int_{\theta_1}^{\theta_2} \sin \theta \, d\theta = g \, \Delta V \bar{\sigma} \quad (23)$$

where $\bar{\sigma}$ corresponds to $\sigma(\theta, R)$ averaged over the size distribution present as well as the unique solid-angle capture of the instrument as shown in Fig. 7.

Because of the broad range of collection optics, sample cell geometries, and light sources, the terms turbidimeter and nephelometer are frequently interchanged. Indeed, as we have seen in Section I,A, diffraction effects preclude the operation of a "true" turbidimeter since the near-forward beam must be blocked out. And just as turbidity and cloudiness refer to the same

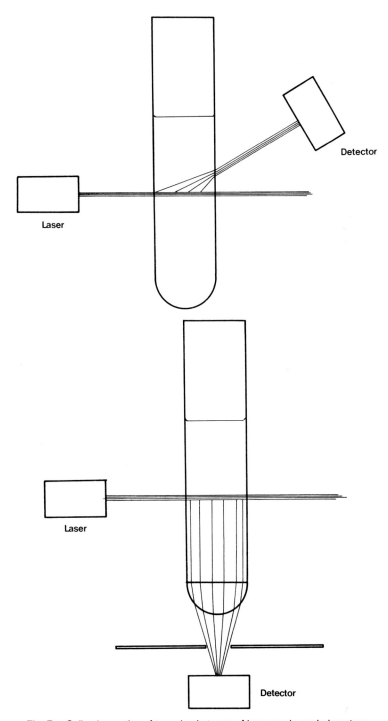

**Fig. 7.** Collection optics of two simple types of large-angle nephelometers.

concepts when applied to a liquid, so too will turbidimeters and nephelometers be measuring the same physical phenomenon, though often from a slightly different geometry.

What, then, do such instruments measure? Even for the simplest situations, it is impossible to extract *absolute* particle size or number density values. Even comparisons with known standards need not yield conclusive results. At best such instruments will detect the presence of particles and *differences* between measurements, but that is really all we can expect. The extraction of particle shapes, size distributions, and refractive index variation from a conventional nephelometric measurement is, of course, impossible. The general unreliability of such measurements, or their mediocre correlation (Sigrist, 1979) with other techniques, should not be surprising. Yet the very simplicity of the nephelometric measurement and its great sensitivity make it difficult to give up for more complex and time-consuming methods. Despite efforts to improve nephelometric results by using lasers and modern electronic processing techniques, the really important applications for the food and beverage industries continue to meet with disappointing results. Certainly the control and prediction of beer haze formation is an important example of such inadequacies, yet the nephelometric methods continue in use. Many types of water-quality measurements also make use of nephelometric devices, again yielding results of negligible significance. Let us now examine two other techniques.

## C. Differential Light-Scattering and Polarization Effects

Figure 5 presented a DLS (Wyatt, 1968) measurement from a water suspension of bacteria cells with a relatively small range of sizes ($\pm 30\%$) and based on the detection technique shown schematically in Fig. 4. This is an interesting type of measurement for several reasons. First, it is equivalent to a simultaneously performed set of different nephelometric measurements; that is, each detector placed at its own unique angle acts as a separate narrow-angle nephelometer. Second, the laser light source is both highly monochromatic and polarized, as we shall discuss. Finally, the effects of the primary scattering particles (bacteria) and secondary particles (debris, agglomerates) are both clearly visible. The bacteria at a density of about $10^6$/ml produce the smooth diffraction-like DLS pattern, while the debris or bacterial agglomerates produce "noise" or irregular features superimposed on the pattern. [The measurements were actually made using a Differential III light-scattering photometer (Stull, 1972) with a scanning photomultiplier rather than a discrete array of detectors. The detected intensity values were digitized at $1°$ intervals for subsequent storage and retrieval for plotting. The newer Differential III units (Differential Light Scattering, 1978), incorporating a detector array of only 15 elements, yield essentially the same patterns via simple mathematical interpolation.]

Since different types of narrow-angle nephelometers make measurements at different angles $\theta_i$, one would expect (referring to Fig. 5, for example) that the combined measurements should yield *additional* information. Indeed, the angular positions of the extrema of the DLS patterns yield very important size and size distribution results, as we shall see shortly. First, however, a few remarks about the polarization of the laser light source should be made.

When making monochromatic light-scattering measurements of particulates in liquid, it is always advisable to use vertically polarized sources, that is, light whose electric field is *perpendicular* to the plane of detection. A different DLS pattern will be obtained if the incident light is horizontally polarized, and an unpolarized source (not to be confused with a *random* polarized source, in which the net polarization may vary in time and/or be elliptic) will yield a resultant DLS pattern that is a superposition of the two. Such measurements tend to wash out the (possibly) sharper patterns associated with each component, and, thereby, one will lose information. For particulates whose refractive index is very close to that of the suspending liquid, such as the bacteria of Fig. 5, which have an average refractive index of about 1.38 (Wyatt, 1970), or the so-called haze particles of protein origin found in beers and other beverages, the intensity variation of scattered horizontally polarized incident light is identical to the pattern for vertically polarized incident light *times* $\cos^2 \theta$, where $\theta$ is the scattering angle. Thus, for such particles, there is no additional information contained in the horizontal DLS pattern, and an unpolarized source will wash out the vertical pattern by a factor of $(1 + \cos^2 \theta)$. So long as the particle's refractive indices are very close to that of the surrounding medium (i.e., $|m - 1| \ll 1$) or if the particles are spherically symmetric, the polarization of the scattered light is the same as that of the incident light. If the particles are of significantly different refractive index or highly irregular shape, then the detector(s) should be covered with a plane-polarizing analyzer. There are, therefore, four general sets of DLS patterns for linearly polarized incident light that one could detect from an arbitrary suspension of particulates: VV, VH, HH, and HV, where the first letter of the pair corresponds to the polarization of the incident source and the second to the polarizing analyzer covering the detector. The VH and HV measurements are useful for the detection of depolarization effects and, thereby, irregular particulates of high refractive index (asbestos fibers, metallic flakes, crystals, etc.). Since the sample-holding cuvette itself may scatter and depolarize some light, it is usually recommended to place analyzers over the detectors, in any event. The data of Fig. 5 did *not* involve such analyzers, however. (The small noise contributions are not shown.)

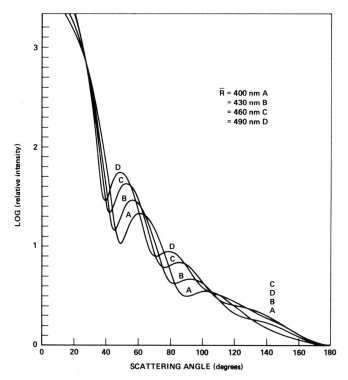

**Fig. 8.** Variation of DLS patterns of homogeneous spherical particles of refractive index 1.39 in water with a mean size distribution of ±13% as a function of $R$. Incident light is vertically polarized ($\lambda$ = 632.8 nm). (Reprinted by permission of Wyatt Technology Company, Santa Barbara, California.)

The very young bacterial cells yielding the DLS patterns of Fig. 5 are short rods of length about 1 μm, width 0.5 – 1 μm, and rms mean diameter about 750 nm. The rms value is easily deduced from plots (Wyatt, 1975) such as shown in Fig. 8 and the simple procedures of Wyatt (1973). The calculation yielding Fig. 8 shows that as the particles' mean radius increases, the DLS patterns (first peak values have all been normalized) shift to *smaller* angles.

The effect of size distribution on the resulting DLS patterns is shown in Fig. 9. This figure shows very clearly how the DLS pattern washes out rapidly as the particle size distribution broadens. This lack of detail in DLS patterns for even the very simplest of homogeneous structures underscores the fact that even the detailed measurements that produced these patterns cannot be expected to yield unequivocal results. Now if we knew that the particles in solution were all homogeneous spheres and also knew their refractive index as well as the smallest and largest size present (Dave, 1971), then we could deduce their size distribution from the recorded DLS pattern.

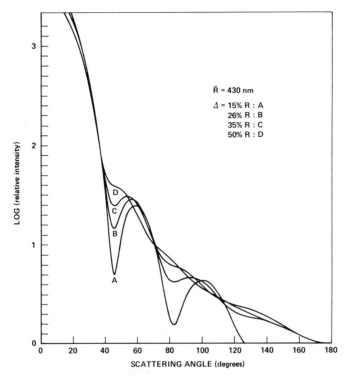

**Fig. 9.** The effects of changing size distributions on DLS patterns. This is the same model of particles as shown in Fig. 8. (Reprinted by permission of Wyatt Technology Company, Santa Barbara, California.)

This is not much information from so many *a priori* facts! If the particles are of unknown shape and composition, the DLS pattern can generally only represent a *fingerprint* with, surprisingly, many features.

Some specific deductions that one can make from these fingerprints are relatively straightforward. We have seen in the last paragraph that a sharp diffraction-like DLS pattern corresponds to a narrow size distribution of relatively homogeneous particles. In addition we have seen that, as the mean particle size increases, these patterns shift toward smaller angles. Thus it is not unexpected that the slope of the DLS pattern at small scattering angles gets steeper as the mean particle size increases. This holds quite generally, even for heterogeneous distributions of particles. Another interesting fact is that as the particles become larger, the predominant mode of scattering is into the forward (small $\theta$) direction. The forward-to-backward scattering ratio can easily exceed factors of $10^3$ for very large particles. Particles very small compared to the wavelength of the incident light tend to scatter light isotropically if it is vertically polarized. For such particles, the forward-to-backward scattering ratio is unity. Mixtures containing large numbers of very small particles as well as significant numbers of large ones produce DLS patterns showing a very steep forward-scattering intensity, yet a relatively flat scattering in the backward direction. In this regard Simms (1972) has proposed that standard nephelometric measurements be varied depending upon the type of particles whose presence is being sought. If the particles are haze particles in beers or worts, for example, the nephelometer should be set at 90° to the forward direction. In order to detect other (larger) particulates, he proposes using near-forward measurements.

If the DLS patterns are plotted in a semilogarithmic mode (Wyatt, 1973), as in the previous figures, then the *vertical* displacement of one curve relative to another must be proportional to the difference of the logarithm of the particle density of each sample, since at any scattering angle

$$\log[I(\theta)] = \log[I_0\sigma(\theta)\mathrm{n}f] = \log n + \log [I_0\sigma(\theta)f] \qquad (24)$$

where $f$ is a constant depending on various geometrical features of the measuring system. Now if the particles of two samples have number densities $n_1$ and $n_2$, respectively, and if their particle makeup is essentially identical, then $\sigma(\theta)$ will be the same for both (as will $I_0$ and $f$), and, therefore,

$$\log[I_2(\theta)] - \log[I_1(\theta)] \approx \log n_2 - \log n_1 \qquad (25)$$

independent of $\theta$. A standard nephelometer operating at one scattering angle can usually monitor particle density changes via Eq. (25). But what if two samples yield DLS patterns with both vertical *and* angular changes such as shown in Fig. 10? A nephelometer operating at, say, 40° would indicate that the control contained more particles than the sample, while at 60° it would indicate the opposite. By measuring mean vertical displacements (Wyatt, 1973) of these patterns, one can establish an estimate of subtle number density changes even though morphological changes would cause erroneous results for a conventional nephelometer.

The major problem of the DLS method relates to the difficulties associated with obtaining quantitative results from so many data points. A nephelometer produces a single number, and, irrespective of its validity, it is easy to use. Indeed this very simplicity is responsible for the popularity of nephelometric measurements, despite their questionable utility. The DLS patterns, on the other hand, may be used to generate a *set* of numbers. Wyatt *et al.* (1977) describe the mathematical decomposition of the DLS patterns in terms of Chebyshev polynomial coefficients. These coefficients may then be used to characterize the curves and make quantitative judgments.

A simpler approach to using DLS patterns for *comparison* purposes consists (Wyatt, 1973) of beginning with DLS semilog plots (or their digital

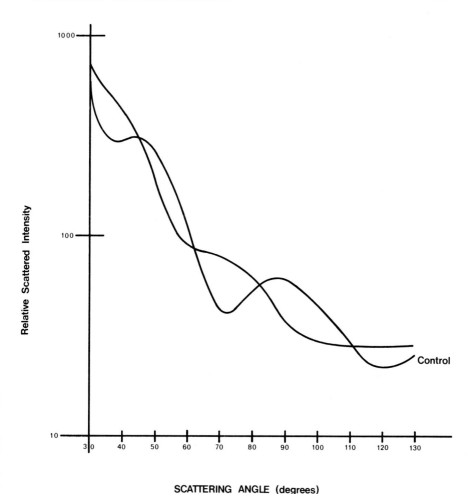

**Fig. 10.**  A comparison of DLS patterns of two particle samples resulting from different environmental effects.

representations) such as shown in Figs. 5 and 10. Two patterns are compared by choosing one as a reference (called R) and then moving the other (called T) vertically a distance $D$ (to T′) until the algebraic sum of the overlap areas between the two curves is 0. This is shown in Fig. 11. If now the *absolute value* of the areas between the curves is calculated and called $M$, a score $S$ of the differences between them may be calculated readily by a formula such as

$$S = D + aM \qquad (26)$$

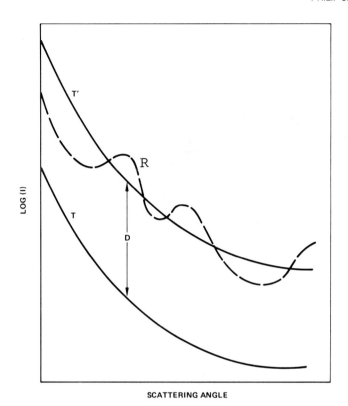

**SCATTERING ANGLE**

**Fig. 11.** The DLS patterns of a reference R and test curve T. T has been displaced a distance $D$ to yield T' and a zero algebraic overlap area between the curves.

where $a$ is a constant. Note that the term $D$ corresponds approximately to the change in number density given by Eq. (25).

### D. Dual-Angle Weighted Nephelometry (DAWN)

The simplicity of nephelometric measurements and the accuracy of DLS measurements can be combined by the technique of dual-angle weighted nephelometry (DAWN*) as incorporated in some new instruments. The technique is straightforward and yields for each suspension measured two numbers, rather than a single nephelometer value or the numerous DLS values. Measurements are made *about* two angular scattering positions as

* DAWN is a trademark of the Wyatt Technology Company.

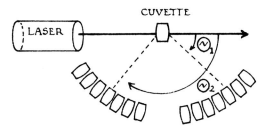

**Fig. 12.** Two sets of detectors, each of which subtends a discrete scattering angle, form the basis of a DAWN measurement. The usual configuration provides for a vertically polarized monochromatic light source.

shown in Fig. 12. Each set is then used to calculate an *average light-scattering value* associated with the corresponding subtended angle $\Theta_1$ or $\Theta_2$. Figure 13 illustrates the basis of the DAWN determinations. A typical pair of detector sets is shown in Fig. 12 and the individual detector locations are shown by the crosses in Fig. 13 as superimposed on the DLS pattern. The weighted values of each set are shown by the open circles. Between these circles is drawn a dotted line whose displacement $D$ and slope $b$ with respect to a fixed reference line can be incorporated into a simple linear equation of

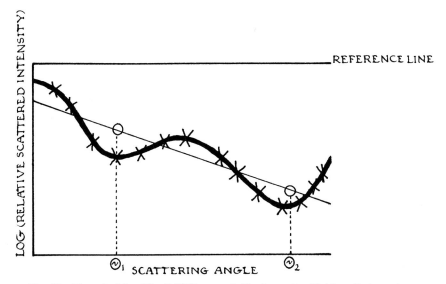

**Fig. 13.** The principle of the DAWN concept. The two sets of light-scattering values are used to generate two weighted values. Angular locations of the individual detectors are shown by the crosses, and their corresponding weighted averages are indicated by the open circles.

the type

$$y = D - b\theta \tag{27}$$

where $y$ is proportional to the logarithm of the relative scattered intensity and $b$ has been chosen as a positive number. If the center of the DAWN line lies below the reference line, then $D$ is a negative number.

If the values of the individual intensities at the various angular locations of each detector element are measured accurately, then the derived parameters $D$ and $b$ will be of equal precision. (Usually each value is measured on the order of two times spanning a time period of several seconds.) For a given reference line (set by the user), each sample is characterized by its two DAWN values $D$ and $b$. These parameters make it particularly easy to monitor changes in solutions. Thus, if $D$ increases, one can assume that the number of particles is increasing. If $b$ increases, then the average particle size is increasing. Conversely, a decrease of these parameters implies that the number density and mean size are decreasing, respectively. The ability to correlate such changes to variations of the particulates' physical properties generally requires that such changes be relatively small. For major changes in particle composition, distribution, and/or shape, the changes of these parameters must be studied phenomenologically, and suitable algorithms must be developed by the user.

Irrespective of the type of instrument used, the light-scattering measurements *must* be performed in a single-scattering environment. Any significant multiple-scattering events will distort and often mask particle changes that might otherwise be detected accurately. Insofar as DLS or DAWN techniques are concerned, this is particularly true, especially at the larger scattering angles.

For particle (including molecular) suspensions that satisfy the single-scattering criteria discussed earlier, it is a relatively simple matter to monitor the amount of debris or foreign matter present. Such agglomerates and adulterants do not generally belong in a suspension such as a beverage, often detract from the product's flavor, and may represent a health hazard for the consumer. Since $D$ and $b$ are determined by repeated measurements, their standard deviations and extreme values can be calculated at the same time. From these latter values one should be able to form an estimate of debris content and some of its physical qualities.

It is important to remember that the light-scattering properties of a suspension of unknown particles can tell something about the particles, but not everything, and certainly not without considerable thought and effort. We must rely heavily on our familiarity with our product, as well as being continually aware of our production objectives.

## II. APPLICATIONS FOR QUALITY CONTROL

In the sections that follow, various quality control features of beverages and foods, and their relationship to their corresponding light-scattering properties, are illustrated by reference to a particular substance. For example, sensory evaluation is discussed in terms of wines. Soft drinks are used to illustrate formulation, and beers are discussed in terms of their formulations and particulate matter content [see also Shah (1975)], the latter being related in many cases to hazes and unwelcome particulates whose origins may even be the filter apparatus itself. Other foodstuffs are discussed quite generally with recommendations as to their sensory analysis via light-scattering techniques.

### A. Wines and Their Sensory Qualities

One of the most recent applications of light-scattering techniques involving DLS or DAWN concepts for the study of beverage quality relates to an attempt at judging taste (Wyatt, 1981). The DLS patterns of several brands of Pinot Noir wines were made and then compared phenomenologically to extract empirical properties that might correlate with subjective tasting opinions. The presence of large particulates in wines is often associated with an unpleasant taste and/or appearance. During the production of most wines, at least in California, considerable effort is expended in inducing the precipitation of or mechanically removing large proteinlike substances. This procedure is also of importance for the beer industry. Centrifugation and filtration procedures are typical of mechanical clarification, while cold chilling (used frequently in beer processing, as well) often accelerates natural precipation phenomena. That filtration can and does change the taste of beverages is easily demonstrated by tasting two aliquots of an inexpensive red wine where one aliquot is unfiltered and the other has been filtered through a 0.5-$\mu$m Nuclepore filter. In the production of wines and other beverages, such filtration and cooling steps add considerably to the cost of production and are often skipped or shortened to yield less expensive (though inferior) products. Alternatively, clarification by aging (such as practiced frequently in France, or for certain wines with high tannin content) often results in excellent wines, but with additional inventory expenses.

In order to measure the light-scattering properties of red wines, they must be diluted to ensure again that their particulates produce single-scattering events only. Using 15-ml sample cuvettes, wine aliquots were diluted with distilled water, one part wine to nine parts water. The 15-ml cuvettes have a

**TABLE 1   Wine Qualities**[a]

| ID | Taste panel rank[b] | Comments[c] |
|----|------|----------|
| A | 4 | NR |
| B | 5 | NR |
| C | 3 | Y,T |
| D | 7 | SP,T,H |
| E | 6 | T,V,H |
| F | 2 | P,N |
| G | 1 | G,N,Y |

[a] Reprinted with permission from Wyatt (1981). Copyright American Association for the Advancement of Science.

[b] The panel consisted of a physicist, a mathematician, an electrical engineer, an office manager, an X-ray technician, a coin-dealer, an electromechanical assembler, and an attorney.

[c] G, good; SP, soda pop aftertaste; T, thin; P, pleasant; N, nice flavor; NR, no resemblance to a Pinot; V, vegetable taste; H, horrible; and Y, young.

path length of about 2.5 cm. The trace rankings of the seven wines tested are given in Table 1. Figure 14 shows the resulting DLS patterns from four of the wines measured (A, B, and F are not shown). Note that a flatter DLS pattern corresponds to a better-tasting wine. Curve G is typical of a good wine ready to drink. Curve C is a thin wine with much debris, and curves D and E show the presence of large particle contributions and too much body. Curve G shows a fine particle distribution with very little debris.

If a smaller cuvette were used, the path length would be reduced and wine concentration could be increased. If suitable single-scattering signatures were obtained at a dilution factor $1 : f_0$ (in this case for red wines, $1 : 10$), then for a smaller cuvette of diameter $d$ the new dilution factor, $1 : f$, would just be calculated from

$$f = f_0 \exp(d/D) \qquad (28)$$

where $D$ is the initial cuvette diameter (in this case 2.5 cm). A change to a 1-cm-diameter cuvette would mean that we could use a new dilution factor of $f = 10 \exp(1/2.5) \sim 1$; that is, no dilution would be needed. A suitable dilution that avoids multiple scattering yet does not eliminate too many of the particles being examined will be obtained by ensuring that, following dilution, the shape of the DLS pattern or (more simply) the DAWN parameter $b$ remains unchanged. Remember that changes in shape (or $b$) following dilution of identical particle suspensions usually means that

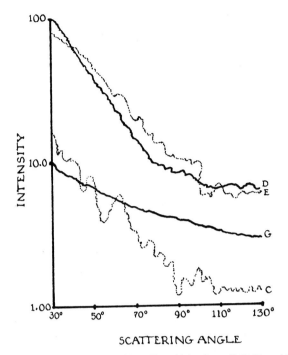

**Fig. 14.** Light-scattering properties of four Pinot Noir wines, C, D, E, and G, diluted 10:1 with water. [Reprinted with permission from Wyatt (1981). Copyright American Association for the Advancement of Science.]

multiple-scattering effects may still be present. Because dilution with distilled water (or any solvent, for that matter) will affect immediately the physical environment of the diluted particles or molecules, measurements must be made as soon as possible with all such samples so prepared.

For the case of white wines or wines exhibiting a "blush" of color, such as the so-called Sauvignon Blanc varieties, *no* dilution is usually required. All wines should be carefully decanted to ensure that sediment contribution (i.e., large agglomerates that have cleared by natural settling) is avoided.

Many wines, especially reds and sauterne types are made for long shelf life. A high tannin content usually ensures (for the case of red wines) that natural clearing via sedimentation will occur for many years and that the wine quality may improve with such aging. Using light-scattering properties to *predict* the outcome of such aging is a difficult and risky business. Some generally accepted guidelines include the following:

1. Wines with good light-scattering properties will probably become too thin after several years of aging. Thus a predominant small-particle distri-

bution at the present time would yield a very bodiless wine after the subsequent agglomeration and precipitation of many of these particles over the years.

2. A young wine with a very heavy body (large vertical displacement of DLS pattern) *and* a relatively large average particle size (such as curves D and E in Fig. 14) will probably age poorly and yield a heavy-bodied wine with an unpleasant taste despite the eventual development of moderate-sized particles.

3. Wines with intermediate body (a little above wine G of Fig. 14) and a modestly large average particle size (curve steepness between curves C and G of Fig. 14), should evolve into a very fine wine with the passage of time.

It is important to note that such wine judgments cannot be made easily by other than the very finest of wine judges or connoisseurs, and even among these there is a great divergence of opinion.

The manufacture of wine requires a great many skills. When we talk about a wine's taste or smell, we are really considering sensory qualities whose quantitation is extremely difficult. An excellent book by Amerine and Roessler (1976) discusses these qualities or properties in considerable detail and addresses the many problems associated with their quantitation. The laser techniques described briefly in this section are but a brief attempt to quantitate sensory qualities that are certainly associated with particulate content and distributions. At the present time, these laser measurements represent a phenomenological approach to the quantitation of wine quality. The user of these techniques would be well advised to examine the sample wines and measure, for example, the two DAWN parameters for each sample. Based upon the producer's (or the judging panel's) preferences, a quality score can be constructed empirically based on the simple $D$ and $b$ parameters or the far more complex Chebyshev decomposition parameters (Wyatt *et al.*, 1977) of the DLS curves. As an example, a quality score might be calculated from a simple linear relation of the type

$$Q = c_1 D + c_2 b \qquad (29)$$

where $c_1$ and $c_2$ are constants chosen so that a *low* value of the quality score, $Q$, corresponds to a *higher* relative overall quality and vice versa.

## B. Soft Drinks and Their Formulation Consistency

The soft drink industry makes considerable use of syrups in their formulations. Individual bottlers receive such syrups and combine them in various proportions with sweetening agents and carbonated water. Depending upon the type of sweetening agent, however (sugar, corn syrup, etc.), the amount of flavoring syrup may have to be varied to produce a consistent product

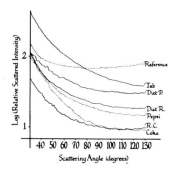

**Fig. 15.** Light-scattering patterns of six diluted cola samples and a reference 30 nm latex curve ($\lambda = 632.8$ nm). [Reprinted with permission from Wyatt (1982). Copyright Optical Society of America.]

throughout the country or region served by several bottlers. If each bottler uses the formulation specified by the manufacturer, the resultant product should meet all quality requirements. Such formulation includes, of course, water of a prescribed quality as well as the specified syrup-to-sweetener ratio. On occasion, these prescriptions may not be followed faithfully by the bottler, to the eventual detriment of the manufacturer or franchiser. The need to maintain formulation quality control throughout regions supplied by different bottlers represents a major problem for the manufacturer. The sheer magnitude of the chemical tests required to maintain a large-scale quality control program often precludes such efforts from practical implementation. As might be expected, DLS or DAWN measurements provide a rapid, inexpensive, and powerful means by which soft drink formulations can be fingerprinted and easily checked in the field.

As an example of such formulation fingerprinting, consider the various cola drinks currently on the market. Figure 15 presents (Wyatt, 1982) the DLS patterns of six such drinks (diluted 1 : 1 with deionized water) as well as a reference curve made from a suspension of small (30-nm) latex particles. Relative to this reference pattern the DLS characterization parameters of Table 2 may be obtained. Alternatively, the two more meaningful DAWN parameters could have been generated using any arbitrary horizontal line as a reference. In Table 2 the parameters $D$ and $S$ are similar to the DAWN parameters $D$ and $-b$, respectively (see Section I,D). Irrespective of the physical meaning of the two parameters of Table 2 [the interested reader should consult Wyatt (1973) for details], the important fact presented is that each formulation yields two significantly distinct and reproducible quantitative parameters. A sample of any of the above colas could easily be identified in a matter of seconds by means of its two light-scattering parameters.

TABLE 2   DLS Cola Parameters[a]

| Cola | D | S |
|------|------|------|
| Coca-Cola (Coke) | $-236 \pm 5$ | $237 \pm 5$ |
| Pepsi-Cola | $-171 \pm 9$ | $159 \pm 4$ |
| RC Cola | $-268 \pm 2$ | $117 \pm 6$ |
| Tab | $8 \pm 11$ | $255 \pm 6$ |
| Diet Pepsi | $-95 \pm 4$ | $88 \pm 14$ |

[a] Reprinted with permission from Wyatt (1982). Copyright Optical Society of America.

Distinguishing between formulations of different manufacturers is relatively straightforward. However, the DAWN method shows an even more selective capability when it is used to distinguish between sucrose and corn syrup formulations of the same drink. Changes of only a few percentage points in the quantity of flavoring syrup, or even relatively minor differences in water quality (as long as they affect taste), usually yield quantitative light-scattering differences readily monitored in an on-line or batch process environment.

A research and development colleague in a beverage-related industry recently remarked, with respect to the formulation consistency of another beverage, that the best way to produce a uniform, high-quality beer was to combine hops extract, malt barley extract, a little alcohol, some carbonated water, and a few minor chemicals. "Unfortunately," he opined, "we couldn't *call* it beer under current federal regulations. But just wait . . . one day we'll do it!" And at that time, of course, light-scattering methods will represent the best way to ensure product quality and consistency.

## C. Beer: An Example of Particle Monitoring

The monitoring of beer and wort hazes (Erber, 1978; Moll and That, 1976; Moll *et al.,* 1981;Tebeau, 1976; Dadic and Belleau, 1980) and particulate impurities (Maurer and Coors, 1975; Bernstein, 1974) represents a significant task for the brewing industry. Like wines, beers will yield precipitable particles during aging that affect appearance and, eventually, taste. Current processing techniques attempt to ensure that no significant hazes will form during the product's normal shelf life. The *prediction* of haze formation [see Moll (Chapter 9, this volume)] has been attempted through various nephelometric techniques, but with only marginal results to date (Sigrist, 1979). (Many of the problems associated with turbidimetric and nephelometric measurements have already been discussed in Section I of this chapter.) It is highly unlikely, therefore, that such measurements will ever prove particularly useful for predicting haze formations, since such measurements do not yield information about average particle size; until haze particles reach a sufficient size, they *cannot* be seen with the naked eye.

Beers often contain other particles besides the precipitating proteinaceous materials, and these particles are basically adulterants whose presence should be unacceptable in the final product. They often represent fragments from filters used in some processes, particulate contaminants of the water supply, boiler scale, sulfate precipitates, or even shards of glass or metal [cf. Shah (1975); Maurer and Coors (1975)]. Unlike their haze companions, these latter particles are often quite large and/or of high density. They scatter light specularly and fall rapidly to the bottom of the beer container. The detection of any such nonhaze particles as early as possible in the brewing process is both desirable from the consumer's point of view and beneficial to the manufacturer. Unfortunately, simple nephelometric measurements can be expected to help very little with the detection of such particulate contaminants.

Haze and particulate contaminant monitoring are rather accurately achieved via DLS or DAWN measuring techniques, and the addition of depolarization analyzers will often disclose fiber contributions to contamination. Although no specific haze *predictions* have as yet been made using DLS or DAWN techniques, one would expect the techniques to be extremely useful in this regard. The formation rate of hazes depends critically upon the average particle size present and its variation with time, the size distribution present and its rate of change, and finally the number density of potentially contributing proteinaceous particles. This assumes that other contributing factors such as temperature and pH remain relatively constant during the period of prediction. Thus the mere presence of suitable haze-forming particles in a beer or wort does not necessarily mean that a haze will form. There must be a sufficient number density of such key particles in the solution to ensure that the larger haze particles form at all within the period of interest. The formation of large particles by the coalescence of smaller ones, like any chemical reation, depends on the collision rate and the adhesion cross sections of the colliding particles. These in turn are directly proportional to the particle density and size. Thus one should recognize from the start that any attempts to monitor haze formation and predict the shelf life of the continuously haze-forming materials requires at the very least that the measurement process include determinations indicative of particle size *and* density simultaneously. The DAWN parameters $b$ and $D$ are suited for this purpose.

The light-scattering properties of *beers* may be measured without dilution for all but the dark varieties. Naturally, like all carbonated beverages, the samples must be degassed before measurement since the small carbon dioxide bubbles can interfere with the accurate recording of the light-scat-

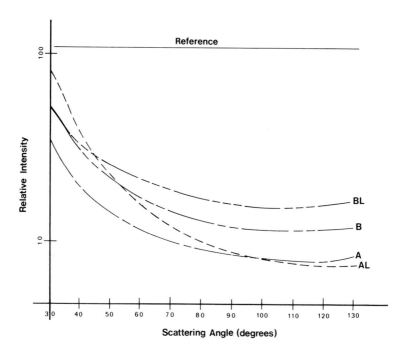

**Fig. 16.** Light-scattering patterns of two national brands of beer, A and B, and their companion light varieties, AL and BL ($\lambda$ = 632.8 nm).

tering properties. Figure 16 shows the DLS patterns of two well-known national beers (A and B) and their companion light varieties (AL and BL). There are several important deductions that can be made from these patterns. First is the somewhat surprising fact that the light beers have more material or molecules than their regular counterparts. This is easily noted in Fig. 16 by the vertical upward displacement of the DLS pattern of the light beer relative to its regular companion. (This difference is well known to the industry, however.) Second, one can see that the light beers seem to yield noisier curves (not shown explicitly in Fig. 16) than their regular companions, indicative of a larger haze or particulate contaminant content. Finally, since the curves are not parallel, the light and regular beers do not have the same particle distributions, are of different formulations, and probably have different tastes.

The results of Fig. 16 have been expanded to include six additional light beers. The DLS parameters for these beers are shown in Table 3, in which A and B are the only regular beers. The standard deviations of $D$ and $S$, based on 10 separate measurements, represent a measure of the noise contributions to the DLS patterns themselves. All values have been calculated

TABLE 3 DLS Properties of 10 Beers

| Beer | $D$ | $S$ |
|------|-----|-----|
| A | $-637 \pm 22$ | $159 \pm 34$ |
| AL | $-622 \pm 34$ | $229 \pm 28$ |
| B | $-599 \pm 9$ | $151 \pm 7$ |
| BL | $-543 \pm 42$ | $132 \pm 32$ |
| CL | $-578 \pm 29$ | $67 \pm 9$ |
| DL | $-488 \pm 4$ | $190 \pm 6$ |
| EL | $-560 \pm 14$ | $166 \pm 4$ |
| FL | $-612 \pm 11$ | $79 \pm 11$ |
| GL | $-629 \pm 17$ | $168 \pm 26$ |
| HL | $-597 \pm 4$ | $93 \pm 18$ |

relative to the reference line indicated in Fig. 16, and $D$ becomes more negative as the curve lies farther below this reference level. Beers with *larger* (less negative) values of $D$, therefore, have the greatest body or particulate and molecular contributions. Thus beer DL has the most body ($D = -488$), and beer A the least ($D = -637$). If one takes into account the standard deviations, then beers A, AL, FL, and GL have about the same body. As has been mentioned earlier, the shape parameter $S$ is very similar to the DAWN parameter $b$, with beer AL having the largest average particle size and beer CL the smallest. In terms of haze content, the standard deviations indicate that beers B and DL have the least, and, as far as light beers are concerned, beer DL is the clearest. It is interesting to note that both regular beer A and its light counterpart contain equivalent haze and particulate matter contributions whereas BL, the light variant of B, is, of the beers tested, one of those with the most particulate matter.

There are many other types of measurements that can be made with beers and beverages in general. Studying the short time variations of $D$ and $b$ (or $S$) can yield important information about larger-particle sedimentation. The long time variations of $D$ and $b$ obviously will be useful for the characterization and prediction of haze formation. Considerable study must be made using the DAWN or DLS formalism before a self-consistent theory can be developed. But insofar as the food and beverage industry is concerned, these light-scattering techniques are still in their infancy. Much work remains, but it should provide the researcher with a few surprises, in addition to considerable insights.

The beer and wine industries share many similar problems but seem for the most part to ignore the other industry. For example, beer hazes are the equivalent of wine sediments, and the treatment to speed up sedimentation by the use of sulfur dioxide is common to both industries. Since the light-scattering methods discussed in this chapter are applicable to all beverage formulations, it is hoped that these methods and their promulgation will bring the various industries closer together, at least from the research point of view, than they have been historically.

## D. Foods

Very little work has been done with DLS methods for the quality control of solid foods, although some applications for chemical and drug residue testing have been successful (see Section III,A). If a soluble exudate of a food or, in the case of dry foods, an extract can be obtained, then such extracts can be resuspended in a suitable solvent and studied by light-scattering techniques.

Some preliminary work (P. J. Wyatt, unpublished, 1982) to determine whether a DAWN measurement might yield differences between various brands of corn flakes was performed with reasonable success but with unsatisfactory reproducibility. The preparation procedures were admittedly crude and consisted in emulsifying equal weights of corn flake samples in equal volumes of deionized water. After the samples settled for 1 hr, aliquots of the supernatant solutions were diluted and compared using a Differential III system (Stull, 1972). The curves (not illustrated here) showed slight differences between brands, but similar differences were often seen between different lots of the same manufacturer.

Despite the marginal success of these first crude tests, one would expect that comparisons of the soluble fractions of processed foods by DAWN or DLS techniques could provide a useful tool for the food chemist. Perhaps a more interesting application would be to monitor bacterial metabolism and inhibition in such soluble extracts. In the next section, a discussion of the use of bacteria to monitor drug and chemical residues is presented. Suffice it to remark at this time that bacteria represent excellent targets for light-scattering measurements, and their responses to their biochemical surroundings can be readily monitored by DLS techniques, often within 60 min or less. Any growth-inhibiting (e.g., preservative) or growth-promoting (e.g., vitamin) components of foods can be monitored by measuring the *changes* of suitable selected bacterial strains exposed to an environment of these components. The next section describes how such measurements can be applied to residues. The methods may, of course, be applied to any food extracts.

## III. BIOASSAY TECHNIQUES

### A. Residues in Food-Producing Animals

A significant article by Oehme (1973) published in the early 1970s reemphasized the potential health hazards in our food supplies resulting from the use of veterinary drugs. Such drugs have been used to treat disease but also to stimulate growth. The controversy still continues as to whether or not such minuscule amounts of drugs in ingested foods have any effects on the consumer. The most often cited example of the dangers of these residues is penicillin-induced anaphylactic shock in allergic individuals. To date, however, most accounts of such food reactions are anecdotal, with no substantive documentation in the medical literature. Be that as it may, there are other residues such as pesticides, herbicides, heavy metals, mycotoxins, and chemical toxins that do from time to time appear in food-producing animals and that should be screened. Although a great variety of analytical procedures (many of which are described in this book) exist for quantitating various residues, their specificity as well as their relatively high cost makes them generally unsuitable for simple screening purposes. Bacteria, on the other hand, when used in conjunction with light-scattering procedures, provide a very general bioassay screen.

Many classes of bacteria are highly susceptible to very broad ranges of antimicrobial agents and other potentially toxic substances. As such, they may be used for a broad screening bioassay (Wyatt *et al.,* 1976, 1977). Figure 17 presents the DLS patterns from suspensions of a susceptible strain of Staphylococcus aureus (SSI41) incubated in the presence of 1, 0.33, and 0 (control) $\mu g/ml$ of the growth-promoting antimicrobial sulfaquinoxaline potentiated with 0.8 $\mu g/ml$ of trimethiprin. Aliquots of exponential-phase bacteria were added to 15 ml of a one-third-strength brain–heart infusion (BHI) broth and incubated for 3 hr. The initial bacterial concentration was about $10^6$ cells/ml. Note that the DLS patterns show a marked shift toward smaller scattering angles with increasing drug concentration, indicative of a slight increase of average cell size per Section I,C.

Consider now how the bioassay technique could be used to screen meat, for example, for the presence of antimicrobial residues. The general procedure (Wyatt *et al.,* 1977) consists of first freezing the tissues, then letting them thaw, and finally collecting the coarsely filtered exudate in small cups. A small volume of this exudate is then combined with an aliquot of exponential-phase, broadly susceptible bacteria and incubated for 60–180 min, the exact time being determined by the sensitivity required. At the same time the exudate of a similar, drug-free tissue is collected and an equal volume combined with an identical aliquot of the test bacteria. After incubation, the control and test samples are diluted (generally with deionized water) and measured in a Differential III or DAWN instrument. Simultaneously, measurements have been made of dilutions from the two exudates without bacteria. These blanks are then subtracted from the patterns recorded with bacteria present prior to scoring. Figure 18 shows the effective dose–response curve for tetracycline-fortified bovine muscle tissue

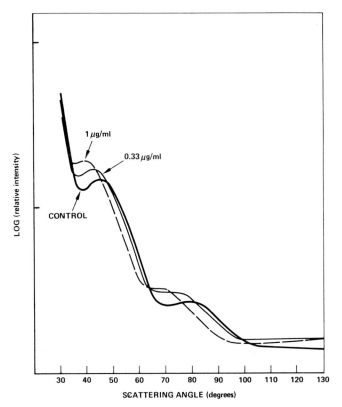

**Fig. 17.** Effect of sulfaquinoxaline on a susceptible strain of *S. aureus* (SSI41). All samples were potentiated with 0.8 μg/ml of trimethiprim. (Reprinted by permission of Wyatt Technology Company, Santa Barbara, California.)

exudate (Wyatt *et al.,* 1977) based on DLS scoring, and Fig. 19 shows dose–response curves for various antimicrobials in milk whey. (Wyatt *et al.,* 1976).

Further details of the above procedures can be found in the cited references. In general such screening tests require about 120 min. For food processors using significant quantities of meats or other items with high probability of residue contamination, such testing can be both useful and practical, especially if the instrumentation is available for other uses.

## B. Water Quality

So many beverages and foods are prepared with water that a major task for the processor will always be the assurance of water quality. Trace organic

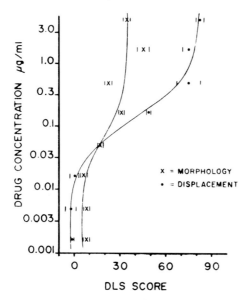

**Fig. 18.** Differential III dose–response curve for chlortetracycline in bovine muscle exudate using *S. aureus* SSI41. [Reprinted with permission from Wyatt *et al.* (1977). Copyright 1977 American Chemical Society.]

and inorganic compounds can ruin a process or a final product if not detected early enough. Various chromatographic techniques have in recent years become standard tools for water-quality assurance (see Chapters 3–5, this volume), yet they are time-consuming and require an *a priori* knowledge of the types of substances sought so that suitable columns are used. But

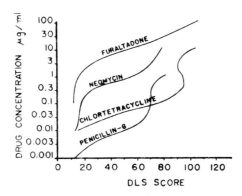

**Fig. 19.** Differential light-scattering dose–response curve for various antimicrobials in drug-fortified whey. [Reprinted with permission from Wyatt *et al.* (1976). Copyright 1976 American Chemical Society.]

how can one assay water for completely unknown and arbitrary toxicants? The obvious way is to use only reverse osmosis and deionized or distilled water and to check and confirm that the so-prepared waters maintain their toxicant-free properties. Sometimes this may not be possible; at other times the incoming untreated waters contain such huge loads of toxicant materials that the treatment facilities themselves cannot handle the levels without some spillover into the final treated water itself. Plant sabotage, especially in the food industry, must always be considered a real threat. Light-scattering techniques have recently (Wyatt, submitted for publication) been applied to the rapid bioassay for toxicants in water supplies with remarkable success.

The main reasons for the success of water bioassay toxicant-detection techniques using light scattering have been the speed of the test (60 min) and the identification of a single nonpathogenic bacterial strain responsive to

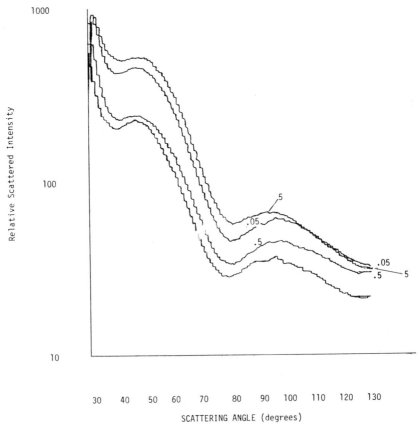

**Fig. 20.** The response of *B. subtilis* TKJ6321 to various levels of paraoxon.

most potential toxicants. The bacterial strain Bacillus subtilis TKJ6321 has been described by Felkner and his co-workers (1978) together with other serotypes of this strain.

The bioassay protocol is very simple:

1. Prepare from an overnight BHI culture a bacterial subculture, using prewarmed (39°C) BHI adjusted to a McFarland standard of 0.5. Incubate 30–40 min.

2. Prepare 15-ml prewarmed (39°C) samples of a control water and a test water and inoculate each with 0.3 ml of the bacterial subculture from step 1.

3. Incubate for 60 min at 39°C and then read in a DAWN or DLS instrument.

Figure 20 presents DLS patterns from the *B. subtilis* strain for water samples contaminated with various levels of the pesticide paraoxon. Note how the compound has *stimulated* bacterial growth. Figure 21 shows corresponding results for T2-mycotoxin, a frequent and dangerous fungal contaminant of grain. Note how different mycotoxin levels produce different bacterial responses: extremely low levels inhibit, while high levels stimulate, growth. Interestingly, there are no other analytical methods currently available that can detect such low mycotoxin levels so easily. Thus the technique might well be applied for the rapid screening of grain itself prior to purchase and use in the brewing–fermentation industry.

## IV. FUTURE DIRECTIONS

In this chapter we have discussed the basis of light-scattering measurements together with some of their limitations and applications. The quantitative aspects of DAWN or DLS measurements present considerable opportunities in the food and beverage industry. Before the application of these newer measurement techniques can become widespread, however, a major reappraisal of the time-honored, yet unreliable, turbidimetric and nephelometric techniques must be made. With the advent of inexpensive high-performance electronic components, much more detailed light-scattering measurements can be made at the same, or even less, cost than traditional single-point determinations. And the potential information they contain is so much greater. Yet this potential has barely been exploited. A new and bold approach is needed in the research laboratory, not only to implement the advanced light-scattering concepts presented in this chapter but to exploit the many other important methods described throughout this book.

Insofar as light-scattering techniques themselves are concerned, their future, especially with the use of laser monochromatic light sources, seems

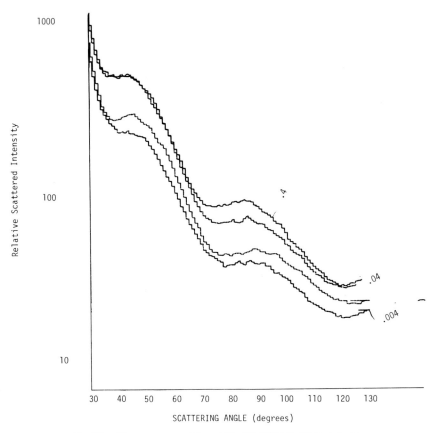

**Fig. 21.** The response of *B. subtilis* TKJ6321 to T2-mycotoxin.

very bright indeed. Among the many applications for these methods, some will certainly receive more attention than others. The author believes that the haze prediction problem will be solved very quickly for a variety of beverages. With this solution will come a better understanding of the complex chemical processes involved in haze formation and the means needed to prevent it. By combining light-scattering measurements at several wavelengths simultaneously with chromatography and other modern methods, many of the sensory aspects such as taste of foods and beverages will at last be amenable to quantitation. Manufacturers will be able to produce products of high consistency over periods of many years without reference to subjective estimations. Bioassays will become more important, especially for screening foods and waters for unexpected or accidental contaminants.

## REFERENCES

Amerine, M. A., and Roessler, E. B. (1976). "Wines: Their Sensory Evaluation." Freeman, San Francisco, California.

Bernstein, L. (1974). *Tech. Q. Master Brew. Assoc. Am.* **11**, 198–203.

Dadic, M., and Belleau, G. (1980). *J. Am. Soc. Brew. Chem.* **38**, 154–158.

Dave, J. V. (1971). *Appl. Opt.* **10**, 2035.

Differential Light Scattering, Rapid Quantitative Antimicrobial Testing, and the New Programmable Differential III (1978). Science Spectrum, Inc., Santa Barbara, California.

Ditchburn, R. W. (1955). "Light." Wiley (Interscience), New York.

Erber, H.-L. (1978). *Brauwelt* **118**, 484–487.

Felkner, I. C. (1982). *In* "Microbial Testers." Dekker, New York.

Kerker, M. (1969). "The Scattering of Light and Other Electromagnetic Radiation." Academic Press, New York.

Maurer, C. L., and Coors, J. H. (1975). *Brew. Dig.* **50**, 54–56, 58, 60, 64.

Moll, M., and That, V. (1976). *Brew. Distilling Int.* **6**, 8.

Moll, M., That, V., Bazard, D., Vincent, L. M., and Andre, J. C. (1981). *J. Am. Soc. Brew. Chem.* **39**, 15–19.

Oehme, F. W. (1973). *Toxicology* **1**, 205.

Sigrist, W. (1979). *Proc. Meet. Swiss Brew. Res. Stn., 1978.*

Shah, S. K. (1975). *J. Inst. Brew.* **81**, 293–295.

Simms, R. J. (1972). *Tech. Q. Master Brew. Assoc. Am.* **9**, 25–30.

Stull, V. R. (1972). *J. Bacteriol.* **109**, 1301.

Tebeau, C. (1976). *J. Am. Soc. Brew. Chem.* **34**, 110–111.

Wyatt, P. J. (1968). *Appl. Opt.* **7**, 1879–1896.

Wyatt, P. J. (1970). *Nature (London)* **226**, 277–279.

Wyatt, P. J. (1973). *In* "Methods of Microbiology" (J. R. Norris and D. W. Ribbons, eds.), pp. 183–263. Academic Press, New York.

Wyatt, P. J., ed. (1975). "Atlas of Light Scattering Curves." Wyatt Technology Co., Santa Barbara, California.

Wyatt, P. J. (1981). *Science* **212**, 1212–1214.

Wyatt, P. J. (1982). *Appl. Opt.* **21**, 2471–2472.

Wyatt, P. J. (submitted for publication).

Wyatt, P. J., Phillips, D. T., and Allen, E. H. (1976). *J. Agric. Food Chem.* **24**, 984–988.

Wyatt, P. J., Phillips, D. T., Scher, M. G., Kahn, M. R., and Allen, E. H. (1977). *J. Agric. Food Chem.* **25**, 1080–1086.

# 19

# Automated Multisample Analysis Using Solution Chemistry

GARY R. BEECHER

*Nutrient Composition Laboratory*
*Beltsville Human Nutrition Research Center*
*United States Department of Agriculture*
*Beltsville, Maryland*


KENT K. STEWART

*Department of Food Science*
*Virginia Polytechnic Institute and State University*
*Blacksburg, Virginia*

ANALYSIS OF FOODS AND BEVERAGES
ISBN 0-12-169160-8

## I. INTRODUCTION

There is an ever-increasing demand for the rapid, precise chemical analysis of components in large numbers of food samples. These samples may be generated for a variety of reasons, including process control, regulatory information, nutrient composition analysis, and research, both basic and applied. In many cases the concentration of the compound of interest is extremely low; however, accuracy and high precision are still required. Mechanized analytical systems capable of analyzing discrete samples at rates of at least 100 samples per hour, with good precision (at least 5% relative standard deviation) and sensitivities similar to manual methods, are absolutely essential to provide the food scientist and food technologist with results from a large number of analyses as quickly as possible.

There may be several reasons to automate chemical procedures in a laboratory. The usual overriding benefit of automation is economic (Stockwell and Foreman, 1979). Another important benefit, although not immediately obvious, is the capability to improve the quality of the resulting analytical data (Ferris, 1980). Analytical precision can often be improved through automation by improving the precision of specific steps, timing, temperature, etc. The increased sample throughput resulting from automation allows quality control samples to be analyzed frequently without negating the economic advantages of automation. These quality control samples can take the form of standards, quality control samples per se, reference materials, etc., all of which provide information to evaluate the accuracy of data from unknown samples. Health and safety are also important reasons to automate particularly difficult procedures.

Maintenance of discrete sample integrity is the most important consideration in automating sample analysis using solution chemistry. As a result, two general approaches have been followed during the design and development of automated instrumentation. One approach has been to develop instrumentation that places each sample and the appropriate reagents in a discrete container much as a chemist does in a manual method of analysis. Several types of analyzers that use this approach have been developed and will be discussed in the next section of this chapter (Section II). The second approach has followed the principle of analyses in flowing streams. This approach was introduced by Skeggs (1957) in the form of air-segmented continuous-flow analysis. Continuous-flow analysis (CFA) has undergone many significant developments and improvements in the intervening one-quarter century. A recent advance in the principle of analyses in flowing streams has been the development of flow-injection analysis (FIA). Instrumentation for FIA was developed simultaneously during the early 1970s by several research groups (Stewart, 1981). Flow-injection analysis takes advantage of the physical property of laminar solution flow in small-bore tubes (ca. 0.5 mm) to maintain sample integrity. The absence of gas bubbles in the flowing stream eliminates the need for sophisticated debubblers or "gated" detectors and permits analysis rates as high as several hundred per hour. Both CFA and FIA will be discussed in Section III.

In keeping with the general theme of this volume, we have elected to present the principal concepts involved in analysis in discrete containers, CFA and FIA. Selected examples of specific applications rather than exhaustive lists of applications will be presented for each type of analysis. It is beyond the scope of this chapter to discuss in depth the many instruments and procedures available. It is assumed that the reader is familiar with, or has access to, commercial literature and scientific publications that describe instruments and procedures in detail.

## II. ANALYSIS IN DISCRETE CONTAINERS

A logical approach to automated analysis is to develop instrumentation that places each sample and the appropriate reagents in a test tube or beaker and processes the reaction to a specific point with subsequent data presentation, that is, direct mechanization of manual methods or a "Human Mimic System" (Malmstadt *et al.,* 1971). This approach has been followed by several manufacturers and has resulted in a variety of instruments.

### A. Titration Systems

Titrations are the simplest category of assays that rely on solution chemistry, are conducted in discrete containers, and have been automated. Basic automated titration systems contain a titrator, a sample transport mechanism, and an instrument controller, usually a microprocessor. In operation, samples are manually introduced into reaction vessels, the vessels placed on the sample transport mechanism, and the titration process started. A titration cycle consists of rinsing electrodes, burette tip, and stirring device; priming the burette tip with fresh reagent; moving the reaction vessel containing the sample into titration position; titrating the sample by adding titrant and acquiring data; and returning the reaction vessel to the sample transport mechanism. Complex titration systems may also include dissolution and reagent-dispensing stations, temperature control, and more than one titrator.

A common feature of automatic titration systems is a feedback system between the electrodes (detector) and burette (reagent dispenser) that slows the rate of reagent delivery as the preset end point is approached and finally stops addition of reagent when the end point is reached. Titration data may be processed and displayed as reagent volumes directly but may also be presented as analog or digital signals that are fed to a recorder or to a computer for additional manipulation. Schematics and photographs of several automatic titration systems have been published (Thomas, 1979; Arndt and Werder, 1979). Pungor *et al.* (1982) have recently reviewed automation of potentiometric and voltametric measurements.

## B. Automatic Discrete Analyzers

Many analyses use specific chemical reactions to generate a product or intermediate that is easily detectable and proportional to the amount of analyte. Many automatic discrete analyzers capable of conducting complex reactions and simulating the same operations used in manual procedures have been developed and described (Foreman and Stockwell, 1975; Bierens de Haan, 1979; Malmstadt *et al.,* 1971, 1979).

A generalized schematic representing these analyzers is shown in Fig. 1. Automatic discrete analyzers are characterized by a sample transport system and an independent analytical transport system. When a sample reaches the sampling unit, a defined aliquot of sample is accurately transferred to an adjacent reaction tube or beaker on the analytical transport system. Reagents are added as part of the sample transfer and at specific "stations" as the sample moves along the transport system. The transfer of sample and reagents is accomplished with pneumatic or motor-driven syringes, or pneumatic burettes. Provisions are made to agitate the sample reagent mixture after reagent addition and to control the temperature of the reaction during analysis.

Photometric or fluorometric measurements are made as the transport

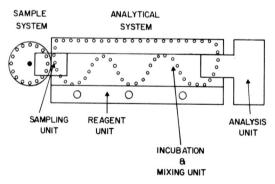

**Fig. 1.** General schematic of automatic discrete analyzers.

system moves each reactive unit through the analytical system and into the analysis unit (Fig. 1). Many discrete analyzer systems also include a wash cycle to clean the reaction vessels and prepare them for a new series of reactions. The test results are usually calculated with the aid of a computer, and suitable analytical reports are generated.

Two discrete analyzer systems have special features that set them apart from other analyzers. The Automatic Clinical Analyzer (ACA) developed by du Pont* is built around disposable plastic analytical test packs that have sealed reagent compartments and that may include a small chromatographic column. The pack also serves as the reaction chamber and observation cell. In operation, an aliquot of sample and a preset volume of diluent are added to an analytical pack, and the pack then proceeds through the chemical-processing section on a transport chain. At strategic locations along the transport chain, reagent compartments within the analytical pack are opened and their contents are mixed with the reagents in the reaction chamber. Upon completion of reagent addition and incubation, the analytical pack advances to the photometer section, where appropriate readings are taken, data are presented, and the pack is subsequently discarded. Several authors have presented detailed descriptions of the ACA (Ferris, 1980; Johnson et al., 1982; Foreman and Stockwell, 1975; Malmstadt et al., 1971).

The second discrete analyzer with special features is the Technicon STAC™ Analyzer†. The STAC system makes use of a patented cell design in which the cell functions as a reagent container, a reaction vessel, and an optical cuvette. Each cell has two compartments, the trigger and optical compartments, each of which can contain lyophilized reagents. These two compartments are interconnected by an inverted U-tube that provides isolation of the reconstituted reagents until reactions are initiated by mixing on the system. The complete automatic system consists of a diluter–dispenser module, an analytical module, and an electronics module. The diluter–dispenser module transfers a selected volume of sample, water, and buffer to the appropriate compartments of the cell. The analytical module processes the cell through a series of stations, each with a specific function, and terminates the analytical cycle at the photometric station. The electronics module controls the operation of the system, manipulates data into a meaningful form, and presents the results. Detailed descriptions of this system have also been published (Rao and Hahn, 1982; Malmstadt et al., 1979).

Discrete analyzers offer the opportunity to automate existing techniques by direct duplication of established manual methods. They also provide the

---

* E. I. du Pont de Nemours and Company, Wilmington, Delaware 19801.
† Technicon Industrial Systems, Tarrytown, New York 10591.

capability of quickly changing a specific procedure by reprogramming the selection, quantity, and sequence of reagents without changing hardware and plumbing. A common disadvantage of discrete analyzers is their mechanical complexity.

## C. Centrifugal Analyzers

The discrete analyzers discussed in the preceding section process samples sequentially past stations where reagent additions, heating, mixing, or other manipulations occur. The interval between sampling and acquisition of results in these analyzers is relatively long. The mechanical design of these systems limits the sample throughput, and it is usually quite difficult to increase the throughput or decrease the time between the addition of sample and the acquisition of results. In the 1960s, N. G. Anderson, at the Oak Ridge National Laboratory, increased sample throughput by developing instrumentation (GeMSAEC) for the analysis of a number of samples (batches) simultaneously (in parallel). GeMSAEC utilizes centrifugal force to mix samples and reagents and transfer them to cuvettes (Anderson, 1970). The absorbance and/or fluorescence of each sample is determined while the sample tray and cuvettes rotate. A number of centrifugal systems have been manufactured (Malmstadt *et al.*, 1979) and a large number of analytical procedures developed (Price and Spencer, 1980; Burtis *et al.*, 1976).

The basic concepts of centrifugal analyzers are shown in Fig. 2. Samples and reagents are manually or automatically pipetted into small individual compartments of the transfer disk. The transfer disk is then appropriately

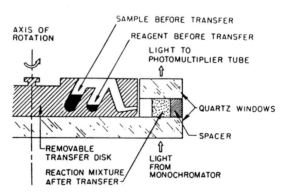

**Fig. 2.** Cross section of a centrifugal analyzer sample disk and rotor. [Reprinted with permission from Coleman *et al.* (1971).]

aligned and locked into the rotor, which ensures proper indexing of the samples. The compartments are radially aligned and angularly arranged so that samples and reagents are transferred to the reaction chamber (the cuvette) only while centrifugal force is applied (the rotor speed is approximately 400 rpm). Data are obtained while the rotor is spinning and are displayed on an oscilloscope and presented as some type of permanent record. The transfer disk and rotor are subsequently processed through a wash cycle in preparation for another batch of samples.

## D. Stopped-Flow Analyzers

Stopped-flow analysis, originally described by Gibson and Roughton (1955), has been used primarily to investigate the kinetics of fast reactions (Malmstadt *et al.*, 1971). This technique consists of transferring reagents from syringes and rapidly mixing reactants by forcing the solutions through a mixing chamber and into an observation cell. The flow of solution is abruptly stopped, creating a small back pressure that augments mixing, and the rate measurement is rapidly made. These systems have a special advantage in that each sample can serve as its own blank.

Recently, Walser and Bartels (1982) described an automated stopped-flow analyzer capable of processing a number of samples in the absence of operator intervention. Two stopped-flow systems based on the principle of FIA have also been described (Ruzicka and Hansen, 1979a; Malmstadt *et al.*, 1980). These systems not only provide the capability of evaluating reaction kinetics but also allow various ratios of sample and reagent volumes to be tested and permit the rapid changeover of analytical procedures.

## E. Applications of Analysis in Discrete Containers for Food Analysts

The various types of analyzers described in this section have the potential of becoming an important laboratory tool for the food analyst. Heretofore, procedures developed for these analyzers have been directed toward clinical analyses. Several of these procedures, however, may be directly applicable to the analysis of foods and beverages for calcium, glucose, total protein, etc. The potential to automate existing techniques by direct duplication of manual methods provides the food analyst with instrumentation to increase analytical productivity. The ability to change procedures easily and quickly without altering hardware makes these analyzers particularly attractive to the laboratory that has several analytical procedures to run on a modest number of samples.

## III. ANALYSIS IN FLOWING STREAMS

### A. Continuous-Flow Analysis (Segmented Continuous-Flow Systems)

#### 1. Background

The heavy analytical load of clinical laboratories prompted L. T. Skeggs to explore the possibility of automatic analytical methods in the mid-1950s. The system that he developed was the first analyzer to perform colorimetric analyses in continuously flowing streams (Skeggs, 1957). Shortly after the development of CFA, the Technicon Corporation introduced a commercial instrument (Autoanalyzer I) that became the first widely used automated equipment for clinical and other analyses. During the ensuing 25 years, a number of new and improved instruments, all based on Skeggs's original design, were developed by the Technicon Corporation (Conetta *et al.,* 1981) as well as by other companies (Salpeter and LaPerch, 1981). The availability of moderately priced instrumentation stimulated the development of a large number of analytical methods capable of quantitating components in many different matrices. As a result, continuous-flow analyzers have become the analytical workhorses of many industrial, clinical, and research laboratories. Several reviews (Snyder *et al.,* 1976; Snyder, 1980; Thomas, 1979) and a book (Furman, 1976) have recently been published describing CFA.

#### 2. System Description

A typical flow diagram of a continuous-flow analyzer, demonstrating the relationship of air segments to liquid segments, is shown in Fig. 3. Samples and intersample wash fluid are intermittently aspirated from a sampler by the pump. The resulting stream is combined with an air-segmented reagent stream and passed through a mixing–reaction coil (reaction bath), where a chromophore, a fluorophore, or another detectable product is formed. The stream is finally fed to the detector (the colorimeter), where the results are recorded. In this way a series of separate steps in an analytical procedure can be carried out in a controlled, reproducible, and fully automatic fashion. These steps include precise dilution of sample, quantitative addition of one or more reagents, reaction of a sample for a specific time at a specific temperature, and measurement of a final reaction product.

An essential feature of CFA systems is the air segmentation of those liquid streams containing samples. This segmentation is retained through the various stages of analysis and removed immediately before the detector, in the case of the older systems, or directed through the detector, in the new instrument models. The introduction of air causes each individual sample to be divided into a number of small segments, which has several advan-

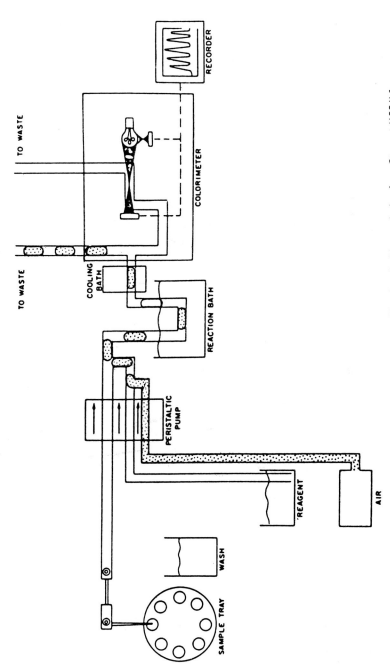

**Fig. 3.** General schematic of a continuous-flow analyzer. [Reprinted with permission from Stewart (1981).]

tages. First, the air segments establish a series of pseudoboundary conditions that reduce sample dispersion; this forces each sample to travel through the system as a distinct, identifiable component. In the absence of air segmentation, dispersion in the tubing of the system (~ 1-mm inside diameter [i.d.]) would be adversely large, resulting in interaction between successive samples or loss of sample integrity completely. Second, the presence of air bubbles promotes mixing within each liquid segment as it rises and falls through each turn of the mixing coil. Last, the wiping action of air segments along the tubing wall reduces the carry-over of residues from preceding liquid segments (Foreman and Stockwell, 1975).

Continuous-flow analyzer systems consist of a series of components or modules. The least complex system (Fig. 3) that might be used to analyze food extracts or beverages requires a sampler, a proportioning pump, a heating (constant temperature) bath or coil, a detector, a manifold (an organized series of tubes and fittings that interconnect all components), and a recorder. Detailed descriptions of the available CFA components have been published by Furman (1976) and Thomas (1979).

The most critical component of a CFA system is the proportioning pump, because it determines and maintains temporal and volumetric relationships throughout the system. Pumping action is attained by a chain-driven roller head assembly that squeezes a series of fluid-carrying flexible tubes against a platen and at the same time moves across the tubes (in a peristaltic-type action). Since the roller head runs at a constant speed and simultaneously squeezes all flexible tubes in the system, flow rates are controlled by the inside diameter of each tube. The precise physical control of tubing diameters, the constant and correct tension on tubes by the rollers, and the constant speed of the rollers all aim at delivering an accurate flow rate (volume) of sample, reagents, and gases in the flow stream. This control assures accuracy of the dynamic proportioning of all materials that enter the measuring and combining procedure and is analogous to accurate pipetting in manual methods of chemical analysis.

A large number of different detectors have been used with CFA, which contributes to the wide applicability of these systems (Furman, 1976). The most commonly used detectors continue to be colorimeters. The first automatic sampler for CFA systems was described by Skeggs (1957); advances since that time include accommodation for a large number of samples and provision to wash the sample probe between each sample. Several additional components have been developed that allow sophisticated in-line processes to be accomplished. These include dialyzers, digesters, and filtration systems, to name only a few.

Continuous-flow analyzer systems were among the first automated systems to use computers for calculating results and generating reports. The sophistication achieved with some of these systems is remarkable. Much of the work done in the interfacing of computers to instrumentation has its origin in the CFA computer system.

## 3. Theoretical Considerations

Air segmentation of liquid streams in CFA systems and the subsequent removal of air segments prior to stream monitoring greatly complicates the mathematical description of dispersion in these systems. Snyder (1980) summarized a "semi-rigorous" theory relating sample dispersion to several variables of the air-segmented flowing stream. The amount of dispersion was found to be related to tubing i.d., liquid flow rate, bubble segmentation rate, liquid viscosity, surface tension, time in the flow network, and sample mass transfer coefficient. Dispersion was found to be mainly a function of flow rate, tubing i.d., bubble segmentation rate, viscosity, and time in the flow network, with dispersion being less for smaller values of time, tubing i.d., and viscosity (Snyder *et al.,* 1976). The complex relationship between dispersion and flow rate – bubble segmentation rate has been summarized by Snyder *et al.* (1976).

The influence of the debubbler (a device to remove air segments) and flow cell on sample dispersion have been summarized by Walker (1976). Under normal operating conditions, the dispersion due to these components conforms closely to an exponential function. Although dispersion contributed by the debubbler and flow cell is somewhat less than disperion in the segmented flowing stream, both sources must be considered when determining dispersion in CFA systems.

These theoretical considerations, even though they are not rigorous definitive mathematical approaches, have practical applications in the design of CFA systems. Minimum sample dispersion is desired, which permits maximum sample throughput and maximum analytical sensitivity. Thus the food analyst, while developing a new analytical procedure, would want to consider using a high flow rate with appropriate bubble segmentation rate, a short time in the flow network, and a detector devoid of a debubbler (a "bubble-gated" detector).

## 4. Applications of CFA for Food Analysts

A large number of CFA systems have been developed and described for the analysis of components in foods and beverages. As an example, the most recent edition of the AOAC (Horwitz, 1980) contains over a dozen CFA systems as official methods; CFA systems for the analysis of many vitamins in foods and beverages have recently been summarized (Roy, 1979; Gregory, 1983). Commercial suppliers of CFA systems have extensive application bibliographies. Both new and modified CFA applications are routinely published in professional journals, such as *Analytical Biochemistry, the Journal of the Association of Official Analytical Chemists,* and others.

## B. Flow-Injection Analysis (Unsegmented Continuous-Flow Systems)

### 1. Background

In the middle of the 1970s several research groups simultaneously developed systems for the sequential analysis of discrete samples by sequential insertion of the samples into unsegmented continuously flowing streams (Stewart, 1981). This procedure now goes under the name of flow-injection analysis. Flow-injection analysis has been defined as "an automated or semi-automated analytical process consisting of the insertion of sequential discrete sample solutions into an unsegmented continuously flowing liquid stream with subsequent detection of the analyte" (Stewart, 1981). This process, which has the potential for rapid and precise analysis of discrete samples, has attracted the attention of many analysts and shows every indication of becoming a significant tool for the analyst of the future. A number of reviews (Betteridge, 1978; Ruzicka and Hansen, 1979a,b,c, 1980; Ranger, 1981; Mottola, 1981; Rocks and Riley, 1982) and one textbook (Ruzicka and Hansen, 1981) have been published that describe the process.

### 2. Description of the System

The basic versions of the semiautomated and automated versions of the flow-injection analyzer are shown in Figs. 4a and 4b, respectively. In the European version of the semiautomated system (Fig. 4a), a sample is injected into a stream of reagent, the analyte is mixed with the reagent by convective and diffusion forces, and the product is monitored as it passes through a flow-through detector, where the peak height or the peak area is measured. In the American version of the automated system (Fig. 4b), a sample is aspirated into the loop of an insertion valve from a cup in an automatic sampler. The valve is actuated and the sample is inserted into a continuously flowing stream of solvent that is subsequently mixed with a continuously flowing reagent stream. The resulting mixture then flows onto the detector, as in the semiautomated system. Both versions have the ability to perform replicate analysis at 100 or more samples per hour. A typical FIA recorder tracing is shown in Fig. 5. Critical to the success of the system is the use of small-bore tubing (0.7-mm i.d. or less), precisely controlled flow rates (1–14 ml/min), and minimization of system mixing volumes. These features result in a minimized and controlled sample dispersion, which is one of the unique aspects of FIA. These simple concepts have proved to be

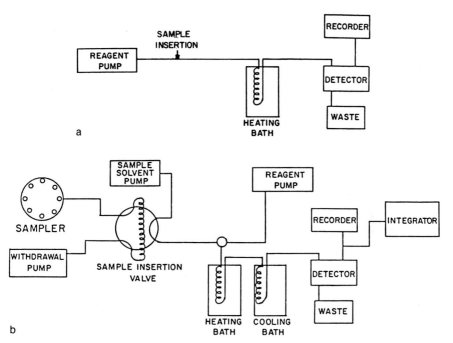

a

b

**Fig. 4.** (a) A simple semiautomated FIA system. (b) A simple automated FIA system. [Reprinted with permission from Stewart *et al.* (1980).]

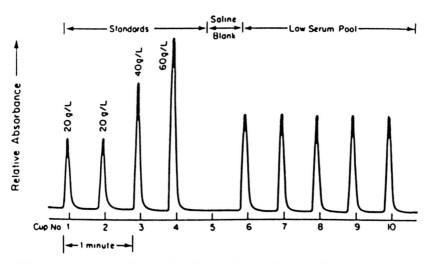

**Fig. 5.** A recorder tracing of the determination of albumin with bromocresol green. [Reprinted with permission from Renoe *et al.* (1980).]

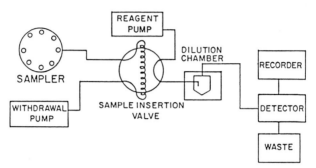

**Fig. 6.** An automated FIA system for pseudotitrations and other exponential dilution systems. [Reprinted with permission from Stewart *et al.* (1980).]

extremely successful in the development of analytical systems for a wide variety of analytes. Flow-injection analysis has been used with colorimetric, fluorometric, potentiometric, flame emission, atomic absorption, plasma emission, ion selective electrode, and conductometric detectors (Ruzicka and Hansen, 1981). Most likely any detector that can be used for high-performance liquid chromatography (HPLC) systems can be used with FIA systems. Obviously many different determinations can be done with FIA.

Another type of FIA system is shown in Fig. 6. This was originally described as an FIA titration system (Ruzicka *et al.*, 1977) but because of technicalities of nomenclature has been more recently called a pseudotitration system (Stewart and Rosenfeld, 1982). In this mode the sample is inserted into the flowing titrant and its dispersion is manipulated so as to create a very large dispersion by means of either a stirred mixing chamber or a suitable length of wide- or narrow-bore tubing. The resulting concentration profiles are shown in Fig. 7a. Under these conditions, the following equation holds, where $\Delta t$ is the time required for the signal to rise from and return to a preset level (see Fig. 7b), $C_{as}^o$ is the original concentration of the analyte, $C_r^o$ is the original concentration of the reagent, and $K_1$ and $K_2$ are constants appropriate to the individual system.

$$\Delta t = K_1(\log C_{as}^o/C_r^o) + K_2 \tag{1}$$

Recent work has shown that stirred mixing chambers may also be used in a scale expansion system for many assay systems (Stewart and Rosenfeld, 1982).

A novel FIA diluter has been described (Stewart *et al.*, 1980) and is shown diagrammatically in Fig. 8. In this system the sample is inserted into a diluent and the stream is collected in a fraction collector. The dilution ratio is determined by the sample size, the flow rate of the diluent, and the timing on the fraction collector. Precisions of 0.8% relative standard deviation have been obtained at 80 samples per hour.

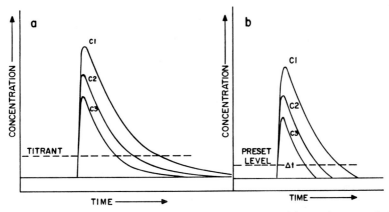

**Fig. 7.** (a) Concentration profiles (solid lines) of three samples of decreasing concentration (C1, C2, C3) after leaving the dilution chamber and concentration of the titrant (broken line). (b) Concentration profiles of the same three samples as in (a) after reacting with titrant.

## 3. Theoretical Considerations

**a. General Theory.** All FIA systems fit Pardue's kinetic analytical classification (Pardue, 1977). The most profound implication of its being a kinetic system is that the only time that an FIA system is in equilibrium is when no sample is in the system. While assays based upon equilibrium assumptions can be (and frequently are) used in FIA systems, any interpretation of the results must be done with caution. Empirical systems in which the concentration of an unknown is obtained by comparison with a standard curve prepared by running a series of standards appear to give excellent results. Systems relying upon absolute measurements have not been developed for FIA.

A striking difference between most assay systems and FIA is the spatial

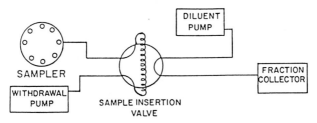

**Fig. 8.** An automated FIA diluter. The sample dilution is determined by the volume of sample inserted into the diluent stream, the flow rate of the diluent stream, and the time between subsequent samples collected in the fraction collector. [Reprinted with permission from Stewart *et al.* (1980).]

distribution of the analytes and the reagents. Traditionally, analytes and reagents are uniformally distributed through a beaker, test tube, or cuvette. In FIA the analyte is not evenly distributed along the entire volume of the system; rather, the sample bolus is inserted into a moving stream, and its concentration distribution along the tubing is a time-dependent function. Often the concentration distribution is not even uniform within an infinitely thin vertical slice of the tubing.

**b. Theory of Standard Systems.** The simplest FIA system can be described as the insertion of a sample bolus into a moving stream inside a segment of narrow-bore tubing and the measurement of the analyte at some point downstream. The theoretical description of the time-dependent concentration profile of analytes, reagents, and products under these rather simple conditions is currently under debate. While there is general agreement that the work of Taylor (1953, 1954) is the starting point of the modern theory, there is no agreement as to what is the best theoretical description. One group prefers the "tanks in series model" (Ruzicka and Hansen, 1978); another group prefers the description of Vanderslice *et al.* (1981). The two key parameters predicted by Vanderslice *et al.* are the time ($t_A$) from injection to the initial appearance of sample bolus at the detector and the baseline-to-baseline time ($\Delta t_B$) for each sample bolus at the detector. Equations (2) and (3) show the relationship of the crucial parameters needed to predict these two times. The definition of the symbols is given in Table 1.

**TABLE 1  Definition of Symbols**

| Symbol | Definition |
|--------|------------|
| $t_A$ | Time from injection of sample until appearance of sample peak in seconds |
| $\Delta t_B$ | Time from initial measurement of peak until final measurement of peak in seconds |
| $a$ | Internal radius of tubing in centimeters |
| $D$ | Diffusion constant of the analyte in centimeters squared per second |
| $L$ | Length of reaction tubing in centimeters |
| $q$ | Flow rate of flowing stream in milliters per minute |
| $C_{as}^{o}$ | Initial concentration of analyte in moles per liter |
| $C_{ag}$ | Time-dependent concentration of analyte in gradient chamber in moles per liter |
| $C_b^{o}$ | Reagent concentration in undiluted flow stream in moles per liter |
| $[A]_g^{ep}$ | Analyte concentration at end point (after reaction) in moles per liter |
| $V_g$ | Volume of dilution chamber in milliliters |
| $V_s$ | Volume of sample in milliliters |

$$t_A = 109a^2D^{0.025}(L/q)^{1.025} \tag{2}$$

$$\Delta t_B = (35.4a^2/D^{0.36})(L/q)^{0.64} \tag{3}$$

Recent studies have suggested that alternatives to the use of narrow-bore tubing as a reaction chamber can be quite useful. Some have suggested that tightly coiled tubes (Tijssen, 1980) should be used and others that packed beds (van der Berg et al., 1980) should be used. One of the most promising reaction chambers appears to be threaded bead reactors (Reijn et al., 1981).

c. **Theory of Pseudotitrations and Exponential Dilution Chambers.** Pardue and Fields (1981) have developed the theory for the sample dispersion when a stirred reactor FIA system is used. In these cases the sample dispersion can be described by Eq. (4), which is a detailed version of Eq. (1).

$$\Delta t = 60(V_g/q) \ln[\exp(V_s/V_g) - 1]C^o_{as} - 60(V_g/q) \ln(C^o_b + [A]^{ep}_g) \tag{4}$$

## 4. Applications of FIA for Food Analysts

Flow-injection analysis has the potential of being an extremely useful tool for food analysts. Examples of its usefulness in other fields are reflected in the application papers listed by Mottola (1981) and Ruzicka and Hansen (1981). J. F. van Staden (personal communication, 1981), has completed an extensive listing of FIA application papers. Such papers are appearing at an ever-increasing rate; however, most of the applications have been developed for assays other than food analysis. Many of these techniques appear to be readily adaptable for food assay systems. See, for example, the phosphate assay system described by Howe and Beecher (1981). In general, experience would indicate that most assay systems developed for CFA can be easily adapted for FIA with only slight modification. In some cases reaction times can be changed from minutes to seconds without loss of sensitivity.

## C. Comparison of CFA and FIA

There are several differences between the classical CFA systems of Skeggs (1957) and FIA. The lack of segmenting air bubbles in the reaction stream is the most obvious, but FIA uses connection tubing of much smaller diameters, can use smaller sample volumes, can get much faster readouts, has the potential for much higher throughputs, and is probably a more precise system for many analytical chemistries. Having said this, the authors believe that the two systems are closely related to each other and are complementary tools for the food analyst. The CFA systems appear to be more suitable for assays that require more than 2 min of reaction time, and/or that require the sequential addition of three or more reagents; FIA systems appear to be more suitable for assays that require 30 sec or less, and that use only one or two sequential reagent additions. Many assay chemistries may be used with either flow system. In many cases the final choice may well depend upon the personal taste of the analyst and upon the availability of instrumentation.

## IV. SUMMARY

Analysis of specific components in foods and beverages relies very heavily on chemical reactions in solution. A large number of automated analytical instruments capable of carrying out reactions in solution is currently available. These instruments can be categorized into two general groups: (1) analyzers that use discrete containers as reaction vessels and (2) analyzers that conduct analyses in flowing streams. The first group of analyzers includes the "Human Mimic Systems," centrifugal analyzers, and stopped-flow systems. Although these analyzers may be mechanically complex, they provide the flexibility of quickly changing analytical procedures without altering hardware.

The second category of analyzers can be subdivided into continuous-flow analyzers and flow-injection analyzers. Continuous-flow analyzers, characterized by air-segmented liquid streams, have found wide analytical applications in all phases of food science and technology. They are particularly useful when a large number of samples requiring long (>100 sec) reaction times is encountered. Flow-injection analyzers, characterized by unsegmented flowing streams in small-bore tubes, are capable of conducting analyses requiring short reaction times at the rate of several hundred per hour. Although a very limited number of applications has been developed for the analysis of components in foods and beverages, these analyzers have the potential of being extremely useful tools for food analysts. The use of one or more of the analyzers discussed in this chapter in combination with appropriate sample preparation techniques will provide the food analyst with high-quality results from a large number of samples in a reasonably short period of time.

## REFERENCES

Anderson, N. G. (1970). *Am. J. Clin. Pathol.* **53**, 778–785.
Arndt, R. W., and Werder, R. D. (1979). *Top. Autom. Chem. Anal.* **1**, 73–94.
Betteridge, D. (1978). *Anal. Chem.* **50**, 832A–846A.
Bierens de Haan, J. (1979). *Top Autom. Chem. Anal.* **1**, 208–234.
Burtis, C. A., Tiffany, T. O., and Scott, C. D. (1976). *In* "Methods of Biochemical Analysis" (D. Glick, ed.), Vol. 33, pp. 190–224. Wiley, New York.
Coleman, R. L., Shults, W. D., Kelley, M. T., and Dean, J. A. (1971). *Am. Lab. (Fairfield, Conn.)* **3**(7), 26–32.

Conetta, A., Dorsheimer, W. T., and Du Cros, M. J. F. (1981). *Am. Lab. (Fairfield, Conn.)* **13**(9), 116–121.

Ferris, C. D. (1980). "Guide to Medical Laboratory Instruments." Little, Brown, Boston, Massachusetts.

Foreman, J. K., and Stockwell, P. B. (1975). "Automatic Chemical Analysis." Halsted Press, New York.

Furman, W. B. (1976). "Continuous Flow Analysis: Theory and Practice." Dekker, New York.

Gibson, Q. H., and Roughton, F. J. W. (1955). *Proc. R. Soc. London, Ser. B* **143**, 310–334.

Gregory, J. F. (1983). *Food Technol. (Chicago)* **37**(1), 75–80.

Horwitz, W., ed. (1980). "Official Methods of Analysis of the Association of Official Analytical Chemists." Assoc. Off. Anal. Chem., Washington, D.C.

Howe, J. C., and Beecher, G. R. (1981). *J. Nutr.* **111**, 708–720.

Johnson, P. E., Smock, P. L., and Rautela, G. S. (1982). *Contemp. Top. Anal. Clin. Chem.* **4**, 55–74.

Malmstadt, H. V., Delaney, C. J., and Cordas, E. A. (1971). *Crit. Rev. Anal. Chem.* **2**, 559–619.

Malmstadt, H. V., Krottinger, D. L., and McCracken, M. S. (1979). *Top. Autom. Chem. Anal.* **1**, 95–134.

Malmstadt, H. V., Walczak, K. M., and Koupparis, M. A. (1980). *Am. Lab. (Fairfield, Conn.)* **12**(9), 27–40.

Mottola, H. A. (1981). *Anal. Chem.* **53**, 1312A–1316A.

Pardue, H. L. (1977). *Clin. Chem.* **23**, 2189–2201.

Pardue, H. L., and Fields, B. (1981). *Anal. Chim. Acta* **124**, 39–63, 65–79.

Price, C. P., and Spencer, K., eds. (1980). "Centrifugal Analyzers in Clinical Chemistry." Praeger, New York.

Pungor, E., Zehér, Z., Nagy, G., and Tóth, K. (1982). *Crit. Rev. Anal. Chem.* **14**, 53–91.

Ranger, C. B. (1981). *Anal. Chem.* **53**, 20A–32A.

Rao, J., and Hahn, B. (1982). *Contemp. Top. Anal. Clin. Chem.* **4**, 1–25.

Reijn, J. M., van der Linden, W. E., and Poppe, H. (1981). *Anal. Chim. Acta* **123**, 229–237.

Renoe, B. W., Stewart, K. K., Beecher, G. R., Wills, M. R., and Savory, J. (1980). *Clin. Chem.* **26**, 331–334.

Rocks, B., and Riley, C. (1982). *Clin. Chem.* **28**, 409–421.

Roy, R. B. (1979). *Top. Autom. Chem. Anal.* **1**, 138–160.

Ruzicka, J., and Hansen, E. H. (1978). *Anal. Chim. Acta* **99**, 37–76.

Ruzicka, J., and Hansen, E. H. (1979a). *NBS Spec. Publ. (U.S.)* No. 519, pp. 501–507.

Ruzicka, J., and Hansen, E. H. (1979b). *CHEMTECH* **9**, 756–764.

Ruzicka, J., and Hansen, E. H. (1979c). *Anal. Chim. Acta* **106**, 207–224.

Ruzicka, J., and Hansen, E. H. (1980). *Anal. Chim. Acta* **114**, 19–44.

Ruzicka, J., and Hansen, E. H. (1981) "Flow Injection Analysis." Wiley, New York.

Ruzicka, J., Hansen, E. H., and Mosbaek, H. (1977). *Anal. Chim. Acta* **92**, 235–249.

Salpeter, J., and LaPerch, F. (1981). *Am. Lab. (Fairfield, Conn.)* **13**(9), 78–85.

Skeggs, L. T. (1957). *Am. J. Clin. Pathol.* **28**, 311–322.

Snyder, L. R. (1980). *Anal. Chim. Acta* **114**, 3–18.

Snyder, L. R., Levine, J., Stoy, R., and Conetta, A. (1976). *Anal. Chem.* **48**, 942A–956A.

Stewart, K. K. (1981). *Talanta* **28**, 789–797.

Stewart, K. K., and Rosenfeld, A. G. (1982). *Anal. Chem.* **54**, 2368–2372.

Stewart, K. K., Brown, J. F., and Golden, B. M. (1980). *Anal. Chim. Acta* **114**, 119–127.

Stockwell, P. B., and Foreman, J. K. (1979). *Top. Autom. Chem. Anal.* **1**, 15–42.

Taylor, G. (1953). *Proc. R. Soc. London, Ser. A* **219**, 186–203.

Taylor, G. (1954). *Proc. R. Soc. London, Ser. A* **225**, 473–477.

Thomas, H. E. (1979). "Handbook of Automated Electronic Clinical Analysis." Reston Publ. Co., Reston, Virginia.

Tijssen, R. (1980). *Anal. Chim. Acta* **114,** 71–89.

van den Berg, J. H. M., Deelder, R. S., and Egberink, H. G. M. (1980). *Anal. Chim. Acta* **114,** 91–104.

Vanderslice, J. T., Stewart, K. K., Rosenfeld, A. G., and Higgs, D. J. (1981). *Talanta* **29,** 11–18.

Walker, W. H. C. (1976). *In* "Continuous Flow Analysis: Theory and Practice" (W. B. Furman, ed.), pp. 207–225. Dekker, New York.

Walser, P. E., and Bartels, H. A. (1982). *Am. Lab (Fairfield, Conn.)* **14**(2), 113–120.

# Index

# FOOD SCIENCE AND TECHNOLOGY
## A SERIES OF MONOGRAPHS

Maynard A. Amerine, Rose Marie Pangborn, and Edward B. Roessler, PRINCIPLES OF SENSORY EVALUATION OF FOOD. 1965.

S. M. Herschdoerfer, QUALITY CONTROL IN THE FOOD INDUSTRY. Volume I — 1967. Volume II — 1968. Volume III — 1972.

Hans Riemann, FOOD-BORNE INFECTIONS AND INTOXICATIONS. 1969.

Irvin E. Leiner, TOXIC CONSTITUENTS OF PLANT FOODSTUFFS. 1969.

Martin Glicksman, GUM TECHNOLOGY IN THE FOOD INDUSTRY. 1970.

L. A. Goldblatt, AFLATOXIN. 1970.

Maynard A. Joslyn, METHODS IN FOOD ANALYSIS, second edition. 1970.

A. C. Hulme (ed.), THE BIOCHEMISTRY OF FRUITS AND THEIR PRODUCTS. Volume 1 — 1970. Volume 2 — 1971.

G. Ohloff and A. F. Thomas, GUSTATION AND OLFACTION. 1971.

C. R. Stumbo, THERMOBACTERIOLOGY IN FOOD PROCESSING, second edition. 1973.

Irvin E. Liener (ed.), TOXIC CONSTITUENTS OF ANIMAL FOODSTUFFS. 1974.

Aaron M. Altschul (ed.), NEW PROTEIN FOODS: Volume 1, TECHNOLOGY, PART A — 1974. Volume 2, TECHNOLOGY, PART B — 1976. Volume 3, ANIMAL PROTEIN SUPPLIES, PART A — 1978. Volume 4, ANIMAL PROTEIN SUPPLIES, PART B — 1981.

S. A. Goldblith, L. Rey, and W. W. Rothmayr, FREEZE DRYING AND ADVANCED FOOD TECHNOLOGY. 1975.

R. B. Duckworth (ed.), WATER RELATIONS OF FOOD. 1975.

Gerald Reed (ed.), ENZYMES IN FOOD PROCESSING, second edition. 1975.

A. G. Ward and A. Courts (eds.), THE SCIENCE AND TECHNOLOGY OF GELATIN. 1976.

John A. Troller and J. H. B. Christian, WATER ACTIVITY AND FOOD. 1978.

A. E. Bender, FOOD PROCESSING AND NUTRITION. 1978.

D. R. Osborne and P. Voogt, THE ANALYSIS OF NUTRIENTS IN FOODS. 1978.

Marcel Loncin and R. L. Merson, FOOD ENGINEERING: PRINCIPLES AND SELECTED APPLICATIONS. 1979.

Hans Riemann and Frank L. Bryan (eds.), FOOD-BORNE INFECTIONS AND INTOXICATIONS, second edition. 1979.

N. A. Michael Eskin, PLANT PIGMENTS, FLAVORS AND TEXTURES: THE CHEMISTRY AND BIOCHEMISTRY OF SELECTED COMPOUNDS. 1979.

J. G. Vaughan (ed.), FOOD MICROSCOPY. 1979.

J. R. A. Pollock (ed.), BREWING SCIENCE, Volume 1 — 1979. Volume 2 — 1980.

Irvin E. Liener (ed.), TOXIC CONSTITUENTS OF PLANT FOODSTUFFS, second edition. 1980.

J. Christopher Bauernfeind (ed.), CAROTENOIDS AS COLORANTS AND VITAMIN A PRECURSORS: TECHNOLOGICAL AND NUTRITIONAL APPLICATIONS. 1981.

Pericles Markakis (ed.), ANTHOCYANINS AS FOOD COLORS. 1982.

Vernal S. Packard, HUMAN MILK AND INFANT FORMULA. 1982.

George F. Stewart and Maynard A. Amerine, INTRODUCTION TO FOOD SCIENCE AND TECHNOLOGY, SECOND EDITION. 1982.

# FOOD SCIENCE AND TECHNOLOGY

## A SERIES OF MONOGRAPHS

Malcolm C. Bourne, FOOD TEXTURE AND VISCOSITY: CONCEPT AND MEASUREMENT. 1982.

R. Macrae (ed.), HPLC IN FOOD ANALYSIS. 1982.

Héctor A. Iglesias and Jorge Chirife, HANDBOOK OF FOOD ISOTHERMS: WATER SORPTION PARAMETERS FOR FOOD AND FOOD COMPONENTS. 1982.

John A. Troller, SANITATION IN FOOD PROCESSING. 1983.

Colin Dennis (ed.), POST-HARVEST PATHOLOGY OF FRUITS AND VEGETABLES. 1983.

George Charalambous (ed.), ANALYSIS OF FOODS AND BEVERAGES: MODERN TECHNIQUES. 1984.

In preparation

David Pimentel and Carl W. Hall, FOOD AND ENERGY RESOURCES. 1984.

Joe M. Regenstein and Carrie E. Regenstein, FOOD PROTEIN CHEMISTRY: AN INTRODUCTION FOR FOOD SCIENTISTS. 1984.

Y. Pomeranz, FUNCTIONAL PROPERTIES OF FOOD COMPONENTS. 1984.